Undergraduate Texts in Mathematics

Undergraduate Texts in Mathematics

Undergraduate Texts in Mathematics are generally aimed at third- and fourth-year undergraduate mathematics students at North American universities. These texts strive to provide students and teachers with new perspectives and novel approaches. The books include motivation that guides the reader to an appreciation of interrelations among different aspects of the subject. They feature examples that illustrate key concepts as well as exercises that strengthen understanding.

More information about this series at http://www.springer.com/series/666

Miklós Laczkovich • Vera T. Sós

Real Analysis

Foundations and Functions of One Variable

First English Edition

 Springer

Miklós Laczkovich
Department of Analysis
Eötvös Loránd University
Budapest, Hungary

Vera T. Sós
Alfréd Rényi Institute of Mathematics
Hungarian Academy of Sciences
Budapest, Hungary

ISSN 0172-6056   ISSN 2197-5604 (electronic)
Undergraduate Texts in Mathematics
ISBN 978-1-4939-4222-0   ISBN 978-1-4939-2766-1 (eBook)
DOI 10.1007/978-1-4939-2766-1

Springer New York Heidelberg Dordrecht London
1st Hungarian edition: T. Sós, Vera, Analízis I/1 © Nemzeti Tankönyvkiadó, Budapest, 1972
2nd Hungarian edition: T. Sós, Vera, Analízis A/2 © Nemzeti Tankönyvkiadó, Budapest, 1976
3rd Hungarian edition: Laczkovich, Miklós & T. Sós, Vera: Analízis I © Nemzeti Tankönyvkiadó, Budapest, 2005
4th Hungarian edition: Laczkovich, Miklós & T. Sós, Vera: Analízis I © Typotex, Budapest, 2012
Translation from the Hungarian language 3rd edition: Valós analízis I by Miklós Laczkovich
& T. Sós, Vera, © Nemzeti Tankönyvkiadó, Budapest, 2005. All rights reserved © Springer 2015.
© Springer New York 2015
Softcover re-print of the Hardcover 1st edition 2015

Printed on acid-free paper

Springer Science+Business Media LLC New York is part of Springer Science+Business Media (www.springer.com)

Preface

Analysis forms an essential basis of mathematics as a whole, as well as of the natural sciences, and more and more of the social sciences too. The theory of analysis (differentiation and integration) was created—after Galileo's insight—for the purposes of describing the universe in the language of mathematics. Working out the precise theory took almost 300 years, with a large portion of this time devoted to definitions that encapsulate the essence of limits and continuity. Mastering these concepts can be a difficult process; this is one of the reasons why analysis is only barely present in most high-school curricula.

At the same time, in postsecondary education where mathematics is part of the program—including various branches of science and mathematics—analysis appears as a basic requirement. Our book is intended to be an introductory analysis textbook; we believe it would be useful in any areas where analysis is a part of the curriculum, in addition to the above, also in teacher education, engineering, or even some high schools. In writing this book, we used the experience we gained from our many decades of lectures at the Eotvos Lorand University, Budapest, Hungary.

We have placed strong emphasis on discussing the foundations of analysis: before we begin the actual topic of analysis, we summarize all that the theory builds upon (basics of logic, sets, real numbers), even though some of these concepts might be familiar from previous studies. We believe that a strong basis is needed not only for those who wish to master higher levels of analysis, but for everyone who wants to apply it, and especially for those who wish to teach analysis at any level.

The central concepts of analysis are limits, continuity, the derivative, and the integral. Our primary goal was to develop the precise concepts gradually, building on intuition and using many examples. We introduce and discuss applications of our topics as much as possible, while ensuring that understanding and mastering of this difficult material is advanced. This, among other reasons, is why we avoided a more abstract or general (topological or multiple variable) discussion.

We would like to emphasize that the—classical, mostly more than 100 year old—results discussed here still inspire active research in different areas. Due to the nature of this book, we cannot delve into this; we only mention a small handful of unsolved problems.

Mastering the material can be only achieved through solving many exercises of various difficulties. We have posed more than 500 exercises in our book, but few of these are routine questions—which can be found in many workbooks and exercise collections. However, we found it important to include questions that call for deeper understanding of results and methods. Of these, several more difficult questions, requiring novel ideas, are marked by (∗). A large number of exercises come with hints for solutions, while many others are provided with complete solutions. Exercises with hints and solutions are denoted by (H) and (S) respectively.

The book contains a much greater amount of material than what is necessary for most curricula. We trust that the organization of the book—namely the structure of subsections—makes the selection of self-contained curricula possible for several levels.

The book drew from Vera T. Sós' university textbook *Analízis*, which has been in print for over 30 years, as well as analysis lecture notes by Miklós Laczkovich. This book, which is the translation of the third edition of the Hungarian original, naturally expands on these sources and the previous editions in both content and presentation.

Budapest, Hungary   Miklós Laczkovich
May 16, 2014   Vera T. Sós

Contents

Chapter 1
A Brief Historical Introduction

The first problems belonging properly to mathematical analysis arose during fifth century BCE, when Greek mathematicians became interested in the properties of various curved shapes and surfaces. The problem of squaring a circle (that is, constructing a square of the same area as a given circle with only a compass and straightedge) was well known by the second half of the century, and Hippias had already discovered a curve called the *quadratix* during an attempt at a solution. Hippocrates was also active during the second half of the fifth century BCE, and he defined the areas of several regions bound by curves ("Hippocratic lunes").

The discovery of the fundamental tool of mathematical analysis, approximating the unknown value with arbitrary precision, is due to Eudoxus (408–355 BCE). Eudoxus was one of the most original thinkers in the history of mathematics. His discoveries were immediately appreciated by Greek mathematicians; Euclid (around 300 BCE) dedicated an entire book (the fifth) of his *Elements* [3] to Eudoxus's theory of proportions of magnitudes. Eudoxus invented the method of *exhaustion* as well, and used it to prove that the volume of a pyramid is one-third that of a prism with the same base and height. This beautiful proof can be found in book XII of the *Elements* as the fifth theorem.

The method of exhaustion is based on the fact that if we take away at least one-half of a quantity, then take away at least one-half of the remainder, and continue this process, then for every given value *v*, sooner or later we will arrive at a value smaller than *v*. One variant of this principle is nowadays called the *axiom of Archimedes*, even though Archimedes admits, in his book *On the Sphere and Cylinder*, that mathematicians before him had already stated this property (and in the form above, it appeared as the first theorem in book X of Euclid's *Elements*). Book XII of the *Elements* gives many applications of the method of exhaustion. It is worth noting the first application, which states that *the ratio of the area of two circles is the same as the ratio of the areas of the squares whose sides are the circles' diameters*. The proof uses the fact that the ratio between the areas of two similar polygons is the same as the ratio of the squares of the corresponding sides (and this fact is proved by Euclid

© Springer New York 2015
M. Laczkovich, V.T. Sós, *Real Analysis*, Undergraduate Texts
in Mathematics, DOI 10.1007/978-1-4939-2766-1_1

in his previous books in detail). Consider a circle C. A square inscribed in the circle contains more than half of the area of the circle, since it is equal to half of the area of the square circumscribed about the circle. A regular octagon inscribed in the circle contains more than half of the remaining area of the circle (as seen in Figure 1.1). This is because the octagon is larger than the square by four isosceles triangles, and each isosceles triangle is larger than half of the corresponding slice of the circle, since it can be inscribed in a rectangle that is twice the triangle. A similar argument tells us that a regular 16-gon covers at least one-half of the circle not covered by the octagon, and so on. The method of exhaustion (the axiom of Archimedes) then tells us that we can inscribe a regular polygon in the circle C such that the areas differ by less than any previously fixed number.

To finish the proof, it is easiest to introduce modern notation. Consider two circles, C_1 and C_2, and let a_i and d_i denote the areas and the diameters of the circle C_i $(i = 1, 2)$. We want to show that $a_1/a_2 = d_1^2/d_2^2$. Suppose that this is not true. Then a_1/a_2 is either larger or smaller than d_1^2/d_2^2. It suffices to consider the case $a_1/a_2 > d_1^2/d_2^2$, since in the second case, $a_2/a_1 > d_2^2/d_1^2$, so we can interchange the roles of the two circles to return to the first case. Now if $a_1/a_2 > d_1^2/d_2^2$, then the value

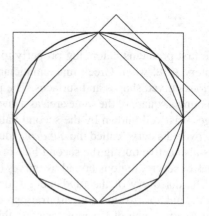

Fig. 1.1

$$\delta = \frac{a_1}{a_2} - \frac{d_1^2}{d_2^2}$$

is positive. Inscribe a regular polygon P_1 into C_1 that differs from the area of C_1 by less than $\delta \cdot a_2$. If a regular polygon P_2 similar to P_1 is inscribed in C_2, then the ratio of the areas of P_1 and P_2 is equal to the ratio of the squares of the corresponding sides, which in turn is equal to d_1^2/d_2^2 (which was shown precisely by Euclid). If the area of P_i is p_i $(i = 1, 2)$, then

$$\frac{a_1}{a_2} - \delta = \frac{d_1^2}{d_2^2} = \frac{p_1}{p_2} > \frac{a_1 - \delta \cdot a_2}{a_2} = \frac{a_1}{a_2} - \delta,$$

which is a contradiction.

Today, we would express the above theorem by saying that the area of a circle is a constant times the square of the diameter of a circle. This constant was determined by Archimedes. In his work *Measurement of a Circle*, he proves that the area of a circle is equal to the area of the right triangle whose side lengths are the radius of the circle and the circumference of the circle. With modern notation (and using the theorem above), this is none other than the formula πr^2, where π is one-half of the circumference of the unit circle.

Archimedes (287–212 BCE) ranks as one of the greatest mathematicians of all time, but is without question the greatest mathematician of antiquity. Although the greater part of his works is lost, a substantial corpus has survived. In his works, he computed the areas of various regions bounded by curves (such as slices of the parabola), determined the surface area and volume of the sphere, the arc length of certain spirals, and studied paraboloids and hyperboloids obtained by rotations. Archimedes also used the method of exhaustion, but he extended the idea by app-

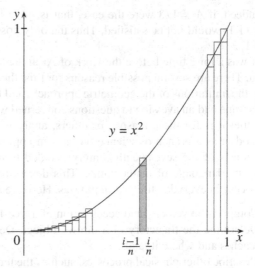

Fig. 1.2

roximating figures not only from the inside, but from the outside as well. Let us see how Archimedes used this method to find the area beneath the parabola. We will use modern notation again.

The area of the region beneath the parabola and above $[0, 1]$, as seen in Figure 1.2 will be denoted by A. The (shaded) region over the ith interval (for any n and $i \leq n$) can be approximated by rectangles from below and from above, which—with the help of Exercise 2.5(b)—gives

$$A > \frac{1}{n} \cdot \left(\left(\frac{1}{n} \right)^2 + \cdots + \left(\frac{n-1}{n} \right)^2 \right) = \frac{(n-1) \cdot n \cdot (2n-1)}{6n^3} > \frac{1}{3} - \frac{1}{n},$$

and

$$A < \frac{1}{n} \cdot \left(\left(\frac{1}{n} \right)^2 + \cdots + \left(\frac{n}{n} \right)^2 \right) = \frac{n \cdot (n+1) \cdot (2n+1)}{6n^3} < \frac{1}{3} + \frac{1}{n}.$$

It follows that

$$\left| A - \frac{1}{3} \right| < \frac{1}{n}. \tag{1.1}$$

This approximation does not give a precise value for A *for a specific n*. However, the approximation (1.1) *for every n* already shows that the area of A can only be $1/3$.

Indeed, if $A \neq 1/3$ were the case, that is, $|A - 1/3| = \alpha > 0$, then if $n \geq 1/\alpha$, then (1.1) would not be satisfied. Thus the only possibility is that $|A - 1/3| = 0$, so $A = 1/3$.

It was a long time before the work of Archimedes was taken up again and built upon. There are several possible reasons for this: the lack of a proper system of notation, the limitations of the geometric approach, and the fact that the mathematicians of the time had an aversion to questions concerned with infinity and with movement. Whether it is for these reasons or others, analysis as a widely applicable general method or as a branch of science on its own appeared only when European mathematicians of the seventeenth century decided to describe movement and change using the language of mathematics. This description was motivated by problems that occur in everyday life and in physics. Here are some examples:

- Compute the velocity and acceleration of a free-falling object.
- Determine the trajectory of a thrown object. Determine what height the object reaches and where it lands.
- Describe other physical processes, such as the temperature of a cooling object. If we know the temperature at two points in time, can we determine the temperature at every time?
- Construct tangent lines to various curves. How, for example, do we draw the tangent line to a parabola at a given point?
- What is the shape of a suspended rope?
- Solve maximum/minimum problems such as the following: What is the cylinder with the largest volume that can be inscribed in a ball? What path between two points can be traversed in the shortest time if velocity varies based on location? (This last question is inspired by the refraction of light.)
- Find approximate solutions of equations.
- Approximate values of powers (e.g., $2^{\sqrt{3}}$) and trigonometric functions (e.g., $\sin 1°$).

It turned out that these questions are strongly linked with determining area, volume, and arc length, which are also problems arising from the real world. Finally, to solve these problems, mathematicians of the seventeenth century devised a theory, called the *differential calculus*, or in today's terms, differentiation, which had three components.

The first component was the coordinate system, which is attributed to René Descartes (1596–1650), even though such a system had been used by Apollonius (262–190 BCE) when he described conic sections. However, Descartes was the first to point out that the coordinate system can help transform geometric problems into algebraic problems.

Consider the parabola, for example. By definition this is the set of points that lie equally distant from a given point and a given line. This geometric definition can be transformed into a simple algebraic condition with the help of the co-ordinate system. Let the given point be $P = (0, p)$, and the given line the horizontal line $y = -p$, where p is a fixed positive number. The distance of the point (x, y) from the point P is $\sqrt{x^2 + (y - p)^2}$, while the distance of the point from the line is $|y + p|$. Thus the point (x, y) is on the parabola if and only if $\sqrt{x^2 + (y - p)^2} = |y + p|$. Squaring both sides gives

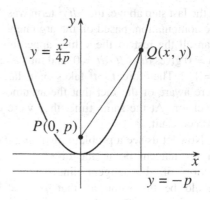

Fig. 1.3

$$x^2 + y^2 - 2py + p^2 = y^2 + 2py + p^2,$$

which we can rearrange to give us that $x^2 = 4py$ or $y = x^2/(4p)$. Thus we get the equation of the parabola, an algebraic condition that describes the points of the parabola precisely: a point (x, y) lies on the parabola if and only if $y = x^2/(4p)$.

The second component of the differential calculus is the concept of a variable quantity. Mathematicians of the seventeenth century considered quantities appearing in physical problems to be variables depending on time, whose values change from one moment to another. They extended this idea to geometric problems as well. Thus every curve was thought of as the path of a continuously moving point. This concept does not interpret the equation $y = x^2/(4p)$ as one in which y depends on x, but as a relation between x and y both depending on time as the point (x, y) traverses the parabola.

The third and most important component of the differential calculus was the notion of a differential of a variable quantity. The essence of this concept is the intuitive notion that every change is the result of the sum of "infinitesimally small" changes. Thus time itself is made up of infinitesimally small time intervals. The differential of the variable quantity x is the infinitesimally small value by which x changes during an infinitesimally small time interval. The differential of x is denoted by dx. Thus the value of x after an infinitesimally small time interval changes to $x + dx$.

How did calculus work? We illustrate the thinking of the wielders of calculus with the help of some simple examples.

The key to solving maximum/minimum problems was the fact that if the variable y reaches its highest value at a point, then $dy = 0$ there (since when a thrown object reaches the highest point along its path, it flies horizontally "for an instant"; thus if the y-coordinate of the object has a maximum, then $dy = 0$ there).

Let us use calculus to determine the largest value of $t - t^2$. Let $x = t - t^2$. Then at the maximum of x, we should have $dx = 0$. Now dx is none other than the change of x as t changes to $t + dt$. From this, we infer that

$$dx = \left[(t + dt) - (t + dt)^2\right] - \left[t - t^2\right] = dt - 2t \cdot dt - (dt)^2 = dt - 2t \cdot dt.$$

In the last step above, the $(dt)^2$ term was "ignored." That is, it was simply left out of the computation, based on the argument that the value $(dt)^2$ is "infinitely smaller" than all the rest of the values appearing in the computation. Thus the condition $dx = 0$ gives $dt - 2t \cdot dt = 0$, and thus after dividing by dt, we obtain $1 - 2t = 0$ and $t = 1/2$. Therefore, $t - t^2$ takes on its largest value at $t = 1/2$. The users of calculus were aware of the fact that the argument above lacks the requisite mathematical precision. At the same time, they were convinced that the argument leads to the correct result.

Now let us see a problem involving the construction of tangent lines to a curve. At the tangent point, the direction of the tangent line should be the same as that of the curve. The direction of the curve at a point (x,y) can be computed by connecting the point by a line to a point "infinitesimally close"; this will tell us the slope of the tangent. After an infinitesimally small change in time, the x-coordinate changes to $x + dx$, while the y-coordinate changes to $y + dy$. The point $(x+dx, y+dy)$ is thus a point of the curve that is "infinitesimally close" to (x,y). The slope of the line connecting the points (x,y) and $(x+dx, y+dy)$ is

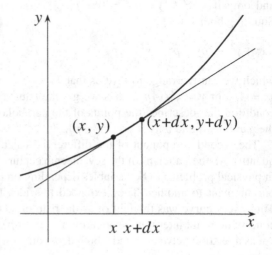

Fig. 1.4

$$\frac{(y+dy) - y}{(x+dx) - x} = \frac{dy}{dx}.$$

This is the quotient of two differentials, thus a *differential quotient*. We get that the slope of the tangent line at the point (x,y) is exactly the differential quotient dy/dx. Computing this quantity is very simple.

Consider, for example, the parabola with equation $y = x^2$. Since the point $(x+dx, y+dy)$ also lies on the parabola, the equation tells us that

$$dy = (y+dy) - y = (x+dx)^2 - x^2 = 2xdx + (dx)^2 = 2xdx,$$

where we "ignore" the term $(dx)^2$ once again. It follows that $dy/dx = 2x$, so at the point (x,y), the slope of the tangent of the parabola given by the equation $y = x^2$ is $2x$. Now consider the point (a, a^2) on the parabola. The slope of the tangent is $2a$ here, so the equation of the tangent is

$$y = 2a \cdot (x - a) + a^2.$$

This intersects the x-axis at $a/2$. Thus—according to mathematicians of the seventeenth century—we can construct the line tangent to the parabola at the point (a, a^2) by connecting the point $(a/2, 0)$ to the point (a, a^2).

Finally, let us return to the problem of computing area that we already addressed. Take the parabola given by the equation $y = x^2$ again, and compute the area of the region R that is bordered by the segment $[0, x]$ on the x-axis, the arc of the parabola between the origin and the point (x, x^2), and the segment connecting the points $(x, 0)$ and (x, x^2). Let A denote the area in question. Then A itself is a variable quantity. After an infinitesimally small change in time, the value of x changes to $x + dx$, so the region R grows by an infinitesimally narrow "rectangle" with width dx and height x. Thus the change of the area A is $dA = y \cdot dx = x^2 \cdot dx$.

Let us look for a variable z whose differential is exactly $x^2 \cdot dx$. We saw before that $d(x^2) = 2x \cdot dx$. A similar computation gives $d(x^3) = 3x^2 \cdot dx$. Thus the choice $z = x^3/3$ works, so $dz = x^2 \cdot dx$. This means that the differentials of the unknowns A and z are the same: $dA = dz$. This, in turn, means that $d(A - z) = dA - dz = 0$, that is, that $A - z$ does not change, so it is constant. If $x = 0$, then A and z are both equal to zero, so the constant $A - z$ is zero, so $A = z = x^3/3$. When $x = 1$, we obtain Archimedes' result. Again, the users of calculus were convinced that they had arrived at the correct value.

We can see that calculus is a very efficient method, and many different types of problems can be tackled with its help. Calculus, as a stand-alone method, was developed by a list of great mathematicians (Barrow, Cavalieri, Fermat, Kepler, and many others), and was completed—by seventeenth-century standards—by Isaac Newton (1643–1727) and G. W. Leibniz (1646–1716). Mathematicians immediately seized on this method and produced numerous results. By the end of the century, it was time for a large-scale monograph that would summarize what had been obtained thus far. This was L'Hospital's book *Infinitesimal Calculus* (1696), which remained the most important textbook on calculus for nearly all of the next century.

From the beginning, calculus was met with heavy criticism, which was completely justified. For the logic of the method was unclear, since it worked with imprecise definitions, and the arguments themselves were sometimes obscure. The great mathematicians of antiquity were no doubt turning over in their graves. The "proofs" outlined above seem to be convincing, but they leave many questions unanswered. What does an infinitely small quantity really mean? Is such a quantity zero, or is it not? If it is zero, we cannot divide by it in the differential quotient dy/dx. But if it is not zero, then we cannot ignore it in our computations. Such a contradiction in a mathematical concept cannot be overlooked. Nor was the method of computing the maximum very convincing. Even if we accept that the differential at the maximum is zero (although the argument for this already raises questions and eyebrows), we would need the converse of the statement: if the differential is zero, then there is a maximum. However, this is not always the case. We know that $d(x^3) = 3x^2 \cdot dx = 0$ if $x = 0$, while x^3 clearly does not have a maximum at 0.

An important part of the criticism of calculus is aimed at the contradictions having to do with infinite series. Adding infinitely many numbers together (or more generally, just the concept of infinity) was found to be problematic much earlier, as the Greek philosopher Zeno[1] had already shown. He illustrated the problem with his famous paradox of Achilles and the tortoise. The paradox states that no matter how much faster Achilles runs than the tortoise, if the tortoise is given a head start, Achilles will never pass it. For Achilles needs some time to catch up to where the tortoise is located when Achilles starts to run. However, once he gets there, the tortoise has already moved away to a further point. Achilles requires some time to get to that further point, but in the meantime, the tortoise travels even farther, and so on. Thus Achilles will never catch up to the tortoise.

Of course, we all know that Achilles will catch up to the tortoise, and we can easily compute when that happens. Suppose that Achilles runs ten yards every second, while the tortoise travels at one yard per second (to make our computations simpler, we let Achilles race against an exceptionally fast tortoise). If the tortoise gets a one-yard advantage, then after x seconds, Achilles will be $10x$ yards from the starting point, while the tortoise will be $1+x$ yards away from that point. Solving the equation $10x = 1+x$, we get that Achilles catches up to the tortoise after $x = 1/9$ seconds.

Zeno knew all of this too; he just wanted to show that an intuitive understanding of summing infinitely many components to produce movement is impossible and leads to contradictions. Zeno's argument expressed in numbers can be summarized as follows: Achilles needs to travel 1 yard to catch up to the starting point of the tortoise, which he does in $1/10$ of a second. During that time, the tortoise moves $1/10$ of a yard. Achilles has to catch up to this point as well, which takes $1/100$ of a second. During that time, the tortoise moves $1/100$ yards, to which Achilles takes $1/1000$ seconds to catch up, and so on. In the end, Achilles needs to travel infinitely many distances, and this requires $(1/10) + (1/100) + (1/1000) + \cdots$ seconds in total. Thus we get that

$$\frac{1}{10} + \frac{1}{100} + \frac{1}{1000} + \cdots = \frac{1}{9}. \tag{1.2}$$

With this, we have reduced Zeno's paradox to the question whether can we put infinitely many segments (distances) next to each other so that we get a bounded segment (finite distance), or in other words, can the sum of infinitely many numbers be finite?

If the terms of an infinite series form a geometric sequence, then its sum can be determined using simple arithmetic—at least formally. Consider the series $1 + x + x^2 + \cdots$, where x is an arbitrary real number. If $1 + x + x^2 + \cdots = A$, then

$$A = 1 + x \cdot (1 + x + x^2 + \cdots) = 1 + x \cdot A,$$

which, when $x \neq 1$, gives us the equality

$$1 + x + x^2 + \cdots = \frac{1}{1-x}. \tag{1.3}$$

---

[1] Zeno (333–262 BCE) Greek philosopher.

If in (1.3) we substitute $x = 1/10$ and subtract 1 from both sides, then we get (1.2). In the special case $x = 1/2$, we get the identity $1 + 1/2 + 1/4 + \cdots = 2$, which is immediate from Figure 1.5 as well.

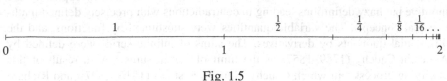

Fig. 1.5

However, the identity (1.3) can produce strange results as well. If we substitute $x = -1$ in (1.3), then we get that

$$1 - 1 + 1 - 1 + \cdots = \tfrac{1}{2}. \tag{1.4}$$

This result is strange from at least two viewpoints. On one hand, we get a fraction as a result of adding integers. On the other hand, if we put parentheses around pairs of numbers, the result is

$$(1 - 1) + (1 - 1) + \cdots = 0 + 0 + \cdots = 0.$$

In fact, if we begin the parentheses elsewhere, we get that

$$1 - (1 - 1) - (1 - 1) - \ldots = 1 - 0 - 0 - \ldots = 1.$$

Thus three different numbers qualify as the sum of the series $1 - 1 + 1 - 1 + \cdots$: $1/2$, 0, and 1.

We run into a different problem if we seek the sum of the series $1 + 1 + 1 + \cdots$. If its value is y, then

$$1 + y = 1 + (1 + 1 + 1 + \cdots) = 1 + 1 + 1 + \cdots = y.$$

Such a number y cannot exist, however. We could say that the sum must be ∞, but can we exclude $-\infty$? We could argue that we are adding positive terms so cannot get a negative number, but are we so sure? If we substitute $x = 2$ in (1.3), then we get that

$$1 + 2 + 4 + \cdots = -1, \tag{1.5}$$

and so, it would seem, the sum of positive numbers can actually be negative.

These strange, impossible, and even contradictory results were the subject of many arguments up until the beginning of the nineteenth century. To resolve the contradictions, some fantastic ideas were born: to justify the equality $1 + 2 + 4 + \cdots = -1$, there were some who believed that the numbers "start over" and that after infinity, the negative numbers follow once again.

The arguments surrounding calculus lasted until the end of the nineteenth century, and they often shifted to philosophical arguments. Berkeley maintained that the statements of calculus are no more scientific than religious beliefs, while Hegel argued that the problems with calculus could be solved only through philosophical arguments.

These problems were eventually solved by mathematicians by replacing the intuitive but hazy definitions leading to contradictions with precisely defined mathematical concepts. The variable quantities were substituted by functions, and the differential quotients by derivatives. The sums of infinite series were defined by Augustin Cauchy (1789–1857) as the limit of partial sums.[2] As a result of this clarifying process—in which Cauchy, Karl Weierstrass (1815–1897), and Richard Dedekind (1831–1916) played the most important roles—by the end of the nineteenth century, the theory of **differentiation and integration** (or **analysis**, for short) reached the logical clarity that mathematics requires.

The creation of the precise theory of analysis was one of the greatest intellectual accomplishments of modern Western culture. We should not be surprised when this theory—especially its basics, and first of all its central concept, the limit—is found to be difficult. We want to facilitate the mastery of this concept as much as possible, which is why we begin with limits of sequences. But before everything else, we must familiarize ourselves with the foundations on which this branch of mathematics, analysis, is based.

---

[2] See the details of this in Chapter 7.

Chapter 2
Basic Concepts

2.1 A Few Words About Mathematics in General

In former times, mathematics was defined as the science concerned with numbers and figures. (This is reflected in the title of the classic book by Hans Rademacher and Otto Toeplitz, *Von Zahlen und Figuren*, literally *On Numbers and Figures* [6].) Nowadays, however, such a definition will not do, for modern algebra deals with abstract structures instead of numbers, and some branches of geometry study objects that barely resemble any figure in the plane or in space. Other branches of mathematics, including analysis, discrete mathematics, and probability theory, also study objects that we would not call numbers or figures. All we can say about the objects studied in mathematics is that generally, they are abstractions from the real world (but not always). In the end, mathematics should be defined not by the objects we study, but how we study them. We can say that mathematics is the science that establishes absolute and irrefutable facts about abstract objects. In mathematics, these truths are called **theorems**, and the logical arguments showing that they are irrefutable are called **proofs**. The method (or language) of proofs is *mathematical logic*.

2.2 Basic Concepts in Logic

Mathematical logic works with *statements*. Statements are declarative sentences that are either *true* or *false*. (That is, we cannot call the wish "If I were a swift cloud to fly with thee" a statement, nor the command "to thine own self be true.") We can link statements together using *logical operators* to yield new statements. The logical operators are *conjunction* (and), *disjunction* (or), *negation* (not), *implication* (if, then), and *equivalence* (if and only if).

© Springer New York 2015
M. Laczkovich, V.T. Sós, *Real Analysis*, Undergraduate Texts
in Mathematics, DOI 10.1007/978-1-4939-2766-1_2

Logical Operators

Conjunction joins two statements together to assert that both are true. That is, the conjunction of statements A and B is the statement "A and B," denoted by $A \wedge B$. The statement $A \wedge B$ is true if both A and B are true. It is false otherwise.

Disjunction asserts that at least one of two statements is true. That is, the disjunction of statements A and B is the statement "A or B," denoted by $A \vee B$. The statement $A \vee B$ is true if at least one of A and B is true, while if both A and B are false, then so is $A \vee B$.

It is important to note that in everyday life, the word "or" can be used with several different meanings:

(i) At least one of two statements is true (**inclusive or**).
(ii) Exactly one of two statements is true (**complementary or**).
(iii) At most one of two statements is true (**exclusive or**).

For example, if Alicia says, "I go to the theater or the opera every week," then she is probably using the inclusive or, since there might be weeks when she goes to both. On the other hand, if a young man talking to his girlfriend says, "Today let's go to the movies or go shopping," then he probably means that they will do exactly one of those two activities, but not both, so he is using the complementary or. Finally, if at a family dinner, a father is lecturing his son by saying, "At the table, one either eats or reads," then he means that his son should do at most one of those things at a time, so he is using the exclusive or.

Let us note that in mathematics, the logical operator *or* is, unless stated otherwise, meant as an *inclusive or*.

Negation is exactly what it sounds like. The negation of statement A is the statement "A is false," or in other words, "A is not true." We can express this by saying "not A," and we denote it by $\overline{A}$. The statement $\overline{A}$ is true exactly when A is false.

We need to distinguish between the concepts of a statement that negates A and a statement that contradicts A. We say that a statement B contradicts A if both A and B cannot be true simultaneously. The negation of a statement A is also a contradiction of A, but the converse is not necessarily true: a contradiction is not necessarily a negation. If we let A denote the statement "the real number x is positive," and we let B denote "the real number x is negative," then B contradicts A, but B is not the negation of A, since the correct negation is "the real number x is negative or zero." Similarly, the statement "these letters are black" does not have the contradiction "these letters are white" as its negation, for while it contradicts our original statement, the correct negation would be "these letters are not black" (since there are many other colors besides white that they could be). *The negation of statement A comprises all cases in which A is not true.*

It is easy to check that if A and B are arbitrary statements, then the identities $\overline{A \wedge B} = \overline{A} \vee \overline{B}$ and $\overline{A \vee B} = \overline{A} \wedge \overline{B}$ hold.

Implications help us express when one statement follows from another. We can express this by the statement "if A is true, then so is B," or "if A, then B" for short.

We denote this by $A \Rightarrow B$. We also say "A implies B." The statement $A \Rightarrow B$ then means that in every case in which A is true, B is true. We can easily see that $A \Rightarrow B$ shares the same value as $\overline{A} \vee B$ in the sense that these statements are true exactly in the same cases. (Think through that the statement "if the victim was not at home Friday night, then his radio was playing" means exactly the same thing as "the victim was at home Friday night or his radio was playing (or both).")

The statement $A \Rightarrow B$ (or $\overline{A} \vee B$) is true in every case in which B is true, and also in every case in which A is false. The only case in which $A \Rightarrow B$ is false is that A is true and B is false. We can convince ourselves that $A \Rightarrow B$ is true whenever A is false with the following example. If we promise a friend "if the weather is nice tomorrow, we will go on a hike," then there is only one case in which we break our promise: if tomorrow it is nice out but we don't go on a hike. In all the other cases—including whatever happens if the weather is not nice tomorrow—we have kept our promise.

In mathematics, just as we saw with the use of the term "or," the "if, then" construction does not always correspond to its use in everyday language. Generally, when we say "if A then B," we mean that there is a cause-and-effect relationship between A and B. And therefore, a statement of the form "if A then B" in which A and B are unrelated sounds nonsensical or comical ("if it's Tuesday, this must be Belgium"). In mathematical logic, we do not concern ourselves with the (philosophical) cause; we care only for the truth values of our statements. The implication "if it is Tuesday, then this is Belgium" is true exactly when the statement "it is not Tuesday or this is Belgium" is true, which depends only on whether it is Tuesday and whether we are in Belgium. If it is Tuesday and this is not Belgium, then the implication is false, and in every other case it is true.

We can express the $A \Rightarrow B$ implication in words as "B follows from A," "A is sufficient for B," and "B is necessary for A."

Equivalence expresses the condition of two statements always being true at the same time. The statement "A is equivalent to B," denoted by $A \iff B$, is true if both A and B are true or both A and B are false. If one of A and B is true but the other false, then $A \iff B$ is false. If $A \iff B$, we can say "A if and only if B," or "A is necessary and sufficient for B."

Quantifiers

We call statements that contain variables *predicates*. Their truth or falsity depends on the values we substitute for the variables. For example, the statement "Mr. N is bald" is a predicate with variable N, and whether it is true depends on whom we substitute for N. Similarly, "x is a perfect square" is a predicate that is true if $x = 4$, but false if $x = 5$. Let $A(x)$ be a predicate, where x is the variable. Then there are two ways to form a "normal" statement (fixed to be either true or false) from it. The first of these is "$A(x)$ is true for all x," which we denote by $(\forall x)A(x)$. The symbol $\forall$ is called the **universal quantifier.** It represents an upside-down A as in

$$\frac{2q-p}{p-q} = \frac{2-(p/q)}{(p/q)-1} = \frac{2-\sqrt{2}}{\sqrt{2}-1} = \sqrt{2}.$$

Since $2p-q$ and $p-q$ are integers and $0 < p-q < q$, we have contradicted the minimality of q again. □

The above examples use a proof by contradiction to show the impossibility or nonexistence of something. Many times, however, we need to show the existence of something using a proof by contradiction. We give two examples.

Theorem 2.2. *If a country has only finitely many cities, and every city has at least two roads leading out of it, then we can make a round trip between some cities.*

Proof. Suppose that we cannot make any round trips. Let C_1 be a city, and suppose we start travelling from C_1 along a road. If we arrive at the city C_2, then we can keep going to another city C_3, as there are at least two roads going from C_2. Also the city C_3 is different than C_1, since otherwise we would have made a round trip already, which we assumed is impossible. We can keep travelling from C_3 to a new city C_4, as there are at least two roads leading out from C_3. This C_4 is different from all the previous cities. Indeed if for example $C_4 = C_2$, then $C_2, C_3, C_4 = C_2$ would be a round trip. From C_4 we can go on to a new city C_5, and then to C_6 and so on continuing indefinitely. However, this is a contradiction, as according to our assumption there are only finitely many cities in the country. □

Theorem 2.3. *There are infinitely many prime numbers.*

Proof. Suppose there are only finitely many primes; call them $p_1, \ldots, p_n$. Consider the number $N = p_1 \cdots p_n + 1$. It is well known—and not hard to prove—that N (like every integer greater than 1) has a prime divisor. Let p be any of the prime divisors of N. Then p is different from all of $p_1, \ldots, p_n$, since otherwise, both N and $N-1$ would be divisible by p, which is impossible. But by our assumption $p_1, \ldots, p_n$ are all the prime numbers, and this is a contradiction. □

Proof by Induction

Another important proof technique is known as **mathematical induction**. Induction arguments prove infinitely many statements at once. In its simplest form, we prove the statements $A_1, A_2, A_3, \ldots$ in two steps: First we show that A_1 is true, and then we show that A_{n+1} follows from A_n for every n. (This second step, namely $A_n \Rightarrow A_{n+1}$, is called the **inductive step**, and A_n is called the **induction hypothesis**.)

These two steps truly prove all of the statements A_n. Indeed, the first statement, A_1, is proved directly. Since we have shown that A_n implies A_{n+1} for each n, it follows that in particular, A_1 implies A_2. That is, since A_1 is true, A_2 must be true as well. But by the inductive step, the truth of A_3 follows from the truth of A_2, and thus A_3 is true. Then from $A_3 \Rightarrow A_4$, we get A_4, and so on.

Let us see a simple example.

Theorem 2.4. *For every positive integer n, $2^n > n$.*

Proof. The statement A_n that we are trying to prove for all positive integers n is $2^n > n$. The statement A_1, namely $2^1 > 1$, is obviously true. Given the induction hypothesis $2^n > n$, it follows that $2^{n+1} = 2 \cdot 2^n = 2^n + 2^n \geq 2^n + 1 > n + 1$. This proves the theorem. □

It often happens that the index of the first statement is not 1, in which case we need to alter our argument slightly. We could, for example have stated the previous theorem for every nonnegative integer n, since $2^0 > 0$ is true, and the inductive step works for $n \geq 0$.

Here is another example: if we want to show that $2^n > n^2$ for $n \geq 5$, then in the first step we have to check this assertion for $n = 5$ ($32 > 25$), and then show, as before, in the inductive step that if the statement holds for n, then it holds for $n + 1$.

Now as an application of the method, we prove an important inequality (from which Theorem 2.4 immediately follows).

Theorem 2.5 (Bernoulli's[1] Inequality). *If $a \geq -1$, then for every positive integer n, we have*

$$(1+a)^n \geq 1 + na.$$

We have equality if and only if $n = 1$ or $a = 0$.

Proof. If $n = 1$, then the statement is clearly true. We want to show that if the statement holds for n, then it is true for $n + 1$. If $a \geq -1$, then $1 + a \geq 0$, so

$$(1+a)^{n+1} = (1+a)^n (1+a) \geq (1+na)(1+a) = 1 + (n+1)a + na^2$$
$$\geq 1 + (n+1)a.$$

The above argument also shows that $(1+a)^{n+1} = 1 + (n+1)a$ can happen only if $na^2 = 0$, that is, if $a = 0$. □

There are times when this simple form of induction does not lead to a proof. To illustrate this, we examine the **Fibonacci[2] numbers**, which are defined as follows. Let $u_0 = 0$ and $u_1 = 1$. We define $u_2, u_3, \ldots$ to be the sum of the two previous (already defined) numbers. That is, $u_2 = 0 + 1 = 1$, $u_3 = 1 + 1 = 2$, $u_4 = 1 + 2 = 3$, $u_5 = 2 + 3 = 5$, and so on. Clearly, the Fibonacci numbers are increasing, that is, $u_0 \leq u_1 \leq u_2 \leq \cdots$. Using induction, we now show that $u_n < 2^n$ for each n. First of all, $u_0 = 0 < 1 = 2^0$. If $u_n < 2^n$ is true, then

$$u_{n+1} = u_n + u_{n-1} \leq u_n + u_n < 2^n + 2^n = 2^{n+1},$$

which proves the statement.

---

[1] Jacob Bernoulli (1654–1705), Swiss mathematician.

[2] Fibonacci (Leonardo of Pisa) (c. 1170–1240), Italian mathematician.

Actually, more than $u_n < 2^n$ is true, namely $u_n < 1.7^n$ holds for all n. However, this cannot be proved using induction in the same form as above, since the most we can deduce from $u_n < 1.7^n$ is $u_{n+1} = u_n + u_{n-1} \leq u_n + u_n < 1.7^n + 1.7^n = 2 \cdot 1.7^n$, which is larger than 1.7^{n+1}. To prove that $u_{n+1} < 1.7^{n+1}$, we need to use not only that $u_n < 1.7^n$, but also that $u_{n-1} < 1.7^{n-1}$. The statement then follows, since

$$u_{n+1} = u_n + u_{n-1} < 1.7^n + 1.7^{n-1} = 1.7 \cdot 1.7^{n-1} + 1.7^{n-1} = 2.7 \cdot 1.7^{n-1}$$
$$< 2.89 \cdot 1.7^{n-1} = 1.7^2 \cdot 1.7^{n-1} = 1.7^{n+1}.$$

The proof then proceeds as follows. First we check that $u_0 < 1.7^0$ and $u_1 < 1.7^1$ hold (both are clear). Suppose $n > 1$, and let us hypothesize that $u_{n-1} < 1.7^{n-1}$ and $u_n < 1.7^n$. Then as the calculations above show, $u_{n+1} < 1.7^{n+1}$ also holds. This proves the inequality for every n. Indeed, let us denote the statement $u_n < 1.7^n$ by A_n. We have shown explicitly that A_0 and A_1 are true. Since we showed $(A_{n-1} \wedge A_n) \Rightarrow A_{n+1}$ for every n, $(A_0 \wedge A_1) \Rightarrow A_2$ is true. Therefore, A_2 is true. Then $(A_1 \wedge A_2) \Rightarrow A_3$ shows that A_3 is true, and so on.

We can use this method of induction to prove relationships between the arithmetic, geometric, and harmonic means. We define the **arithmetic mean** of the numbers $a_1, \dots, a_n$ to be

$$A = \frac{a_1 + \cdots + a_n}{n}.$$

For nonnegative $a_1, \dots, a_n$, the **geometric mean** is

$$G = \sqrt[n]{a_1 \cdots a_n},$$

and for positive $a_1, \dots, a_n$, the **harmonic mean** is

$$H = n \cdot \left(\frac{1}{a_1} + \cdots + \frac{1}{a_n} \right)^{-1}.$$

The term "mean" indicates that each of these numbers falls between the largest and the smallest of the a_i. If the largest of the a_i is M and the smallest is m, then $A \geq n \cdot m / n = m$, and similarly, $A \leq M$. If the numbers are not all equal, then $m < A < M$ also holds. This is so because one of the numbers is then larger than m, so $a_1 + \cdots + a_n > n \cdot m$ and $A > m$. We can similarly obtain $A < M$.

The same method can be used to see that for $m > 0$, we have $m \leq G \leq M$ and $m \leq H \leq M$; moreover, if the numbers are not all equal, then the inequalities are strict.

Another important property of the three means above is that if we append the arithmetic mean A to $a_1, \dots, a_n$, then the resulting extended set of numbers has the same arithmetic mean, namely A. That is,

$$\frac{a_1 + \cdots + a_n + A}{n+1} = \frac{n \cdot A + A}{n+1} = A.$$

The same is true (and can be proved similarly) for the geometric and harmonic means.

Theorem 2.6. *If $a_1, \ldots, a_n$ are arbitrary nonnegative numbers, then*

$$\sqrt[n]{a_1 \cdots a_n} \leq \frac{a_1 + \cdots + a_n}{n}.$$

Equality holds if and only if $a_1 = \cdots = a_n$.

Proof. Let $a_1, \ldots, a_n$ be arbitrary nonnegative numbers, and let A and G denote their respective arithmetic and geometric means. Suppose that k of the numbers $a_1, \ldots, a_n$ are different from A. We prove the statement using induction on k. (The number of terms, n, is arbitrary.) If $k = 0$, then the statement clearly holds, since then, all terms are equal to A, so the value of each mean is also A. Since $k = 1$ can never happen (if all terms but one of them are equal to A, then the arithmetic mean cannot be equal to A), the statement holds for $k = 1$ as well.

Now let $k > 1$, and suppose that our statement is true for all sets of numbers such that the number of terms different from A is either $k - 1$ or $k - 2$. We prove that $G < A$. We may suppose that $a_1 \leq \cdots \leq a_n$, since the order of the numbers plays no role in the statement. Since there are terms not equal to A, we see that $a_1 < A < a_n$. This also implies that $k \geq 2$.

If 0 occurs among the terms, then $G = 0$ and $A > 0$, so the statement is true. So we can suppose that the terms are positive. Replace a_n with A, and a_1 with $a_1 + a_n - A$. This new set of numbers will have the same arithmetic mean A, since the replacements did not affect the value of their sum. Also, the new geometric mean after the replacement must have increased, since $A(a_1 + a_n - A) > a_1 a_n$, which is equivalent to $(a_n - A)(A - a_1) > 0$, which is true.

In this new set of numbers, the number of terms different from A decreased by one or two. Then, by our induction hypothesis, $G' \leq A$, where G' denotes the geometric mean of the new set. Since $G < G'$, we have $G < A$, which concludes the proof. □

Theorem 2.7. *If $a_1, \ldots, a_n$ are arbitrary positive numbers, then*

$$n \cdot \left(\frac{1}{a_1} + \cdots + \frac{1}{a_n} \right)^{-1} \leq \sqrt[n]{a_1 \cdots a_n}.$$

Equality holds if and only if $a_1 = \cdots = a_n$.

Proof. Apply the previous theorem to the numbers $1/a_1, \ldots, 1/a_n$, and then take the reciprocal of both sides of the resulting inequality. □

The more general form of induction (sometimes called *strong induction*, while the previous form is sometimes known as *weak induction*) is to show that A_1 holds, then in the induction step to show that if $A_1, \ldots, A_n$ are all true, then A_{n+1} is also true. This proves that every A_n is true just as our previous arguments did.

Exercises

2.5. Prove that the following identities hold for every positive integer n:

(a) $\dfrac{x^n - y^n}{x - y} = x^{n-1} + x^{n-2} \cdot y + \cdots + x \cdot y^{n-2} + y^{n-1}$;

(b) $1^2 + \cdots + n^2 = \dfrac{n \cdot (n+1) \cdot (2n+1)}{6}$;

(c) $1^3 + \cdots + n^3 = \left(\dfrac{n \cdot (n+1)}{2} \right)^2$;

(d) $1 - \dfrac{1}{2} + \dfrac{1}{3} - \cdots - \dfrac{1}{2n} = \dfrac{1}{n+1} + \cdots + \dfrac{1}{2n}$;

(e) $\dfrac{1}{1 \cdot 2} + \cdots + \dfrac{1}{(n-1) \cdot n} = \dfrac{n-1}{n}$.

2.6. Express the following sums in simpler terms:

(a) $\dfrac{1}{1 \cdot 2 \cdot 3} + \cdots + \dfrac{1}{n \cdot (n+1) \cdot (n+2)}$;

(b) $1 \cdot 2 + \cdots + n \cdot (n+1)$;

(c) $1 \cdot 2 \cdot 3 + \cdots + n \cdot (n+1) \cdot (n+2)$.

2.7. Prove that the following inequalities hold for every positive integer n:

(a) $\sqrt{n} \le 1 + \dfrac{1}{\sqrt{2}} + \cdots + \dfrac{1}{\sqrt{n}} < 2\sqrt{n}$;

(b) $1 + \dfrac{1}{2 \cdot \sqrt{2}} + \cdots + \dfrac{1}{n \cdot \sqrt{n}} \le 3 - \dfrac{2}{\sqrt{n}}$.

2.8. Denote the nth Fibonacci number by u_n. Prove that $u_n > 1.6^n/3$ for every $n \ge 1$.

2.9. Prove the following equalities:

(a) $u_n^2 - u_{n-1} u_{n+1} = (-1)^{n+1}$;

(b) $u_1^2 + \cdots + u_n^2 = u_n u_{n+1}$.

2.10. Express the following sums in simpler terms:

(a) $u_0 + u_2 + \cdots + u_{2n}$;

(b) $u_1 + u_3 + \cdots + u_{2n+1}$;

(c) $u_0 + u_3 + \cdots + u_{3n}$;

(d) $u_1 u_2 + \cdots + u_{2n-1} u_{2n}$.

2.11. Find the flaw in the following argument: We want to prove that no matter how we choose n lines in the plane such that no two are parallel, with $n > 1$, they will always intersect at a single point. The statement is clearly true for $n = 2$. Let $n \ge 2$, and suppose the statement is true for n lines. Let $l_1, \ldots, l_{n+1}$ be lines such that no two are parallel. By the induction hypothesis, the lines $l_1, \ldots, l_n$ intersect at the single point P, while the lines $l_2, \ldots, l_{n+1}$ intersect at the point Q. Since each of the lines $l_2, \ldots, l_n$ passes through both P and Q, necessarily $P = Q$. Thus we see that each line goes through the point P. (S)

2.12. If we have n lines such that no two are parallel and no three intersect at one point, into how many regions do they divide the plane? (H)

2.13. Prove that finitely many lines (or circles) partition the plane into regions that can be colored with two colors such that the borders of two regions with the same color do not share a common segment or circular arc.

2.14. Deduce Bernoulli's inequality from the inequality of arithmetic and geometric means. (S)

2.15. Prove that if $a_1, \ldots, a_n \geq 0$, then

$$\frac{a_1 + \cdots + a_n}{n} \leq \sqrt{\frac{a_1^2 + \cdots + a_n^2}{n}}.$$

(Use the arguments of the proof of Theorem 2.7.)

2.16. Find the largest value of $x^2 \cdot (1 - x)$ for $x \in [0, 1]$. (H)

2.17. Find the cylinder of maximal volume contained in a given right circular cone.

2.18. Find the cylinder of maximal volume contained in a given sphere.

2.4 Sets, Functions, Sequences

Sets

Every branch of mathematics involves inspecting some set of objects or elements defined in various ways. In geometry, those elements are points, lines, and planes; in analysis, they are numbers, sequences, and functions; and so on. Therefore, we need to clarify some basic notions regarding sets.

What is a set? Intuitively, a set is a collection, family, system, assemblage, or aggregation of specific things. It is important to note that these terms are redundant and are rather synonyms than definitions. In fact, we could not even accept them as definitions, since we would first have to define a collection, family, system, assemblage, or aggregation, putting us back right where we started. Sometimes, a set is defined as a collection of things sharing some property. Disregarding the fact that we again used the term "collection," we can raise another objection: what do we mean by a shared property? That is a very subjective notion, which we cannot allow in a formal definition. (Take, for example, the set that consists of all natural numbers and all circles in the plane. Whether there is a common property here could be a matter of debate.) So we cannot accept this definition either.

It seems that we are unable to define sets using simpler language. In fact, we have to accept that there are concepts that we cannot express in simpler terms (since otherwise, the process of defining things would never end), so we need to have some fundamental concepts that we do not define. *A set is one of those fundamental concepts. All that we suppose is that every item in our universe of discourse is either an element of a given set or not an element of that set.* (Here we are using the word "or" to mean the complementary or.)

If x is an element of the set H (we also say that x belongs to H, or is in H), we denote this relationship by $x \in H$. If x is not an element of the set (we also say that x does not belong to H, or is not in H), we denote this fact by $x \notin H$.

We can represent sets themselves in two different ways. The ostensibly simpler way is to list the set's elements inside a pair of curly braces: $A = \{1,2,3,4,6,8, 12,24\}$. If a set H contains only one element, x, then we denote it by $H = \{x\}$. We can even list an infinite set of elements if the relationship among its elements is unambiguous, for example $\mathbb{N} = \{0,1,2,3,\dots\}$. The other common way of denoting sets is that after a letter or symbol denoting a generic element we put a colon, and then provide a rule that defines such elements:

$\mathbb{N} = \{n : n \text{ is a nonnegative integer}\} = \{0,1,2,\dots\}$,

$\mathbb{N}^+ = \{n : n \text{ is a positive integer}\} = \{1,2,\dots\}$,

$B = \{n : n \text{ is an odd natural number}\} = \{2n-1 : n = 1,2,\dots\} = \{1,3,5,\dots\}$,

$C = \left\{\dfrac{1}{n} : n = 1,2,\dots\right\} = \left\{1, \dfrac{1}{2}, \dfrac{1}{3}, \dots\right\}$,

$D = \{n : n \mid 24 \text{ and } n > 0\} = \{1,2,3,4,6,8,12,24\}$.

What do we mean by the equalities here? By agreement, we consider two sets A and B to be equal if they have the same elements, that is, for every object x, $x \in A$ if and only if $x \in B$. Using notation:

$$A = B \iff (\forall x)(x \in A \iff x \in B).$$

In other words, $A = B$ if every element of A is an element of B and vice versa. Let us note that if we list an element multiple times, it does not affect the set. For example, $\{1,1,2,2,3,3,3\} = \{1,2,3\}$, and $\{n^2 : n \text{ is an integer}\} = \{n^2 : n \geq 0 \text{ is an integer}\} = \{0,1,4,9,16,\dots\}$.

As we have seen, a set can consist of a single element, for example $\{5\}$. Such a set is called a *singleton*; this set, of course, is not equal to the number 5 (which is not a set), nor to $\{\{5\}\}$ which is indeed a set (and a singleton), but its only element is not 5 but $\{5\}$. We see that an element of a set can be a set itself.

Is it possible for a set to have no elements? Consider the following examples:

$G = \{p : p \text{ is prime, and } 888 \leq p \leq 906\}$,

$H = \{n^2 : n \in \mathbb{N} \text{ and the sum of the decimal digits of } n^2 \text{ is } 300\}$.

At first glance, these sets seem just as legitimate as all the previously listed ones. But if we inspect them more carefully, we find out that they are empty, that is, they have no elements. Should we exclude them from being sets? Were we to do so, then before specifying any set, we would have to make sure that it had an element. Other than making life more difficult, there would be another drawback: we are not always able to check this. *It is a longstanding open problem whether or not there is an odd prefect number* (a number is perfect if it is equal to the sum of its positive proper divisors). If we exclude sets with no elements from being sets, we would not be able to determine whether $\{n : n$ is an odd perfect number$\}$ is a well-defined set.

Thus it seems practical to agree that the above definitions form sets, and so (certainly in the case of G or H above) we accept sets that do not have any elements. How many such sets are there? By our previous notion of equality, there is just one, since if neither A nor B has any elements, then it is vacuously true that every element of A is an element of B, and vice versa, since there are no such elements. We call this single set the **empty set** and denote it by $\emptyset$.

If every element of a set B is an element of a set A, then we say that B is a **subset** of A, or that B is contained in A. Notation: $B \subset A$ or $A \supset B$; we can say equivalently that B is a subset of A and that A contains B.[3]

It is clear that $A = B$ if and only if $A \subset B$ and $B \subset A$. If $B \subset A$, but $B \neq A$, then we say that B is a **proper subset** of A. We denote this by $B \subsetneq A$.

Just as we define operations between numbers (such as addition and multiplication), we can define operations between sets. For two sets A and B, their **union** is the set of elements that belong to at least one of A and B. We denote the union of A and B by $A \cup B$, so

$$A \cup B = \{x : x \in A \vee x \in B\}.$$

We can define the union of any number of sets, even infinitely many: $A_1 \cup A_2 \cup \cdots \cup A_n$ denotes the set of elements that belong to at least one of $A_1, \ldots, A_n$. This same set can also be denoted more concisely by $\bigcup_{i=1}^{n} A_i$. Similarly, $A_1 \cup A_2 \cup \cdots$ or $\bigcup_{i=1}^{\infty} A_i$ denotes the set of all elements that belong to at least one of $A_1, A_2, \ldots$.

For two sets A and B, their **intersection** is the set of elements that belong to both A and B. We denote the intersection of A and B by $A \cap B$, so

$$A \cap B = \{x : x \in A \wedge x \in B\}.$$

The intersection of any finite number of sets or infinitely many sets is defined similarly as for union.

We say that two sets A and B are **disjoint** if $A \cap B = \emptyset$.

For two sets A and B, their **set-theoretic difference** is the set of elements that are in A but not in B. We denote the difference of sets A and B by $A \setminus B$, so

$$A \setminus B = \{x : x \in A \wedge x \notin B\}.$$

Let H be a fixed set, and let $X \subset H$. We call $H \setminus X$ the **complement** (with respect to H) of X, and we denote it by $\overline{X}$. It is easy to see that

$$\overline{\overline{A}} = A,$$

---

[3] Sometimes, containment is denoted by $\subseteq$.

and the so-called De Morgan identities are also straightforward:

$$\overline{A \cap B} = \overline{A} \cup \overline{B}, \quad \overline{A \cup B} = \overline{A} \cap \overline{B}.$$

Here are some further identities:

$$A \setminus A = \emptyset,$$
$$A \setminus \emptyset = A,$$
$$A \cup A = A,$$
$$A \cap A = A,$$
$$A \cup B = B \cup A,$$
$$A \cap B = B \cap A,$$
$$A \cup (B \cup C) = (A \cup B) \cup C = A \cup B \cup C,$$
$$A \cap (B \cap C) = (A \cap B) \cap C = A \cap B \cap C,$$
$$A \cup (B \cap C) = (A \cup B) \cap (A \cup C),$$
$$A \cap (B \cup C) = (A \cap B) \cup (A \cap C),$$
$$A \subset A,$$
$$A \subset B, \; B \subset C \Rightarrow A \subset C,$$
$$\emptyset \subset A. \tag{2.1}$$

Functions

Consider a mathematical expression that contains the variable x, such as

$$x^2 + 1 \quad \text{or} \quad \frac{x+3}{x-2}.$$

To evaluate such an expression for a particular numeric value of x, we have to compute the value of the expression when the given number is substituted for x (this is possible, of course, only if the expression makes sense; in the second example, $x = 2$ is impermissible). These formulas thereby associate certain numbers with other numbers. There are many ways of creating such correspondences between numbers. For example, we could consider the number of positive divisors of an integer n, the sum of the digits of n, etc. We can even associate numbers with other things, for example to each person their weight or how many strands of hair they have. Even more generally, we could associate to each person their name. These examples define functions, or mappings. These two terms are synonymous, and they mean the following.

Consider two sets A and B. Suppose that for each $a \in A$, we somehow associate with a an element $b \in B$. Then this association is called a **function** or **mapping**.

If this function is denoted by f, then we say that f is a function from A to B, and we write $f : A \to B$. If the function f associates $a \in A$ to $b \in B$, we say that f maps a to b (or a is mapped to b under f), or that the value of f at a is b, and we denote this by $b = f(a)$. We call A the **domain** of f.

When we write down a formula, for example $n^2 + 1$, and we want to emphasize that we are not talking about the *number* $n^2 + 1$ but the *mapping* that maps n to $n^2 + 1$, then we can denote this by

$$n \mapsto n^2 + 1 \quad (n \in \mathbb{N}).$$

Sequences

Writing arbitrary elements sequentially, that is, in a particular order, gives us a **sequence**. If the number of elements in the list is n, then we say that we have a sequence of length n. To define a sequence, we need to specify what the first, second, and generally the kth element is, for each $k = 1, \ldots, n$. A sequence of length n is usually denoted by $(a_1, a_2, \ldots, a_n)$, where, of course, instead of a we can use any other letter or symbol. We often call the elements $a_1, \ldots, a_n$ the **terms** of the sequence; the number representing each term's position is called its **index**. We consider two sequences of length n to be equal if their kth terms agree for each $k = 1, \ldots, n$, so the order of the terms matters. That is, $(a_1, a_2, a_3) = (\text{Chicago}, \sqrt{2}, \text{Shakespeare})$ if and only if $a_1 = \text{Chicago}$, $a_2 = \sqrt{2}$, and $a_3 = \text{Shakespeare}$. The terms of the sequence need not be distinct, so $(2, 3, 3, 3)$ is a sequence of length 4.

We sometimes also call sequences of length n **ordered n-tuples**. For $n = 2$, instead of ordered two-tuples we say **ordered pairs**, and for $n = 3$, we usually say **ordered triples**.

If we want to make the ambiguous "sequentially" above more precise, we may say that a sequence of length n is a mapping with domain $\{1, 2, \ldots, n\}$ that maps k to the kth term of the sequence. So the sequence defined by the map $a : \{1, 2, \ldots, n\} \to B$ is $(a(1), a(2), \ldots, a(n))$, or as we previously denoted it, $(a_1, a_2, \ldots, a_n)$.

We will often work with infinite sequences. We obtain these by writing infinitely many elements sequentially. More precisely, we define an **infinite sequence** to be a function with domain $\mathbb{N}^+ = \{1, 2, \ldots\}$. That is, the function $a : \mathbb{N}^+ \to B$ defines a sequence $(a(1), a(2), \ldots)$, which we can also denote by $(a_1, a_2, \ldots)$ or by $(a_i)_{i=1}^{\infty}$.

We also consider functions with domain $\mathbb{N}$ as infinite sequences. These can be denoted by $(a(0), a(1), \ldots)$, or $(a_0, a_1, \ldots)$, or even $(a_i)_{i=0}^{\infty}$. More generally, every function from a set of the form $\{k, k + 1, \ldots\}$ is called an infinite sequence; the notation is straightforward.

Exercises

2.19. Prove the identities in (2.1).

2.20. Prove that $(A \cup B) \setminus (A \cap B) = (A \setminus B) \cup (B \setminus A)$.

2.21. Let $A \triangle B$ denote the set $(A \setminus B) \cup (B \setminus A)$. (We call $A \triangle B$ the **symmetric difference** of A and B.) Show that for arbitrary sets A, B, C,

(a) $A \triangle \emptyset = A$,
(b) $A \triangle A = \emptyset$,
(c) $(A \triangle B) \triangle C = A \triangle (B \triangle C)$.

2.22. Prove that $x \in A_1 \triangle A_2 \triangle \cdots \triangle A_n$ holds if and only if x is an element of an odd number of $A_1, \ldots, A_n$.[4]

2.23. Determine which of the following statements are true and which are false:[5]

(a) $(A \cup B) \setminus A = B$,
(b) $(A \cup B) \setminus C = A \cup (B \setminus C)$,
(c) $(A \setminus B) \cap C = (A \cap C) \setminus B = (A \cap C) \setminus (B \cap C)$,
(d) $A \setminus B = A \setminus (A \cap B)$.

2.24. Let $U(A_1, \ldots, A_n)$ and $V(A_1, \ldots, A_n)$ be expressions that are formed from the operations $\cup$, $\cap$, and $\setminus$ on the sets $A_1, \ldots, A_n$. (For example, $U = A_1 \cap (A_2 \cup A_3)$ and $V = (A_1 \cap A_2) \cup (A_1 \cap A_3)$.) Prove that $U(A_1, \ldots, A_n) = V(A_1, \ldots, A_n)$ holds for every $A_1, \ldots, A_n$ if and only if it holds for all sets $A_1, \ldots, A_n$ such that the nonempty ones are all equal. (∗H)

---

[4] We can write $A_1 \triangle A_2 \triangle \cdots \triangle A_n$, since we showed in part (c) of Exercise 2.21 that different placements of parentheses define the same set (so we may simply omit them).

[5] That is, either prove that the identity holds for every A, B, C, or give some A, B, C for which it does not hold.

Chapter 3
Real Numbers

What are the real numbers? The usual answer is that they comprise the rational and irrational numbers. That is correct, but what are the irrational numbers? They are the numbers whose infinite decimal expansions are infinite and nonrepeating. But for this, we need to know precisely what an infinite decimal expansion is. We obtain an infinite decimal expansion by

1. executing a division algorithm indefinitely, e.g.,

$$\frac{1}{7} = 0,142857142857\ldots,$$

   or
2. locating a point on the number line, e.g.,

$$1 < \sqrt{2} < 2$$
$$1,4 < \sqrt{2} < 1,5$$
$$1,41 < \sqrt{2} < 1,42$$

   etc. Based on this, we can say the decimal expansion of $\sqrt{2}$ is $1.41\ldots$.

We note that the decimal expansion in 1. above also determines the location of the respective (rational) point on the number line. That is, a decimal expansion like $1.41421356\ldots$, always locates a point on the number line.

Now the question is the following: is the decimal expansion the *number itself*, or just a representation of it? The latter hypothesis is supported by the fact that we can write the number in different numeral systems to obtain different representations. For example, the number $1/2$ has the decimal form 0.5, while in binary, $1/2 = 0.1$; in ternary, $1/2 = 0.111\ldots$. But if decimal expansions are just a representation of the number, then what is the number itself? Perhaps it is the *point* that is represented by the decimal expansion?

© Springer New York 2015
M. Laczkovich, V.T. Sós, *Real Analysis*, Undergraduate Texts
in Mathematics, DOI 10.1007/978-1-4939-2766-1_3

We can imagine the real numbers in various ways. But they will always be objects that we can use to measure distance, area, etc., and on which we can have operations (addition, multiplication, etc.). In the end, there are two ways in which we can clarify what real numbers are:

I. **Constructive approach.** We state that one of the above (or another) concepts defines the real numbers. For example, we can declare that the set of real numbers is the set of all infinite decimal expansions. The advantage of such a construction is that it answers the question as to what the real numbers are, albeit arbitrarily. But a disadvantage also arises: those who had a different image of real numbers will have a hard time following.

In a constructive approach, operations are usually not easy to define. (For example, it is not really clear what the product $2 \cdot 0.898899888999\ldots$ should be.) Properties of an operation, such as the distributive law $a(b+c) = ab + ac$, can also be inconvenient to check.

II. **Axiomatic approach.** In the axiomatic construction, we do not state *what* the real numbers are, but what properties they satisfy. Everyone can imagine what they want, but we fix some basic properties, and we refer only to these. Whoever accepts that the real numbers are *like* this also must accept any conclusions that we logically draw from the basic properties. Of course, these properties should be something that we should generally expect from the real numbers.

The axiomatic approach does not make direct constructions useless, since the question arises whether there exists such a construct satisfying the basic properties that we have stated. Several direct constructions are known. The construction by infinite decimal expansions can be seen in the book [1], while another construction will be touched upon in Remark 6.14. A third construction can be seen in Walter Rudin's textbook [7].

In the following sections, we follow an axiomatic approach to the real numbers. In particular, the notion of real numbers will be a fundamental concept. The basic properties that we accept without proof are called the *axioms* of the real numbers. We state the axioms in four groups. The first group consists of axioms regarding operations (addition and multiplication), while the second is made up of properties about ordering (less than, greater than). The third and fourth groups are just one axiom each. These will express that there are "arbitrarily large" natural numbers, as well as the fact that the set of real numbers is "complete".

I. Field Axioms

The first group of axioms requires some further fundamental concepts. We denote the set of real numbers by $\mathbb{R}$. We assume that there are two operations defined on the real numbers, called *addition* and *multiplication*. By this, we mean that for any two (not necessarily distinct) real numbers $a, b \in \mathbb{R}$, there correspond a number denoted by $a + b$ (the sum of a and b) as well as a number denoted by $a \cdot b$ (the product of *a* and *b*). We also suppose that two distinct numbers 0 and 1 are specified. The first group of axioms deals with these concepts.

1. *Commutativity of addition: $a + b = b + a$ for each $a, b \in \mathbb{R}$.*
2. *Associativity of addition: $(a+b)+c = a+(b+c)$ for each $a, b, c \in \mathbb{R}$.*
3. *$a + 0 = a$ for each $a \in \mathbb{R}$.*
4. *For each $a \in \mathbb{R}$, there exists a $b \in \mathbb{R}$ such that $a + b = 0$.*
5. *Commutativity of multiplication: $a \cdot b = b \cdot a$ for each $a, b \in \mathbb{R}$.*
6. *Associativity of multiplication: $(a \cdot b) \cdot c = a \cdot (b \cdot c)$ for each $a, b, c \in \mathbb{R}$.*
7. *$a \cdot 1 = a$ for each $a \in \mathbb{R}$.*
8. *For each $a \in \mathbb{R}$, $a \neq 0$, there exists a $b \in \mathbb{R}$ such that $a \cdot b = 1$.*
9. *Distributivity: $a \cdot (b + c) = a \cdot b + a \cdot c$ for each $a, b, c \in \mathbb{R}$.*

If two operations that satisfy the nine axioms above are defined on a set, then we say that the set with the two operations is a **field**. (This includes the specification of the distinct elements 0 and 1.) The first group of axioms thus tells us that the real numbers form a field. This is why we call axioms 1–9 the **field axioms**.

It is easy to show that for each $a \in \mathbb{R}$, there is exactly one $b \in \mathbb{R}$ such that $a + b = 0$ (see Theorem 3.28 in the first appendix). We denote this unique b by $-a$. Similarly, for each $a \neq 0$ there is exactly one $b \in \mathbb{R}$ such that $a \cdot b = 1$ (see Theorem 3.30 in the first appendix). This element b will be denoted by $\frac{1}{a}$ or $1/a$. If $c \neq 0$, then we denote $a \cdot (1/c)$ by $\frac{a}{c}$ or a/c.

We can show that all the usual properties of arithmetic we are used to and use without hesitation in solving algebraic expressions follow from the field axioms. We list these properties in the first appendix of the chapter, proving the most important ones.

II. Order axioms

The second group of axioms requires another fundamental concept. We suppose that there is a so-called **order relation**, denoted by $<$ (less than), defined on the real numbers. By this, we mean that for any two real numbers a and b, $a < b$ is either true or false. (We could also formulate this as follows: we have a map from the ordered pairs of real numbers to the logical statements "true" and "false." If (a, b) maps to "true," then we denote this by $a < b$.) The order axioms fix some properties of this order relation.

10. *Trichotomy: For any two real numbers a and b, exactly one of $a < b$, $a = b$, $b < a$ is true.*
11. *Transitivity: For each $a, b, c \in \mathbb{R}$, if $a < b$ and $b < c$, then $a < c$.*
12. *For each $a, b, c \in \mathbb{R}$, if $a < b$, then $a + c < b + c$.*
13. *For each $a, b, c \in \mathbb{R}$, if $a < b$ and $0 < c$, then $a \cdot c < b \cdot c$.*

To talk about consequences of the order axioms, we need to introduce some notation. We say that $a \le b$ (a is less than or equal to b) if $a < b$ or $a = b$. The notation $a > b$ is equivalent to $b < a$; $a \ge b$ is equivalent to $b \le a$.

We call the number a **positive** if $a > 0$; **negative** if $a < 0$; **nonnegative** if $a \ge 0$; **nonpositive** if $a \le 0$.

We call numbers that we get from adding 1 repeatedly **natural numbers**:

$$1,\ 1+1 = 2,\ 1+1+1 = 2+1 = 3, \ldots.$$

The set of natural numbers is denoted by $\mathbb{N}^+$.

The **integers** are the natural numbers, these times -1, and 0. The set of integers is denoted by $\mathbb{Z}$.

We denote the set of nonnegative integers by $\mathbb{N}$. That is, $\mathbb{N} = \mathbb{N}^+ \cup \{0\}$.

A real number is **rational** if we can write it as p/q, where p and q are integers and $q \ne 0$. We will denote the set of rational numbers by $\mathbb{Q}$.

A real number is called **irrational** if it is not rational.

As with the case of field axioms, any properties and rules we have previously used to deal with inequalities follow from the order axioms above. We will discuss these further in the second appendix of this chapter. We recommend that the reader look over the Bernoulli inequality as well as the relationship between the arithmetic and geometric means, as well as the harmonic and arithmetic means (Theorems 2.5, 2.6, 2.7), and check to make sure that the properties used in the proofs all follow from the order axioms.

We highlight one consequence of the order axioms, namely that the natural numbers are positive and distinct; more precisely, they satisfy the inequalities $0 < 1 < 2 < \cdots$ (see Theorem 3.38 in the second appendix). Another important fact is that there are no neighboring real numbers. That is, *for arbitrary real numbers $a < b$, there exists a c such that $a < c < b$,* for example, $c = (a+b)/2$ (see Theorem 3.40).

Definition 3.1. The *absolute value* of a real number a, denoted by $|a|$, is defined as follows:

$$|a| = \begin{cases} a, & \text{if } a \ge 0, \\ -a, & \text{if } a < 0. \end{cases}$$

From the definition of absolute value and the previously mentioned consequences of the order axioms, it is easy to check the statements below for arbitrary real numbers a and b:

$|a| \ge 0$, and $|a| = 0$ if and only if $a = 0$;

$|a| = |-a|$;

$|a \cdot b| = |a| \cdot |b|$;

If $b \ne 0$, then $\left| \dfrac{1}{b} \right| = \dfrac{1}{|b|}$ and $\left| \dfrac{a}{b} \right| = \dfrac{|a|}{|b|}$;

Triangle Inequality: $|a+b| \le |a| + |b|$; $||a| - |b|| \le |a - b|$.

III. The Axiom of Archimedes

14. *For an arbitrary real number b, there exists a natural number n larger than b.*

If we replace b by b/a (where a and b are positive numbers) in the above axiom, we get the following consequence:

If a and b are arbitrary positive numbers, then there exists a natural number n such that $n \cdot a > b$.

And if instead of b we write $1/\varepsilon$, where $\varepsilon > 0$, then we get:

If ε is an arbitrary positive number, then there exists a natural number n such that $1/n < \varepsilon$.

An important consequence of the axiom of Archimedes is that the rational numbers are "everywhere dense" within the real numbers.

Theorem 3.2. *There exists a rational number between any two real numbers.*

Proof. Let $a < b$ be real numbers, and suppose first that $0 \le a < b$. By the axiom of Archimedes, there exists a positive integer n such that $1/n < b - a$. By the same axiom, there exists a positive integer m such that $a < m/n$. Let k be the smallest positive integer for which $a < k/n$. Then

$$\frac{k-1}{n} \le a < \frac{k}{n},$$

and thus

$$\frac{k}{n} - a \le \frac{k}{n} - \frac{k-1}{n} = \frac{1}{n} < b - a.$$

So what we have obtained is that $a < k/n < b$, so we have found a rational number between a and b.

Now suppose that $a < b \le 0$. Then $0 \le -b < -a$, so by the above argument, there is a rational number r such that $-b < r < -a$. Then $a < -r < b$, and the statement is again true.

Finally, if $a < 0 < b$, then there is nothing to prove, since 0 is a rational number. $\square$

Remark 3.3. The axiom of Archimedes is indeed necessary, since it does not follow from the previous field and order axioms. We can show this by giving an example of an ordered field (a set with two operations and an order relation where the field and order axioms hold) where the axiom of Archimedes is not satisfied. We outline the construction.

We consider the set of polynomials in a single variable with integer coefficients, that is, expressions of the form $a_n x^n + a_{n-1} x^{n-1} + \cdots + a_1 x + a_0$, where $n \ge 0$, $a_0, \ldots, a_n$ are integers, and $a_n \ne 0$, and also the expression 0. We denote the set of polynomials with integer coefficients by $\mathbb{Z}[x]$. We say that the polynomial $a_n x^n + a_{n-1} x^{n-1} + \cdots + a_1 x + a_0$ is nonzero if at least one of its coefficients a_i is nonzero. In this case, the leading coefficient of the polynomial is a_k, where k is the largest index with $a_k \neq 0$. Expressions of the form p/q, where $p, q \in \mathbb{Z}[x]$ and $q \neq 0$, are called rational expressions with integer coefficients (often simply rational expressions). We denote the set of rational expressions by $\mathbb{Z}(x)$. We consider the rational expressions p/q and r/s to be equal if $ps = qr$; that is, if ps and qr expand to the same polynomial.

We define addition and multiplication of rational expressions as

$$\frac{p}{q} + \frac{r}{s} = \frac{ps + qr}{qs} \quad \text{and} \quad \frac{p}{q} \cdot \frac{r}{s} = \frac{pr}{qs}.$$

One can show that with these operations, the set $\mathbb{Z}(x)$ becomes a field (where $0/1$ and $1/1$ play the role of the zero and the identity element respectively). We say that the rational expression p/q is positive if the signs of the leading coefficients of p and q agree. We say that the rational expression f is less than the rational expression g if $g - f$ is positive. We denote this by $f < g$. It is easy to check that $\mathbb{Z}(x)$ is an ordered field; that is, the order axioms are also satisfied. In this structure, the natural numbers will be the constant rational expressions $n/1$.

Now we show that in the above structure, the axiom of Archimedes is not met. We must show that there exists a rational expression that is greater than every natural number. We claim that the expression $x/1$ has this property. Indeed, $(n/1) < (x/1)$ for each n, since the difference $(x/1) - (n/1) = (x - n)/1$ is positive.

IV. Cantor's Axiom

The properties listed up until now (the 14 axioms and their consequences) still do not characterize the real numbers, since it is clear that the rational numbers satisfy those same properties. On the other hand, there are properties that we expect from the real numbers but are not true for rational numbers. For example, we expect a solution in the real numbers to the equation $x^2 = 2$, but we know that a rational solution does not exist (Theorem 2.1).

The last, so-called Cantor's axiom,[1] plays a central role in analysis. It expresses the fact that the set of real numbers is in some sense "complete."

To state the axiom, we need some definitions and notation. Let $a < b$. We call the set of real numbers x for which $a \leq x \leq b$ is true a **closed interval**, and we denote it by $[a, b]$. The set of x for which $a < x < b$ holds is denoted by (a, b), and we call it an **open interval**.

Let $I_n = [a_n, b_n]$ be a closed interval for every natural number n. We say that $I_1, I_2, \ldots$ form a **sequence of nested closed intervals** if $I_1 \supset I_2 \supset \ldots \supset I_n \supset \ldots$, that is, if

$$a_n \leq a_{n+1} < b_{n+1} \leq b_n$$

holds for each n. We can now formulate Cantor's axiom.

---

[1] Georg Cantor (1845–1918) German mathematician.

15. *Every sequence of nested closed intervals $I_1 \supset I_2 \supset \ldots$ has a common point, that is, there exists a real number x such that $x \in I_n$ for each n.*

Remark 3.4. It is important that the intervals I_n be closed: a sequence of nested open intervals does not always contain a common element. If, for example, $J_n = (0, 1/n)$ $(n = 1, 2, \ldots)$, then $J_1 \supset J_2 \supset \ldots$, but the intervals J_n do not contain a common element. For if $x \in J_1$, then $x > 0$. Then by the axiom of Archimedes, there exists an n for which $1/n < x$, and for this n, we see that $x \notin J_n$.

Let us see when the nested closed intervals $I_1, I_2, \ldots$ have exactly one common point.

Theorem 3.5. *The sequence of nested closed intervals $I_1, I_2, \ldots$ has exactly one common point if and only if there is no positive number that is smaller than every $b_n - a_n$, that is, if for every $\delta > 0$, there exists an n such that $b_n - a_n < \delta$.*

Proof. If x and y are both shared elements and $x < y$, then $a_n \leq x < y \leq b_n$, and so $y - x \leq b_n - a_n$ for each n. In other words, if the closed intervals $I_1, I_2, \ldots$ contain more than one common element, then there exists a positive number smaller than each $b_n - a_n$.

Conversely, suppose that $b_n - a_n \geq \delta > 0$ for all n. Let x be a common point of the sequence of intervals. If $b_n \geq x + (\delta/2)$ for every n, then $x + (\delta/2)$ is also a common point. Similarly, if $a_n \leq x - (\delta/2)$ for all n, then $x - (\delta/2)$ is also a common point. One of these two cases must hold, since if $b_n < x + (\delta/2)$ for some n and $a_m > x - (\delta/2)$ for some m, then for $k \geq \max(n, m)$, we have that $x - (\delta/2) < a_k < b_k < x + (\delta/2)$ and $b_k - a_k < \delta$, which is impossible. □

Cantor's axiom concludes the axioms of the real numbers. With this axiomatic construction, by the real numbers we mean a structure that satisfies axioms 1–15. We can also express this by saying that

the real numbers form an Archimedean ordered field in which Cantor's axiom is satisfied.

As we mentioned before, such a field exists. A sketch of the construction of a field satisfying the conditions will be given in Remark 6.14.

Before beginning our detailed exposition of the theory of analysis, we give an important example of the application of Cantor's axiom. If $a \geq 0$ and k is a positive integer, then $\sqrt[k]{a}$ denotes the nonnegative number whose kth power is a. But it is not at all obvious that such a number exists. As we saw, the first 14 axioms do not even guarantee the existence of $\sqrt{2}$. We show that from the complete axiom system, the existence of $\sqrt[k]{a}$ follows.

Theorem 3.6. *If $a \geq 0$ and k is a positive integer, then there exists exactly one non-negative real number b for which $b^k = a$.*

Proof. We can suppose that $a > 0$. We give the proof only for the special case $k = 2$; the general case can be proved similarly. (Later, we give another proof of the theorem; see Corollary 10.59.)

We will find the b we want as the common point of a sequence of nested closed intervals. Let u_1 and v_1 be nonnegative numbers for which $u_1^2 \le a \le v_1^2$. (Such are, for example, $u_1 = 0$ and $v_1 = a + 1$, since $(a+1)^2 > 2 \cdot a > a$.)

Suppose that $n \ge 1$ is an integer, and we have already defined the numbers u_n and v_n such that

$$u_n^2 \le a \le v_n^2 \tag{3.1}$$

holds. We distinguish two cases. If

$$\left(\frac{u_n + v_n}{2} \right)^2 < a,$$

then let

$$u_{n+1} = \frac{u_n + v_n}{2} \quad \text{and} \quad v_{n+1} = v_n.$$

But if

$$\left(\frac{u_n + v_n}{2} \right)^2 \ge a,$$

then let

$$u_{n+1} = u_n \quad \text{and} \quad v_{n+1} = \frac{u_n + v_n}{2}.$$

It is clear that in both cases, $[u_{n+1}, v_{n+1}] \subset [u_n, v_n]$ and

$$u_{n+1}^2 \le a \le v_{n+1}^2.$$

With this, we have defined u_n and v_n for each n. It follows from the definition that the intervals $[u_n, v_n]$ are nested closed intervals, so by Cantor's axiom, there exists a common point.

If b is a common point, then $u_n \le b \le v_n$, so

$$u_n^2 \le b^2 \le v_n^2 \tag{3.2}$$

holds for each n. Thus by (3.1), a and b^2 are both common points of the interval system $[u_n^2, v_n^2]$. We want to see that $b^2 = a$. By Theorem 3.5, it suffices to show that for every $\delta > 0$, there exists n such that $v_n^2 - u_n^2 < \delta$.

We obtained the $[u_{n+1}, v_{n+1}]$ interval by "halving" the interval $[u_n, v_n]$ and taking one of the halves. From this, we see clearly that $v_{n+1} - u_{n+1} = (v_n - u_n)/2$. Of course, we can also conclude this from the definition, since

$$v_n - \frac{u_n + v_n}{2} = \frac{u_n + v_n}{2} - u_n = \frac{v_n - u_n}{2}.$$

Then by induction, we see that $v_n - u_n = (v_1 - u_1)/2^{n-1}$ for each n. From this, we see that

$$v_n^2 - u_n^2 = (v_n - u_n) \cdot (v_n + u_n) \leq \frac{v_1 - u_1}{2^{n-1}} \cdot (v_1 + v_1) \leq \frac{2 \cdot v_1^2}{2^{n-1}} = \frac{4 \cdot v_1^2}{2^n} \leq \frac{4 \cdot v_1^2}{n} \quad (3.3)$$

for each n. Let δ be an arbitrary positive number. By the axiom of Archimedes, there exists n for which $4 \cdot v_1^2/n$ is smaller than δ. This shows that for suitable n, we have $v_n^2 - u_n^2 < \delta$, so by Theorem 3.5, $b^2 = a$.

The uniqueness of b is clear. For if $0 < b_1 < b_2$, then $b_1^2 < b_2^2$, so only one of b_1^2 and b_2^2 can be equal to a.   □

Let us note that Theorem 2.1 became truly complete only now. In Chapter 1, when stating the theorem, we did not care whether the number denoted by $\sqrt{2}$ actually *exists*. In the proof of Theorem 2.1, we proved only that *if $\sqrt{2}$ exists*, then it cannot be rational. It is true, however, that in this proof we used only the field and order axioms (check).

Exercises

3.1. Consider the set $\{0, 1, \ldots, m-1\}$ with addition and multiplication modulo m. (By this, we mean that $i + j \equiv k$ if the remainders of $i + j$ and k on dividing by m are the same, and similarly $i \cdot j \equiv k$ if the remainders of $i \cdot j$ and k on dividing by m agree.) Show that this structure satisfies the field axioms if and only if m is prime.

3.2. Give an addition and a multiplication rule on the set $\{0, 1, a, b\}$ that satisfy the field axioms.

3.3. Let F be a subset of the real numbers such that $1 \in F$ and $F \neq \{0, 1\}$. Suppose that if $a, b \in F$ and $a \neq 0$, then $(1/a) - b \in F$. Prove that F is a field.

3.4. Prove that a finite field cannot be ordered in a way that it satisfies the order axioms. (H)

3.5. Using the field and order axioms, deduce the properties of the absolute value listed.

3.6. Check that the real numbers with the operation $(a, b) \mapsto a + b + 1$ satisfy the first four axioms. What is the zero element? Define a multiplication with which we get a field.

3.7. Check that the positive real numbers with the operation $(a, b) \mapsto a \cdot b$ satisfy the first four axioms. What is the zero element? Define a multiplication with which we get a field.

3.8. Check that the positive rational numbers with the operation $(a, b) \mapsto a \cdot b$ satisfy the first four axioms. What is the zero element? Prove that *there is no* multiplication here that makes it a field.

3.9. Check that the set of rational expressions with the operations and ordering given in Remark 3.3 satisfy the field and order axioms.

3.10. Is it true that the set of rational expressions with the given operations satisfies Cantor's axiom? (H)

3.11. In Cantor's axiom, we required the sequence of nested intervals to be made up of closed, bounded, and nonempty intervals. Check that the statement of Cantor's axiom is no longer true if any of these conditions are omitted.

3.1 Decimal Expansions: The Real Line

As we mentioned before, the decimal expansion of a real number "locates a point on the number line." We will expand on this concept now. First of all, we will give the exact definition of the decimal expansion of a real number. We will need the conventional notation for finite decimal expansions: if n is a nonnegative integer and each $a_1, \ldots a_k$ is one of $0, 1, \ldots, 9$, then $n.a_1 \ldots a_k$ denotes the sum

$$n + \frac{a_1}{10} + \frac{a_2}{10^2} + \cdots + \frac{a_k}{10^k}.$$

Let x be an arbitrary nonnegative real number. We say that the **decimal expansion** of x is $n.a_1 a_2 \ldots$ if

$$n \leq x \leq n+1,$$

$$n.a_1 \leq x \leq n.a_1 + \frac{1}{10},$$

$$n.a_1 a_2 \leq x \leq n.a_1 a_2 + \frac{1}{10^2}, \tag{3.4}$$

and so on hold.

In other words, the decimal expansion of $x \geq 0$ is $n.a_1 a_2 \ldots$ if

$$n.a_1 \ldots a_k \leq x \leq n.a_1 \ldots a_k + \frac{1}{10^k} \tag{3.5}$$

holds for each positive integer k.

Several questions arise from the above definitions. Does every real number have a decimal expansion? Is the expansion unique? Is every decimal expansion the decimal expansion of a real number? (That is, for a given decimal expansion, does there exist a real number whose decimal expansion agrees with it?) The following theorems give answers to these questions.

Theorem 3.7. *Every nonnegative real number has a decimal expansion.*

Proof. Let $x \geq 0$ be a given real number. By the axiom of Archimedes, there exists a positive integer larger than x. If k is the smallest positive integer larger than x, and

$n = k - 1$, then $n \leq x < n + 1$. Since $n + (a/10)$ $(a = 0, 1, \ldots, 10)$ is not larger than x for $a = 0$ but larger for $a = 10$, there is an $a_1 \in \{0, 1, \ldots, 9\}$ for which $n.a_1 \leq x < (n.a_1) + 1/10$. Since $(n.a_1) + (a/10^2)$ $(a = 0, 1, \ldots, 10)$ is not larger than x for $a = 0$ but larger for $a = 10$, there is an $a_2 \in \{0, 1, \ldots, 9\}$ for which $n.a_1 a_2 \leq x < (n.a_1 a_2) + 1/10^2$. Repeating this process, we get the digits $a_1, a_2, \ldots$, which satisfy (3.5) for each k. $\qquad\square$

Let us note that in the above theorem, we saw that for each $x \geq 0$, there is a decimal expansion $n.a_1 a_2 \ldots$ for which the stronger inequality

$$n.a_1 \ldots a_k \leq x < n, a_1 \ldots a_k + \frac{1}{10^k} \qquad (3.6)$$

holds for each positive integer k. Decimal expansions with this property are unique, since there is only one nonnegative integer for which $n \leq x < n + 1$, only one digit a_1 for which $n.a_1 \leq x < (n.a_1) + \frac{1}{10}$, and so on.

Now if x has another decimal expansion $m.b_1 b_2 \ldots$, then this cannot satisfy (3.6), so either $x = m + 1$ or $x = m.b_1 \ldots b_k + (1/10^k)$ for some k. It is easy to check that in the case of $x = m + 1$, we have $n = m + 1$, $a_i = 0$ and $b_i = 9$ for each i; in the case of $x = m.b_1 \ldots b_k + (1/10^k)$ we get $a_i = 0$ and $b_i = 9$ for each $i > k$. We have then the following theorem.

Theorem 3.8. *Positive numbers with a finite decimal expansions have two infinite decimal expansions: one has all 0 digits from a certain point on, while the other repeats the digit 9 after some point. Every other nonnegative real number has a unique decimal expansion.*

The next theorem expresses the fact that the decimal expansions of real numbers contain all formal decimal expansions.

Theorem 3.9. *For arbitrary $n \in \mathbb{N}$ and $(a_1, a_2, \ldots)$ consisting of digits $\{0, 1, \ldots, 9\}$, there exists exactly one nonnegative real number whose decimal expansion is $n.a_1 a_2 \ldots$.*

Proof. The condition (3.5) expresses the fact that x is an element of

$$I_k = \left[n, a_1 \ldots a_k, \ n, a_1 \ldots a_k + \frac{1}{10^k} \right]$$

for each k. Since these are nested closed intervals, Cantor's axiom implies that there exists a real number x such that $x \in I_k$ for each k. The fact that this x is unique follows from Theorem 3.5, since by the axiom of Archimedes, for each $\delta > 0$, there exists a k such that $1/k < \delta$, and then $1/10^k < 1/k < \delta$. $\qquad\square$

Remarks 3.10. **1.** Being able to write real numbers in decimal form has several interesting consequences. The most important corollary concerns the question of how accurately the axioms of the real numbers describe the real numbers. The question is whether we "forgot" something from the axioms; is it not possible that further properties are needed? To understand the answer, let us recollect the idea behind the axiomatic approach. Recall that we pay no attention to what the real numbers actually are, just the properties they satisfy. Instead of the set $\mathbb{R}$ we are used to, we could take another set $\mathbb{R}'$, assuming that there are two operations and a relation defined on it that satisfy the 15 axioms.

The fact that we were able to deduce our results about decimal expansions using only the 15 axioms means that these are true in both $\mathbb{R}$ and $\mathbb{R}'$. So we can pair each nonnegative element of x with the $x' \in \mathbb{R}'$ that has the same decimal expansion. We extend this association to $\mathbb{R}$ by setting $(-x)' = -x'$. By the above theorems regarding decimal expansions, we have a one-to-one correspondence[2] between $\mathbb{R}$ and $\mathbb{R}'$.

It can also be shown (although we do not go into detail here) that this correspondence commutes with the operations, that is, $(x+y)' = x' + y'$ and $(x \cdot y)' = x' \cdot y'$ for each $x, y \in \mathbb{R}$, and moreover, $x < y$ holds if and only if $x' < y'$. The existence of such a correspondence (or isomorphism, for short) shows us that $\mathbb{R}$ and $\mathbb{R}'$ are "indistinguishable": if a statement holds in one, then it holds in the other as well. This fact is expressed in mathematical logic by saying that *any two models of the axioms of real numbers are isomorphic*. So if our goal is to describe the properties of the real numbers as precisely as possible, then we have reached this goal; including other axioms cannot restrict the class of models satisfying the axioms of real numbers any further.

2. Another important consequence of being able to express real numbers as infinite decimals is that we get *a one-to-one correspondence between the set of real numbers and points on a line*.

Let e be a line, and let us pick two different points P and Q that lie on e. Call the direction $\overrightarrow{PQ}$ positive, and the opposite direction negative. In our correspondence we assign the number 0 to P, and 1 to Q. From the point P, we can then measure multiples of the distance $\overline{PQ}$ in both positive and negative directions, and assign the integers to the corresponding points. If we subdivide each segment determined by these integer points into k smaller equal parts (for each k), we get the points that we can map to rational numbers. Let x be a nonnegative real number, and let its decimal expansion be $n.a_1 a_2 \dots$. Let A_k and B_k be the points that we mapped to $n.a_1 \dots a_k$, and $(n.a_1 \dots a_k) + (1/10^k)$ respectively. It follows from the properties of the line that there is exactly one common point of the segments $A_k B_k$. This will be the point A that we map to x. Finally, measuring the segment PA in the negative direction from P yields us the point that will be mapped to $-x$.

It can be shown that this has given us a one-to-one correspondence between the points of e and the real numbers. The real number corresponding to a certain point on the line is called its **coordinate**, and the line itself is called a **number line**, or the **real line**. In the future, when we talk about a number x, we will sometimes refer to it as the point with the coordinate x, or simply the point x.

---

[2] We say that a function f is a **one-to-one correspondence** (or a **bijective map**, or a **bijection**) between sets A and B if different points of A get mapped to different points of B (that is, if $a_1 \neq a_2$ then $f(a_1) \neq f(a_2)$), and for every $b \in B$ there exists an $a \in A$ such that $f(a) = b$.

The benefit of this correspondence is that we can better see or understand certain statements and properties of the real numbers when they are viewed as the number line. Many times, we get ideas for proofs by looking at the real line. It is, however, important to note that properties that we can "see" on the number line cannot be taken for granted, or as proved; in fact, statements that seem to be true on the real line might turn out to be false. In our proofs, we can refer only to the fundamental principles of real numbers (that is the axioms) and the theorems already established.

Looking at the real numbers as the real line suggests many concepts that prove to be important regardless of our visual interpretation. Such is, for example, the notion of an everywhere dense set.

Definition 3.11. We say that a set of real numbers H is *everywhere dense* in $\mathbb{R}$ if every open interval contains elements of H; that is, if for every $a < b$, there exists an $x \in H$ for which $a < x < b$.

So for example, by Theorem 3.2, the set of rational numbers is everywhere dense in $\mathbb{R}$. We now show that the same is true for the set of irrational numbers.

Theorem 3.12. *The set of irrational numbers is everywhere dense in $\mathbb{R}$.*

Proof. Let $a < b$ be arbitrary. Since the set of rational numbers is everywhere dense and $a - \sqrt{2} < b - \sqrt{2}$, there is a rational number r such that $a - \sqrt{2} < r < b - \sqrt{2}$. Then $a < r + \sqrt{2} < b$, and the open interval (a,b) contains the irrational number $r + \sqrt{2}$. The irrationality of $r + \sqrt{2}$ follows from the fact that if it were rational, then $\sqrt{2} = (r + \sqrt{2}) - r$ would also be rational, whereas it is not. □

Motivated by the visual representation of the real line, we shall sometimes use the word *segment* instead of *interval*. Now we will expand the concept of intervals (or segments).

Let H be a closed or open interval. Then clearly, H contains every segment whose endpoints are contained in H (this property is called convexity). It is reasonable to call every set that satisfies this property an interval. Other than closed and open intervals, the following are intervals as well.

Let $a < b$. The set of all x for which $a \leq x < b$ is denoted by $[a,b)$ and is called a **left-closed, right-open interval**. The set of all x that satisfy $a < x \leq b$ is denoted by $(a,b]$ and is called a **left-open, right-closed interval**. That is,

$$[a,b) = \{x : a \leq x < b\} \text{ and } (a,b] = \{x : a < x \leq b\}.$$

Intervals of the form $[a,b]$, (a,b), $[a,b)$, and $(a,b]$ are called **bounded** (or finite) intervals. We also introduce the notation

$$\begin{array}{ll} (-\infty, a] = \{x : x \leq a\}, & [a, \infty) = \{x : x \geq a\}, \\ (-\infty, a) = \{x : x < a\}, & (a, \infty) = \{x : x > a\}, \end{array} \tag{3.7}$$

as well as $(-\infty, \infty) = \mathbb{R}$.

Intervals of the form $(-\infty, a]$, $(-\infty, a)$, $[a, \infty)$, (a, ∞) as well as $(-\infty, \infty) = \mathbb{R}$ itself are called **unbounded** (or infinite) intervals. Out of these, $(-\infty, a]$ and $[a, \infty)$ are **closed half-lines** (or closed rays), while $(-\infty, a)$ and (a, ∞) are **open half-lines** (or open rays).

We consider the empty set and a set consisting of a single point (a singleton) as intervals too; these are the **degenerate** intervals. Singletons are considered to be closed intervals, which is expressed by the notation $[a, a] = \{a\}$.

Remark 3.13. The symbol ∞ appearing in the unbounded intervals does not represent a specific object. The notation should be thought of as an abbreviation. For example, $[a, \infty)$ is merely a shorter (and more expressive) way of writing the set $\{x : x \geq a\}$. The symbol ∞ will pop up many more times. In every case, the same is true, so we give meaning (in each case clearly defined) only to the whole expression.

Exercises

3.12. Prove that

(a) If x and y are rational, then $x + y$ is rational.
(b) If x is rational and y is irrational, then $x + y$ is irrational.

Is it true that if x and y are irrational then $x + y$ is irrational?

3.13. Prove that the decimal expansion of a positive real number is periodic if and only if the number is rational.

3.14. Prove that the set of numbers having finite decimal expansions and their negatives is everywhere dense.

3.15. Partition the number line into the union of infinitely many pairwise disjoint everywhere dense sets.

3.16. Let $H \subset \mathbb{R}$ be a nonempty set that contains the difference of every pair of its (not necessarily distinct) elements. Prove that either there is a real number a such that $H = \{n \cdot a : n \in \mathbb{Z}\}$, or H is everywhere dense. $(*H)$

3.17. Prove that if α is irrational, then the set $\{n \cdot \alpha + k : n, k \in \mathbb{Z}\}$ is everywhere dense.

3.2 Bounded Sets

If a set of real numbers A is finite and nonempty, then there has to be a largest element among them (we can easily check this by induction on the number of elements in A). If, however, a set has infinitely many elements, then there is not necessarily a largest element. Clearly, none of the sets $\mathbb{R}$, $[a,\infty)$, and $\mathbb{N}$ have a largest element. These sets all share the property that they have elements larger than every given real number (in the case of $\mathbb{N}$, this is the axiom of Archimedes). We see that such a set can never have a largest element.

Sets without the above property are more interesting, that is, sets whose elements are all less than some real number. Such is, for example, the open interval (a,b), whose elements are all smaller than b. However the set (a,b) does not have a largest element: if $x \in (a,b)$, then $x < b$, so by Theorem 3.40, there exists a real number y such that $x < y < b$, so automatically $y \in (a,b)$. Another example is the set

$$B = \left\{ \frac{1}{2}, \frac{2}{3}, \ldots, \frac{n-1}{n}, \ldots \right\}. \tag{3.8}$$

Every element of this is less than 1, but there is no largest element, since for every element $(n-1)/n$, the next one, $n/(n+1)$ is larger.

To understand these phenomena better, we introduce some definitions and notation. If a set A has a largest (or in other words, maximal) element, we denote it by $\max A$. If a set A has a smallest (or in other words, minimal) element, we denote it by $\min A$. If the set A is finite, $A = \{a_1, \ldots a_n\}$, then instead of $\max A$, we can write $\max_{1 \le i \le n} a_i$, and similarly, we can write $\min_{1 \le i \le n} a_i$ instead of $\min A$.

Definition 3.14. We say that a set of real numbers A is *bounded from above* if there exists a number b such that for every element x of A, $x \le b$ holds (that is no element of A is larger than b). Every number b with this property is called an *upper bound* of A. So the set A is bounded from above exactly when it has an upper bound.

The set A is *bounded from below* if there exists a number c such that for every element x of A, the inequality $x \ge c$ holds (that is, no element of A is smaller than c). Every number c with such a property is called a *lower bound* of A. The set A is bounded from below if and only if it has a lower bound.

We say A is *bounded* when it is bounded both from above and from below.

Let us look at the previous notions using these new definitions. If $\max A$ exists, then it is clearly also an upper bound of A, so A is bounded from above. The following implications hold:

$$(A \text{ is finite and nonempty}) \Rightarrow (\max A \text{ exists}) \Rightarrow (A \text{ is bounded from above}).$$

The reverse implications are not usually true, since, for example, the closed interval $[a,b]$ has a largest element but is not finite, while (a,b) is bounded from above but does not have a largest element.

Further examples: The set $\mathbb{N}^+$ is bounded from below (in fact, it has a smallest element) but is not bounded from above.

$\mathbb{Z}$ is not bounded from above or from below.

Every finite interval is a bounded set.

Half-lines are not bounded.

The set $C = \{1, \frac{1}{2}, \ldots, \frac{1}{n} \ldots\}$ is bounded, since the greatest element is 1, so it is also an upper bound. On the other hand, every element of C is nonnegative, so 0 is a lower bound. The set does not have a smallest element.

Let us note that 0 is the *greatest* lower bound of C. Clearly, if $\varepsilon > 0$, then by the axiom of Archimedes, there is an n such that $1/n < \varepsilon$. Since $1/n \in C$, this means that ε is not a lower bound.

We can see similar behavior in the case of the open interval (a,b). As we saw, the set does not have a greatest element. But we can again find a *least* upper bound: it is easy to see that the number b is the least upper bound of (a,b). Or take the set B defined in (3.8). This does not have a largest element, but it is easy to see that among the upper bounds, there is a smallest one, namely 1. The following important theorem expresses the fact that this is true for every (nonempty) set bounded from above. Before stating the theorem, let us note that if b is an upper bound of a set H, then every number greater than b is also an upper bound. Thus a set bounded from above always has infinitely many upper bounds.

Theorem 3.15. *Every nonempty set that is bounded from above has a least upper bound.*

Proof. Let A be a nonempty set that is bounded from above. We will obtain the least upper bound of A as a common point of a sequence of nested closed intervals (similarly to the proof of Theorem 3.6).

Let v_1 be an upper bound of A. Let, moreover, a_0 be an arbitrary element of A (this exists, for we assumed $A \neq \emptyset$), and pick an arbitrary number $u_1 < a_0$. Then u_1 is not an upper bound of A, and $u_1 < v_1$ (since $u_1 < a_0 \leq v_1$).

Let us suppose that $n \geq 1$ is an integer and we have defined the numbers $u_n < v_n$ such that u_n is not an upper bound, while v_n is an upper bound of A. We distinguish two cases. If $(u_n + v_n)/2$ is not an upper bound of A, then let

$$u_{n+1} = \frac{u_n + v_n}{2} \text{ and } v_{n+1} = v_n.$$

However, if $(u_n + v_n)/2$ is an upper bound of A, then let

$$u_{n+1} = u_n \text{ and } v_{n+1} = \frac{u_n + v_n}{2}.$$

It is clear that in both cases, $[u_{n+1}, v_{n+1}] \subset [u_n, v_n]$, and u_{n+1} is not an upper bound, while v_{n+1} is an upper bound of A.

With the above, we have defined u_n and v_n for every n. It follows from the definition that the sequence of intervals $[u_n, v_n]$ forms a sequence of nested closed intervals, so by Cantor's axiom, they have a common point. We can also see that there is only one common point. By Theorem 3.5, it suffices to see that for every number $\delta > 0$, there exists an n for which $v_n - u_n < \delta$. But it is clear that (just as in the proof of Theorem 3.6) $v_n - u_n = (v_1 - u_1)/2^{n-1}$ for each n. So if n is large enough that $(v_1 - u_1)/2^{n-1} < \delta$ (which will happen for some n by the axiom of Archimedes), then $v_n - u_n < \delta$ will also hold.

So we see that the sequence of intervals $[u_n, v_n]$ has one common point. Let this be denoted by b. We want to show that b is an upper bound of A. Let $a \in A$ be arbitrary, and suppose that $b < a$. Then $u_n \leq b < a \leq v_n$ for each n. (Here the third inequality follows from the fact that v_n is an upper bound.) This means that a is a common point of the sequence of intervals $[u_n, v_n]$, which is impossible. This shows that $a \leq b$ for each $a \in A$, making b an upper bound.

Finally, we show that b is the least upper bound. Let c be another upper bound, and suppose that $c < b$. Then $u_n < c < b \leq v_n$ for each n. (Here the first inequality follows from the fact that $u_n \geq c$ would mean that u_n is an upper bound, but that cannot be.) This means that c is a common point of the interval sequence $[u_n, v_n]$, which is impossible. Therefore, we have $b \leq c$ for each upper bound c, making b the least upper bound.   □

A straightforward modification of the above proof yields Theorem 3.15 for lower bounds.

Theorem 3.16. *Every nonempty set that is bounded from below has a greatest lower bound.*

We can reduce Theorem 3.16 to Theorem 3.15. For let A be a nonempty set bounded from below. It is easy to check that the set $B = \{-x : x \in A\}$ is nonempty and bounded from above. By Theorem 3.15, B has a least upper bound. If b is the least upper bound of B, then it is easy to see that $-b$ will be the greatest lower bound of A.

The above argument uses the field axioms (since even the existence of the numbers $-x$ requires the first four axioms). It is worth noting that Theorem 3.16 has a proof that uses only the order axioms in addition to Theorem 3.15 (see Exercise 3.25).

Definition 3.17. We call the least upper bound of a nonempty set A that is bounded from above the *supremum* of A, and denote it by $\sup A$. The greatest lower bound of a nonempty set A that is bounded from below is called the *infimum* of A, and is denoted by $\inf A$.

For completeness, we extend the notion of infimum and supremum to sets that are not bounded.

Definition 3.18. If the set A is not bounded from above, then we say that the least upper bound or supremum of A is infinity, and we denote it by $\sup A = \infty$. If the set A is not bounded from below, then we say that the greatest lower bound or infimum of A is negative infinity, and we denote it by $\inf A = -\infty$.

Remark 3.19. Clearly, $\sup A \in A$ (respectively $\inf A \in A$) holds if and only if A has a largest (respectively smallest) element, and then $\max A = \sup A$ (respectively $\min A = \inf A$). The infimum and supremum of a nonempty set A agree if and only if A is a singleton.

For any two sets $A, B \subset \mathbb{R}$, the **sumset**, denoted by $A + B$, is the set $\{a + b : a \in A, \ b \in B\}$. The following relationships hold between the supremum and infimum of sets and their sumsets.

Theorem 3.20. *If A, B are nonempty sets, then* $\sup(A + B) = \sup A + \sup B$ *and* $\inf(A + B) = \inf A + \inf B$.

If either one of $\sup A$ and $\sup B$ is infinity, then what we mean by the statement is that $\sup(A + B)$ is infinity. Similarly, if either one of $\inf A$ and $\inf B$ is negative infinity, then the statement is to be understood to mean that $\inf(A + B)$ is negative infinity.

Proof. We prove only the statement regarding the supremum. If $\sup A = \infty$, then A is not bounded from above. It is clear that $A + B$ is then also not bounded from above, so $\sup(A + B) = \infty$.

Now suppose that both $\sup A$ and $\sup B$ are finite. If $a \in A$ and $b \in B$, then we know that $a + b \leq \sup A + \sup B$, so $\sup A + \sup B$ is an upper bound of $A + B$.

On the other hand, if c is an upper bound of $A + B$, then for arbitrary $a \in A$ and $b \in B$, we have $a + b \leq c$, that is, $a \leq c - b$, so $c - b$ is an upper bound for A. Then $\sup A \leq c - b$, that is, $b \leq c - \sup A$ for each $b \in B$, meaning that $c - \sup A$ is an upper bound of B. From this, we get that $\sup B \leq c - \sup A$, that is, $\sup A + \sup B \leq c$, showing that $\sup A + \sup B$ is the least upper bound of the set $A + B$. $\qquad\square$

Exercises

3.18. Let H be a set of real numbers. Which properties of H do the following statements express?

(a) $(\forall x \in \mathbb{R})(\exists y \in H)(x < y)$;
(b) $(\forall x \in H)(\exists y \in \mathbb{R})(x < y)$;
(c) $(\forall x \in H)(\exists y \in H)(x < y)$.

3.19. Prove that

$$\max(a, b) = \frac{|a - b| + a + b}{2} \quad \text{and} \quad \min(a, b) = \frac{-|a - b| + a + b}{2}.$$

3.20. Let $A \cap B \neq \emptyset$. What can we say about the relationships between $\sup A$, $\sup B$; $\sup(A \cup B)$, $\sup(A \cap B)$; and $\sup(A \setminus B)$?

3.21. Let $A = (0, 1)$, $B = [-\sqrt{2}, \sqrt{2}]$ and

$$C = \left\{ \frac{1}{2^n} + \frac{1}{2^m} : n \in \mathbb{N}^+, \ m \in \mathbb{N}^+ \right\}.$$

Find—if they exist—the supremum, infimum, maximum, and minimum of the above sets.

3.22. Let $A \cdot B = \{a \cdot b : a \in A, \ b \in B\}$ for arbitrary sets $A, B \subset \mathbb{R}$. What kind of relationships can we find between $\sup A$, $\sup B$, $\inf A$, $\inf B$ and $\inf(A \cdot B)$, $\sup(A \cdot B)$? What if we suppose $A, B \subset (0, \infty)$?

3.23. Let A be an arbitrary set of numbers, and let

$$B = \{-b : b \in A\}, \quad C = \{1/c : c \in A, \ c \neq 0\}.$$

What kind of relationships can we find between $\sup A$, $\inf A$, $\sup B$, $\inf B$, $\sup C$, $\inf C$?

3.24. Prove that if $a > 0$, $k \in \mathbb{N}^+$, $H_- = \{x > 0 : x^k < a\}$, and $H_+ = \{x > 0 : x^k > a\}$, then $\sup H_- = \inf H_+ = \sqrt[k]{a}$, that is, $(\sup H_-)^k = (\inf H_+)^k = a$. (This can also show us the existence of $\sqrt[k]{a}$ for $a > 0$.)

3.25. Let X be an ordered set. (This means that we have a relation $<$ given on X that satisfies the first two of the order axioms, trichotomy and transitivity.) Suppose that whenever a nonempty subset of X has an upper bound, it has a least upper bound. Show that if a nonempty subset of X has a lower bound, then it has a greatest lower bound. (H)

3.26. We say a set $H \subset \mathbb{R}$ is convex if $[x, y] \subset H$ whenever $x, y \in H$, $x < y$. Prove that a set is convex if and only if it is an interval. (Use the theorem about the existence of a least upper bound and greatest lower bound.) Show that without assuming Cantor's axiom, this would not be true.

3.27. Assume the field axioms, the order axioms, and the statement that if a non-empty set has an upper bound, then it has a least upper bound. Deduce from this the axiom of Archimedes and Cantor's axiom. (∗ S)

3.3 Exponentiation

We denote the n-fold product $a \cdot \ldots \cdot a$ by a^n, and call it the nth **power** of a (we call n the **exponent** and a the **base**). It is clear that for all real numbers a, b and positive integers x, y, the equations

$$(ab)^x = a^x \cdot b^x, \qquad a^{x+y} = a^x \cdot a^y, \qquad (a^x)^y = a^{xy} \tag{3.9}$$

hold. Our goal is to extend the notion of exponentiation in a way that satisfies the above identities. If $a \neq 0$, the equality $a^{x+y} = a^x \cdot a^y$ can hold if and only if we declare a^0 to be 1, and a^{-n} to be $1/a^n$ for every positive integer n. Accepting this definition, we see that the three identities of (3.9) hold for each $a, b \neq 0$ and $x, y \in \mathbb{Z}$. In the following, we concern ourselves only with powers of numbers different from zero; *we define only the positive integer powers of zero (for now).*

To extend exponentiation to rational exponents, we use Theorem 3.6, guaranteeing the existence of roots. If $a < 0$ and k is odd, then we denote the number $-\sqrt[k]{|a|}$ by $\sqrt[k]{a}$. Then $(\sqrt[k]{a})^k = a$ clearly holds.

Let r be rational, and suppose that $r = p/q$, where p, q are relatively prime integers, and $q > 0$. If $a \neq 0$, then by $(a^x)^y = a^{xy}$, if $a^r = b$, then $b^q = a^p$ holds. If q is odd, then this uniquely determines b: the only possible value is $b = \sqrt[q]{a^p}$. If q is even, then p is odd, and $a^p = b^q$ must be positive. This is possible only if $a > 0$, and then $b = \pm\sqrt[q]{a^p}$. Since it is natural that every power of a positive number should be positive, the logical conclusion is that $b = \sqrt[q]{a^p}$. We accept the following definition.

Definition 3.21. Let p and q be relatively prime integers, and suppose $q > 0$. If $a > 0$, then the value of $a^{p/q}$ is defined to be $\sqrt[q]{a^p}$. We define $a^{p/q}$ similarly if $a < 0$ and q is odd.

In the following, we deal only with powers of positive numbers.

Theorem 3.22. *If $a > 0$, n, m are integers, and $m > 0$, then $a^{n/m} = \sqrt[m]{a^n}$. (Note that we did not require n and m to be relatively prime.)*

Proof. Let $n/m = p/q$, where p, q are relatively prime integers and $q > 0$. We have to show that $\sqrt[m]{a^n} = \sqrt[q]{a^p}$. Since both sides are positive, it suffices to show that $\left(\sqrt[m]{a^n}\right)^{mq} = \left(\sqrt[q]{a^p}\right)^{mq}$, that is, $a^{nq} = a^{mp}$. However, this is clear, since $n/m = p/q$, so $nq = mp$.   □

Theorem 3.23. *The identities in (3.9) hold for each positive a, b and rational x, y.*

Proof. We prove only the first identity, since the rest can be proved similarly. Let $x = p/q$, where p, q are relatively prime integers and $q > 0$. Then $(ab)^x = \sqrt[q]{(ab)^p}$ and $a^x \cdot b^x = \sqrt[q]{a^p} \cdot \sqrt[q]{b^p}$. Since these are all positive numbers, it suffices to show that

$$\left(\sqrt[q]{(ab)^p}\right)^q = \left(\sqrt[q]{a^p} \cdot \sqrt[q]{b^p}\right)^q.$$

The left side equates to $(ab)^p$. To compute the right side, we apply the first identity in (3.9) to the integer powers q and then p.

We get that

$$\left(\sqrt[q]{a^p} \cdot \sqrt[q]{b^p}\right)^q = \left(\sqrt[q]{a^p}\right)^q \cdot \left(\sqrt[q]{b^p}\right)^q = a^p \cdot b^p = (ab)^p.$$

□

Theorem 3.24. *$a^r > 0$ for each positive a and rational r. If $r_1 < r_2$, then $a > 1$ implies $a^{r_1} < a^{r_2}$, while $0 < a < 1$ leads to $a^{r_1} > a^{r_2}$.*

Proof. The inequality $a^r > 0$ is clear from the definition. If $a > 1$ and p, q are positive integers, then $a^{p/q} = \sqrt[q]{a^p} > 1$, and so $a^r > 1$ for each positive rational r. So if $r_1 < r_2$, then $a^{r_2} = a^{r_1} a^{r_2 - r_1} > a^{r_1}$. The statement for $0 < a < 1$ can be proved the same way.   □

To extend exponentiation to irrational powers, we will keep in mind the monotone property shown in the previous theorem. Let $a > 1$. If we require that whenever $x \leq y$, we also have $a^x \leq a^y$, then a^x needs to satisfy $a^r \leq a^x \leq a^s$ whenever s and r are rational numbers such that $r \leq x \leq s$. We show that this restriction uniquely defines a^x.

Theorem 3.25. *If $a > 1$, then for an arbitrary real number x, we have*

$$\sup\{a^r : r \in \mathbb{Q}, \ r < x\} = \inf\{a^s : s \in \mathbb{Q}, \ s > x\}. \tag{3.10}$$

If $0 < a < 1$, then for an arbitrary real number x, we have

$$\inf\{a^r : r \in \mathbb{Q}, \ r < x\} = \sup\{a^s : s \in \mathbb{Q}, \ s > x\}. \tag{3.11}$$

Proof. Let $a > 1$. The set $A = \{a^r : r \in \mathbb{Q}, \ r < x\}$ is nonempty and bounded from above, since a^s is an upper bound for each rational s greater than x. Thus $\alpha = \sup A$ is finite, and $\alpha \leq a^s$ whenever $s > x$ for rational s. Then α is a lower bound of the set $B = \{a^s : s \in \mathbb{Q}, \ s > x\}$, so $\beta = \inf B$ is finite and $\alpha \leq \beta$. We show that $\alpha = \beta$.

Suppose that $\beta > \alpha$, and let $(\beta/\alpha) = 1 + h$, where $h > 0$. For each positive integer n, there exists an integer k for which $k/n \leq x < (k+1)/n$. Then we know that $(k-1)/n < x < (k+1)/n$, and so $a^{(k-1)/n} \leq \alpha < \beta \leq a^{(k+1)/n}$. We get

$$\frac{\beta}{\alpha} \leq \frac{a^{(k+1)/n}}{a^{(k-1)/n}} = a^{2/n},$$

and by applying Bernoulli's inequality,

$$a^2 \geq \left(\frac{\beta}{\alpha}\right)^n = (1 + h)^n \geq 1 + nh.$$

However, this is impossible if $n > a^2/h$. Thus $\alpha = \beta$, which proves (3.10). The second statement can be proved similarly. $\qquad\square$

Definition 3.26. Let $a > 1$. For an arbitrary real number x, the number a^x denotes $\sup\{a^r : r \in \mathbb{Q}, \ r < x\} = \inf\{a^s : s \in \mathbb{Q}, \ s > x\}$. If $0 < a < 1$, then the value of a^x is $\inf\{a^r : r \in \mathbb{Q}, \ r < x\} = \sup\{a^s : s \in \mathbb{Q}, \ s > x\}$. We define the exponent 1^x to be 1 for each x.

Let us note that if x is rational, then the above definition agrees with the previously defined value by Theorem 3.25.

Theorem 3.27. $a^x > 0$ *for each positive a and real number x. If $x_1 < x_2$, then $a > 1$ implies $a^{x_1} < a^{x_2}$, while $0 < a < 1$ leads to $a^{x_1} > a^{x_2}$.*

Proof. Let $a > 1$. For an arbitrary x, we can pick a rational $r < x$ such that $a^x \geq a^r > 0$. If $x_1 < x_2$, then let r_1 and r_2 be rational numbers for which $x_1 < r_1 < r_2 < x_2$. Then $a^{x_1} \leq a^{r_1} < a^{r_2} \leq a^{x_2}$. We can argue similarly in the case $0 < a < 1$. $\qquad\square$

Later, we will see (refer to Theorem 11.4) that all three identities in (3.9) hold for every $a, b > 0$ and $x, y \in \mathbb{R}$.

Exercises

3.28. Prove that if $n \in \mathbb{N}$, then $\sqrt{n}$ is either an integer or irrational.

3.29. Prove that if $n, k \in \mathbb{N}^+$, then $\sqrt[k]{n}$ is either an integer or irrational.

3.30. Let $a > 0$ be rational. Prove that if a^a is rational, then a is an integer.

3.31. Let a and b be rational numbers, $0 < a < b$. Prove that $a^b = b^a$ holds if and only if there exists an $n \in \mathbb{N}^+$ such that

$$a = \left(1 + \frac{1}{n}\right)^n \quad \text{and} \quad b = \left(1 + \frac{1}{n}\right)^{n+1}. \tag{H}$$

3.32. Prove that if $0 < a \le b$ are real numbers and $r > 0$ is rational, then $a^r \le b^r$. (S)

3.33. Prove that if $x > -1$ and $b \ge 1$, then $(1+x)^b \ge 1 + bx$. But if $x > -1$ and $0 \le b \le 1$, then $(1+x)^b \le 1 + bx$. (H S)

3.4 First Appendix: Consequences of the Field Axioms

Theorem 3.28. *If $a + b = 0$ and $a + c = 0$, then $b = c$.*

Proof. Using the first three axioms, we get that

$$c = c + 0 = c + (a+b) = (c+a) + b = (a+c) + b = 0 + b = b + 0 = b.$$

$\square$

If we compare the result of the previous theorem with axiom 4, then we get that for every $a \in \mathbb{R}$, there is exactly one b such that $a + b = 0$. We denote this unique b by $-a$.

Theorem 3.29. *For every a and b, there is exactly one x such that $a = b + x$.*

Proof. If $x = (-b) + a$, then

$$b + x = b + ((-b) + a) = (b + (-b)) + a = 0 + a = a + 0 = a.$$

On the other hand, if $a = b + x$, then

$$x = x + 0 = 0 + x = ((-b) + b) + x = (-b) + (b + x) = (-b) + a.$$

$\square$

From now on, we will denote the element $(-b) + a = a + (-b)$ by $a - b$.

Theorem 3.30. *If $a \cdot b = 1$ and $a \cdot c = 1$, then $b = c$.*

This can be proven in the same way as Theorem 3.28, using axioms 5–8 here. Comparing Theorem 3.30 with axiom 8, we obtain that for every $a \neq 0$, there exists exactly one b such that $a \cdot b = 1$. We denote this unique element b by $\frac{1}{a}$ or $1/a$.

Theorem 3.31. *For any a and $b \neq 0$, there is exactly one x such that $a = b \cdot x$.*

Proof. Mimicking the proof of Theorem 3.29 will show us that $x = a \cdot (1/b)$ is the unique real number that satisfies the condition of the theorem. $\qquad \square$

If $a, b \in \mathbb{R}$ and $b \neq 0$, then we denote the number $a \cdot (1/b)$ by $\frac{a}{b}$ or a/b.

Theorem 3.32. *Every real number a satisfies $a \cdot 0 = 0$.*

Proof. Let $a \cdot 0 = b$. By axioms 3 and 9,

$$b = a \cdot 0 = a \cdot (0 + 0) = (a \cdot 0) + (a \cdot 0) = b + b.$$

Since $b + 0 = b$ also holds, Theorem 3.29 implies $b = 0$. $\qquad \square$

It is easy to check that each of the following identities follows from the field axioms:

$$-a = (-1) \cdot a, \qquad (a - b) - c = a - (b + c),$$

$$(-a) \cdot b = -(a \cdot b), \qquad \frac{1}{a/b} = \frac{b}{a} \ (a, b \neq 0), \qquad \frac{a}{b} \cdot \frac{c}{d} = \frac{a \cdot c}{b \cdot d} \ (b, d \neq 0).$$

With the help of induction, it is easy to justify that putting parentheses anywhere in a sum or product of several terms does not change the value of the sum or product. For example, $(a + b) + (c + d) = (a + (b + c)) + d$ and $(a \cdot b) \cdot (c \cdot d) = a \cdot ((b \cdot c) \cdot d)$. Therefore, we can omit parentheses in sums or products; by the sum $a_1 + \cdots + a_n$ and the product $a_1 \cdot \ldots \cdot a_n$, we mean the number that we would get by putting parentheses at arbitrary places in the sum (or product), thereby adding or multiplying two numbers at a time.

3.5 Second Appendix: Consequences of the Order Axioms

Theorem 3.33. *If $a < b$ and $c < d$, then $a + c < b + d$. If $a \leq b$ and $c \leq d$, then $a + c \leq b + d$.*

Proof. Let $a < b$ and $c < d$. Applying axiom 12 and commutativity twice, we get that $a + c < b + c = c + b < d + b = b + d$. The second statement is a clear consequence of this. $\qquad \square$

We can similarly show that if $0 < a < b$ and $0 < c < d$, then $a \cdot c < b \cdot d$; moreover, if $0 \leq a \leq b$ and $0 \leq c \leq d$, then $a \cdot c \leq b \cdot d$. (We need to use Theorem 3.32 for the proof of the second statement.)

Theorem 3.34. *If $a < b$ then $-a > -b$.*

Proof. By axiom 12, we have $-b = a + (-a - b) < b + (-a - b) = -a$. $\square$

Theorem 3.35. *If $a < b$ and $c < 0$, then $a \cdot c > b \cdot c$. If $a \le b$ and $c \le 0$, then $a \cdot c \ge b \cdot c$.*

Proof. Let $a < b$ and $c < 0$. By the previous theorem, $-c > 0$, so by axiom 13, we have $-a \cdot c = a \cdot (-c) < b \cdot (-c) = -b \cdot c$. Thus using the previous theorem again, we obtain $a \cdot c > b \cdot c$. The second statement is a simple consequence of this one, using Theorem 3.32. $\square$

Theorem 3.36. $1 > 0$.

Proof. By axiom 10, it suffices to show that neither the statement $1 = 0$ nor $1 < 0$ holds. We initially assumed the numbers 0 and 1 to be distinct, so all we need to exclude is the case $1 < 0$. Suppose that $1 < 0$. Then by Theorem 3.35, we have $1 \cdot 1 > 0 \cdot 1$, that is, $1 > 0$. However, this contradicts our assumption. Thus we conclude that our assumption was false, and so $1 > 0$. $\square$

Theorem 3.37. *If $a > 0$, then $1/a > 0$. If $0 < a < b$, then $1/a > 1/b$. If $a \ne 0$, then $a^2 > 0$.*

Proof. Let $a > 0$. If $1/a \le 0$, then by Theorems 3.35 and 3.32, we must have $1 = (1/a) \cdot a \le 0 \cdot a = 0$, which is impossible. Thus by axiom 10, only $1/a > 0$ is possible.

Now suppose that $0 < a < b$. If $1/a \le 1/b$, then by axiom 13, $b = (1/a) \cdot a \cdot b \le (1/b) \cdot a \cdot b = a$, which is impossible.

The third statement is clear by axiom 13 and Theorem 3.35. $\square$

Theorem 3.38. *The natural numbers are positive and distinct.*

Proof. By Theorem 3.36, we have that $0 < 1$. Knowing this, axiom 12 then implies $1 = 0 + 1 < 1 + 1 = 2$. Applying axiom 12 again yields $2 = 1 + 1 < 2 + 1 = 3$, and so on. Finally, we get that $0 < 1 < 2 < \cdots$, and using transitivity, both statements of the theorem are clear. $\square$

Theorem 3.39. *If n is a natural number, then for every real number a, we have*

$$\underbrace{a + \cdots + a}_{n \text{ terms}} = n \cdot a.$$

Proof. $\underbrace{a + \cdots + a}_{n \text{ terms}} = \underbrace{1 \cdot a + \cdots + 1 \cdot a}_{n \text{ terms}} = \underbrace{(1 + \cdots + 1)}_{n \text{ terms}} \cdot a = n \cdot a.$ $\square$

Theorem 3.40. *There are no neighboring real numbers. That is, for arbitrary real numbers $a < b$, there exists a real number c such that $a < c < b$. One such number, for example, is $c = (a + b)/2$.*

Proof. By Theorem 3.39 and axiom 12, $2 \cdot a = a + a < a + b$. If we multiply this inequality by $1/2$ (which is positive by Theorems 3.37 and 3.38), then we get that $a < c$. A similar argument shows that $c < b$. $\square$

Chapter 4
Infinite Sequences I

In this chapter, we will be dealing with sequences of real numbers. For brevity, by a sequence we shall mean an infinite sequence whose terms are all real numbers.

We can present a sequence in various different ways. Here are a few examples (each one is defined for $n \in \mathbb{N}^+$):

Examples 4.1. 1. $a_n = \frac{1}{n}$: $(a_n) = \left(1, \frac{1}{2}, \ldots, \frac{1}{n}, \ldots\right)$;

2. $a_n = (-1)^{n+1} \cdot \frac{1}{n}$: $(a_n) = \left(1, -\frac{1}{2}, \frac{1}{3}, -\frac{1}{4}, \ldots\right)$;

3. $a_n = (-1)^n$: $(a_n) = (-1, 1, \ldots, -1, 1, \ldots)$;

4. $a_n = (n+1)^2$: $(a_n) = (4, 9, 16, \ldots)$;

5. $a_n = \sqrt{n+1} - \sqrt{n}$: $(a_n) = \left(\sqrt{2} - 1, \sqrt{3} - \sqrt{2}, \ldots\right)$;

6. $a_n = \frac{n+1}{n}$: $(a_n) = \left(2, \frac{3}{2}, \frac{4}{3}, \frac{5}{4}, \ldots\right)$;

7. $a_n = (-1)^n \cdot n^2$: $(a_n) = (-1, 4, -9, 16, \ldots)$;

8. $a_n = n + \frac{1}{n}$: $(a_n) = \left(2, \frac{5}{2}, \frac{10}{3}, \frac{17}{4}, \ldots\right)$;

9. $a_n = \sqrt{n+10}$: $(a_n) = \left(\sqrt{11}, \sqrt{12}, \ldots\right)$;

10. $a_n = \left(1 + \frac{1}{n}\right)^n$;

11. $a_n = \left(1 + \frac{1}{n}\right)^{n^2}$;

12. $a_n = \left(1 + \frac{1}{n^2}\right)^n$;

13. $a_1 = -1$, $a_2 = 2$, $a_n = (a_{n-1} + a_{n-2})/2$ $(n \geq 3)$: $(a_n) = \left(-1, 2, \frac{1}{2}, \frac{5}{4}, \frac{7}{8}, \frac{17}{16}, \ldots\right)$;

14. $a_1 = 1$, $a_2 = 3$, $a_{n+1} = \sqrt[n]{a_1 \cdots a_n}$ $(n \geq 2)$: $(a_n) = (1, 3, \sqrt{3}, \sqrt{3}, \sqrt{3}, \ldots)$;

15. $a_1 = 0$, $a_{n+1} = \sqrt{2 + a_n}$ $(n \geq 1)$: $(a_n) = \left(0, \sqrt{2}, \sqrt{2 + \sqrt{2}}, \ldots\right)$;

16. $a_n = \begin{cases} n, & \text{if } n \text{ is even,} \\ 1, & \text{if } n \text{ is odd} \end{cases}$: $(a_n) = (1, 2, 1, 4, 1, 6, 1, 8, \ldots)$;

17. $a_n =$ the nth prime number: $(a_n) = (2, 3, 5, 7, 11, \ldots)$;

18. $a_n =$ the nth digit of the infinite decimal expansion of $\sqrt{2}$:
$(a_n) = (4, 1, 4, 2, 1, 3, 5, 6, \ldots)$.

© Springer New York 2015
M. Laczkovich, V.T. Sós, *Real Analysis*, Undergraduate Texts
in Mathematics, DOI 10.1007/978-1-4939-2766-1_4

In the first 12 sequences, the value of a_n is given by an "explicit formula." The terms of (13)–(15) are given **recursively**. This means that we give the first few, say k, terms of the sequence, and if $n > k$, then the nth term is given by the terms with index less than n. In (16)–(18), the terms are not given with a specific "formula." There is no real difference, however, concerning the validity of the definitions; they are all valid sequences. As we will later see, whether a_n (or generally a function) can be expressed with some kind of formula depends only on how frequently such an expression occurs and how important it is. For if it defines an important map that we use frequently, then it is worthwhile to define some notation and nomenclature that converts the long definition (of a_n or the function) into a "formula." If it is not so important or frequently used, we usually leave it as a lengthy description. For example, in the above sequence (9), a_n is the positive number whose square is $n + 10$. However, the map expressed here (the square root) occurs so frequently and with such importance that we create a new symbol for it, and the definition of a_n becomes a simple formula.

Exercises

4.1. Give a closed formula for the nth term of sequence (13) in Example 4.1. (S)

4.2. Let $p(x) = x^k - c_1 x^{k-1} - c_2 x^{k-2} - \ldots - c_k$, let $\alpha_1, \ldots, \alpha_m$ be roots of the polynomial p (not necessarily all of the roots), and let $\beta_1, \ldots, \beta_m$ be arbitrary real numbers. Show that the sequence

$$a_n = \beta_1 \cdot \alpha_1^n + \cdots + \beta_m \cdot \alpha_m^n \quad (n = 1, 2, \ldots)$$

satisfies the recurrence relation (recursion) $a_n = c_1 a_{n-1} + c_2 a_{n-2} + \cdots + c_k a_{n-k}$ for all $n > k$. (H)

4.3. Give a closed formula for the nth term of the following sequences, given by recursion.

(a) $u_0 = 0$, $u_1 = 1$, $u_n = u_{n-1} + u_{n-2}$ $(n \geq 2)$ (HS);
(b) $a_0 = 0$, $a_1 = 1$, $a_n = a_{n-1} + 2 \cdot a_{n-2}$ $(n \geq 2)$;
(c) $a_0 = 0$, $a_1 = 1$, $a_n = 2 \cdot a_{n-1} + a_{n-2}$ $(n \geq 2)$.

4.4. Let $p(x) = x^k - c_1 x^{k-1} - c_2 x^{k-2} - \ldots - c_k$, and let α be a double root of the polynomial p (this means that $(x - \alpha)^2$ can be factored from p). Show that the sequence $a_n = n \cdot \alpha^n$ $(n = 1, 2, \ldots)$ satisfies the following recurrence for all $n > k$: $a_n = c_1 a_{n-1} + c_2 a_{n-2} + \cdots + c_k a_{n-k}$.

4.5. Give a closed formula for the nth term of the sequence

$$a_0 = 0, \ a_1 = 0, \ a_2 = 1, \ a_n = a_{n-1} + a_{n-2} - a_{n-3} \ (n \geq 3)$$

given by a recurrence relation.

4.1 Convergent and Divergent Sequences

When we make measurements, we are often faced with a value that we can only express through approximation—although arbitrarily precisely. For example, we define the circumference (or area) of a circle in terms of the perimeter (or area) of the inscribed or circumscribed regular n-gons. According to this definition, the circumference (or area) of the circle is the number that the perimeter (or area) of the inscribed regular n-gon "approximates arbitrarily well as n increases past every bound."

Some of the above sequences also have this property that the terms "tend" to some "limit value." The terms of the sequence (1) "tend" to 0 in the sense that if n is large, the value of $1/n$ is "very small," that is it is very close to 0. More precisely, no matter how small an interval we take around 0, if n is large enough, then $1/n$ is inside this interval (there are only finitely many n such that $1/n$ is outside the interval).

The terms of the sequence (2) also "tend" to 0, while the terms of (6) "tend" to 1 by the above notion.

For the sequences (3), (4), (7), (8), (9), (16), (17), and (18), no number can be found that the terms of any of these sequences "tend" to, by the above notion.

The powers defining the a_n in the sequences (10), (11), and (12) all behave differently as n increases: the base approaches 1, but the exponent gets very big. Without detailed examination, we cannot "see" whether they tend to some value, and if they do, to what. We will see later that these three sequences each behave differently from the point of view of limits.

To define limits precisely, we use the notion outlined in the examination of (1). We give two definitions, which we promptly show to be equivalent.

Definition 4.2. The sequence (a_n) *tends to b* (or *b is the limit of the sequence*) if for every $\varepsilon > 0$, there are only finitely many terms falling outside the interval $(b - \varepsilon, b + \varepsilon)$. In other words, the limit of (a_n) is b if for every $\varepsilon > 0$, the terms of the sequence satisfy, with finitely many exceptions, the inequality $b - \varepsilon < a_n < b + \varepsilon$.

Definition 4.3. The sequence (a_n) *tends to b* (or *b is the limit of the sequence*), if for every $\varepsilon > 0$ there exists a number n_0 (depending on ε) such that

$$|a_n - b| < \varepsilon \text{ for all indices } n > n_0. \tag{4.1}$$

Let us show that the *two definitions are equivalent*. Suppose first that the sequence (a_n) tends to b by Definition 4.2. Consider an arbitrary $\varepsilon > 0$. Then only finitely many terms of the sequence fall outside $(b - \varepsilon, b + \varepsilon)$. If there are no terms of the sequence outside the interval, then (4.1) holds for any choice of n_0. If terms of the sequence fall outside $(b - \varepsilon, b + \varepsilon)$, then out of those finitely many terms, there is one with maximal index. Denote this index by n_0. Then for each $n > n_0$, the a_n are in the interval $(b - \varepsilon, b + \varepsilon)$, that is, $|a_n - b| < \varepsilon$ if $n > n_0$. Thus we see that (a_n) satisfies Definition 4.3.

Secondly, suppose that (a_n) tends to b by Definition 4.3. Let $\varepsilon > 0$ be given. Then there exists an n_0 such that if $n > n_0$, then a_n is in the interval $I = (b - \varepsilon, b + \varepsilon)$. Thus only among the terms a_i $(i \leq n_0)$ can there be terms that do not fall in the interval I. The number of these is at most n_0, thus finite. It follows that the sequence satisfies Definition 4.2.

If the sequence (a_n) tends to b, then we denote this by

$$\lim_{n \to \infty} a_n = b \qquad \text{or} \qquad a_n \to b, \text{ as } n \to \infty \text{ (or just } a_n \to b).$$

If there is a real number b such that $\lim_{n \to \infty} a_n = b$, then we say that the sequence (a_n) is **convergent**. If there is no such number, then the sequence (a_n) is **divergent**.

Examples 4.4. **1.** By Definition 4.3, it is easy to see that the sequence (1) in 4.1 truly tends to 0, that is, $\lim_{n \to \infty} 1/n = 0$. For if ε is an arbitrary positive number, then $n > 1/\varepsilon$ implies $1/n < \varepsilon$, and thus $|(1/n) - 0| = 1/n < \varepsilon$. That is, we can pick the n_0 in the definition to be $1/\varepsilon$. (The definition did not require that n_0 be an integer. But it is clear that if some n_0 satisfies the conditions of the definition, then every number larger than n_0 does as well, and among these—by the axiom of Archimedes—is an integer.)

2. In the same way, we can see that the sequence (2) in 4.1 tends to 0. Similarly,

$$\lim_{n \to \infty} \frac{n+1}{n} = 1,$$

or (6) is also convergent, and it has limit 1. Clearly, $(n+1)/n \in (1 - \varepsilon, 1 + \varepsilon)$ if $1/n < \varepsilon$, that is, the only a_n outside the interval $(1 - \varepsilon, 1 + \varepsilon)$ are those for which $1/n \geq \varepsilon$, that is, $n \leq 1/\varepsilon$.

3. Now we show that the sequence (5) tends to 0, that is,

$$\lim_{n \to \infty} (\sqrt{n+1} - \sqrt{n}) = 0. \tag{4.2}$$

Since

$$\sqrt{n+1} - \sqrt{n} = \frac{1}{\sqrt{n+1} + \sqrt{n}} < \frac{1}{2\sqrt{n}},$$

if $n > 1/(4\varepsilon^2)$, then $1/(2\sqrt{n}) < \varepsilon$ and $a_n \in (-\varepsilon, \varepsilon)$.

Remarks 4.5. **1.** It is clear that if a threshold n_0 is good, that is it satisfies the conditions of (4.1), then every number larger than n_0 is also a good index. Generally when finding a threshold n_0 we do not strive to find the smallest one.

2. It is important to note the following regarding Definition 4.2. If the infinite sequence (a_n) has only finitely many terms *outside* the interval $(a - \varepsilon, a + \varepsilon)$, then naturally there are infinitely many terms *inside* the interval $(a - \varepsilon, a + \varepsilon)$. It is clear, however, that for the sequence (a_n), if for every $\varepsilon > 0$ there are infinitely many terms *inside* the interval $(a - \varepsilon, a + \varepsilon)$ it does not necessarily mean that there are only

finitely many terms *outside* the interval; that is, it does not follow that $\lim\limits_{n\to\infty} a_n = a$.
For example, in Example 4.1, for the sequence (3) there are infinitely many terms
inside $(1-\varepsilon, 1+\varepsilon)$ for every $\varepsilon > 0$ (and the same holds for $(-1-\varepsilon, -1+\varepsilon)$), but if
$\varepsilon < 2$, then it is not true that there are only finitely many terms outside $(1-\varepsilon, 1+\varepsilon)$.
That is, 1 is not a limit of the sequence, and it is easy to see that the sequence is di-
vergent.

3. Denote the set of numbers occurring in the sequence (a_n) by $\{a_n\}$. Let us examine
the relationship between (a_n) and $\{a_n\}$. On one hand, we know that for the set, a
number is either an element of it or it is not, and there is no meaning in saying that it
appears multiple times, while in a sequence, a number can occur many times. In fact,
for the infinite sequence (a_n) in Example (3), the corresponding set $\{a_n\} = \{-1, 1\}$
is finite. (This distinction is further emphasized by talking about *elements* of sets,
and *terms* of sequences.) Consider the following two properties:

I. For every $\varepsilon > 0$, there are finitely many terms of (a_n) outside the interval $(a - \varepsilon, a+\varepsilon)$.
II. For every $\varepsilon > 0$, there are finitely many elements of $\{a_n\}$ outside the interval $(a-\varepsilon, a+\varepsilon)$.

Property I. means that $\lim\limits_{n\to\infty} a_n = a$. The same cannot be said of Property II., since
we can take our example above, where $\{a_n\}$ had only finitely many elements outside
every interval $(a-\varepsilon, a+\varepsilon)$, but the sequence was still divergent. Therefore, it is
clear that I implies II, but II does not imply I. In other words, I is a stronger property
than II.

Exercises

4.6. Find the limits of the following sequences, and find an n_0 (not necessarily the
smallest) for a given $\varepsilon > 0$ as in Definition 4.3.

(a) $1/\sqrt{n}$;

(b) $(2n+1)/(n+1)$;

(c) $(5n-1)/(7n+2)$;

(d) $1/(n-\sqrt{n})$;

(e) $(1+\cdots+n)/n^2$;

(f) $(\sqrt{1}+\sqrt{2}+\cdots+\sqrt{n})/n^{4/3}$;

(g) $n\cdot\left(\sqrt{1+(1/n)}-1\right)$;

(h) $\sqrt{n^2+1}+\sqrt{n^2-1}-2n$;

(i) $\sqrt[3]{n+2}-\sqrt[3]{n-2}$;

(j) $\dfrac{1}{1\cdot 2}+\dfrac{1}{2\cdot 3}+\cdots+\dfrac{1}{(n-1)\cdot n}$.

4.7. Consider the definition of $a_n \to b$: $(\forall \varepsilon > 0)(\exists n_0)(\forall n \geq n_0)(|a_n - b| < \varepsilon)$. Per-
muting and changing the quantifiers yields the following statements:

(a) $(\forall \varepsilon > 0)(\exists n_0)(\exists n \geq n_0)(|a_n - b| < \varepsilon)$;
(b) $(\forall \varepsilon > 0)(\forall n_0)(\forall n \geq n_0)(|a_n - b| < \varepsilon)$;
(c) $(\forall \varepsilon > 0)(\forall n_0)(\exists n \geq n_0)(|a_n - b| < \varepsilon)$;

(d) $(\exists \varepsilon > 0)(\forall n_0)(\forall n \geq n_0)(|a_n - b| < \varepsilon)$;
(e) $(\exists \varepsilon > 0)(\forall n_0)(\exists n \geq n_0)(|a_n - b| < \varepsilon)$;
(f) $(\exists \varepsilon > 0)(\exists n_0)(\forall n \geq n_0)(|a_n - b| < \varepsilon)$;
(g) $(\exists \varepsilon > 0)(\exists n_0)(\exists n \geq n_0)(|a_n - b| < \varepsilon)$;
(h) $(\exists n_0)(\forall \varepsilon > 0)(\forall n \geq n_0)(|a_n - b| < \varepsilon)$;
(i) $(\exists n_0)(\forall \varepsilon > 0)(\exists n \geq n_0)(|a_n - b| < \varepsilon)$;
(j) $(\forall n_0)(\exists \varepsilon > 0)(\forall n \geq n_0)(|a_n - b| < \varepsilon)$;
(k) $(\forall n_0)(\exists \varepsilon > 0)(\exists n \geq n_0)(|a_n - b| < \varepsilon)$.

What properties of (a_n) do these statements express? For each property, give a sequence (if exists) with the given property.

4.8. Prove that a convergent sequence always has a smallest or a largest term.

4.9. Give examples such that $a_n - b_n \to 0$, but a_n/b_n does not tend to 1, as well as $a_n/b_n \to 1$ but $a_n - b_n$ does not tend to 0.

4.10. Prove that if (a_n) is convergent, then $(|a_n|)$ is convergent. Is this statement true the other way around?

4.11. If $a_n^2 \to a^2$, does it follow that $a_n \to a$?   If $a_n^3 \to a^3$, does it follow that $a_n \to a$?

4.12. Prove that if $a_n \to a > 0$, then $\sqrt{a_n} \to \sqrt{a}$.

4.13. For the sequence (a_n), consider the corresponding sequence of arithmetic means, $s_n = (a_1 + \cdots + a_n)/n$. Prove that if $\lim_{n \to \infty} a_n = a$, then $\lim_{n \to \infty} s_n = a$. Give a sequence for which (s_n) is convergent but (a_n) is divergent. (S)

4.2 Sequences That Tend to Infinity

It is easy to see that the sequences (3), (4), (7), (8), (9), (16), and (17) in Example 4.1 are divergent. Moreover, terms of the sequences (4), (8), (9), and (17)—aside from diverging—share the trend that for "large" n, the a_n terms are "large"; more precisely, for an arbitrarily large number P, there are only finitely many terms of the sequence that are smaller than P. We say that sequences like this "diverge to ∞." This is expressed precisely be the definition below. As in the case of convergent sequences, we give two definitions and then show that they are equivalent.

Definition 4.6. We say that the *limit* of the sequence (a_n) is ∞ (or that (a_n) tends to infinity) if for arbitrary P, there are only finitely many terms of the sequence outside the interval (P, ∞).[1]

---

[1] That is, to the left of P on the number line.

Definition 4.7. We say that the *limit* of the sequence (a_n) is ∞ (or that (a_n) tends to infinity) if for arbitrary P, there exists a number n_0 (depending on P) for which the statement

$$a_n > P, \quad \text{if} \quad n > n_0 \tag{4.3}$$

holds.

We can show the equivalence of the above definitions in the following way. If there are no terms outside the interval (P, ∞), then (4.3) holds for every n_0. If there are terms outside the interval (P, ∞), but only finitely many, then among these indices, call the largest n_0, which will make (4.3) hold.

Conversely, if (4.3) holds, then there are at most n_0 terms of the sequence, finitely many, outside the interval (P, ∞).

If the sequence (a_n) tends to infinity, then we denote this by $\lim_{n \to \infty} a_n = \infty$, or by $a_n \to \infty$ as $n \to \infty$, or just $a_n \to \infty$ for short. In this case, we say that the sequence (a_n) **diverges to infinity**.

We define the concept of tending to $-\infty$ in the same manner.

Definition 4.8. We say that the *limit* of the sequence (a_n) is $-\infty$ (or that (a_n) tends to negative infinity) if for arbitrary P, there are only finitely many terms of the sequence outside the interval $(-\infty, P)$.[2]

The following is equivalent.

Definition 4.9. We say that the *limit* of the sequence (a_n) is $-\infty$ (or that (a_n) tends to negative infinity) if for arbitrary P, there exists a number n_0 (dependent on P) for which if $n > n_0$, then $a_n < P$ holds.

If the sequence (a_n) tends to negative infinity, then we denote this by $\lim_{n \to \infty} a_n = -\infty$, or by $a_n \to -\infty$ if $n \to \infty$, or just $a_n \to -\infty$ for short. In this case, we say that the sequence (a_n) **diverges to negative infinity**.

There are several sequences in Example 4.1 that tend to infinity. It is clear that (4) and (8) are such sequences, for in both cases, if $n > P$, then $a_n > P$. The sequence (9) also tends to infinity, for if $n > P^2$, then $a_n > P$.

Now we show that the sequence (11) tends to infinity as well. Let P be given. Then Bernoulli's inequality (Theorem 2.5) implies

$$\left(1 + \frac{1}{n}\right)^{n^2} > 1 + n^2 \cdot \frac{1}{n} = 1 + n,$$

so for arbitrary $n > P$, we have

$$\left(1 + \frac{1}{n}\right)^{n^2} > 1 + n > P.$$

---

[2] That is, to the right of P on the number line.

Exercises

4.14. For a fixed P, find an n_0 (not necessarily the smallest one) satisfying Definition 4.7 for the following sequences:

(a) $n - \sqrt{n}$;

(b) $(1 + \cdots + n)/n$;

(c) $(\sqrt{1} + \sqrt{2} + \cdots + \sqrt{n})/n$;

(d) $\dfrac{n^2 - 10n}{10n + 100}$;

(e) $2^n/n$.

4.15. Consider the definition of $a_n \to \infty$: $(\forall P)(\exists n_0)(\forall n \geq n_0)(a_n > P)$. Permuting or changing the quantifiers yields the following statements:

(a) $(\forall P)(\exists n_0)(\exists n \geq n_0)(a_n > P)$;
(b) $(\forall P)(\forall n_0)(\forall n \geq n_0)(a_n > P)$;
(c) $(\forall P)(\forall n_0)(\exists n \geq n_0)(a_n > P)$;
(d) $(\exists P)(\forall n_0)(\forall n \geq n_0)(a_n > P)$;
(e) $(\exists P)(\forall n_0)(\exists n \geq n_0)(a_n > P)$;
(f) $(\exists P)(\exists n_0)(\forall n \geq n_0)(a_n > P)$;
(g) $(\exists P)(\exists n_0)(\exists n \geq n_0)(a_n > P)$;
(h) $(\exists n_0)(\forall P)(\forall n \geq n_0)(a_n > P)$;
(i) $(\exists n_0)(\forall P)(\exists n \geq n_0)(a_n > P)$;
(j) $(\forall n_0)(\exists P)(\forall n \geq n_0)(a_n > P)$;
(k) $(\forall n_0)(\exists P)(\exists n \geq n_0)(a_n > P)$.

What properties of (a_n) do these statements express? For each property, give a sequence (if exists) with the given property.

4.16. Prove that a sequence tending to infinity always has a smallest term.

4.17. Find the limit of $(n^2 + 1)/(n + 1) - an$ for each a.

4.18. Find the limit of $\sqrt{n^2 - n + 1} - an$ for each a.

4.19. Find the limit of $\sqrt{(n + a)(n + b)} - n$ for each a, b.

4.20. Prove that if $a_{n+1} - a_n \to c$, where $c > 0$, then $a_n \to \infty$.

4.3 Uniqueness of Limit

If the sequence (a_n) is convergent, or tends to plus or minus infinity, then we say that (a_n) **has a limit**. Instead of saying that a sequence is *convergent*, we can say that **the sequence has a finite limit**. If (a_n) doesn't have a limit, we can say that it **oscillates at infinity**. The following table illustrates the classification above.

$$\left.\begin{array}{ll} \text{convergent} & a_n \to b \in \mathbb{R} \\ \left\{\begin{array}{l} \\ \text{divergent} \left\{\begin{array}{l} a_n \to \infty \\ a_n \to -\infty \\ \text{has no limit} \end{array}\right. \end{array}\right. \end{array}\right.$$

convergent $\quad a_n \to b \in \mathbb{R}$

divergent $\left\{\begin{array}{l} a_n \to \infty \\ a_n \to -\infty \end{array}\right\}$ has limit

has no limit $\quad$ oscillates at infinity

To justify the above classification, we need to show that the properties in the middle column are mutually exclusive. We will show more than this: in Theorem 4.14, we will show that a sequence can have at most one limit. For this, we need the following two theorems.

Theorem 4.10. *If the sequence* (a_n) *is convergent, then it is a bounded sequence.*[3]

Proof. Let $\lim_{n\to\infty} a_n = b$. Pick, according to the notation in Definition 4.2, ε to be 1. We get that there are only finitely many terms of the sequence outside the interval $(a-1, a+1)$. If there are no terms greater than $a+1$, then it is an upper bound. If there are terms greater than $a+1$, there are only finitely many of them. Out of these, the largest one is the largest term of the whole sequence, and as such, is an upper bound too. A bound from below can be found in the same fashion. $\square$

Remark 4.11. The converse of the statement above is not true: the sequence $(-1)^n$ is bounded, but not convergent.

Theorem 4.12. *If the sequence* (a_n) *tends to infinity, then it is bounded from below and not bounded from above. If the sequence* (a_n) *tends to negative infinity, then it is bounded from above and not bounded from below.*

Proof. Let $\lim_{n\to\infty} a_n = \infty$. Comparing Definition 4.6 with the definition of being bounded from above (3.14) clearly shows that (a_n) cannot be bounded from above.

Pick, according to the notation in Definition 4.6, P to be 0. The sequence has only finitely many terms outside the interval $(0, \infty)$. If there are no terms outside the interval, then 0 is a lower bound. If there are terms outside the interval, then there are only finitely many. Out of these, the smallest one is the smallest term of the whole sequence, and thus a lower bound too. This shows that (a_n) is bounded from below. The case $a_n \to -\infty$ can be dealt with in a similar way. $\square$

Remark 4.13. The converses of the statements of the theorem are not true. It is clear that in Example 4.1, the sequence (16) is bounded from below (the number 1 is a lower bound), not bounded from above, but the sequence doesn't tend to infinity.

Theorem 4.14. *Every sequence has at most one limit.*

Proof. By Theorems 4.10 and 4.12, it suffices to show that every convergent sequence has at most one limit. Suppose that $a_n \to b$ and $a_n \to c$ both hold, where b and c are distinct real numbers. Then for each $\varepsilon > 0$, only finitely many terms of the sequence lie outside the interval $(b - \varepsilon, b + \varepsilon)$, so there are infinitely many terms inside $(b - \varepsilon, b + \varepsilon)$. Let ε be so small that the intervals $(b - \varepsilon, b + \varepsilon)$ and $(c - \varepsilon, c + \varepsilon)$ are disjoint, that is, do not have any common points. (Such is, for

---

[3] By this, we mean that the set $\{a_n\}$ is bounded.

example, $\varepsilon = |c - b|/2$.) Then there are infinitely many terms of the sequence outside the interval $(c - \varepsilon, c + \varepsilon)$, which is impossible, since then c cannot be a limit of a_n.   □

Exercises

4.21. Let

   S be the set of all sequences;
   C be the set of convergent sequences;
   D be the set of divergent sequences;
   D_∞ be the set of sequences diverging to ∞;
   $D_{-\infty}$ be the set of sequences diverging to $-\infty$;
   O be the set of sequences oscillating at infinity;
   K be the set of bounded sequences.

Prove the following statements:

(a) $S = C \cup D$.
(b) $D = D_\infty \cup D_{-\infty} \cup O$.
(c) $C \subset K$.
(d) $K \cap D_\infty = \emptyset$.

4.22. Give an example for each possible behavior of a sequence (a_n) (convergent, tending to infinity, tending to negative infinity, oscillating at infinity), while $a_{n+1} - a_n \to 0$ also holds. (H)

4.23. Give an example for each possible behavior of a sequence (a_n) (convergent, tending to infinity, tending to negative infinity, oscillating at infinity), while $a_{n+1}/a_n \to 1$ also holds.

4.24. Give an example of a sequence (a_n) that

(a) is convergent,
(b) tends to infinity,
(c) tends to negative infinity,

while $a_n < (a_{n-1} + a_{n+1})/2$ holds for each $n > 1$.

4.25. Prove that if $a_n \to \infty$ and (b_n) is bounded, then $(a_n + b_n) \to \infty$.

4.26. Is it true that if (a_n) oscillates at infinity and is unbounded, and (b_n) is bounded, then $(a_n + b_n)$ oscillates at infinity and is unbounded?

4.27. Let (a_n) be sequence (18) from Example 4.1, that is, let a_n be the nth term in the decimal expansion of $\sqrt{2}$. Prove that the sequence (a_n) oscillates at infinity. (H)

4.4 Limits of Some Specific Sequences

Theorem 4.15.

(i) For every fixed integer p,

$$\lim_{n\to\infty} n^p = \begin{cases} \infty, & \text{if } p > 0, \\ 1, & \text{if } p = 0, \\ 0, & \text{if } p < 0. \end{cases} \tag{4.4}$$

(ii) If $p > 0$, then $\lim_{n\to\infty} \sqrt[p]{n} = \infty$.

Proof. (i) Let first $p > 0$. For arbitrary $P > 0$, if $n > P$, then $n^p \geq n > P$. Then $n^p \to \infty$. If $p = 0$, then $n^p = 1$ for every n, so $n^p \to 1$. Finally, if p is a negative integer and $\varepsilon > 0$, then $n > 1/\varepsilon$ implies $0 < n^p \leq 1/n < \varepsilon$, which proves that $n^p \to 0$. (ii) If $n > P^p$, then $\sqrt[p]{n} \geq P$, so $\sqrt[p]{n} \to \infty$. $\quad\square$

Theorem 4.16. *For every fixed real number a,*

$$\lim_{n\to\infty} a^n = \begin{cases} \infty, & \text{if } a > 1, \\ 1, & \text{if } a - 1, \\ 0, & \text{if } |a| < 1. \end{cases} \tag{4.5}$$

If $a \leq -1$, then (a^n) oscillates at infinity.

Proof. If $a > 1$, then by Bernoulli's inequality, we have that

$$a^n = (1 + (a-1))^n \geq 1 + n \cdot (a-1)$$

holds for every n. So for an arbitrary real number P, if $n > (P-1)/(a-1)$, then $a^n > P$, so $a^n \to \infty$.

If $a = 1$, then $a^n = 1$ for each n, and so $a^n \to 1$.

If $|a| < 1$, then $1/|a| > 1$. If an $\varepsilon > 0$ is given, then by the already proved statement, there is an n_0 such that for $n > n_0$,

$$\frac{1}{|a|^n} = \left(\frac{1}{|a|}\right)^n > \frac{1}{\varepsilon}$$

holds, that is, $|a^n| = |a|^n < \varepsilon$. This means that $a^n \to 0$ in this case.

Finally, if $a \leq -1$, then for even n, we have $a^n \geq 1$, while for odd n, we have $a^n \leq -1$. It is clear that a sequence with these properties can have neither a finite nor an infinite limit. $\quad\square$

Theorem 4.17.

(i) For every fixed positive real number a, we have $\lim_{n\to\infty} \sqrt[n]{a} = 1$.
(ii) $\lim_{n\to\infty} \sqrt[n]{n} = 1$.

Proof. Let $a > 0$ be fixed. If $0 < \varepsilon \leq 1$, then by Theorem 4.16, $(1+\varepsilon)^n \to \infty$ and $(1-\varepsilon)^n \to 0$. Thus there exist n_1 and n_2 such that when $n > n_1$, we have $(1+\varepsilon)^n > a$, and when $n > n_2$, we have $(1-\varepsilon)^n < a$. So if $n > \max(n_1, n_2)$, then $(1-\varepsilon)^n < a < (1+\varepsilon)^n$, that is, $1-\varepsilon < \sqrt[n]{a} < 1+\varepsilon$. This shows that if $0 < \varepsilon \leq 1$, then there is an n_0 such that when $n > n_0$ holds, we also have $|\sqrt[n]{a} - 1| < \varepsilon$. It follows from this that for arbitrary positive ε, we can find an n_0, since if $\varepsilon \geq 1$, then the n_0 corresponding to 1 also works. We have shown (i).

(ii) Let $0 < \varepsilon < 1$ be fixed. If $n > 4/\varepsilon^2$ is even, then by Bernoulli's inequality, we have $(1+\varepsilon)^{n/2} > n\varepsilon/2$, so

$$(1+\varepsilon)^n > \left(\frac{n}{2}\varepsilon\right)^2 > n.$$

If $n > 16/\varepsilon^2$ is odd, then $(1+\varepsilon)^{(n-1)/2} > (n-1)\varepsilon/2 > n\varepsilon/4$, which gives

$$(1+\varepsilon)^n > (1+\varepsilon)^{n-1} > \left(\frac{n-1}{2}\varepsilon\right)^2 > \left(\frac{n}{4}\varepsilon\right)^2 > n.$$

Therefore, if $n > 16/\varepsilon^2$, then we have $\sqrt[n]{n} < 1+\varepsilon$. $\square$

Chapter 5
Infinite Sequences II

Finding the limit of a sequence is generally a difficult task. Sometimes, just determining whether a sequence has a limit is tough. Consider the sequence (18) in Example 4.1, that is, let a_n be the nth digit in the decimal expansion of $\sqrt{2}$. We know that (a_n) does not have a limit. But does the sequence $c_n = \sqrt[n]{a_n}$ have a limit? First of all, let us note that $a_n \geq 1$, and thus $c_n \geq 1$ for infinitely many n. Now if there are infinitely many zeros among the terms a_n, then $c_n = 0$ also holds for infinitely many n, so the sequence (c_n) is divergent. However, if there are only finitely many zeros among the terms a_n, that is, $a_n \neq 0$ for all $n > n_0$, then $1 \leq a_n \leq 9$, and so $1 \leq c_n \leq \sqrt[n]{9}$ also holds if $n > n_0$. By Theorem 4.17, $\sqrt[n]{9} \to 1$. Thus for a given $\varepsilon > 0$, there is an n_1 such that $\sqrt[n]{9} < 1 + \varepsilon$ for all $n > n_1$. So if $n > \max(n_0, n_1)$, then $1 \leq c_n < 1 + \varepsilon$, and thus $c_n \to 1$.

This reasoning shows that the sequence (c_n) has a limit if and only if there are only finitely many zeros among the terms a_n. However, *the question whether the decimal expansion of $\sqrt{2}$ has infinitely many zeros is a famous open problem in number theory.* Thus with our current knowledge, we are unable to determine whether the sequence (c_n) has a limit.

Fortunately, the above example is atypical; we can generally determine the limits of sequences that we encounter in practice. In most cases, we use the method of decomposing the given sequence into simpler sequences whose limits we know. Of course, to determine the limit, we need to know how to find the limit of a sequence that is constructed from simpler sequences. In the following, this will be explored.

5.1 Basic Properties of Limits

Definition 5.1. For a sequence $(a_1, a_2, \ldots, a_n, \ldots)$, we say that a *subsequence* is a sequence of the form

$$(a_{n_1}, a_{n_2}, \ldots, a_{n_k}, \ldots),$$

where $n_1 < n_2 < \cdots < n_k < \ldots$ are positive integers.

© Springer New York 2015
M. Laczkovich, V.T. Sós, *Real Analysis*, Undergraduate Texts
in Mathematics, DOI 10.1007/978-1-4939-2766-1_5

A subsequence, then, is formed by deleting some (possibly infinitely many) terms from the original sequence, keeping infinitely many.

Theorem 5.2. *If the sequence (a_n) has a limit, then every subsequence (a_{n_k}) does too, and* $\lim_{k\to\infty} a_{n_k} = \lim_{n\to\infty} a_n$.

Proof. Suppose first that (a_n) is convergent, and let $\lim_{n\to\infty} a_n = b$ be a finite limit. This means that for every positive ε, there are only finitely many terms of the sequence outside the interval $(b - \varepsilon, b + \varepsilon)$. Then clearly, the same holds for terms of the subsequence, which means exactly that $\lim_{k\to\infty} a_{n_k} = b$.

The statement can be proved similarly if (a_n) tends to infinity or negative infinity.

$\square$

We should note that the existence of a limit for a subsequence does not imply the existence of a limit for the original sequence, as can be seen in sequences (3) and (16) in Example 4.1. However, if we already know that (a_n) has a limit, then (by Theorem 5.2) every subsequence will have this same limit.

Definition 5.3. We say that the sequences (a_n) and (b_n) have *identical convergence behavior* if it is the case that (a_n) has a limit if and only if (b_n) has a limit, in which case the limits agree.

To determine limits of sequences, it is useful to inspect what changes we can make to a sequence that results in a new sequence whose convergence behavior is identical to that of the old one.

We will list a few such changes below:

I. We can "rearrange" a sequence, that is, we can change the order of its terms. A rearranged sequence contains the same numbers, and moreover, each one is listed the same number of times as in the previous sequence. (The formal definition is as follows: the sequence $(a_{n_1}, a_{n_2}, \ldots)$ is a rearrangement of the sequence $(a_1, a_2, \ldots)$ if the map $f(k) = n_k$ ($k \in \mathbb{N}^+$) is a permutation of $\mathbb{N}^+$, which means that f is a one-to-one correspondence from $\mathbb{N}^+$ to itself.)
II. We can repeat certain terms (possibly infinitely many) of a sequence finitely many times.
III. We can add finitely many terms to a sequence.
IV. We can take away finitely many terms from a sequence.

The above changes can, naturally, change the indices of the terms.

Examples 5.4. Consider the following sequences:

(1) $a_n = n$: \qquad\qquad\qquad\qquad\qquad $(a_n) = (1, 2, \ldots, n, \ldots)$;

(2) $a_n = n + 2$: \qquad\qquad\qquad\qquad $(a_n) = (3, 4, 5, \ldots)$;

(3) $a_n = n - 2$: \qquad\qquad\qquad\qquad $(a_n) = (-1, 0, 1, \ldots)$;

(4) $a_n = k$, if $k(k-1)/2 < n \le k(k+1)/2$:

$(a_n) = (1, 2, 2, 3, 3, 3, 4, 4, 4, 4, \ldots, k, \ldots)$;

(5) $a_n = 2n - 1$: $(a_n) = (1, 3, 5, 7 \ldots)$;

(6) $a_n = \begin{cases} 0, & \text{if } n = 2k+1, \\ k, & \text{if } n = 2k \end{cases}$: $(a_n) = (0, 1, 0, 2, 0, 3, \ldots)$;

(7) $a_n = \begin{cases} n+1, & \text{if } n = 2k+1, \\ n-1, & \text{if } n = 2k \end{cases}$: $(a_n) = (2, 1, 4, 3, 6, 5, \ldots)$.

In the above sequences, starting from the sequence (1), we can get

(2) by a type IV change,
(3) by a type III change,
(4) by a type II change,
(7) by a type I change.

Moreover, (6) cannot be a result of type I–IV changes applied to (1), since there is only one new term in (6), 0, but it appears infinitely many times. (Although we gained infinitely many new numbers in (4), too, we can get there using II.)

Theorem 5.5. *The sequences (a_n) and (b_n) have identical convergence behavior if one can be reached from the other by applying a finite combination of type I–IV changes.*

Proof. The property that there are finitely or infinitely many terms outside an interval remains unchanged by each one of the type I–IV changes. From this, by Definition 4.2 and Definition 4.6, the statement of the theorem is clear.   □

Exercises

5.1. Prove that if every subsequence of (a_n) has a subsequence that tends to b, then $a_n \to b$.

5.2. Prove that if the sequence (a_n) does not have a subsequence tending to infinity, then (a_n) is bounded from above.

5.3. Prove that if (a_{2n}), (a_{2n+1}), and (a_{3n}) are convergent, then (a_n) is as well.

5.4. Give an example for an (a_n) that is divergent, but for each $k > 1$, the subsequence (a_{kn}) is convergent. (H)

5.2 Limits and Inequalities

First of all, we introduce some terminology. Let (A_n) be a sequence of statements. We say that A_n *holds for all n sufficiently large* if there exists an n_0 such that A_n is true for all $n > n_0$. So, for example, we can say that $2^n > n^2$ for all n sufficiently large, since this inequality holds for all $n > 4$.

Theorem 5.6. *If* $\lim_{n\to\infty} a_n = \infty$ *and* $b_n \geq a_n$ *for all n sufficiently large, then* $\lim_{n\to\infty} b_n = \infty$. *If* $\lim_{n\to\infty} a_n = -\infty$ *and* $b_n \leq a_n$ *for all n sufficiently large, then* $\lim_{n\to\infty} b_n = -\infty$.

Proof. Suppose that $b_n \geq a_n$ if $n > n_0$. If $a_n > P$ for all $n > n_1$, then $b_n \geq a_n > P$ holds for all $n > \max(n_0, n_1)$. From this, it is clear that when $\lim_{n\to\infty} a_n = \infty$ holds, $\lim_{n\to\infty} b_n = \infty$ also holds. The second statement follows in the same way. □

The following theorem is often known as the **squeeze theorem** (or **sandwich theorem**).

Theorem 5.7. *If* $a_n \leq b_n \leq c_n$ *for all n sufficiently large and*

$$\lim_{n\to\infty} a_n = \lim_{n\to\infty} c_n = a,$$

then $\lim_{n\to\infty} b_n = a$.

Proof. By the previous theorem, it suffices to restrict ourselves to the case that a is finite. Suppose that $a_n \leq b_n \leq c_n$ for all $n > n_0$. It follows from our assumption that for every $\varepsilon > 0$, there exist n_1 and n_2 such that

$$a - \varepsilon < a_n < a + \varepsilon, \quad \text{if} \quad n > n_1$$

and

$$a - \varepsilon < c_n < a + \varepsilon, \quad \text{if} \quad n > n_2.$$

Then for $n > \max(n_0, n_1, n_2)$,

$$a - \varepsilon < a_n \leq b_n \leq c_n < a + \varepsilon,$$

that is, $b_n \to a$. □

We often use the above theorem for the special case $a_n \equiv 0$: if $0 \leq b_n \leq c_n$ and $\lim_{n\to\infty} c_n = 0$, then $\lim_{n\to\infty} b_n = 0$.

The following theorems state that strict inequality between limits is inherited by terms of sufficiently large index, while not-strict inequality between terms is inherited by limits.

Theorem 5.8. *Let* (a_n) *and* (b_n) *be convergent sequences, and let* $\lim_{n\to\infty} a_n = a$, $\lim_{n\to\infty} b_n = b$. *If* $a < b$, *then* $a_n < b_n$ *holds for all n sufficiently large.*

Proof. Let $\varepsilon = (b - a)/2$. We know that for suitable n_1 and n_2, $a_n < a + \varepsilon$ if $n > n_1$, and $b_n > b - \varepsilon$ if $n > n_2$. Let $n_0 = \max(n_1, n_2)$. If $n > n_0$, then both inequalities hold, that is, $a_n < a + \varepsilon = b - \varepsilon < b_n$. □

Remark 5.9. Note that from the weaker assumption $a \leq b$, we generally do not get that $a_n \leq b_n$ holds even for a single index. If, for example, $a_n = 1/n$ and $b_n = -1/n$, then $\lim_{n\to\infty} a_n = 0 \leq 0 = \lim_{n\to\infty} b_n$, but $a_n > b_n$ for all n.

Theorem 5.10. *Let* (a_n) *and* (b_n) *be convergent sequences, and let* $\lim_{n\to\infty} a_n = a$, $\lim_{n\to\infty} b_n = b$. *If* $a_n \leq b_n$ *holds for all sufficiently large n, then* $a \leq b$.

Proof. Suppose that $a > b$. By Theorem 5.8, it follows that $a_n > b_n$ for all n sufficiently large, which contradicts our assumption.   □

Remark 5.11. Note that even the assumption $a_n < b_n$ does not imply $a < b$. If, for example, $a_n = -1/n$ and $b_n = 1/n$, then $a_n < b_n$ for all n, but $\lim_{n\to\infty} a_n = 0 = \lim_{n\to\infty} b_n$.

Exercises

5.5. Prove that if $a_n \to a > 1$, then $(a_n)^n \to \infty$.

5.6. Prove that if $a_n \to a$, where $|a| < 1$, then $(a_n)^n \to 0$.

5.7. Prove that if $a_n \to a > 0$, then $\sqrt[n]{a_n} \to 1$.

5.8. $\lim_{n\to\infty} \sqrt[n]{2^n - n} = ?$

5.9. Prove that if $a_1, \dots, a_k \geq 0$, then

$$\lim_{n\to\infty} \sqrt[n]{a_1^n + \cdots + a_k^n} = \max_{1\leq i\leq k} a_i. \tag{S}$$

5.3 Limits and Operations

We say that the **sum of the sequences** (a_n) and (b_n) is the sequence $(a_n + b_n)$. The following theorem states that in most cases, the order of taking sums and taking limits can be switched, that is, the sum of the limits is equal to the limit of the sum of terms.

Theorem 5.12.

(i) *If the sequences (a_n) and (b_n) are convergent and $a_n \to a$, $b_n \to b$, then the sequence $(a_n + b_n)$ is convergent and $a_n + b_n \to a + b$.*

(ii) *If the sequence (a_n) is convergent, $a_n \to a$ and $b_n \to \infty$, then $a_n + b_n \to \infty$.*

(iii) *If the sequence (a_n) is convergent, $a_n \to a$ and $b_n \to -\infty$, then $a_n + b_n \to -\infty$.*

(iv) *If $a_n \to \infty$ and $b_n \to \infty$, then $a_n + b_n \to \infty$.*

(v) *If $a_n \to -\infty$ and $b_n \to -\infty$, then $a_n + b_n \to -\infty$.*

Proof. (i) Intuitively, if a_n is close to a and b_n is close to b, then $a_n + b_n$ is close to $a + b$. Basically, we only need to make this idea precise using limits.

If $a_n \to a$ and $b_n \to b$, then for all $\varepsilon > 0$, there exist n_1 and n_2 for which $|a_n - a| < \varepsilon/2$ holds if $n > n_1$, and $|b_n - b| < \varepsilon/2$ holds if $n > n_2$. It follows from this, using the triangle inequality, that

$$|(a_n + b_n) - (a + b)| \leq |a_n - a| + |b_n - b| < \varepsilon,$$

if $n > \max(n_1, n_2)$. Since ε was arbitrary, this proves that $a_n + b_n \to a + b$.

(ii) If (a_n) is convergent, then by Theorem 4.10, it is bounded. This means that for suitable $K > 0$, $|a_n| \le K$ for all n. Let P be arbitrary. Since $b_n \to \infty$, there exists an n_0 such that when $n > n_0$, $b_n > P + K$. Then $a_n + b_n > (-K) + (P + K) = P$ for all $n > n_0$. Since P was arbitrary, this proves that $a_n + b_n \to \infty$. Statement (iii) can be proved in the same way.

(iv) Suppose that $a_n \to \infty$ and $b_n \to \infty$. Let P be arbitrary. Then there are n_1 and n_2 such that when $n > n_1$, $a_n > P/2$, and when $n > n_2$, $b_n > P/2$. If $n > \max(n_1, n_2)$, then $a_n + b_n > (P/2) + (P/2) = P$. Since P was chosen arbitrarily, this proves that $a_n + b_n \to \infty$. Statement (v) can be proved in the same way. $\qquad\square$

If (a_n) is convergent and $a_n \to a$, then by applying statement (i) in Theorem 5.12 to the constant sequence $b_n = -a$, we get that $a_n - a \to 0$. Conversely, if $a_n - a \to 0$, then $a_n = (a_n - a) + a \to a$. This shows the following.

Corollary 5.13. *A sequence (a_n) tends to a finite limit a if and only if $a_n - a \to 0$.*

The statements of Theorem 5.12 can be summarized in the table below.

		$\lim b_n$		
		b	∞	$-\infty$
	a	$a+b$	∞	$-\infty$
$\lim a_n$	∞	∞	∞	?
	$-\infty$	$-\infty$	?	$-\infty$

The question marks appearing in the table mean that the given values of $\lim a_n$ and $\lim b_n$ do not determine the value of $\lim(a_n + b_n)$. Specifically, if $\lim a_n = \infty$ and $\lim b_n = -\infty$ (or the other way around), then using only this information, we cannot say what value $\lim(a_n + b_n)$ takes. Let us see a few examples.

$$a_n = n + c, \qquad b_n = -n, \qquad a_n + b_n = c \to c \in \mathbb{R}$$
$$a_n = 2n, \qquad b_n = -n, \qquad a_n + b_n = n \to \infty$$
$$a_n = n, \qquad b_n = -2n, \quad a_n + b_n = -n \to -\infty$$
$$a_n = n + (-1)^n, \quad b_n = -n, \quad a_n + b_n = (-1)^n \text{ oscillates at infinity.}$$

We see that $(a_n + b_n)$ can be convergent, can diverge to positive or negative infinity, and can oscillate at infinity. We express this by saying that the limit $\lim(a_n + b_n)$ is a **critical limit** if $\lim a_n = \infty$ and $\lim b_n = -\infty$. More concisely, we can say that limits of the type $\infty - \infty$ are critical.

Now we look at the limit of a product. We call the **product of two sequences** (a_n) and (b_n) the sequence $(a_n \cdot b_n)$.

Theorem 5.14.

(i) *If the sequences (a_n) and (b_n) are convergent and $a_n \to a$, $b_n \to b$, then the sequence $(a_n \cdot b_n)$ is convergent, and $a_n \cdot b_n \to a \cdot b$.*

(ii) *If the sequence (a_n) is convergent, $a_n \to a$ where $a > 0$, and $b_n \to \pm\infty$, then $a_n \cdot b_n \to \pm\infty$.*

(iii) *If the sequence (a_n) is convergent, $a_n \to a$ where $a < 0$, and $b_n \to \pm\infty$, then $a_n \cdot b_n \to \mp\infty$.*

(iv) *If $a_n \to \pm\infty$ and $b_n \to \pm\infty$, then $a_n \cdot b_n \to \infty$.*

(v) *If $a_n \to \pm\infty$ and $b_n \to \mp\infty$, then $a_n \cdot b_n \to -\infty$.*

Lemma 5.15. *If $a_n \to 0$ and (b_n) is bounded, then $a_n \cdot b_n \to 0$.*

Proof. Since (b_n) is bounded, there is a $K > 0$ such that $|b_n| \leq K$ for all n. Let $\varepsilon > 0$ be given. From the assumption that $a_n \to 0$, it follows that $|a_n| < \varepsilon/K$ for all n sufficiently large. Thus $|a_n \cdot b_n| < (\varepsilon/K) \cdot K = \varepsilon$ for all n sufficiently large. Since ε was chosen arbitrarily, this proves that $a_n \cdot b_n \to 0$. $\qquad\square$

Proof (Theorem 5.14). (i) Since by our assumption, (a_n) and (b_n) are convergent, by Theorem 4.10, both are bounded. If $a_n \to a$ and $b_n \to b$, then by Corollary 5.13, $a_n - a \to 0$ and $b_n - b \to 0$. Moreover,

$$a_n \cdot b_n - a \cdot b = (a_n - a) \cdot b_n + a \cdot (b_n - b). \tag{5.1}$$

By Lemma 5.15, both terms on the right-hand side tend to 0, so by Theorem 5.12, the limit of the right-hand side is 0. Then $a_n \cdot b_n - a \cdot b \to 0$, so by Corollary 5.13, $a_n \cdot b_n \to a \cdot b$.

(ii) Suppose that $a_n \to a > 0$ and $b_n \to \infty$. Let P be an arbitrary positive number. Since $a/2 < a$, there exists an n_1 such that $a_n > a/2$ for all $n > n_1$. By the assumption that $b_n \to \infty$, there exists an n_2 such that $b_n > 2P/a$ for all $n > n_2$. Then for $n > \max(n_1, n_2)$, $a_n \cdot b_n > (a/2) \cdot (2P/a) = P$. Since P was chosen arbitrarily, this proves that $a_n \cdot b_n \to \infty$. It can be shown in the same way that if $a_n \to a > 0$ and $b_n \to -\infty$, then $a_n \cdot b_n \to -\infty$. Statement (iii) can also be proved in the same manner.

(iv) If $a_n \to \infty$ and $b_n \to \infty$, then for all $P > 0$, there exist n_1 and n_2 such that for all $n > n_1$, $a_n > P$, and for all $n > n_2$, $b_n > 1$. Then for $n > \max(n_1, n_2)$, we have $a_n \cdot b_n > P \cdot 1 = P$. It can be shown in the same way that if $a_n \to -\infty$ and $b_n \to -\infty$, then $a_n \cdot b_n \to \infty$. Statement (v) can also be proved in the same manner. $\qquad\square$

The statements of Theorem 5.14 are summarized in the table below.

		$\lim b_n$			
	$b > 0$	0	$b < 0$	∞	$-\infty$
$a > 0$	$a \cdot b$	0	$a \cdot b$	∞	$-\infty$
0	0	0	0	$?$	$?$
$a < 0$	$a \cdot b$	0	$a \cdot b$	$-\infty$	∞
∞	∞	$?$	$-\infty$	∞	$-\infty$
$-\infty$	$-\infty$	$?$	∞	$-\infty$	∞

$\lim a_n$ labels the left column group.

The question marks indicate the critical limits again. As the examples below show, $\lim(a_n \cdot b_n)$ is critical if $a_n \to 0$ and $b_n \to \infty$. (In short, the limit of type $0 \cdot \infty$ is critical.)

$$a_n = c/n, \qquad b_n = n, \qquad a_n \cdot b_n = c \to c \in \mathbb{R}$$

$$a_n = 1/n, \qquad b_n = n^2, \qquad a_n \cdot b_n = n \to \infty$$

$$a_n = -1/n, \qquad b_n = n^2, \qquad a_n \cdot b_n = -n \to -\infty$$

$$a_n = (-1)^n/n, \quad b_n = n, \qquad a_n \cdot b_n = (-1)^n \text{ oscillates at infinity.}$$

Similar examples show that the limit of type $0 \cdot (-\infty)$ is also critical.

We now turn to defining quotient limits. Suppose that $b_n \neq 0$ for all n. We sometimes call the sequence (a_n/b_n) the **quotient sequence**.

Theorem 5.16. *Suppose that the sequences (a_n) and (b_n) have limits, and that $b_n \neq 0$ for all n. Then the limit of the sequence (a_n/b_n) is given by the table below.*

		$\lim b_n$				
		$b > 0$	0	$b < 0$	∞	$-\infty$
	$a > 0$	a/b	?	a/b	0	0
	0	0	?	0	0	0
$\lim a_n$	$a < 0$	a/b	?	a/b	0	0
	∞	∞	?	$-\infty$	?	?
	$-\infty$	$-\infty$	?	∞	?	?

Lemma 5.17. *If (b_n) is convergent and $b_n \to b \neq 0$, then $1/b_n \to 1/b$.*

Proof. Let $\varepsilon > 0$ be given; we need to show that $|1/b_n - 1/b| < \varepsilon$ if n is sufficiently large. Since

$$\frac{1}{b_n} - \frac{1}{b} = \frac{b - b_n}{b \cdot b_n},$$

we have to show that if n is large, then $b - b_n$ is very small, while $b \cdot b_n$ is not too small. By the assumption that $b_n \to b$, there exists an n_1 such that for $n > n_1$, $|b_n - b| < \varepsilon b^2/2$. Since $|b|/2 > 0$, we can find an n_2 such that $|b_n - b| < |b|/2$ for $n > n_2$. Then for $n > n_2$, $|b_n| > |b|/2$, since if $b > 0$, then $b_n > b - (b/2) = b/2$, while if $b < 0$, then $b_n < b + (|b|/2) = b/2 = -|b|/2$. Then for $n > \max(n_1, n_2)$,

$$\left| \frac{1}{b_n} - \frac{1}{b} \right| = \frac{|b - b_n|}{|b \cdot b_n|} < \frac{\varepsilon b^2/2}{|b| \cdot |b|/2} = \varepsilon.$$

Since ε was arbitrary, this proves that $1/b_n \to 1/b$. □

Lemma 5.18. *If $|b_n| \to \infty$, then $1/b_n \to 0$.*

Proof. Let $\varepsilon > 0$ be given. Since $|b_n| \to \infty$, there is an n_0 such that for $n > n_0$, $|b_n| > 1/\varepsilon$. Then when $n > n_0$, $|1/b_n| = 1/|b_n| < \varepsilon$, so $1/b_n \to 0$. □

Corollary 5.19. *If $b_n \to \infty$ or $b_n \to -\infty$, then $1/b_n \to 0$.*

Proof. It is easy to check that if $b_n \to \infty$ or $b_n \to -\infty$, then $|b_n| \to \infty$. □

Proof (Theorem 5.16). Suppose first that $a_n \to a \in \mathbb{R}$ and $b_n \to b \in \mathbb{R}$, $b \neq 0$. By Theorem 5.14 and Lemma 5.17,

$$\frac{a_n}{b_n} = a_n \cdot \frac{1}{b_n} \to a \cdot \frac{1}{b} = \frac{a}{b}.$$

If (a_n) is convergent and $b_n \to \infty$ or $b_n \to -\infty$, then by Theorem 5.14 and Corollary 5.19,

$$\frac{a_n}{b_n} = a_n \cdot \frac{1}{b_n} \to a \cdot 0 = 0.$$

Now suppose that $a_n \to \infty$ and $b_n \to b \in \mathbb{R}$, $b > 0$. Then by Theorem 5.14 and Lemma 5.17,

$$\frac{a_n}{b_n} = a_n \cdot \frac{1}{b_n} \to \infty,$$

since $1/b_n \to 1/b > 0$. It can be seen similarly that in the case of $a_n \to \infty$ and $b_n \to b < 0$, we have $a_n/b_n \to -\infty$; in the case of $a_n \to -\infty$ and $b_n \to b > 0$, we have $a_n/b_n \to -\infty$; while in the case of $a_n \to -\infty$ and $b_n \to b < 0$, we have $a_n/b_n \to \infty$. With this, we have justified every (non-question-mark) entry in the table. $\square$

The question marks in the table of Theorem 5.16 once more denote the critical limits. Here, however, we need to distinguish two levels of criticality. As the examples below show, the limit of type $0/0$ is critical in the same sense that, for example, the limit of type $0 \cdot \infty$ is.

$$a_n = c/n, \qquad b_n = 1/n, \qquad a_n/b_n = c \to c \in \mathbb{R}$$
$$a_n = 1/n, \qquad b_n = 1/n^2, \qquad a_n/b_n = n \to \infty$$
$$a_n = -1/n, \qquad b_n = 1/n^2, \qquad a_n/b_n = -n \to -\infty$$
$$a_n = (-1)^n/n, \qquad b_n = 1/n, \qquad a_n/b_n = (-1)^n \text{ oscillates at infinity.}$$

We can see that if $a_n \to 0$ and $b_n \to 0$, then (a_n/b_n) can be convergent, can tend to infinity or negative infinity, and can oscillate at infinity as well.

The situation is different with the other question marks in the table of theorem 5.16. Consider the case that $a_n \to a > 0$ and $b_n \to 0$. The examples $a_n = 1$, $b_n = 1/n$; $a_n = 1$, $b_n = -1/n$; and $a_n = 1$, $b_n = (-1)^n/n$ show that a_n/b_n can tend to infinity or negative infinity, but can oscillate at infinity as well. However, we do not find an example in which a_n/b_n is convergent. This follows immediately from the following theorem.

Theorem 5.20.

(i) *Suppose that $b_n \to 0$ and $b_n \neq 0$ for all n. Then $1/|b_n| \to \infty$.*
(ii) *Suppose that $a_n \to a \neq 0$, $b_n \to 0$, and $b_n \neq 0$ for all n. Then $|a_n/b_n| \to \infty$.*

Proof. It is enough to prove (ii). Let $P > 0$ be given. There exists an n_0 such that for $n > n_0$, $|a_n| > |a|/2$ and $|b_n| < |a|/(2P)$. Then for $n > n_0$, $|a_n/b_n| > P$, that is, $|a_n/b_n| \to \infty$. $\square$

Finally let us consider the case that $a_n \to \infty$ and $b_n \to \infty$. The examples $a_n = b_n = n$ and $a_n = n^2$, $b_n = n$ show that a_n/b_n can be convergent and can tend to infinity as well. Now let

$$a_n = \begin{cases} n^2, & \text{if } n \text{ is even,} \\ n, & \text{if } n \text{ is odd} \end{cases}$$

and $b_n = n$. It is clear that $a_n \to \infty$, and a_n/b_n agrees with sequence (16) from Example 4.1, which oscillates at infinity. However, we cannot find an example in which a_n/b_n tends to negative infinity. Since both $a_n \to \infty$ and $b_n \to \infty$, a_n and b_n are both positive for all sufficiently large n. Thus a_n/b_n is positive for all sufficiently large n, so it cannot tend to negative infinity. Similar observations can be made for the three remaining cases, when (a_n) and (b_n) tend to infinity or negative infinity.

Exercises

5.10. Prove that if $(a_n + b_n)$ is convergent and (b_n) is divergent, then (a_n) is divergent.

5.11. Is it true that if $(a_n \cdot b_n)$ is convergent and (b_n) is divergent, then (a_n) is also divergent?

5.12. Is it true that if (a_n/b_n) is convergent and (b_n) is divergent, then (a_n) is also divergent?

5.13. Prove that if $\lim_{n \to \infty} (a_n - 1)/(a_n + 1) = 0$, then $\lim_{n \to \infty} a_n = 1$.

5.14. Let $\lim_{n \to \infty} a_n = a$, $\lim_{n \to \infty} b_n = b$. Prove that $\max(a_n, b_n) \to \max(a, b)$.

5.15. Prove that if $a_n < 0$ and $a_n \to 0$, then $1/a_n \to -\infty$.

5.4 Applications

First of all, we need generalizations of 5.12 (i) and 5.14 (i) for more summands and terms, respectively.

Theorem 5.21. *Let $\left(a_n^1\right), \ldots, \left(a_n^k\right)$ be convergent sequences,[1] and let $\lim_{n \to \infty} a_n^i = b_i$ for all $i = 1, \ldots, k$. Then the sequences $\left(a_n^1 + \cdots + a_n^k\right)$ and $\left(a_n^1 \cdot \ldots \cdot a_n^k\right)$ are also convergent, and their limits are $b_1 + \cdots + b_k$ and $b_1 \cdot \ldots \cdot b_k$ respectively.*

The statement can be proved easily by induction on k, using that the $k = 2$ case was already proved in Theorems 5.12 and 5.14.

---

[1] here a_n^i denotes the nth term of the ith sequence

It is important to note that Theorem 5.21 *holds only for a fixed number of sequences*, that is, the assumptions of the theorem do not allow the number of sequences (k) to depend on n. Consider

$$1 = \frac{1}{n} + \cdots + \frac{1}{n},$$

if the number of summands on the right-hand side is exactly n. Despite $1/n$ tending to 0, the sum on the left-hand side—the constant-1 sequence—still tends to 1. Similarly,

$$2 = \sqrt[n]{2} \cdot \ldots \cdot \sqrt[n]{2},$$

if the number of terms on the right side is exactly n. Even though $\sqrt[n]{2}$ tends to 1, the product—the constant-2 sequence—still tends to 2.

As a first application, we will determine the limits of sequences that can be obtained from the index n and constants, using the four elementary operations.

Theorem 5.22. *Let*

$$c_n = \frac{a_0 + a_1 n + \cdots + a_k n^k}{b_0 + b_1 n + \cdots + b_\ell n^\ell} \quad (n = 1, 2, \ldots),$$

where $a_k \neq 0$ and $b_\ell \neq 0$. Then

$$\lim_{n \to \infty} c_n = \begin{cases} 0, & \text{if } \ell > k, \\ \infty, & \text{if } \ell < k \text{ and } a_k/b_\ell > 0, \\ -\infty, & \text{if } \ell < k \text{ and } a_k/b_\ell < 0, \\ a_k/b_\ell, & \text{if } \ell = k. \end{cases}$$

Proof. If we take out an n^k from the numerator and n^ℓ from the denominator of the fraction representing c_k, then we get that

$$c_n = \frac{n^k}{n^\ell} \cdot \frac{a_k + \dfrac{a_{k-1}}{n} + \cdots + \dfrac{a_0}{n^k}}{b_\ell + \dfrac{b_{\ell-1}}{n} + \cdots + \dfrac{b_0}{n^\ell}}. \tag{5.2}$$

By Theorem 4.15, in the second term, except for the first summands, everything in both the numerator and the denominator tends to 0.

Then by applying Theorems 5.21 and 5.16, we get that the second term on the right-hand side of (5.2) tends to a_k/b_ℓ. From this, based on the behavior of $n^{k-\ell}$ as $n \to \infty$ (Theorem 4.15), the statement immediately follows.   □

An important sufficient condition for convergence to 0 is given by the next theorem.

Theorem 5.23. *Suppose that there exists a number $q < 1$ such that $a_n \neq 0$ and $|a_{n+1}/a_n| \leq q$ for all n sufficiently large. Then $a_n \to 0$.*

Proof. If $a_n \neq 0$ and $|a_{n+1}/a_n| \leq q$ for every $n \geq n_0$, then

$$|a_{n_0+1}| \leq q \cdot |a_{n_0}|,$$
$$|a_{n_0+2}| \leq q \cdot |a_{n_0+1}| \leq q^2 \cdot |a_{n_0}|,$$
$$|a_{n_0+3}| \leq q \cdot |a_{n_0+2}| \leq q^3 \cdot |a_{n_0}|, \tag{5.3}$$

and so on; all inequalities $|a_n| \leq q^{n-n_0} \cdot |a_{n_0}|$ for $n > n_0$ hold. Since $q^n \to 0$ by Theorem 4.16, $a_n \to 0$.   □

Corollary 5.24. *Suppose that $a_n \neq 0$ for all sufficiently large n, and $a_{n+1}/a_n \to c$, where $|c| < 1$. Then $a_n \to 0$.*

Proof. Fix a number q, for which $|c| < q < 1$. Since $|a_{n+1}/a_n| \to |c|$, $|a_{n+1}/a_n| < q$ for all sufficiently large n, so we can apply Theorem 5.23.   □

Remark 5.25. The assumptions required for Theorem 5.23 and Corollary 5.24 are generally *not necessary conditions* to deduce $a_n \to 0$. The sequence $a_n = 1/n$ tends to zero, but $a_{n+1}/a_n = n/(n+1) \to 1$, so neither the assumptions of Theorem 5.23 nor the assumptions of Corollary 5.24 are satisfied.

Corollary 5.24 can often be applied to sequences that are given as a product. The following theorem introduces two important special cases. The notation $n!$ in the second statement denotes the product $1 \cdot 2 \cdot \ldots \cdot n$. We call this product n **factorial**.

Theorem 5.26.

(i) *For an arbitrary real number $a > 1$ and positive integer k, we have $n^k/a^n \to 0$.*
(ii) *For an arbitrary real number a, $a^n/n! \to 0$.*

Proof. (i) Let $a_n = n^k/a^n$. Then

$$\frac{a_{n+1}}{a_n} = \frac{(n+1)^k \cdot a^n}{n^k \cdot a^{n+1}} = \frac{\left(1 + \frac{1}{n}\right)^k}{a}.$$

Here the numerator tends to 1 by Theorem 5.21, and so $a_{n+1}/a_n \to 1/a$. Since $a > 1$ by our assumption, $0 < 1/a < 1$, so we can apply Corollary 5.24.
(ii) If $b_n = a^n/n!$, then

$$\frac{b_{n+1}}{b_n} = \frac{a^{n+1} \cdot n!}{a^n \cdot (n+1)!} = \frac{a}{n+1}.$$

Then $b_{n+1}/b_n \to 0$, so $b_n \to 0$ by Corollary 5.24.   □

By Theorem 4.16, $a^n \to \infty$ if $a > 1$. We also know that $n^k \to \infty$ if k is a positive integer. By statement (i) of the above theorem, we can determine that $a^n/n^k \to \infty$ (see Theorem 5.20), which means that the sequence (a^n) is "much bigger" than the sequence (n^k). To make this phenomenon clear, we introduce new terminology and notation below.

Definition 5.27. Let (a_n) and (b_n) be sequences tending to infinity. We say that the sequence (a_n) *tends to infinity faster* than the sequence (b_n) if $a_n/b_n \to \infty$. We can also express this by saying that (a_n) *has a larger order of magnitude than* (b_n), and we denote this by $(b_n) \prec (a_n)$.

By Theorems 4.15, 4.16, and 5.26, we can conclude that the following order-of-magnitude relations hold:

$$(n) \prec (n^2) \prec (n^3) \prec \ldots \prec (2^n) \prec (3^n) \prec (4^n) \prec \ldots \prec (n!) \prec n^n.$$

Definition 5.28. If $a_n \to \infty$, $b_n \to \infty$, and $a_n/b_n \to 1$, then we say that the sequences (a_n) and (b_n) are *asymptotically equivalent*, and we denote this by $a_n \sim b_n$.

Thus, for example, $(n^2 + n) \sim n^2$, since $(n^2 + n)/n^2 = 1 + (1/n) \to 1$.

Exercises

5.16. Prove that if $a_n > 0$ and $a_{n+1}/a_n > q > 1$, then $a_n \to \infty$.

5.17. Prove that if $a_n > 0$ and $a_{n+1}/a_n \to c$, where $c > 1$, then $a_n \to \infty$.

5.18. Show that if $a_n > 0$ and $a_{n+1}/a_n \to q$, then $\sqrt[n]{a_n} \to q$.

5.19. Give an example of a positive sequence (a_n) for which $\sqrt[n]{a_n} \to 1$, but a_{n+1}/a_n does not tend to 1.

5.20. Prove that $\lim_{n \to \infty} 2^{\sqrt{n}}/n^k = \infty$ for all k.

5.21. Let $(a_n^1), (a_n^2), \ldots$ be an arbitrary sequence of sequences tending to infinity. (Here a_n^k denotes the nth term of the kth sequence.) Prove that there exists a sequence $b_n \to \infty$, whose order of magnitude is larger than the order of magnitude of every (a_n^k). (H)

5.22. Suppose that
$$(a_n^1) \prec (a_n^2) \prec \ldots \prec (b_n^2) \prec (b_n^1).$$
Prove that there exists a sequence (c_n) for which $(a_n^k) \prec (c_n) \prec (b_n^k)$ for all k.

5.23. Let $p(n) = a_0 + a_1 n + \cdots + a_k n^k$, where $a_k > 0$. Prove that $p(n+1) \sim p(n)$.

Chapter 6
Infinite Sequences III

6.1 Monotone Sequences

In Theorem 4.10, we proved that for a sequence to converge, a necessary condition is the boundedness of the sequence, and in our example of the sequence $(-1)^n$, we saw that boundedness is not a sufficient condition for convergence.

In the following, we prove, however, that for a major class of sequences, the so-called monotone sequences, boundedness already implies convergence. Since boundedness is generally easier to check than convergence directly, the theorem is often a very useful tool in determining whether a sequence converges. Moreover, as we will see, the theorem is important from a conceptual viewpoint and is closely linked to Cantor's axiom.

Definition 6.1. We say the sequence (a_n) is *monotone increasing* if

$$a_1 \leq a_2 \leq \cdots \leq a_n \leq a_{n+1} \leq \ldots.$$

If the statement above with $\leq$ replaced by $\geq$ holds, we say that the sequence is *monotone decreasing*, while in the case of $<$ and $>$ we say *strictly monotone increasing* and *strictly monotone decreasing* respectively.

We call the sequence (a_n) *monotone* if one of the above cases applies.

Theorem 6.2. *If the sequence (a_n) is monotone and bounded, then it is convergent. If (a_n) is monotone increasing, then*

$$\lim_{n \to \infty} a_n = \sup\{a_n\},$$

and if it is monotone decreasing, then

$$\lim_{n \to \infty} a_n = \inf\{a_n\}.$$

© Springer New York 2015
M. Laczkovich, V.T. Sós, *Real Analysis*, Undergraduate Texts
in Mathematics, DOI 10.1007/978-1-4939-2766-1_6

Proof. Let, for example, (a_n) be monotone increasing, bounded, and set

$$\alpha = \sup\{a_n\}.$$

Since α is the least upper bound of the set $\{a_n\}$, it follows that for all $\varepsilon > 0$, $\alpha - \varepsilon$ is not an upper bound, that is, there exists an n_0 for which

$$a_{n_0} > \alpha - \varepsilon.$$

The sequence is monotone increasing, so

$$a_n \geq a_{n_0}, \ \text{if} \ \ n > n_0,$$

and thus

$$0 \leq \alpha - a_n \leq \alpha - a_{n_0} < \varepsilon, \ \text{if} \ \ n > n_0.$$

We see that for all $\varepsilon > 0$, there exists an n_0 for which

$$|\alpha - a_n| < \varepsilon, \ \text{if} \ \ n > n_0$$

holds, which means exactly that $\lim_{n \to \infty} a_n = \alpha$.   □

We can extend Theorem 6.2 with the following result.

Theorem 6.3. *If the sequence (a_n) is monotone increasing and unbounded, then* $\lim_{n \to \infty} a_n = \infty$; *if it is monotone decreasing and unbounded, then* $\lim_{n \to \infty} a_n = -\infty$.

Proof. Let (a_n) be monotone increasing and unbounded. If (a_n) is monotone increasing, then it is bounded from below, since a_1 is a lower bound. Then the assumption that (a_n) is unbounded means that (a_n) is unbounded from above. Then for all P, there exists an n_0 (depending on P) for which $a_{n_0} > P$. But by monotonicity, $a_n \geq a_{n_0} > P$ if $n > n_0$, that is, $\lim_{n \to \infty} a_n = \infty$. The statement for monotone decreasing sequences can be proved similarly.   □

In the previous theorem, it suffices to assume that the sequence is monotone for $n > n_1$, since finitely many terms do not change the convergence behavior of the sequence.

We often denote that (a_n) is monotone increasing (or decreasing) and tends to a by

$$a_n \nearrow a; \quad (a_n \searrow a).$$

With the help of Theorem 6.2, we can justify the convergence of a few frequently occurring sequences.

Theorem 6.4. *The sequence* $a_n = \left(1 + \dfrac{1}{n}\right)^n$ *is strictly monotone increasing and bounded, thus convergent.*

Proof. By the inequality of arithmetic and geometric means,

$$\sqrt[n+1]{1 \cdot \left(1 + \frac{1}{n}\right)^n} < \frac{1 + n \cdot \left(1 + \frac{1}{n}\right)}{n+1} = \frac{n+2}{n+1} = 1 + \frac{1}{n+1}.$$

Raising both sides to the $(n+1)$th power yields

$$\left(1 + \frac{1}{n}\right)^n < \left(1 + \frac{1}{n+1}\right)^{n+1},$$

showing us that the sequence is strictly monotone increasing.

For boundedness from above, we will show that for all positive integers n and m,

$$\left(1 + \frac{1}{n}\right)^n < \left(1 + \frac{1}{m}\right)^{m+1}. \tag{6.1}$$

Using the inequality of the arithmetic and geometric means again, we see that

$$\sqrt[n+m]{\left(1 + \frac{1}{n}\right)^n \cdot \left(1 - \frac{1}{m}\right)^m} < \frac{n \cdot \left(1 + \frac{1}{n}\right) + m \cdot \left(1 - \frac{1}{m}\right)}{n+m} = \frac{n+m}{n+m} = 1,$$

that is,

$$\left(1 + \frac{1}{n}\right)^n \cdot \left(1 - \frac{1}{m}\right)^m < 1. \tag{6.2}$$

Since for $m > 1$, the reciprocal of $(1 - (1/m))^m$ is

$$\left(\frac{m}{m-1}\right)^m = \left(1 + \frac{1}{m-1}\right)^m,$$

dividing both sides of (6.2) by $(1 - (1/m))^m$ yields (6.1). This proves that each $(1 + (1/m))^{m+1}$ is an upper bound of the sequence. $\square$

We denote[1] the limit of the sequence $\left(1 + \frac{1}{n}\right)^n$ by e, that is,

$$e = \lim_{n \to \infty} \left(1 + \frac{1}{n}\right)^n. \tag{6.3}$$

As we will later see, this constant plays an important role in analysis and other branches of mathematics.

---

[1] This notation was introduced by Leonhard Euler (1707–1783), Swiss mathematician.

It can easily be shown that the limit of a strictly increasing sequence is larger than any term in the sequence. If we compare this to (6.1) and Theorem 5.10, then we get that

$$\left(1+\frac{1}{n}\right)^n < e \le \left(1+\frac{1}{n}\right)^{n+1} \tag{6.4}$$

for all n. This then implies $e > 1.1^{10} > 2.5$ and $e \le 1.2^6 < 3$. Further, by (6.4),

$$0 < e - \left(1+\frac{1}{n}\right)^n < \left(\left(1+\frac{1}{n}\right)-1\right)\left(1+\frac{1}{n}\right)^n < \frac{3}{n},$$

and this (theoretically) provides an opportunity to estimate e with arbitrary fixed precision.[2] One can show (although perhaps not with the help of the above approximation), that $e = 2.718281828459045\ldots$. It can be proven that e is an irrational number.[3]

Using the approximation (6.4), we can get more accurate information of the order of magnitude of factorials.

Theorem 6.5.

(i) $n! > (n/e)^n$ for all n, and moreover,

(ii) $n! < n \cdot (n/e)^n$ for all $n \ge 7$.

Proof. We will prove both statements by induction. Since $e > 1$, $1! = 1 > 1/e$. Suppose that $n \ge 1$ and $n! > (n/e)^n$. To prove the inequality $(n+1)! > ((n+1)/e)^{n+1}$, it suffices to show that

$$(n+1)\cdot\left(\frac{n}{e}\right)^n > \left(\frac{n+1}{e}\right)^{n+1},$$

which is equivalent to $(1+1/n)^n < e$. This proves (i).

In proving (ii), first of all, we check that $7! = 5040 < 7\cdot(7/e)^7$. It is easy to check that $720 < (2.56)^7$, from which we get $5040 < 7\cdot(2.56)^7 < 7\cdot(7/e)^7$. Suppose that $n \ge 7$ and $n! < n\cdot(n/e)^n$. To prove $(n+1)! < (n+1)\cdot((n+1)/e)^{n+1}$, it suffices to show that

$$(n+1)\cdot n\cdot\left(\frac{n}{e}\right)^n \le (n+1)\cdot\left(\frac{n+1}{e}\right)^{n+1},$$

which is equivalent to $(1+1/n)^{n+1} \ge e$. This proves (ii). $\square$

The exact order of magnitude of $n!$ is given by **Stirling's formula**,[4] which states that $n! \sim (n/e)^n \cdot \sqrt{2\pi n}$. This can be proved as an application of integrals (see Theorem 15.15).

---

[2] In practice, (6.4) is not a very useful approximation. If we wanted to approximate e to 10 decimal points, we would have to compute a 10^{10}th power. We will later give a much faster approximation method.

[3] In fact, one can also show that e is what is called a *transcendental number*, meaning that it is not a root of any nonzero polynomial with integer coefficients. We will prove irrationality in Exercises 12.87 and 15.23, while one can prove transcendence as an application of integration.

[4] James Stirling (1692–1770) Scottish mathematician.

With the help of Theorem 6.2, we can conclude the convergence of several sequences given by recurrence relations.

Example 6.6. Consider sequence (15) in Example 4.1, that is, the sequence (a_n) given by the recurrence $a_1 = 0$, $a_{n+1} = \sqrt{2+a_n}$. We show that the sequence is monotone increasing. We prove $a_n < a_{n+1}$ using induction. Since $a_1 = 0 < \sqrt{2} = a_2$, the statement is true for $n = 1$. If it holds for n, then

$$a_{n+1} = \sqrt{2+a_n} < \sqrt{2+a_{n+1}} = a_{n+2},$$

so it holds for $n+1$ as well. This shows that the sequence is (strictly) monotone increasing.

Now we show that the sequence is bounded from above, namely that 2 is an upper bound. The statement $a_n \leq 2$ follows from induction as well, since $a_1 = 0 < 2$, and if $a_n \leq 2$, then $a_{n+1} = \sqrt{2+a_n} < \sqrt{2+2} = 2$. This shows that the sequence is monotone and bounded, thus convergent. We can find the limit with the help of the recurrence. Let $\lim_{n\to\infty} a_n = a$. Since $a_{n+1}^2 = 2 + a_n$ for all n,

$$a^2 = \lim_{n\to\infty} a_{n+1}^2 = \lim_{n\to\infty}(2+a_n) = 2+a,$$

which gives $a = 2$ or $a = -1$. The second option is impossible, since the terms of the sequence are nonnegative. So $a = 2$, that is, $\lim_{n\to\infty} a_n = 2$.

Exercises

6.1. Prove that if $A \subset \mathbb{R}$ and $\sup A = \alpha \notin A$, then there exists a sequence (a_n) for which $\{a_n\} \subset A$, (a_n) is strictly monotone increasing, and $a_n \to \alpha$. Is the statement true if $\alpha \in A$?

6.2. A sequence, in terms of monotonicity, boundedness, and convergence, can behave (theoretically) in eight different ways (each property is either present or not). In reality, how many of the eight cases can actually occur?

6.3. Suppose that the terms of the sequence (a_n) satisfy the inequality $a_n \leq (a_{n-1} + a_{n+1})/2$ for all $n > 1$. Prove that (a_n) cannot oscillate at infinity. (H)

6.4. Prove that the following sequences, defined recursively, are convergent, and find their limits.

(a) $a_1 = 0$, $a_{n+1} = \sqrt{a+a_n}$ $(n = 1,2,\ldots)$, where $a > 0$ is fixed;
(b) $a_1 = 0$, $a_{n+1} = 1/(2-a_n)$ $(n = 1,2,\ldots)$;
(c) $a_1 = 0$, $a_{n+1} = 1/(4-a_n)$ $(n = 1,2,\ldots)$;
(d) $a_1 = 0$, $a_{n+1} = 1/(1+a_n)$ $(n = 1,2,\ldots)$ (H);
(e) $a_1 = \sqrt{2}$, $a_{n+1} = \sqrt{2}\sqrt{a_n}$ $(n = 1,2,\ldots)$.

6.5. Let $a > 0$ be given, and define the sequence (a_n) by the recurrence $a_1 = a$, $a_{n+1} = \left(a_n + \dfrac{a}{a_n} \right) / 2$. Prove that $a_n \to \sqrt{a}$. (H)

6.6. Prove that the sequence $\left(1 + \frac{1}{n} \right)^{n+1}$ is monotone decreasing.

6.7. Prove that $n + 1 < e^{1 + \frac{1}{2} + \cdots + \frac{1}{n}} < 3n$ for all $n = 1, 2, \ldots$. (H)

6.8. Let a and b be positive numbers. Suppose that the sequences (a_n) and (b_n) satisfy $a_1 = a$, $b_1 = b$, and $a_{n+1} = (a_n + b_n)/2$, $b_{n+1} = \sqrt{a_n b_n}$ for all $n \geq 1$. Prove that $\lim_{n \to \infty} a_n = \lim_{n \to \infty} b_n$. (This value is the so-called **arithmetic–geometric mean** of a and b.)

6.9. Prove that if (a_n) is convergent and $(a_{n+1} - a_n)$ is monotone, then it follows that $n \cdot (a_{n+1} - a_n) \to 0$. Give an example for a convergent sequence (a_n) for which $n \cdot (a_{n+1} - a_n)$ does not tend to 0. (∗H)

6.10. Suppose that (b_n) is strictly monotone increasing and tends to infinity. Prove that if $\dfrac{a_n - a_{n-1}}{b_n - b_{n-1}} \to c$, then $\dfrac{a_n}{b_n} \to c$.

6.2 The Bolzano–Weierstrass Theorem and Cauchy's Criterion

We saw that the monotone sequences behave simply from the point of view of convergence. We also know (see Theorem 5.5) that rearrangement of the terms of a sequence does not affect whether it converges, and if it does, its limit is the same. It would be useful to see which sequences can be rearranged to give a monotone sequence.

Every finite set of numbers can be arranged in ascending or descending order to give a finite monotone sequence. It is clear however, that not every infinite sequence can be rearranged to give a monotone sequence. For example, the sequence $((-1)^n/n)$ cannot be rearranged into a monotone sequence. The conditions for rearranging a sequence into one that is strictly monotone is given by the following theorem.

Theorem 6.7. *A sequence can be rearranged to give a strictly monotone increasing sequence if and only if its terms are distinct and the sequence either tends to infinity or converges to a value that is larger than all of its terms.*

Proof. The terms of a strictly monotone sequence are distinct, so this is a necessary condition for being able to rearrange a sequence into one that is strictly monotone. We know that every monotone increasing sequence either tends to infinity or is convergent (Theorems 6.2 and 6.3). It is also clear that if a strictly increasing monotone sequence is convergent, then its terms are smaller than the limit. This proves the "only if" part of the statement.

Now suppose that we have a sequence (a_n) whose terms are pairwise distinct and $a_n \to \infty$; we show that (a_n) can be rearranged into a strictly increasing monotone sequence. Consider the intervals

$$I_0 = (-\infty, 0], \ I_1 = (0, 1], \ \ldots, \ I_k = (k-1, k], \ldots.$$

Each of these intervals contains only finitely many terms of the sequence. If we list the terms in I_0 in monotone increasing order, followed by the ones in I_1, and so on, then we get a rearrangement of the sequence into one that is strictly monotone increasing.

Finally let us suppose that we have a sequence (a_n) whose terms are distinct, $a_n \to a$ finite, and $a_n < a$ for all n. We show that (a_n) can be rearranged into a strictly increasing monotone sequence. Consider the intervals

$$J_1 = (-\infty, a-1], \ J_2 = \left(a-1, a-\tfrac{1}{2}\right], \ \ldots, \ J_k = \left(a-\tfrac{1}{k-1}, a-\tfrac{1}{k}\right], \ \ldots.$$

Each of these contains only finitely many of the sequence, and every term of the sequence is in one of the J_k intervals. If we then list the terms in J_0 in monotone increasing order, followed by those in J_1, and so on, then we get a rearrangement of the sequence into one that is strictly monotone increasing. □

Using a similar argument, one can show that if a sequence is convergent and its terms are smaller than its limit (but not necessarily distinct), then it can be rearranged into a (not necessarily strictly) increasing monotone sequence.

The following combinatorial theorem, while interesting in its own right, has important consequences.

Theorem 6.8. *Every sequence has a monotone subsequence.*

Proof. For a sequence (a_n), we will call the terms a_k *peaks*, if for all $m > k$, $a_m \leq a_k$. We distinguish two cases.

I. The sequence (a_n) has infinitely many peaks. In this case, the peaks form a monotone decreasing subsequence.

II. The sequence (a_n) has finitely many peaks. Then there is an n_0 such that whenever $n \geq n_0$, a_n is not a peak. Since a_{n_0} is not a peak, then according to the definition of a peak, there exists an $n_1 > n_0$ such that $a_{n_1} > a_{n_0}$. Since a_{n_1} is also not a peak, there exists $n_2 > n_1$ such that $a_{n_2} > a_{n_1}$, and so on. We then get an infinite sequence of indices $n_0 < n_1 < \cdots < n_k < \ldots$ for which

$$a_{n_0} < a_{n_1} < \cdots < a_{n_k} < \ldots.$$

That is, in this case, the sequence has a (strictly) increasing monotone subsequence. □

A theorem of fundamental importance follows.

Theorem 6.9 (Bolzano–Weierstrass[5] Theorem). *Every bounded sequence has a convergent subsequence.*

Proof. If a sequence is bounded, then every one of its subsequences is also bounded. Then by our previous theorem, every bounded sequence has a bounded monotone subsequence. From Theorem 6.2, it follows that this is convergent.   □

We can extend Theorem 6.8 with the following.

Theorem 6.10. *If a sequence is not bounded from above, then it has a monotone subsequence that tends to ∞; if it is not bounded from below, then it has a monotone subsequence that tends to $-\infty$.*

Proof. If $\{a_n\}$ is not bounded from above, then there is an n_1 such that $a_{n_1} > 0$. Also, by not being bounded from above, we can find an index n_2 for which

$$a_{n_2} > \max(a_1, \dots, a_{n_1}, 1).$$

Then necessarily $n_2 > n_1$. Following this procedure, we can find indices $n_1, n_2, \dots$ such that

$$a_{n_{k+1}} > \max(a_1, \dots, a_{n_k}, k)$$

for all k. Then $n_1 < n_2 < \cdots$, $a_{n_1} < a_{n_2} < \cdots$, and $a_{n_k} > k - 1$. Our constructed subsequence (a_{n_k}) is monotone increasing, diverging to ∞. The second statement of the theorem can be proven in the same manner.   □

Since every monotone sequence has a limit, by Theorem 6.8, *every sequence has a subsequence that tends to a limit*. In the following theorems, we will show that the subsequences with limits determine the convergence behavior of the original sequence.

Theorem 6.11. *Suppose that whenever a subsequence of (a_n) has a limit, then it tends to b (where b can be finite or infinite). Then (a_n) also tends to b.*

Proof. Let first $b = \infty$, and let K be given. We will show that the sequence can have only finitely many terms smaller than K. Suppose this is not the case, and let $a_{n_1}, a_{n_2}, \dots$ all be smaller than K. As we saw before, the sequence (a_{n_k}) has a subsequence that tends to a limit. This subsequence, however, cannot tend to infinity (since each one of its terms is smaller than K), which is impossible. Thus we see that $a_n \to \infty$. We can argue similarly if $b = -\infty$.

Now suppose that b is finite, and let $\varepsilon > 0$ be given. We will see that the sequence can have only finitely many terms outside the interval $(b - \varepsilon, b + \varepsilon)$. Suppose that this is not true, and let $a_{n_1}, a_{n_2}, \dots$ be terms that do not lie inside $(b - \varepsilon, b + \varepsilon)$. As we saw before, the sequence (a_{n_k}) has a subsequence that tends to a limit. This limit, however, cannot be b (since none of its terms is inside $(b - \varepsilon, b + \varepsilon)$), which is impossible. Thus we see that $a_n \to b$.   □

---

[5] Bernhard Bolzano (1781–1848), Italian–German mathematician, and Karl Weierstrass (1815–1897), German mathematician.

By Theorem 5.2, the statement of the previous theorem can be reversed: if $a_n \to b$, then all subsequences of (a_n) that converge to a limit tend to b, since all subsequences of (a_n) tend to b. The following important consequence follows from Theorems 5.2 and 6.11.

Theorem 6.12. *A sequence oscillates at infinity if and only if it has two subsequences that tend to distinct (finite or infinite) limits.*

The following theorem—called Cauchy's[6] criterion—gives a necessary and sufficient condition for a sequence to be convergent. The theorem is of fundamental importance, since it gives an opportunity to check convergence without knowing the limit; it places conditions only on differences between terms, as opposed to differences between terms and the limit.

Theorem 6.13 (Cauchy's Criterion). *The sequence (a_n) is convergent if and only if for every $\varepsilon > 0$, there exists an N such that for all $n, m \geq N$, $|a_n - a_m| < \varepsilon$.*

Proof. If (a_n) is convergent and $\lim_{n \to \infty} a_n = b$, then for every $\varepsilon > 0$, there exists an N for which $|a_n - b| < \varepsilon/2$ and $|a_m - b| < \varepsilon/2$ hold whenever $n \geq N$ and $m \geq N$. Then by the triangle inequality, $|a_n - a_m| < \varepsilon$ if $m, n \geq N$. This shows that the assumption in our theorem is a necessary condition for convergence.

Now we prove that convergence follows from the assumptions. First we show that if the assumptions hold, the sequence is bounded. Certainly, considering the assumption for $\varepsilon = 1$ yields an N such that $|a_n - a_m| < 1$ whenever $n, m \geq N$. Here if we set m to be N, we get that $|a_n - a_N| < 1$ for all $n \geq N$, which clearly says that the sequence is bounded.

By the Bolzano–Weierstrass theorem, the fact that (a_n) has a convergent subsequence follows. Let $\lim_{n \to \infty} a_{n_k} = b$. We prove that (a_n) is convergent and tends to b. We give two different proofs of this.

I. Let $\varepsilon > 0$ be given. There exists a k_0 such that $|a_{n_k} - b| < \varepsilon/2$ whenever $k > k_0$. On the other hand, by the assumption, there exists an N such that $|a_n - a_m| < \varepsilon/2$ whenever $n, m \geq N$. Let us fix an index $k > k_0$ for which $n_k \geq N$. Then for arbitrary $n \geq N$,

$$|a_n - b| \leq |a_n - a_{n_k}| + |a_{n_k} - b| < (\varepsilon/2) + (\varepsilon/2) = \varepsilon,$$

which shows that $a_n \to b$.

II. By Theorem 6.11, it suffices to show that every subsequence of (a_n) that has a limit tends to b. Let (a_{m_i}) be such a subsequence having a limit, and let $\lim_{i \to \infty} a_{m_i} = c$. Since (a_n) is bounded, c is finite.

---

[6] Augustin Cauchy (1789–1857), French mathematician.

Let $\varepsilon > 0$ be given. By our assumption, there exists an N such that $|a_n - a_m| < \varepsilon$ whenever $n, m \geq N$. Now, $a_{n_k} \to b$ and $a_{m_i} \to c$, so there are indices $n_k > N$ and $m_i > N$ for which $|a_{n_k} - b| < \varepsilon$ and $|a_{m_i} - c| < \varepsilon$. Then

$$|b - c| \leq |a_{n_k} - b| + |a_{m_i} - c| + |a_{n_k} - a_{m_i}| < 3\varepsilon.$$

Since ε was arbitrary, $b = c$.   □

The statement of the Cauchy criterion can also be stated (less precisely but more intuitively) by saying that a sequence is convergent if and only if its large-index terms are close to each other. It is important to note that for convergence, it is necessary for terms with large indices to be close to their neighbors, but this in itself is not sufficient. More precisely, the condition that $a_{n+1} - a_n \to 0$ is necessary, but not sufficient, for (a_n) to be convergent. Clearly, if $a_n \to a$, then $a_{n+1} - a_n \to a - a = 0$. Moreover, by (4.2), $\sqrt{n+1} - \sqrt{n} \to 0$, but the sequence $(\sqrt{n})$ is not convergent, but tends to infinity. There even exists a sequence (a_n) such that $a_{n+1} - a_n \to 0$, but the sequence oscillates at infinity (see Exercise 4.22).

Remark 6.14. When we introduced the real numbers, we mentioned alternative constructions, namely the construction of structures that satisfy the axioms of the real numbers. With the help of the Cauchy criterion, we briefly sketch such a structure, taking for granted the existence and properties of the rational numbers. For brevity, we call a sequence (a_n) **Cauchy** if it satisfies the conditions of the Cauchy criterion, that is, for every $\varepsilon > 0$, there exists an N such that $|a_n - a_m| < \varepsilon$ for all $n, m \geq N$.

Let us call the Cauchy sequences consisting of rational numbers C-numbers. We consider the C-numbers (a_n) and (b_n) to be equal if $a_n - b_n \to 0$, that is, if for every $\varepsilon > 0$, there exists an N such that $|a_n - b_n| < \varepsilon$ for all $n \geq N$.

We define the operations of addition and multiplication among C-numbers by the formulas $(a_n) + (b_n) = (a_n + b_n)$ and $(a_n) \cdot (b_n) = (a_n \cdot b_n)$ respectively. (Of course, we need to check whether these operations are well defined.) In the construction, the roles of 0 and 1 are fulfilled by (u_n) and (v_n), which are the constant 0 and 1 sequences, respectively.

One can show that the structure defined as above satisfies the field axioms. The less-than relation is understood in the following way: $(a_n) < (b_n)$ if $(a_n) \neq (b_n)$ (that is, $a_n - b_n \not\to 0$) and there exists an N such that $a_n < b_n$ for all $n \geq N$. It can be shown that with this ordering, we get a structure that satisfies the axiom system of the real numbers. For details of this construction, see [4, 8].

Exercises

6.11. Prove that if the sequence (a_n) does not have a convergent subsequence, then $|a_n| \to \infty$.

6.12. Is it possible that (a_n) does not have convergent subsequences, but $(|a_n|)$ is convergent?

6.13. Prove that if (a_n) is bounded and every one of its convergent subsequences tends to b, then $a_n \to b$.

6.14. Prove that if the sequence (a_n) does not have two convergent subsequences that tend to distinct limits, then (a_n) either has a limit or can be broken into the union of two subsequences, one of which is convergent, while the absolute value of the other tends to infinity.

6.15. Assume the field and ordering axioms, and the Bolzano–Weierstrass theorem. Deduce from these the axiom of Archimedes and Cantor's axiom. (∗S)

6.16. Prove that if a sequence has infinitely many terms smaller than a, and infinitely many terms larger than a, then it cannot be reordered to form a monotone sequence.

6.17. What is the exact condition for a sequence (a_n) to be rearrangeable into a monotone increasing sequence? (H)

6.18. Prove that every convergent sequence breaks into at most three (finite or infinite) subsequences, each of which can be rearranged to give a monotone sequence.

6.19. Prove that if $|a_{n+1} - a_n| \le 2^{-n}$ for all n, then (a_n) is convergent.

6.20. Suppose that $a_{n+1} - u_n \to 0$. Does it follow from this that $a_{2n} - a_n \to 0$?

6.21. Give examples of a sequence (a_n) such that $a_n \to \infty$, and

(a) $a_{2n} - a_n \to 0$ (S);
(b) $a_{n^2} - a_n \to 0$;
(c) $a_{2^n} - a_n \to 0$;
(d) for a fixed sequence $s_n > n$ made up of positive integers, $a_{s_n} - a_n \to 0$ (S).

6.22. Give an example of a sequence for which $a_{n^k} - a_n \to 0$ for all $k > 1$, but $a_{2^n} - a_n$ does not tend to 0. (∗H)

6.23. Is it true, that the following statements are equivalent?

(a) $(\forall \varepsilon > 0)(\exists N)(n, m \ge N \Rightarrow |a_n - a_m| < \varepsilon)$, and
(b) whenever s_n, t_n are positive integers for which $s_n \to \infty$ and $t_n \to \infty$, then $a_{s_n} - a_{t_n} \to 0$. (H)

Chapter 7
Rudiments of Infinite Series

If we add infinitely many numbers (more precisely, if we take the sum of an infinite sequence of numbers), then we get an infinite series.

Mathematicians in India were dealing with infinite series as early as the fifteenth century, while European mathematics caught up with them only in the seventeenth. But then, however, the study of infinite series underwent rapid development, because mathematicians realized that certain quantities and functions can be computed more easily if they are expressed as infinite series.

There were ideas outside of mathematics that led to infinite series as well. One of the earliest such ideas was Zeno's paradox about Achilles and the tortoise (as seen in our *brief historical introduction*). So-called "elementary" mathematics and recreational mathematics also give rise to questions that lead to infinite series:

1. What is the area of the shaded region in Figure 7.1, formed by infinitely many triangles?

If the area of the big triangle is 1, then the area we seek is clearly $(1/4) + (1/4^2) + (1/4^3) + \cdots$. On he other hand, the big triangle is the union of three copies of the shaded region, so the area of the region is $1/3$. We get that

$$\frac{1}{4} + \frac{1}{4^2} + \frac{1}{4^3} + \cdots = \frac{1}{3}.$$

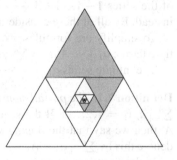

Fig. 7.1

2. A simple puzzle asks that if the weight of a brick is 1 pound plus the weight of half a brick, then how heavy is one brick? Since the weight of half of a brick is $1/2$ pound plus the weight of a quarter of a brick, while the weight of a quarter of a brick is $1/4$ pound plus the weight of an eighth of a brick, and so on, the weight of one brick is $1 + (1/2) + (1/4) + (1/8) + \cdots$. On the other hand,

© Springer New York 2015
M. Laczkovich, V.T. Sós, *Real Analysis*, Undergraduate Texts in Mathematics, DOI 10.1007/978-1-4939-2766-1_7

if we subtract half of a brick from both sides of the equation 1 brick = 1 pound $+\frac{1}{2}$ brick, we get that half a brick weighs 1 pound. Thus the brick weighs 2 pounds, and so

$$1 + \frac{1}{2} + \frac{1}{4} + \frac{1}{8} + \cdots = 2.$$

In the *brief historical introduction*, we saw several examples of how the sum of an infinite series can result in some strange or contradictory statements (for example, (1.4) and (1.5)). We can easily see that these strange results come from faulty reasoning, namely from the assumption that every infinite series has a well-defined "predestined" sum. This is false, because only the axioms can be assumed as "predestined" (once we accept them), and every other concept needs to be created by us. We have to give up on every infinite series having a sum: we have to decide ourselves which series should have a sum, and then decide what that sum should be. The concept that we create should satisfy certain expectations and should mirror any intuition we might already have for the concept.

To define infinite series, let us start with finite sums. Even sums of multiple terms are not "predestined," since the axioms talk only about sums of two numbers. We defined the sum of n numbers by adding parentheses to the sum (in the first appendix of Chapter 3), which simply means that we get the sum of n terms after $n - 1$ summations. It is only natural to define the infinite sum $a_1 + a_2 + \cdots$ with the help of the running sums a_1, $a_1 + a_2$, $a_1 + a_2 + a_3, \ldots$, which are called the partial sums.[1]

It is easy to check (and we will soon do so in Example 7.3) that the partial sums of the series $1 + 1/2 + 1/4 + 1/8 + \cdots$ tend to 2, and the partial sums of $3/10 + 3/100 + 3/1000 + \cdots$ tend to $1/3$. Both results seem to be correct; the second is simply the decimal expansion of the number $1/3$. On the other hand, the partial sums of the problematic series $1 + 2 + 4 + \cdots$ do not tend to -1, while the partial sums of the series $1 - 1 + 1 - 1 + \cdots$ does not tend to anything, but oscillate at infinity instead. By all of these considerations, the definition below follows naturally.

To simplify our formulas, we introduce the following notation for sums of multiple terms: $a_1 + \cdots + a_n = \sum_{i=1}^{n} a_i$.

The infinite series $a_1 + a_2 + \cdots$ will also get new notation, which is $\sum_{n=1}^{\infty} a_n$.

Definition 7.1. The *partial sums* of the infinite series $\sum_{n=1}^{\infty} a_n$ are the numbers $s_n = \sum_{i=1}^{n} a_i$ $(n = 1, 2, \ldots)$. If the sequence of partial sums (s_n) is convergent with limit A, then we say that the *infinite series $\sum_{n=1}^{\infty} a_n$ is convergent, and its sum is A.* We denote this by $\sum_{n=1}^{\infty} a_n = A$.

If the sequence of partial sums (s_n) is divergent, then we say that the *series $\sum_{n=1}^{\infty} a_n$ is divergent.*

If $\lim_{n \to \infty} s_n = \infty$ (or $-\infty$), then we say that *the sum of the series $\sum_{n=1}^{\infty} a_n$ is ∞ (or $-\infty$).* We denote this by $\sum_{n=1}^{\infty} a_n = \infty$ (or $-\infty$).

---

[1] This viewpoint does not treat an infinite sum as an expression whose value is already defined, but instead, the value is "gradually created" with our method. From a philosophical viewpoint, the infinitude of the series is viewed not as "actual infinity" but "potential infinity."

Remark 7.2. Strictly speaking, the expression $\sum_{n=1}^{\infty} a_n$ alone has no meaning. The phrase "consider the infinite series $\sum_{n=1}^{\infty} a_n$" simply means "consider the sequence (a_n)," with the difference that we usually are more concerned about the partial sums[2] $a_1 + \cdots + a_n$. For our purposes, the statement $\sum_{n=1}^{\infty} a_n = A$ is simply a shorthand way of writing that $\lim_{n \to \infty}(a_1 + \cdots + a_n) = A$.

Examples 7.3. **1.** The nth partial sum of the series $1 + 1/2 + 1/4 + 1/8 + \cdots$ is $s_n = \sum_{i=0}^{n-1} 2^{-i} = 2 - 2^{-n+1}$. Since $\lim_{n \to \infty} s_n = 2$, the series is convergent, and its sum is 2.

2. The nth partial sum of the series $3/10 + 3/100 + 3/1000 + \cdots$ is

$$s_n = \sum_{i=1}^{n} 3 \cdot 10^{-i} = \frac{3}{10} \cdot \frac{1 - 10^{-n}}{1 - (1/10)}.$$

Since $\lim_{n \to \infty} s_n = 3/9 = 1/3$, the series is convergent, and its sum is $1/3$.

3. The nth partial sum of the series $1 + 1 + 1 + \cdots$ is $s_n = n$. Since $\lim_{n \to \infty} s_n = \infty$, the sequence is divergent (and its sum is ∞).

4. The $(2k)$th partial sum of the series $1 - 1 + 1 - \ldots$ is zero, while the $(2k+1)$th partial sum is 1 for all $k \in \mathbb{N}$. Since the sequence (s_n) is oscillating at infinity, the series is divergent (and has no sum).

5. The kth partial sum of the series

$$1 - \frac{1}{2} + \frac{1}{3} - \frac{1}{4} + \cdots \tag{7.1}$$

is

$$s_k = 1 - \frac{1}{2} + \frac{1}{3} - \frac{1}{4} + \cdots + (-1)^{k-1} \cdot \frac{1}{k}.$$

If $n < m$, then we can see that

$$|s_m - s_n| = \left| \frac{1}{n+1} - \frac{1}{n+2} + \cdots + (-1)^{n-m+1} \cdot \frac{1}{m} \right| < \frac{1}{n}.$$

It follows that the sequence (s_n) satisfies Cauchy's criterion (Theorem 6.13), so it is convergent. This shows that the series (7.1) is convergent. We will later see that the sum of the series is equal to the natural logarithm of 2 (see Exercise 12.92 and Remark 13.16).

The second example above is a special case of the following theorem, which states that infinite decimal expansions can be thought of as convergent series.

Theorem 7.4. *Let the infinite decimal expansion of x be $m.a_1 a_2 \ldots$. Then the infinite series $m + \frac{a_1}{10} + \frac{a_2}{10^2} + \cdots$ is convergent, and its sum is x.*

---

[2] There are some who actually mean the series (s_n) when they write $\sum_{n=1}^{\infty} a_n$. We do not follow this practice, since then, the expression $\sum_{n=1}^{\infty} a_n = A$ would state equality between a sequence and a number, which is not a good idea.

Proof. By the definition of the decimal expansion,

$$m.a_1 \ldots a_n \leq x \leq m.a_1 \ldots a_n + \frac{1}{10^n}$$

for all n, so $\lim_{n\to\infty} m.a_1 \ldots a_n = x$. We see that $m.a_1 \ldots a_n$ is the $(n+1)$th partial sum of the series

$$m + \frac{a_1}{10} + \frac{a_2}{10^2} + \cdots,$$

which makes the statement of the theorem clear.   □

The sums appearing in Examples 7.3.1. and 7.3.2. are special cases of the following theorem.

Theorem 7.5. *The series* $1 + x + x^2 + \cdots$ *is convergent if and only if* $|x| < 1$, *and then its sum is* $1/(1-x)$.

Proof. We already saw that in the case $x = 1$, the series is divergent, so we can assume that $x \neq 1$. Then the nth partial sum of the series is $s_n = \sum_{i=0}^{n-1} x^i = (1 - x^n)/(1-x)$. If $|x| < 1$, then $x^n \to 0$ and $s_n \to 1/(1-x)$. Thus the series is convergent with sum $1/(1-x)$.

If $x > 1$, then $s_n \to \infty$, so the series is divergent (and its sum is ∞). If, however, $x \leq -1$, then the sequence (s_n) oscillates at infinity, so the series is also divergent (with no sum).   □

The next theorem outlines an important property of convergent series.

Theorem 7.6. *If the series* $\sum_{n=1}^{\infty} a_n$ *is convergent, then* $\lim_{n\to\infty} a_n = 0$.

Proof. Let the sum of the series be A. Since

$$a_n = (a_1 + \cdots + a_n) - (a_1 + \cdots + a_{n-1}) = s_n - s_{n-1},$$

we have $a_n \to A - A = 0$.   □

Remark 7.7. The theorem above states that for $\sum_{n=1}^{\infty} a_n$ to be convergent, it is necessary for $a_n \to 0$ to hold. It is important to note that this condition is in no way sufficient, since there are many divergent sequences whose terms tend to zero. A simple example: The terms of the series $\sum_{i=0}^{\infty} \left(\sqrt{i+1} - \sqrt{i}\right)$ tend to zero by Example 4.4.3. On the other hand, the nth partial sum is $\sum_{i=0}^{n-1} \left(\sqrt{i+1} - \sqrt{i}\right) = \sqrt{n} \to \infty$ as $n \to \infty$, so the series is divergent.

Another famous example of a divergent series whose terms tend to zero is the series $\sum_{n=1}^{\infty} 1/n$, which is called the **harmonic series**.[3]

---

[3] The name comes from the fact that the wavelengths of the overtones of a vibrating string of length h are h/n ($n = 2, 3, \ldots$). The wavelengths $h/2, h/3, \ldots, h/8$ correspond to the octave, octave plus a fifth, second octave, second octave plus a major third, second octave plus a fifth, second octave plus a seventh, and the third octave, respectively. Thus the series $\sum_{n=1}^{\infty} (h/n)$ contains the tone and its overtones, which are often called harmonics.

Theorem 7.8. *The series $\sum_{n=1}^{\infty} \frac{1}{n}$ is divergent.*

We give two proofs of the statement.

Proof. **1.** If the nth partial sum of the series is s_n, then

$$s_{2n} - s_n = \left(1 + \frac{1}{2} + + \cdots + \frac{1}{2n}\right) - \left(1 + \frac{1}{2} + + \cdots + \frac{1}{n}\right) =$$

$$= \left(\frac{1}{n+1} + + \cdots + \frac{1}{2n}\right) \geq$$

$$\geq n \cdot \frac{1}{2n} = \frac{1}{2}$$

for all n. Suppose that the series is convergent and its sum is A. Then if $n \to \infty$, then $s_{2n} - s_n \to A - A = 0$, which is impossible.
2. If $n > 2^k$, then

$$s_n \geq 1 + \frac{1}{2} + + \cdots + \frac{1}{2^k} =$$

$$= 1 + \frac{1}{2} + \left(\frac{1}{3} + \frac{1}{4}\right) + \left(\frac{1}{5} + \cdots + \frac{1}{8}\right) + \cdots + \left(\frac{1}{2^{k-1}+1} + \cdots + \frac{1}{2^k}\right) \geq$$

$$\geq 1 + \frac{1}{2} + 2 \cdot \frac{1}{4} + 4 \cdot \frac{1}{8} + \cdots + 2^{k-1} \cdot \frac{1}{2^k} =$$

$$= 1 + k \cdot \frac{1}{2}.$$

Thus $\lim_{n \to \infty} s_n = \infty$, so the series is divergent, and its sum is ∞. $\qquad\qquad \square$

Remark 7.9. Since the harmonic series contains the reciprocal of every positive integer, one could expect the behavior of the series to have number-theoretic significance. This is true. Using the divergence of the harmonic series, we can give a new proof of the fact that *there exist infinitely many prime numbers.* Suppose that this were not true, and that there were only finitely many prime numbers. Let these be $p_1, \ldots, p_k$. For all i and N, the relations

$$1 + \frac{1}{p_i} + \cdots + \frac{1}{p_i^N} = \frac{1 - p_i^{-(N+1)}}{1 - \frac{1}{p_i}} < \frac{1}{1 - \frac{1}{p_i}}$$

hold. Multiplying these together, we get that

$$\prod_{i=1}^{k} \left(1 + \frac{1}{p_i} + \cdots + \frac{1}{p_i^N}\right) < \prod_{i=1}^{k} \frac{1}{1 - \frac{1}{p_i}} \qquad\qquad (7.2)$$

for all N. (Here we use the notation $\prod_{i=1}^{k} a_i = a_1 \cdots a_k$.) If we expand the multiplication on the left-hand side, then we get the reciprocal of every number whose prime factorization does not contain any prime with power greater than or equal to N

(since we assumed that there are no prime numbers other than $p_1, \ldots, p_k$). It is clear that every number up to N is there, so $\sum_{n=1}^{N} 1/n = s_N$ is smaller than the right-hand side of (7.2). This, however, is impossible, since $s_N \to \infty$ as $N \to \infty$.

With a refinement of the proof above, one can show that the series consisting of the reciprocals of all primes is divergent. We also know more precisely that the sum of reciprocals of primes smaller than x is greater than $\log\log x - 1$ for all $x \geq 2$ (see Chapter 5 of [2] or Corollary 18.16 and Theorem 18.17 of this book).

The second proof of Theorem 7.8 seems to tell us more than the first, because it says not only that the series is divergent, but that its sum is infinite too. By the following simple theorem, we see that the sum of a divergent series consisting of nonnegative terms is always infinite.

Theorem 7.10.

(i) *A series consisting of nonnegative terms is convergent if and only if the sequence of its partial sums is bounded (from above).*

(ii) *If a series consisting of nonnegative terms is divergent, then its sum is infinite.*

Proof. By the assumption that the terms of the series are nonnegative, we clearly get that the sequence of partial sums of the series is monotone increasing. If this sequence is bounded from above, then it is convergent by Theorem 6.2. Then the series in question is convergent. If, however, the sequence of partial sums is not bounded from above, then by Theorem 6.3, it tends to infinity, so the series will be divergent, and its sum will be infinity.   □

We emphasize that by the above theorem, *a series consisting of nonnegative terms always has a sum: either a finite number (if the series converges) or infinity (if the series diverges).*

Examples 7.11. **1.** The series $\sum_{i=1}^{\infty} 1/i^2$ is convergent, because its nth partial sum is

$$\sum_{i=1}^{n} \frac{1}{i^2} \leq 1 + \sum_{i=2}^{n} \frac{1}{(i-1) \cdot i} = 1 + \sum_{i=2}^{n} \left(\frac{1}{i-1} - \frac{1}{i} \right) = 2 - \frac{1}{n} < 2.$$

During the seventeenth century and the beginning of the eighteenth, many mathematicians tried to determine the value of the series $\sum_{i=1}^{\infty} 1/i^2$. Finally, Johann Bernoulli[4] and Euler independently found that $\sum_{i=1}^{\infty} 1/i^2 = \pi^2/6$. This fact now has dozens of proofs (see http://mathworld.wolfram.com/RiemannZetaFunction Zeta2. html). For a relatively elementary proof, see Exercise 11.36.

2. By part (b) of Exercise 2.7, the partial sums of the series $\sum_{i=1}^{\infty} 1/i^{3/2}$ are less than 3. Thus the series is convergent, and its sum is at most 3.

3. We generally call series of the form $\sum_{i=1}^{\infty} 1/i^c$ **hyperharmonic series.** It is easy to see that if $b > 0$, then

$$1 + \frac{1}{2^{b+1}} + \cdots + \frac{1}{n^{b+1}} \leq 1 + \frac{1}{b} - \frac{1}{b \cdot n^b} \tag{7.3}$$

---

[4] Johann Bernoulli (1667–1748) Swiss mathematician, brother of Jacob Bernoulli.

for all n (see Exercise 7.5). It then follows that *the hyperharmonic series* $\sum_{i=1}^{\infty} 1/i^c$
is convergent for all $c > 1$ (see also Exercise 7.6). We denote the sum of the series
by $\zeta(c)$. By equality (7.3), it follows that every partial sum of the series is less than
$c/(c-1)$, and so $1 < \zeta(c) \leq c/(c-1)$ for all $c > 1$.

The theorems of Johann Bernoulli and Euler can be summarized using this notation as $\zeta(2) = \pi^2/6$.

Remark 7.12. The series $\sum_{n=1}^{\infty} 1/n^2$ is an example of the rare occurrence whereby
we can specifically determine the sum of a series (such as the series appearing in
Exercises 7.2 and 7.3). These can be thought of as exceptions, for we generally
cannot express the sum of an arbitrary series in closed form. The sums $\sum_{n=1}^{\infty} 1/n^c$
(that is, the values of $\zeta(c)$) have closed formulas only for some special values of c.
We already saw that $\zeta(2) = \pi^2/6$. Bernoulli and Euler proved that if k is an even
positive integer, then $\zeta(k)$ is equal to a rational multiple of π^k. However, to this day,
we do not know whether this is also true if k is an odd integer. In fact, *nobody has
found closed expressions for the values of* $\zeta(3), \zeta(5),$ *and so on for the past 300
years*, and it is possible that no such closed form exists.

By the transcendence of the number π, we know that the values $\zeta(2k)$ are irrational for every positive integer k. In the 1970s, it was proven that the value $\zeta(3)$
is also irrational. *Whether the numbers* $\zeta(5), \zeta(7),$ *and so on are rational or not,
however, is still an open question.*

In the general case, the following theorem gives us a precise condition for the
convergence of a series.

Theorem 7.13 (Cauchy's Criterion). *The infinite series* $\sum_{n=1}^{\infty} a_n$ *is convergent if
and only if for all* $\varepsilon > 0$, *there exists an index* N *such that for all* $N \leq n < m$,

$$|a_{n+1} + a_{n+2} + \cdots + a_m| < \varepsilon.$$

Proof. Since $a_{n+1} + a_{n+2} + \cdots + a_m = s_m - s_n$, the statement is clear by Cauchy's
criterion for sequences (Theorem 6.13). □

Exercises

7.1. For a fixed $\varepsilon > 0$, give threshold indices above which the partial sums of the
following series differ from their actual sum by less than ε.

(a) $\sum_{n=0}^{\infty} 1/2^n$;   (b) $\sum_{n=0}^{\infty} (-2/3)^n$;

(c) $1 - 1/2 + 1/3 - 1/4 + \cdots$;   (d) $1/1 \cdot 2 + 1/2 \cdot 3 + 1/3 \cdot 4 + \cdots$.

7.2.
(a) $\sum_{n=1}^{\infty} 1/(n^2 + 2n) = ?$ (b) $\sum_{n=1}^{\infty} 1/(n^2 + 4n + 3) = ?$

(c) $\sum_{n=2}^{\infty} 1/(n^3 - n) = ?$ (H S)

7.3. Give a general method for determining the sums $\sum_{n=a}^{\infty} p(n)/q(n)$ in which p and q are polynomials, $\mathrm{gr}\, p \le \mathrm{gr}\, q - 2$, and $q(x) = (x - a_1)\cdots(x - a_k)$, where $a_1,\ldots,a_k$ are distinct integers smaller than a. (H)

7.4. Are the following series convergent?

    (a) $\sum_1^{\infty} 1/\sqrt[n]{2}$;   (b) $\sum_2^{\infty} 1/\sqrt[n]{\log n}$;

    (c) $\sum_2^{\infty} n\log((n+1)/n)$;   (d) $\sum_1^{\infty} (3^{\sqrt{n}} - 2^n)/(3^{\sqrt{n}} + 2^n)$.

7.5. Prove inequality (7.3) for all $b > 0$. (H S)

7.6. Show that

$$\left(1 + \frac{1}{2^c} + \cdots + \frac{1}{n^c}\right)\left(1 - \frac{2}{2^c}\right) < 1$$

for all $n = 1, 2, \ldots$ and $c > 0$. Deduce from this that the series $\sum_{n=1}^{\infty} 1/n^c$ is convergent for all $c > 1$.

7.7. Prove that $\lim_{n\to\infty} \zeta(1 + (1/n)) = \infty$.

7.8. Let $a_1, a_2, \ldots$ be a listing of the positive integers that do not contain the digit 7 (in decimal representation). Prove that $\sum_{n=1}^{\infty} 1/a_n$ is convergent. (H)

7.9. Let $\sum_{n=1}^{\infty} a_n$ be convergent. $\lim_{n\to\infty} (a_{n+1} + a_{n+2} + \cdots + a_{n^2}) = ?$

7.10. Let the series (a_n) satisfy

$$\lim_{n\to\infty} |a_n + a_{n+1} + \cdots + a_{n+2^n}| = 0.$$

Does it then follow that the series $\sum_{n=1}^{\infty} a_n$ is convergent? (H)

7.11. Let the series (a_n) satisfy

$$\lim_{n\to\infty} |a_n + a_{n+1} + \cdots + a_{n+i_n}| = 0$$

for every sequence of positive integers (i_n). Does it then follow that the series $\sum_{n=1}^{\infty} a_n$ is convergent? (H)

7.12. Prove that if $|a_{n+1} - a_n| < 1/n^2$ for all n, then the series $\sum_{n=1}^{\infty} a_n$ is convergent.

7.13. Suppose that $a_n \le b_n \le c_n$ for all n. Prove that if the series $\sum_{n=1}^{\infty} a_n$ and $\sum_{n=1}^{\infty} c_n$ are convergent, then the series $\sum_{n=1}^{\infty} b_n$ is convergent as well.

7.14. Prove that if the series $\sum_{n=1}^{\infty} a_n$ is convergent, then

$$\lim_{n\to\infty} \frac{a_1 + 2a_2 + \cdots + na_n}{n} = 0. \text{ (S)}$$

Chapter 8
Countable Sets

While we were talking about sequences, we noted that care needs to be taken in distinguishing the sequence (a_n) from the set of its terms $\{a_n\}$. We will say that the sequence (a_n) **lists** the elements of H if $H = \{a_n\}$. (The elements of the set H and thus the terms of (a_n) can be arbitrary; we do not restrict ourselves to sequences of real numbers.) If there is a sequence that lists the elements of H, we say that H **can be listed** by that sequence. Clearly, every finite set can be listed, since if $H = \{a_1, \ldots, a_k\}$, then the sequence $(a_1, \ldots, a_k, a_k, a_k, \ldots)$ satisfies $H = \{a_n\}$. Suppose now that H is infinite. We show that if there is a sequence (a_n) that lists the elements of H, then there is another sequence listing H whose terms are all distinct. Indeed, for each element x of H, we can choose a term a_n for which $a_n = x$. The terms chosen in this way (in the order of their indices) form (a_{n_k}), a subsequence of (a_n). If $c_k = a_{n_k}$ for all k, then the terms of the sequence (c_k) are distinct, and $H = \{c_k\}$.

Like every sequence, (c_k) is also a function defined on the set $\mathbb{N}^+$. Since its terms are distinct and $H = \{c_k\}$, this means that the map $k \mapsto c_k$ is a bijection between the sets $\mathbb{N}^+$ and H (see the footnote on p. 38).

We have shown that the elements of a set H can be listed if and only if H is finite or there is a bijection between H and $\mathbb{N}^+$.

Definition 8.1. We say the set H is *countably infinite*, if there exists a bijective map between $\mathbb{N}^+$ and H. The set H is *countable* if it is finite or countably infinite.

With our earlier notation, we can summarize our argument above by saying that a set can be listed if and only if it is countable.

It is clear that $\mathbb{N}$ is countable: consider the sequence $(0, 1, 2, \ldots)$. It is also easy to see that $\mathbb{Z}$ is countable, since the sequence $(0, 1, -1, 2, -2, \ldots)$ contains every integer. More surprising is the following theorem.

© Springer New York 2015
M. Laczkovich, V.T. Sós, *Real Analysis*, Undergraduate Texts
in Mathematics, DOI 10.1007/978-1-4939-2766-1_8

Theorem 8.2. *The set of all rational numbers is countable.*

Proof. We have to convince ourselves that there is a sequence that lists the rational numbers. One example of such a sequence is

$$\frac{0}{1}, \frac{-1}{1}, \frac{0}{2}, \frac{1}{1}, \frac{-2}{1}, \frac{-1}{2}, \frac{0}{3}, \frac{1}{2}, \frac{2}{1}, \frac{-3}{1}, \frac{-2}{2}, \frac{-1}{3}, \frac{0}{4}, \frac{1}{3},$$

$$\frac{2}{2}, \frac{3}{1}, \frac{-4}{1}, \frac{-3}{2}, \frac{-2}{3}, \frac{-1}{4}, \frac{0}{5}, \frac{1}{4}, \frac{2}{3}, \frac{3}{2}, \frac{4}{1}, \frac{-5}{1}, \frac{-4}{2}, \dots. \tag{8.1}$$

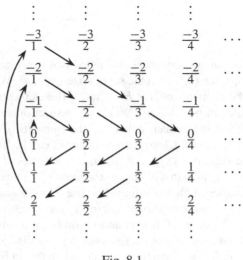

Fig. 8.1

Here fractions of the form p/q (where p, q are integers and $q > 0$) are listed by the size of $|p| + q$. For all n, we list all fractions for which $|p| + q = n$ in some order, and then we attach the finite sequences we get for $n = 1, 2, \dots$ to get an infinite sequence (Figure 8.1). It is clear that we have listed every rational number in this way.   □

Using similar techniques, we can show that sets much "larger" than the set of rational numbers are also countable. We say that the complex number α is **algebraic** if it is the root of a not identically zero polynomial with integer coefficients. It is clear that every rational number is algebraic, since p/q is the root of the polynomial $qx - p$. The number $\sqrt{2}$ is also algebraic, since it is a root of $x^2 - 2$.

Theorem 8.3. *The set of algebraic numbers is countable.*

Proof. We define the *weight* of the polynomial $a_k x^k + \cdots + a_0$ to be the number $k + |a_k| + |a_{k-1}| + \cdots + |a_0|$. It is clear that for every n, there are only finitely many integer-coefficient polynomials whose weight is n. It then follows that there is a sequence $f_1, f_2, \dots$, that contains every nonconstant polynomial. We can get this by first listing all the integer-coefficient nonconstant polynomials whose weight is 2, followed by all those whose weight is 3, and so on. We have then listed every integer-coefficient nonconstant polynomial, since the weight of such a polynomial cannot be 0 or 1. Each polynomial f_i can have only finitely many roots (see Lemma 11.1). List the roots of f_1 in some order, followed by the roots of f_2, and so on. The sequence we get lists every algebraic number, which proves our theorem. $\square$

According to the following theorem, set-theoretic operations (union, intersection, complement) do not lead us out of the class of countable sets.

Theorem 8.4.

(i) *Every subset of a countable set is countable.*

(ii) *The union of two countable sets is countable.*

Proof. (i) Let A be countable, and $B \subset A$. Suppose that the sequence (a_n) lists the elements of A. For each element x of B, choose a term a_n for which $a_n = x$. The terms chosen in this way (in the order of their indices) form a subsequence of (a_n) that lists the elements of B.

(ii) If the sequences (a_n) and (b_n) list the elements of the sets A and B, then the sequence $(a_1, b_1, a_2, b_2, \ldots)$ lists the elements of $A \cup B$. $\square$

An immediate consequence of statement (ii) above (by induction) is that the union of finitely many countable sets is also countable. By the following theorem, more is true.

Theorem 8.5. *The union of a countable number of countable sets is countable.*

Proof. Let $A_1, A_2, \ldots$ be countable sets, and for each k, let $\left(a_n^k\right)$ be a sequence that lists the elements of A_k. Then the sequence

$$\left(a_1^1, a_2^1, a_1^2, a_3^1, a_2^2, a_1^3, a_4^1, a_3^2, a_2^3, a_1^4, \ldots\right)$$

lists the elements of $\bigcup_{k=1}^{\infty} A_k$. We get the above sequence by writing all the finite sequences $\left(a_n^1, a_{n-1}^2, \ldots, a_2^{n-1}, a_1^n\right)$ one after another for each n. $\square$

Based on the previous theorems, the question whether there are uncountable sets at all arises naturally. The following theorem gives an answer to this.

Theorem 8.6. *The set of real numbers is uncountable.*

Proof. Suppose that $\mathbb{R}$ is countable, and let (x_n) be a sequence of real numbers that contains every real number. We will work toward a contradiction by constructing a real number x that is not in the sequence. We outline two constructions.

I. The first construction is based on the following simple observation: if I is a closed interval and c is a given number, then I has a closed subinterval that does not contain c. This is clear: if we choose two disjoint closed subintervals, at least one will not contain c.

Let I_1 be a closed interval that does not contain x_1. Let I_2 be a closed subinterval of I_1 such that $x_2 \notin I_2$. Following this procedure, let I_n be a closed subinterval of

I_{n-1} such that $x_n \notin I_n$. According to Cantor's axiom, the intervals I_n have a shared point. If $x \in \bigcap_{n=1}^{\infty} I_n$, then $x \neq x_n$ for all n, since $x \in I_n$, but $x_n \notin I_n$. Thus x cannot be a term in the sequence, which is what we were trying to show.

II. A second construction for a similar x is the following: Consider the decimal expansion of $x_1, x_2, \ldots$:

$$x_1 = \pm n_1.a_1^1 a_2^1 \ldots$$
$$x_2 = \pm n_2.a_1^2 a_2^2 \ldots$$
$$\vdots$$

Let $x = 0.b_1 b_2 \ldots$, where $b_i = 5$ if $a_i^i \neq 5$, and $b_i = 4$ if $a_i^i = 5$. Clearly, x is different from each x_n. $\qquad \square$

Theorems 8.4 and 8.6 imply that the set of irrational numbers is uncountable. If it were countable, then—since $\mathbb{Q}$ is also countable—$\mathbb{R} = \mathbb{Q} \cup (\mathbb{R} \setminus \mathbb{Q})$ would also be countable, whereas it is not. We call a number **transcendental** if it is not algebraic. Repeating the above argument—using the fact that the set of algebraic numbers is countable—we get that *the set of transcendental numbers is uncountable*.

Definition 8.7. If there exists a bijection between two sets A and B, then we say that the two sets are *equivalent*, or that A and B have the same *cardinality*, and we denote this by $A \sim B$.

By the above definition, a set A is countably infinite if and only if $A \sim \mathbb{N}^+$. It can be seen immediately that if $A \sim B$ and $B \sim C$, then $A \sim C$; if f is a bijection from A to B and g is a bijection from B to C, then the map $x \mapsto g(f(x))$ ($x \in A$) is a bijection from A to C.

Definition 8.8. We say that a set H has the *cardinality of the continuum* if it is equivalent to $\mathbb{R}$.

We show that both the set of irrational numbers and the set of transcendental numbers have the cardinality of the continuum. For this, we need the following simple lemma.

Lemma 8.9. *If A is infinite and B is countable, then $A \cup B \sim A$.*

Proof. First of all, we show that A contains a countably infinite subsequence. Since A is infinite, it is nonempty, and we can choose an $x_1 \in A$. If we have already chosen $x_1, \ldots, x_n \in A$, then $A \neq \{x_1, \ldots, x_n\}$ (since then A would be finite), so we can choose an element $x_{n+1} \in A \setminus \{x_1, \ldots, x_n\}$. Thus by induction, we have chosen distinct x_n for each n. Then $X = \{x_n : n = 1, 2, \ldots\}$ is a countably infinite subset of A.

To prove the lemma, we can suppose that $A \cap B = \emptyset$, since we can substitute B with $B \setminus A$ (which is also countable). By Theorem 8.4, $X \cup B$ is countable. Since it is also infinite, $X \cup B \sim \mathbb{N}^+$, and so $X \cup B \sim X$, since $\mathbb{N}^+ \sim X$. Let f be a bijection from X to $X \cup B$. Then

$$g(x) = \begin{cases} x, & \text{if } x \in A \setminus X, \\ f(x), & \text{if } x \in X \end{cases}$$

is a bijection from A to $A \cup B$.   □

Theorem 8.10. *Both the set of irrational numbers and the set of transcendental numbers have the cardinality of the continuum.*

Proof. By the previous theorem, $\mathbb{R} \setminus \mathbb{Q} \sim (\mathbb{R} \setminus \mathbb{Q}) \cup \mathbb{Q} = \mathbb{R}$, so $\mathbb{R} \setminus \mathbb{Q}$ has the cardinality of the continuum. By a straightforward modification of the argument, we find that the set of transcendental numbers also has the cardinality of the continuum.   □

Theorem 8.11. *Every nondegenerate interval has the cardinality of the continuum.*

Proof. The interval $(-1, 1)$ has the cardinality of the continuum, since the map $f(x) = x/(1 + |x|)$ is a bijection from $\mathbb{R}$ to $(-1, 1)$. (The inverse of f is $f^{-1}(x) = x/(1 - |x|)$ $(x \in (-1, 1))$.) Since every open interval is equivalent to $(0, 1)$ (the function $(b - a)x + a$ maps $(0, 1)$ to (a, b)), every open interval has the cardinality of the continuum.

Moreover, Lemma 8.9 gives $[a, b] \sim (a, b)$, $(a, b] \sim (a, b)$ and $[a, b) \sim (a, b)$, so we get that every bounded nondegenerate interval has the cardinality of the continuum.

The proof for rays (or half-lines) having the cardinality of the continuum is left as an exercise for the reader.   □

Exercises

8.1. Let a_n denote the nth term of sequence (8.1). What is the smallest n for which $a_n = -17/39$?

8.2. Prove that the set of finite sequences with integer terms is countable. (H)

8.3. Show that the set of finite-length English texts is countable.

8.4. Prove that every set of disjoint intervals is countable. (H)

8.5. Prove that a set is infinite if and only if it is equivalent to a proper subset of itself.

8.6. Prove that every ray (half-line) has the cardinality of the continuum.

8.7. Prove that if both A and B have the cardinality of the continuum, then so does $A \cup B$.

8.8. Prove that every circle has the cardinality of the continuum.

8.9. Give a function that maps $(0,1]$ bijectively to the set of infinite sequences made up of positive integers. (H)

8.10. Prove that the set of all subsets of $\mathbb{N}$ has the cardinality of the continuum.

8.11. Prove that the plane (that is, the set $\{(x,y): x,y \in \mathbb{R}\}$) has the cardinality of the continuum. (H)

Chapter 9
Real-Valued Functions of One Real Variable

9.1 Functions and Graphs

Consider a function $f\colon A \to B$. As we stated earlier, by this we mean that for every element a of the set A, there exists a corresponding $b \in B$, which is denoted by $b = f(a)$.

We call the set A the **domain** of f, and we denote it by $A = D(f)$. The set of $b \in B$ that correspond to some $a \in A$ is called the **range** of f, and is denoted by $R(f)$. That is, $R(f) = \{f(a)\colon a \in D(f)\}$. The set $R(f)$ is inside B, but generally it is not required to be equal to B.

If $C \subset A$, then $f(C)$ denotes all the $b \in B$ that correspond to some $c \in C$; $f(C) = \{f(a)\colon a \in C\}$. By this notation, $R(f) = f(D(f))$.

We consider two functions f and g to be equal if $D(f) = D(g)$, and $f(x) = g(x)$ holds for all $x \in D(f) = D(g)$.

Definition 9.1. For a function $f\colon A \to B$, if for every $a_1, a_2 \in A$, $a_1 \neq a_2$, we have $f(a_1) \neq f(a_2)$, then we say that f is *one-to-one* or *injective*.

If $f\colon A \to B$ and $R(f) = B$, then we say that f is *onto* or *surjective*.

If $f\colon A \to B$ is both one-to-one and onto, then we say that f is a *one-to-one correspondence* between A and B. (In other words, f is a *bijection, a bijective map*.)

Definition 9.2. If $f\colon A \to B$ is a one-to-one correspondence between A and B, then we write f^{-1} to denote the map that takes every $b \in B$ to the corresponding $a \in A$ for which $b = f(a)$. Then $f^{-1}\colon B \to A$, $D(f^{-1}) = B$, and $R(f^{-1}) = A$. We call the map f^{-1} the *inverse function* (or the *inverse map* or simply the *inverse*) of f.

Theorem 9.3. *Let $f\colon A \to B$ be a bijection between A and B. Then the map $g = f^{-1}\colon B \to A$ has an inverse, and $g^{-1} = f$.*

Proof. The statement is clear from the definition of an inverse function. □

© Springer New York 2015
M. Laczkovich, V.T. Sós, *Real Analysis*, Undergraduate Texts
in Mathematics, DOI 10.1007/978-1-4939-2766-1_9

We can define various operations between functions that map pairs of functions to functions. One such operations is composition.

Definition 9.4. We can define the *composition* of the functions $f\colon A \to B$ and $g\colon B \to C$ as a new function, denoted by $g \circ f$, that satisfies $D(g \circ f) = A$ and $(g \circ f)(x) = g(f(x))$ for all $x \in A$. If f and g are arbitrary functions, then we define $g \circ f$ for those values x for which $g(f(x))$ is defined. That is, we would have $D(g \circ f) = \{x \in D(f)\colon f(x) \in D(g)\}$, and $(g \circ f)(x) = g(f(x))$ for all such x.

Remark 9.5 (Composition Is Not Commutative). Consider the following example. Let $\mathbb{Z}$ denote the set of integers, and let $f(x) = x + 1$ for all $x \in \mathbb{Z}$. Moreover, let $g(x) = 1/x$ for all $x \in \mathbb{Z} \setminus \{0\}$. Then $D(g \circ f) = \mathbb{Z} \setminus \{-1\}$, and $(g \circ f)(x) = 1/(x+1)$ for all $x \in \mathbb{Z} \setminus \{-1\}$. On the other hand, $D(f \circ g) = \mathbb{Z} \setminus \{0\}$, and $(f \circ g)(x) = (1/x) + 1$ for all $x \in \mathbb{Z} \setminus \{0\}$. We can see that $f \circ g \neq g \circ f$, since they are defined on different domains. But in fact, it is easy to see that $(f \circ g)(x) \neq (g \circ f)(x)$ for each x where both sides are defined.

From this point on, we will deal with functions whose domain and range are both subsets of the real numbers. We call such functions **real-valued function of a real variable** (or simply real functions for short).

We can also define addition, subtraction, multiplication, and division among real functions.

Let f and g be real-valued functions. Their sum $f + g$ and their difference $f - g$ are defined by the formulas $(f + g)(x) = f(x) + g(x)$ and $(f - g)(x) = f(x) - g(x)$ respectively for all $x \in D(f) \cap D(g)$. Then $D(f + g) = D(f - g) = D(f) \cap D(g)$.

Similarly, their product $f \cdot g$ is defined on the set $D(f) \cap D(g)$, having the value $f(x) \cdot g(x)$ at $x \in D(f) \cap D(g)$. Finally, their quotient, f/g, is defined by $(f/g)(x) = f(x)/g(x)$ at every x for which $x \in D(f) \cap D(g)$ and $g(x) \neq 0$. That is, $D(f/g) = \{x \in D(f) \cap D(g)\colon g(x) \neq 0\}$.

Like sequences, real functions can be represented in a variety of ways. Consider the following examples.

Examples 9.6. 1. $f(x) = x^2 + 3 \quad (x \in \mathbb{R})$;

2. $f(x) = \begin{cases} 1, & \text{if } x \text{ is rational} \\ 0, & \text{if } x \text{ is irrational} \end{cases} \quad (x \in \mathbb{R})$;

3. $f(x) = \lim_{n \to \infty} (1 + x + \cdots + x^n) \quad (x \in (-1, 1))$;

4. $f(0.a_1 a_2 \ldots) = 0.a_2 a_4 a_6 \ldots$, where we exclude the forms $0.a_1 \ldots a_n 999 \ldots$.

We gave the function in (1) with a "formula"; in the other examples, the map was defined in other ways. Just as for sequences, the way the function is defined is irrelevant: using a formula to define a function is no better or worse than the other ways (except perhaps shorter).

We can illustrate real functions in the plane using Cartesian coordinates.[1] Let $f\colon A \to B$ be a real function, that is, such that $A \subset \mathbb{R}$ and $B \subset \mathbb{R}$. Consider points of

---

[1] We summarize the basic definitions of coordinate geometry in the appendix of this chapter.

the form $(x, 0)$ on the x-axis, where $x \in A$. At each of these points raise a perpendicular line to the x-axis, and measure $f(x)$ on this perpendicular ("upward" from the x-axis if $f(x) \geq 0$, and "downward" if $f(x) \leq 0$). Then we get the points $(x, f(x))$, where $x \in A$. We call the set of these points the **graph** of f, which we denote by graph f. To summarize:

$$\text{graph} f = \{(x, f(x)) : x \in A\}. \tag{9.1}$$

Examples 9.7.   1. $f(x) = ax + b$;   2. $f(x) = x^2$;

3. $f(x) = (x - a)(x - b)$;   4. $f(x) = x^3$;

5. $f(x) = 1/x$;   6. $f(x) = |x|$;

7. $f(x) = [x]$, where $[x]$ denotes the greatest integer that is less than or equal to x (the **floor** of x);
8. $f(x) = \{x\}$, where $\{x\} = x - [x]$;

9. $f(x) = \begin{cases} 1, & \text{if } x \text{ is rational} \\ 0, & \text{if } x \text{ is irrational;} \end{cases}$ 10. $f(x) = \begin{cases} x, & \text{if } x \text{ is rational} \\ 0, & \text{if } x \text{ is irrational.} \end{cases}$

Let us see the graphs of functions (1)–(10) (see Figure 9.1)!

The function (9), which we now see for the second time, will continue to appear in the sequel to illustrate various phenomena. This function is named—after its discoverer—the Dirichlet[2] function.

We note that illustrating functions by a graph is similar to our use of the number line, and what we said earlier applies: the advantage of these illustrations in the plane are that some statements can be expressed graphically in a more understandable form, and often we can get an idea for a proof from the graphical interpretation. However, we must again emphasize that something that we can "see" from inspecting a graph cannot be taken as a proof, and some things that we "see" turn out to be false. As before, we can rely only on the axioms of the real numbers and the theorems that we have proved from those.

Remark 9.8. Our more careful readers might notice that when we introduced functions, we were not as strict as we were when we defined sets. Back then, we noted that writing sets as collections, classes, or systems does not solve the problem of the definition, so we said that the definition of a set is a fundamental concept. We reduced defining functions to correspondences between sets, but we did not define what we meant by a correspondence. It is a good idea to choose the same solution as in the case of sets, that is, we consider the notion of a function to be a fundamental concept as well, where "correspondence" and "map" are just synonyms.

---

[2] Lejeune Dirichlet (1805–1859), German mathematician.

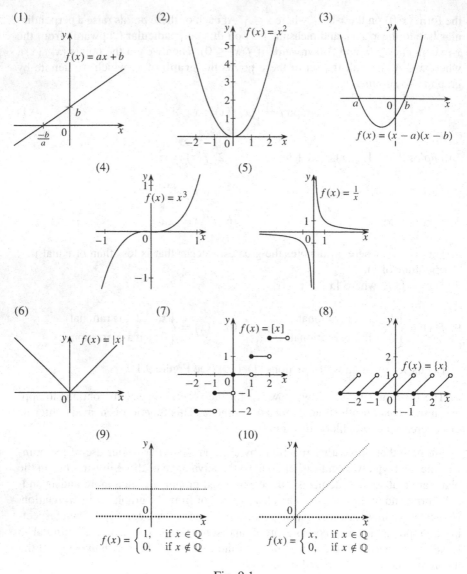

Fig. 9.1

We should note, however, that the concept of a function can be reduced to the concept of sets. We can do this by generalizing what we defined as graphs. Graphs, as defined in (9.1), can easily be generalized to maps between arbitrary sets. Let A and B be arbitrary sets. The set of ordered pairs (a,b) where $a \in A$ and $b \in B$ is called the **Cartesian product** of the sets A and B, denoted by $A \times B$. That is,

$$A \times B = \{(a,b) : a \in A, \ b \in B\}.$$

If $f: A \to B$ is a map, then the graph of f can be defined by (9.1). The set graph f is then a subset of the Cartesian product $A \times B$. Then every function from A to B has a corresponding set graph $f \subset A \times B$. It is clear that different functions have different graphs.

Clearly, the set graph $f \subset A \times B$ has the property that for every element $a \in A$, there is exactly one element $b \in B$ for which $(a,b) \in$ graph f, namely $b = f(a)$. Conversely, suppose that $H \subset A \times B$, and for every $a \in A$, there is exactly one element $b \in B$ for which $(a,b) \in H$. Denote this b by $f(a)$. This defines a map $f: A \to B$, and of course $H =$ graph f. This observation is what makes it possible to reduce the concept of graphs to the concept of sets. In axiomatic set theory, functions are *defined to be* subsets of the Cartesian product $A \times B$ that have the above property. We do not follow this path (since the discussion of axiomatic set theory is not our goal); instead, we continue to consider the notion of a function a fundamental concept.

Remark 9.9. We end this section with a note on mathematical notation. Throughout the centuries, a convention has formed regarding notation of mathematical concepts and objects. According to this, functions are most often denoted by f; this comes from the first letter of the Latin word *functio* (which is the root of our English word *function*). If several functions appear in an argument, they are usually denoted by f, g, h. For similar reasons, natural numbers are generally denoted by n, which is the first letter of the word *naturalis*. Of course, to denote natural numbers, we often also use the letters i, j, k, l, m preceding n. Constants and sequences are mostly denoted by letters from the beginning of the alphabet, while variables (that is, the values to which we apply functions) are generally denoted by the letters x, y, z at the end of the alphabet.

Of course there is no theoretical requirement that a particular letter or symbol be used for a certain object, and it can happen that we have to, or choose to, call a function something other than f, g, h. However, following the conventions above makes reading mathematical texts much easier, since we can discern the nature of the objects appearing very quickly.

Exercises

9.1. Decide which of the following statements hold for all functions $f: A \to B$ and all subsets $H, K \subset A$

(a) $f(H \cup K) = f(H) \cup f(K)$;
(b) $f(H \cap K) = f(H) \cap f(K)$;
(c) $f(H \setminus K) = f(H) \setminus f(K)$.

9.2. For an arbitrary function $f: A \to B$ and set $Y \subset B$, let $f^{-1}(Y)$ denote the set of $x \in A$ for which $f(x) \in Y$. (We do not suppose that f has an inverse. To justify our notation, we note that *if* an inverse of f^{-1} exists, then the two meanings of $f^{-1}(Y)$ denote the same set.)

Prove that for arbitrary sets $Y, Z \subset B$,

(a) $f^{-1}(Y \cup Z) = f^{-1}(Y) \cup f^{-1}(Z)$;
(b) $f^{-1}(Y \cap Z) = f^{-1}(Y) \cap f^{-1}(Z)$;
(c) $f^{-1}(Y \setminus Z) = f^{-1}(Y) \setminus f^{-1}(Z)$.

9.3. Find the inverse functions for the following maps.

(a) $(x+1)/(x-1)$, $x \in \mathbb{R} \setminus \{1\}$;
(b) $1/(2x+3)$, $x \in \mathbb{R} \setminus \{-3/2\}$;
(c) $x/(1+|x|)$, $x \in \mathbb{R}$.

9.4. Does there exist a function $f \colon \mathbb{R} \to \mathbb{R}$ for which $(f \circ f)(x) = -x$ for all $x \in \mathbb{R}$? (H)

9.5. For every real number c, give a function $f_c \colon \mathbb{R} \to \mathbb{R}$ such that $f_{a+b} = f_a \circ f_b$ holds for all $a, b \in \mathbb{R}$. Can this be done even if f_1 is an arbitrary fixed function? ($*$H)

9.6. (a) Give two real functions f_1, f_2 for which there is no g such that both f_1 and f_2 are of the form $g \circ \ldots \circ g$. (H)
(b) Let $f_1, \ldots, f_k$ be arbitrary real functions defined on $\mathbb{R}$. Show that three functions g_1, g_2, g_3 can be given such that each $f_1, \ldots, f_k$ can be written in the form $g_{i_1} \circ g_{i_2} \circ \ldots \circ g_{i_s}$, (where $i_1, \ldots, i_s = 1, 2, 3$). ($*$S)
(c) Can this be solved with two functions g_i instead of three? ($*$HS)
(d) Can this be solved for an infinite sequence of f_i? ($*$H S)

9.2 Global Properties of Real Functions

The graphs of functions found in Examples 9.7 admit symmetries and other properties that are present in numerous other graphs as well. The graphs of functions (2), (6), and (9) are symmetric with respect to the y-axis, the graphs of (4) and (5) are symmetric with respect to the origin, and the graph of the function (8) is made up of periodically repeating segments. The graphs of the functions (2), (4), and (6) also admit other properties. In these graphs, the part above $[0, \infty)$ "moves upward," which corresponds to larger $f(x)$ for increasing x. The graphs of (2) and (3), as well as the part over $(0, \infty)$ in the graphs of (4) and (5), are "concave up," meaning that the line segment between any two points on the graph is always above the graph.

We will define these properties precisely below, and give tools and methods to determine whether a function satisfies some of the given properties.

Definition 9.10. The function f is said to be *even* if for all $x \in D(f)$, we have $-x \in D(f)$ and $f(x) = f(-x)$.

The function f is *odd* if for all $x \in D(f)$, we have $-x \in D(f)$ and $f(x) = -f(-x)$.

Remark 9.11. It is clear that for an arbitrary even integer n, the function $f(x) = x^n$ is even, while if n is odd, then x^n is odd (which is where the nomenclature comes from in the first place). It is also clear that the graph of an even function is symmetric with respect to the y-axis. This follows from the fact that the point (x, y) reflected across the y-axis gives $(-x, y)$. Graphs of odd functions, however, are symmetric with respect to the origin, since the point (x, y) reflected through the origin gives $(-x, -y)$.

Definition 9.12. The function f is *periodic* if there is a $d \neq 0$ such that for all $x \in D(f)$, we have $x + d \in D(f)$, $x - d \in D(f)$, and $f(x + d) = f(x)$. We call the number d a *period* of the function f.

Remark 9.13. It is easy to see that if d is a period of f, then $k \cdot d$ is also a period for each integer k. Thus a periodic function always has infinitely many periods.

Not every periodic function has a smallest positive period. For example the periods of the Dirichlet function are exactly the rational numbers, and there is no smallest positive rational number.

Definition 9.14. The function f is *bounded from above* (or *below*) *on the set* $A \subset \mathbb{R}$ if $A \subset D(f)$ and there exists a number K such that $f(x) \leq K$ ($f(x) \geq K$) for all $x \in A$.

The function f is *bounded on the set* $A \subset \mathbb{R}$ if it is bounded from above and from below on A.

It is easy to see that f is bounded on A if and only if $A \subset D(f)$ and there is a number K such that $|f(x)| \leq K$ for all $x \in A$. The boundedness of f on the set A graphically means that for suitable K, the graph of f lies inside the region bordered by the lines $y = -K$ and $y = K$ over A. The function $f(x) = 1/x$ is bounded on $(\delta, 1)$ for all $\delta > 0$, but is not bounded on $(0, 1)$. In the interval $(0, 1)$, the function is bounded from below but not from above.

The function

$$f(x) = \begin{cases} x, & \text{if } x \text{ is rational} \\ 0, & \text{if } x \text{ is irrational} \end{cases}$$

is not bounded on $(-\infty, +\infty)$. However, f is bounded on the set $\mathbb{R} \setminus \mathbb{Q}$.

Definition 9.15. The function f is *monotone increasing (monotone decreasing) on the set* $A \subset \mathbb{R}$ if $A \subset D(f)$ and for all $x_1 \in A$ and $x_2 \in A$ such that $x_1 < x_2$,

$$f(x_1) \leq f(x_2) \qquad (f(x_1) \geq f(x_2)). \tag{9.2}$$

If in (9.2) we replace $\leq$ or $\geq$ by $<$ and $>$ respectively, then we say that f is *strictly monotone increasing (decreasing)*. We call monotone increasing or decreasing functions *monotone* functions for short.

Let us note that if f is constant on the set A, then f is both monotone increasing and monotone decreasing on A.

The Dirichlet function is neither monotone increasing nor monotone decreasing on any interval. However, on the set of rational numbers, the Dirichlet function is both monotone increasing and decreasing (since it is constant there).

Convex and Concave Functions. Consider the points $(a, f(a))$ and $(b, f(b))$ on the graph graph f. We call the line segment connecting these points a chord of graph f. Let the linear function defining the line of the chord be $h_{a,b}$, that is, let

$$h_{a,b}(x) = \frac{f(b) - f(a)}{b - a}(x - a) + f(a).$$

Definition 9.16. The function f is *convex* on the interval I if for every $a, b \in I$ and $a < x < b$,

$$f(x) \leq h_{a,b}(x). \tag{9.3}$$

If in (9.3) we write $<$ in place of $\leq$, then we say that f is *strictly convex* on I; if we write $\geq$ or $>$, then we say that f is *concave*, or respectively *strictly concave* on I.

The property of f being convex on I can be phrased in a more visual way by saying that for all $a, b \in I$, the part of the graph corresponding to the interval (a, b) is "below" the chord connecting $(a, f(a))$ and $(b, f(b))$. In the figure, $I = (\alpha, \beta)$ (Figure 9.2).

It is clear that f is convex on I if and only if $-f$ is concave on I. We note that sometimes "concave down" is said instead of concave, and "concave up" instead of convex.

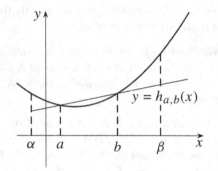

Fig. 9.2

If the function f is linear on the interval I, that is, $f(x) = cx + d$ for some constants c and d, then in (9.3), equality holds for all x. Thus a linear function is both concave and convex.

For applications, it is useful to give the inequality defining convexity in an alternative way.

Let $a < b$ and $0 < t < 1$. Then the number $x = ta + (1 - t)b$ is an element of (a, b), and moreover, x is the point that splits the interval $[a, b]$ in a $(1 - t) : t$ ratio. Indeed,

$$a = ta + (1 - t)a < ta + (1 - t)b =$$
$$= x < tb + (1 - t)b = b,$$

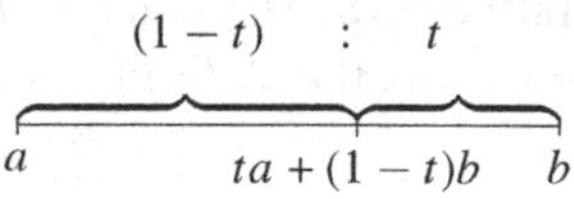

Fig. 9.3

so $x \in (a, b)$. On the other hand, a simple computation shows that $(x - a)/(b - x) = (1 - t)/t$.

By reversing the computation, we can see that every element of (a, b) occurs in the form $ta + (1 - t)b$, where $0 < t < 1$. If $x \in (a, b)$, then the choice $t = (b - x)/(b - a)$ works (Figure 9.3).

Now if $a < x < b$ and $x = ta + (1 - t)b$, then

$$h_{a,b}(x) = \frac{f(b)-f(a)}{b-a} \cdot (ta+(1-t)b-a)+f(a) = tf(a)+(1-t)f(b).$$

If we substitute this into (9.3), then we get the following equivalent condition for convexity.

Lemma 9.17. *The function f is convex on the interval I if and only if for all numbers $a,b \in I$ and $0 < t < 1$,*

$$f(ta+(1-t)b) \leq tf(a)+(1-t)f(b). \tag{9.4}$$

Remark 9.18. By the above argument, we also see that the strict convexity of the function f is equivalent to a strict inequality in (9.4) whenever $a \neq b$.

Theorem 9.19 (Jensen's[3] Inequality). *The function f is convex on the interval I if and only if whenever $a_1,\ldots,a_n \in I$, $t_1,\ldots,t_n > 0$, and $t_1 + \cdots + t_n = 1$, then*

$$f(t_1 a_1 + \cdots + t_n a_n) \leq t_1 f(a_1) + \cdots + t_n f(a_n). \tag{9.5}$$

If f is strictly convex, then strict inequality holds, assuming that the a_i are not all equal.

Proof. By Lemma 9.17, the function f is convex on an interval I if and only if (9.5) holds for $n = 2$. Thus we need to show only that if the inequality holds for $n = 2$, then it also holds for $n > 2$. We prove this by induction.

Let $k > 2$, and suppose that (9.5) holds for all $2 \leq n < k$ and $a_1,\ldots,a_n \in I$, as well as for $t_1,\ldots,t_n > 0$, $t_1 + \cdots + t_n = 1$. Let $a_1,\ldots a_k \in I$ and $t_1,\ldots t_k > 0$, $t_1 + \cdots + t_k = 1$. We see that

$$f(t_1 a_1 + \cdots + t_k a_k) \leq t_1 f(a_1) + \cdots + t_k f(a_k). \tag{9.6}$$

Let

$$t = t_1 + \cdots + t_{k-1},$$
$$\alpha = (t_1/t)a_1 + \cdots + (t_{k-1}/t)a_{k-1}$$

and

$$\beta = (t_1/t)f(a_1) + \cdots + (t_{k-1}/t)f(a_{k-1}).$$

By the induction hypothesis, $f(\alpha) \leq \beta$. If we now use $t_k = 1-t$, then we get

$$\begin{aligned}
f(t_1 a_1 + \cdots + t_k a_k) = f(t \cdot \alpha + (1-t)a_k) &\leq \\
\leq t \cdot f(\alpha) + (1-t) \cdot f(a_k) &\leq \\
\leq t \cdot \beta + (1-t) \cdot f(a_k) &= \\
= t_1 f(a_1) + \cdots + t_k f(a_k), \tag{9.7}
\end{aligned}$$

which proves the statement. The statement for strict convexity follows in the same way. □

---

[3] Johan Ludwig William Valdemar Jensen (1859–1925), Danish mathematician.

The following theorem links convexity and monotonicity.

Theorem 9.20. *The function f is convex on an interval I if and only if for all $a \in I$, the function $x \mapsto (f(x) - f(a))/(x - a)$ $(x \in I \setminus \{a\})$ is monotone increasing on the set $I \setminus \{a\}$.*

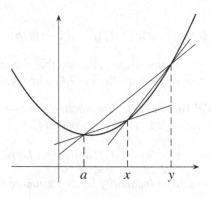

Fig. 9.4

Proof. Suppose that f is convex on I, and let a, x, $y \in I$, where $a < x < y$ (Figure 9.4). By (9.3),

$$f(x) \le \frac{f(y) - f(a)}{y - a}(x - a) + f(a).$$

A simple rearrangement of this yields the inequality

$$\frac{f(x) - f(a)}{x - a} \le \frac{f(y) - f(a)}{y - a}. \tag{9.8}$$

If $x < a < y$, then by (9.3),

$$f(a) \le \frac{f(y) - f(x)}{y - x}(a - x) + f(x),$$

which we can again rearrange to give us (9.8).

If $x < y < a$, (9.8) follows in the same way. Thus we have shown that the function

$$\frac{f(x) - f(a)}{x - a}$$

is monotone increasing.

Now let us suppose that the function $(f(x) - f(a))/(x - a)$ is monotone increasing for all $a \in I$. Let $a, b \in I$, where $a < x < b$. Then

$$\frac{f(x)-f(a)}{x-a} \le \frac{f(b)-f(a)}{b-a},$$

which can be rearranged into

$$f(x) \le \frac{f(b)-f(a)}{b-a}(x-a)+f(a).$$

Thus f satisfies the conditions of convexity. $\qquad\square$

Let us see some applications.

Example 9.21. First of all, we show that the function x^2 is convex on $\mathbb{R}$. By Theorem 9.20, we have to show that the function $(x^2 - a^2)/(x-a) = x+a$ is monotone increasing for all a, which is clear. Thus we can apply Jensen's inequality to $f(x) = x^2$. Choosing $t_1 = \ldots = t_n = 1/n$, we get that

$$\left(\frac{a_1 + \cdots + a_n}{n}\right)^2 \le \frac{a_1^2 + \cdots + a_n^2}{n},$$

or, taking square roots,

$$\frac{a_1 + \cdots + a_n}{n} \le \sqrt{\frac{a_1^2 + \cdots + a_n^2}{n}} \qquad (9.9)$$

for all $a_1, \ldots, a_n \in \mathbb{R}$. This is known as the **root-mean-square–arithmetic mean inequality** (which we have encountered in Exercise 2.15).

Example 9.22. Now we show that the function $1/x$ is convex on the half-line $(0, \infty)$. We want to show that for every $a > 0$, the function

$$\frac{(1/x)-(1/a)}{x-a} = -\frac{1}{a \cdot x}$$

is monotone increasing on the set $(0, \infty) \setminus \{a\}$, which is again clear.

If we now apply Jensen's inequality to $f(x) = 1/x$ with $t_1 = \ldots = t_n = 1/n$, we get the inequality between the arithmetic and harmonic means (Theorems 2.6 and 2.7).

Exercises

9.7. Let f and g be defined on $(-\infty, +\infty)$ and suppose

(a) f is even, g is odd;
(b) f is even, g is even;
(c) f is odd, g is odd.

What can we say in each of the three cases about $f + g$, $f - g$, $f \cdot g$, and $f \circ g$, in terms of being odd or even?

9.8. Repeat the previous question, replacing even and odd with monotone increasing and monotone decreasing.

9.9. Prove that every function $f: \mathbb{R} \to \mathbb{R}$ can be written as the sum of an odd and an even function.

9.10. Suppose that the function $f: \mathbb{R} \to \mathbb{R}$ never takes on the value of -1. Prove that if

$$f(x+1) = \frac{f(x) - 1}{f(x) + 1}$$

for all x, then f is periodic.

9.11. Suppose that $f: \mathbb{R} \to \mathbb{R}$ is periodic with the rational numbers as its periods. Is it true that there exists a function $g: \mathbb{R} \to \mathbb{R}$ such that $g \circ f$ is the Dirichlet function?

9.12. Prove that the function

$$f(x) = \begin{cases} 0, & \text{if } x \text{ is irrational} \\ q, & \text{if } x = p/q, \ p \in \mathbb{Z}, \ q \in \mathbb{N}^+, \ (p,q) = 1 \end{cases}$$

is not bounded on any interval I.

9.13. Let f be defined in the following way. Let $x \in (0,1]$, and let its infinite decimal expansion be

$$x = 0.a_1 a_2 a_3 \ldots a_{2n-1} a_{2n} \ldots,$$

where, for the sake of unambiguity, we exclude decimal expansions of the form $0.a_1 \ldots a_n 0 \ldots 0 \ldots$. We define a function whose value depends on whether the number

$$0.a_1 a_3 \ldots a_{2n+1} \ldots$$

is rational. Let

$$f(x) = \begin{cases} 0, & \text{if } 0.a_1 a_3 a_5 \ldots a_{2n+1} \ldots \text{ is irrational,} \\ 0.a_{2n} a_{2n+2} \ldots a_{2n+2k} \ldots, & \text{if } 0.a_1 a_3 \ldots a_{2n+1} \ldots \text{ is rational} \\ & \text{and its first period starts at } a_{2n-1}. \end{cases}$$

Prove that f takes on every value in $(0,1)$ on each subinterval of $(0,1)$. (It then follows that it takes on each of those values infinitely often on every subinterval of $(0,1)$.)

9.14. Prove that x^k is strictly convex on $[0, \infty)$ for all integers $k > 1$.

9.15. Prove that if $a_1, \ldots, a_n \geq 0$ and $k > 1$ is an integer, then

$$\frac{a_1 + \cdots + a_n}{n} \leq \sqrt[k]{\frac{a_1^k + \cdots + a_n^k}{n}}.$$

9.16. Prove that

(a) $\sqrt{x}$ is strictly concave on $[0, \infty)$;

(b) $\sqrt[k]{x}$ is strictly concave on $[0, \infty)$ for all integers $k > 1$.

9.17. Prove that if g is convex on $[a, b]$, the range of f over $[a, b]$ is $[c, d]$, f is convex and monotone increasing on $[c, d]$, then $f \circ g$ is convex on $[a, b]$.

9.18. Prove that if f is strictly convex on the interval I, then every line intersects the graph of f in at most two points. (S)

9.3 Appendix: Basics of Coordinate Geometry

Let us consider two perpendicular lines in the plane, the first of which we call the x-axis, the second the y-axis. We call the intersection of the two axes the origin. We imagine each axis to be a number line, that is, for every point on each axis, there is a corresponding real number that gives the distance of the point from the origin; positive in one direction and negative in the other.

If we have a point P in the plane, we get its projection onto the x-axis by taking a line that contains P and is parallel to the y-axis, and taking the point on this line that crosses the x-axis. The value of this point as seen on the number line is called the first coordinate of P. We get the projection onto the y-axis similarly, as well as the second coordinate of P. If the first and second coordinates of P are a and b respectively, then we denote this by $P = (a, b)$. In this sense, we assign an ordered pair of real numbers to every point in the plane. It follows from the geometric properties of the plane that this mapping is bijective. Thus from now on, we identify points in the plane with the ordered pair of their coordinates, and the plane itself with the set $\mathbb{R} \times \mathbb{R} = \mathbb{R}^2$. Instead of saying "the point in the plane whose coordinates are a and b respectively," we say "the point (a, b)."

We can also call points in the plane **vectors**. The length of the vector $u = (a, b)$ is the number $|u| = \sqrt{a^2 + b^2}$. Among vectors, the operations of addition, subtraction, and multiplication by a real number (scalar) are defined: the sum of the vectors (a_1, a_2) and (b_1, b_2) is the vector $(a_1 + b_1, a_2 + b_2)$, their difference is $(a_1 - b_1, a_2 - b_2)$, and the multiple of the vector (a_1, a_2) by a real number t is (ta_1, ta_2).

Addition by a given vector $c \in \mathbb{R}^2$ appears as a translation of the coordinate plane: translating a vector u by the vector c takes us to the vector $u + c$. If $A \subset \mathbb{R}^2$ is a set of vectors, then the set $\{u + c : u \in A\}$ is the set A translated by the vector c.

If $c = (c_1, c_2)$ is a fixed vector that is not equal to 0, then the vectors $t \cdot c = (tc_1, tc_2)$ (where t is an arbitrary real number) cover the points on the line connecting the origin and c. If we translate this line by a vector a, then we get the set $\{a + tc : t \in \mathbb{R}\}$; this is a line going through a.

Let a and b be distinct points. By our previous observations, we know that the set $E = \{a + t(b - a) : t \in \mathbb{R}\}$ is a line that crosses the point a as well as b, since $a + 1 \cdot (b - a) = b$. That is, this is exactly the line that passes through a and b. Let $a = (a_1, a_2)$ and $b = (b_1, b_2)$, where $a_1 \neq b_1$. A point (x, y) is an element of E if and only if

$$x = a_1 + t(b_1 - a_1) \quad \text{and} \quad y = a_2 + t(b_2 - a_2) \tag{9.10}$$

for a suitable real number t. If we isolate t in the first expression and substitute it into the second, we get that

$$y = a_2 + \frac{b_2 - a_2}{b_1 - a_1} \cdot (x - a_1). \tag{9.11}$$

Conversely, if (9.11) holds, then (9.10) also holds with $t = (x - a_1)/(b_1 - a_1)$. This means that $(x, y) \in E$ if and only if (9.11) holds. In short we express this by saying that (9.11) is the equation of the line E.

If $a_2 = b_2$, then (9.11) takes the form $y = a_2$. This coincides with the simple observation that the point (x, y) is on the horizontal line crossing $a = (a_1, a_2)$ if and only if $y = a_2$. If $a_1 = b_1$, then the equation of the line crossing points a and b is clearly $x = a_1$.

For $a, b \in \mathbb{R}^2$ given, every point of the segment $[a, b]$ can be obtained by measuring a vector of length at most $|b - a|$ from the point a in the direction of $b - a$. In other words, $[a, b] = \{a + t(b - a) : t \in [0, 1]\}$.

That is, $(x, y) \in [a, b]$ if and only if there exists a number $t \in [0, 1]$ for which (9.10) holds. In the case that $a_1 < a_2$, the exact condition is that $a_1 \leq x \leq a_2$ and that (9.11) holds.

Chapter 10
Continuity and Limits of Functions

If we want to compute the value of a specific function at some point a, it may happen that we can compute only the values of the function near a. Consider, for example, the distance a free-falling object covers. This is given by the equation $s(t) = g \cdot t^2/2$, where t is the time elapsed, and g is the gravitational constant. Knowing this equation, we can easily compute the value of $s(t)$. If, however, we want to calculate $s(t)$ at a particular time $t = a$ by measuring the time, then we will not be able to calculate the *precise* distance corresponding to this given time; we will obtain only a better or worse approximation—depending on the precision of our instruments. However, if we are careful, we will hope that if we use the value of t that we get from the measurement to recover $s(t)$, the result will be close to the original $s(a)$. In essence, such difficulties always arise when we are trying to find some data with the help of another measured quantity. At those times, we assume that if our measured quantity differs from the real quantity by a very small amount, then the value computed from it will also be very close to its actual value.

In such cases, we have a function f, and we assume that $f(t)$ will be close to $f(a)$ as long as t deviates little from a. We call this property continuity. The precise definition of this idea is the following.

Definition 10.1. Let f be defined on an open interval containing a. The function f is *continuous* at a if for all $\varepsilon > 0$, there exists a $\delta > 0$ (dependent on ε) such that

$$|f(x) - f(a)| < \varepsilon, \ \text{if} \ |x - a| < \delta. \tag{10.1}$$

Continuity of the function f at a graphically means that the graph $G = \operatorname{graph} f$ has the following property: given an arbitrary (narrow) strip

$$\{(x,y) : f(a) - \varepsilon < y < f(a) + \varepsilon\},$$

© Springer New York 2015
M. Laczkovich, V.T. Sós, *Real Analysis*, Undergraduate Texts
in Mathematics, DOI 10.1007/978-1-4939-2766-1_10

there exists an interval $(a - \delta, a + \delta)$ such that the part of G over this interval lies inside the given strip (Figure 10.1).

It is clear that if (10.1) holds with some $\delta > 0$, then it holds for all $\delta' \in (0, \delta)$ as well. In other words, if for some $\varepsilon > 0$ a $\delta > 0$ is "good," then every positive $\delta' < \delta$ is also "good." On the other hand, if a $\delta > 0$ is "good" for some ε, then it is good for all $\varepsilon' > \varepsilon$ as well. If our goal is to deduce whether a function is continuous at a point, then generally there is no need to find the

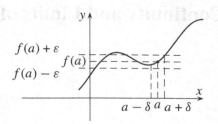

Fig. 10.1

best (i.e., largest) δ for each $\varepsilon > 0$. It suffices to find just *one* such δ. (The situation is analogous to determining the convergence of sequences: we do not have to find the smallest threshold for a given ε; it is enough to find one.)

Examples 10.2. **1.** The constant function $f(x) \equiv c$ is continuous at all a. For every $\varepsilon > 0$, every positive δ is good.
2. The function $f(x) = x$ is continuous at all a. For all $\varepsilon > 0$, the choice of $\delta = \varepsilon$ is good.
3. The function $f(x) = x^2$ is continuous at all a. If $0 < \delta \leq 1$ and $|x - a| < \delta$, then

$$|x^2 - a^2| = |x - a| \cdot |x + a| = |x - a| \cdot |x - a + 2a| < |x - a| \cdot (2|a| + 1).$$

Thus, if

$$\delta = \min\left(1, \frac{\varepsilon}{2|a| + 1}\right),$$

then $|x^2 - a^2| < \delta \cdot (2|a| + 1) < \varepsilon$ whenever $|x - a| < \delta$.
4. The function $f(x) = 1/x$ is continuous at all $a \neq 0$. To see this, for a given $\varepsilon > 0$ we will choose the largest δ, and in fact, we will determine all x where $|f(x) - f(a)| < \varepsilon$. Let, for simplicity's sake, $0 < a < 1$ and $0 < \varepsilon < 1$. Then

$$\frac{1}{x} = \frac{1}{a} + \varepsilon, \text{ if } x = \frac{a}{1 + \varepsilon a},$$

$$\frac{1}{x} = \frac{1}{a} - \varepsilon, \text{ if } x = \frac{a}{1 - \varepsilon a}.$$

Then, because $1/x$ is strictly monotone on $(0, \infty)$,

$$|f(x) - f(a)| < \varepsilon, \text{ if } \frac{a}{1 + \varepsilon a} < x < \frac{a}{1 - \varepsilon a},$$

$$|f(x) - f(a)| \geq \varepsilon, \text{ if } x \notin \left(\frac{a}{1 + \varepsilon a}, \frac{a}{1 - \varepsilon a}\right).$$

It then follows that for a given a and ε,

$$\delta = \min\left(\frac{a}{1-\varepsilon a} - a, a - \frac{a}{1+\varepsilon a}\right) = a - \frac{a}{1+\varepsilon a} = \frac{\varepsilon a^2}{1+\varepsilon a} \qquad (10.2)$$

is the largest δ for which

$$|f(x) - f(a)| < \varepsilon, \quad \text{if} \quad x \in (a - \delta, a + \delta)$$

holds (Figure 10.2).

Later a more general theorem (10.44) will show that the continuity of $f(x) = x^2$ and $f(x) = 1/x$ follows directly from the continuity of $f(x) = x$ without extra work.

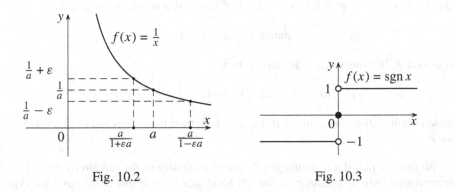

Fig. 10.2   Fig. 10.3

5. The function

$$f(x) = \operatorname{sgn} x = \begin{cases} 1, & \text{if } x > 0 \\ 0, & \text{if } x = 0 \\ -1, & \text{if } x < 0 \end{cases}$$

is continuous at all $a \neq 0$, but at 0, it is not continuous. Since $f \equiv 1$ on the half-line $(0, \infty)$, it follows that for all $a > 0$ and $\varepsilon > 0$, the choice $\delta = a$ is good. Similarly, if $a < 0$, then $\delta = |a|$ is a good δ for all ε. However, $|f(x) - f(0)| = 1$ if $x \neq 0$, so for $a = 0$, if $0 < \varepsilon < 1$, then we cannot choose a good delta (Figure 10.3).

6. The function

$$f(x) = \begin{cases} x, & \text{if } x \text{ is rational} \\ -x, & \text{if } x \text{ is irrational} \end{cases}$$

is continuous at 0, but not continuous at any $x \neq 0$. It is continuous at 0, since for all x, $|f(x) - f(0)| = |x|$, so

$$|f(x) - f(0)| < \varepsilon, \quad \text{if} \quad |x - 0| < \varepsilon,$$

that is, for all $\varepsilon > 0$, the choice $\delta = \varepsilon$ is good.

Now we show that the function is discontinuous at $a \neq 0$. Let a be a rational number. Then at all irrational x (sharing the same sign), $|f(x) - f(a)| > |a|$. This, in turn, means that for $0 < \varepsilon < |a|$, we cannot choose a good δ for that ε. The proof is similar if a is irrational.

(We know that for all $(\delta > 0)$, there exist a rational number and an irrational number in the interval $(a - \delta, a + \delta)$. See Theorems 3.2 and 3.12.) This example is worth noting, for it shows that a function can be continuous at one point but not continuous anywhere else. This situation is perhaps less intuitive than a function that is continuous everywhere except at one point.

7. Let $f(x) = \{x\}$ be the fractional-part function (see Figure 9.1, graph (8)). We see that f is continuous at a if a is not an integer, and it is discontinuous at a if a is an integer. Indeed, $f(a) = a - [a]$ and $f(x) = x - [a]$ if $[a] \leq x < [a+1]$, so $|f(x) - f(a)| = |x - a|$ if $[a] \leq x < [a+1]$. Clearly, if a is not an integer, then

$$\delta = \min(\varepsilon, a - [a], [a] + 1 - a)$$

is a good δ. If, however, a is an integer, then

$$|f(x) - f(a)| = |x - (a-1)| > \tfrac{1}{2}, \quad \text{if} \quad a - \tfrac{1}{2} < x < a,$$

which implies that, for example, if $0 < \varepsilon < 1/2$, then there does not exist a good δ for ε.

We can see that if a is an integer, then the continuity of the function $f(x) = \{x\}$ is prevented only by behavior on the left-hand side of a. If this is the case, we say the function is continuous from the right. We make this more precise below.

Definition 10.3. Let f be defined on an interval $[a, b)$. We say that the function f is *continuous from the right* at a if for all $\varepsilon > 0$, there exists a $\delta > 0$ such that

$$|f(x) - f(a)| < \varepsilon, \quad \text{if} \quad 0 \leq x - a < \delta. \tag{10.3}$$

The function f is *continuous from the left* at a if it is defined on an interval $(c, a]$, and if for all $\varepsilon > 0$, there exists a $\delta > 0$ such that

$$|f(x) - f(a)| < \varepsilon, \quad \text{if} \quad 0 \leq a - x < \delta. \tag{10.4}$$

Exercises

10.1. Show that the function f is continuous at a if and only if it is continuous both from the right and from the left there.

10.2. Show that the function $[x]$ is continuous at a if a is not an integer, and continuous from the right at a if a is an integer.

10.3. For a given $\varepsilon > 0$, find a good δ (according to Definition 10.1) in the following functions.

(a) $f(x) = (x+1)/(x-1)$, $a = 3$;
(b) $f(x) = x^3$, $a = 2$;
(c) $f(x) = \sqrt{x}$, $a = 2$.

10.4. Continuity of the function $f: \mathbb{R} \to \mathbb{R}$ at a can be written using the following expression:

$(\forall \varepsilon > 0)(\exists \delta > 0)(\forall x)(|x - a| < \delta \Rightarrow |f(x) - f(a)| < \varepsilon).$

Consider the following expressions:

$(\forall \varepsilon > 0)(\forall \delta > 0)(\forall x)(|x - a| < \delta \Rightarrow |f(x) - f(a)| < \varepsilon);$
$(\exists \varepsilon > 0)(\forall \delta > 0)(\forall x)(|x - a| < \delta \Rightarrow |f(x) - f(a)| < \varepsilon);$
$(\exists \varepsilon > 0)(\exists \delta > 0)(\forall x)(|x - a| < \delta \Rightarrow |f(x) - f(a)| < \varepsilon);$
$(\forall \delta > 0)(\exists \varepsilon > 0)(\forall x)(|x - a| < \delta \Rightarrow |f(x) - f(a)| < \varepsilon);$
$(\exists \delta > 0)(\forall \varepsilon > 0)(\forall x)(|x - a| < \delta \Rightarrow |f(x) - f(a)| < \varepsilon).$

What properties of f do these express?

10.5. Prove that if f is continuous at a point, then $|f|$ is also continuous there. Conversely, does the continuity of f follow from the continuity of $|f|$?

10.6. Prove that if f and g are continuous at a point a, then $\max(f, g)$ and $\min(f, g)$ are also continuous at a.

10.7. Prove that if the function $f: \mathbb{R} \to \mathbb{R}$ is monotone increasing and assumes every rational number as a value, then it is continuous everywhere.

10.8. Prove that if $f: \mathbb{R} \to \mathbb{R}$ is not constant, continuous, and periodic, then it has a smallest positive period. (H)

10.1 Limits of Functions

Before explaining limits of functions, we discuss three problems that shed light on the need for defining limits and actually suggest a good definition for them. The first two problems raise questions that are of fundamental importance; one can almost say that the theory of analysis was developed to answer these questions. The third problem is a concrete exercise, but it also demonstrates the concept of a limit well.

1. The first problem is defining speed. For constant speed, the value of velocity is $v = s/t$, where s is the length of the path traveled at a time t. Now consider movement with a variable speed, and let $s(t)$ denote the length of the path traveled at a time t. The problem is defining and computing *instantaneous* velocity at a given time t_0. Let $\omega(t)$ denote the average speed on the time interval $[t_0, t]$, that, is let

$$\omega(t) = \frac{s(t) - s(t_0)}{t - t_0}.$$

This is the velocity that in the case of constant speed a point would have moving a distance $s(t) - s(t_0)$ in time $t - t_0$. If, for example,

$$s(t) = t^3 \text{ and } t_0 = 2, \qquad \text{then} \qquad \omega(t) = \frac{t^3 - 8}{t - 2} = (t^2 + 2t + 4).$$

In this case, if t is close to 2, then the average velocity in the interval $[2, t]$ is close to 12. It is clear that we should decide that 12 should be the instantaneous velocity at $t_0 = 2$.

Generally, if we know that there is a value v to which $\omega(t)$ is "very" close for all t close enough to t, then we will call this v the instantaneous velocity at t_0.

2. The second problem concerns defining and finding a tangent line to the graph of a function. Consider the graph of f and a fixed point $P = (a, f(a))$ on it. Let $h_a(x)$ denote the chord crossing P and $(x, f(x))$ (Figure 10.4). We wish to define the tangent line to the curve, that is, the—yet to be precisely defined—line to which these lines tend. Since these lines cross P, their slopes uniquely determine them. The slope of the chord $h_a(x)$ is given by

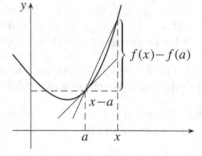

Fig. 10.4

$$m_a(x) = \frac{f(x) - f(a)}{x - a} \qquad (x \neq a).$$

For example, in the case $f(x) = 1/x$,

$$m_a(x) = \frac{\dfrac{1}{x} - \dfrac{1}{a}}{x - a} = -\frac{1}{xa}.$$

It can be seen that if x tends to a, then $m_a(x)$ tends to $-1/a^2$, in a sense yet to be defined. Then it is only natural to say that the tangent at the point P is the line whose slope is $-1/a^2$ that intersects the point $(a, 1/a)$. Thus the equation of the tangent is

$$y = -\frac{1}{a^2}(x - a) + \frac{1}{a}.$$

Generally, if the values

$$m_a(x) = \frac{f(x) - f(a)}{x - a}$$

tend to some value m while x tends to a (again, in a yet to be defined way), then the line crossing P with slope m will be called the tangent of the graph of f at the point P.

3. The third problem is finding the focal length of a spherical mirror. Consider a spherically curved concave mirror with radius r. Light rays traveling parallel to the principal axis at distance x from it will reflect off the mirror and intersect the principal axis at a point P_x. The problem is to find the limit of P_x as x tends to 0 (Figure 10.5).

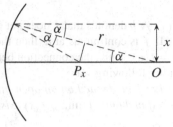

Assuming knowledge of the law of reflection, we have

Fig. 10.5

$$\frac{\overline{OP_x}}{(r/2)} = \frac{r}{\sqrt{r^2 - x^2}}, \quad \text{that is } \overline{OP_x} = \frac{r^2}{2\sqrt{r^2 - x^2}}.$$

We see that if x is close enough to 0, then $\overline{OP_x}$ gets arbitrarily close to the value $r/2$. Thus the focal length of the spherically curved mirror is $r/2$.

In all three cases, the essence of the problem is the following: how do we define what it means to say that "as x tends to a, the values $f(x)$ tend to a number b" or that "the limit of the function f at a is b"? The above three problems indicate that we should define the limit of a function in a way that does not depend on the value of f at $x = a$ or the fact that the function f might not even be defined at $x = a$.

Continuity of f is the precise definition that at places "close" to a, the values of f are "close" to $f(a)$. Then by conveniently modifying the definition of continuity, we get the following definition for the limit of a function.

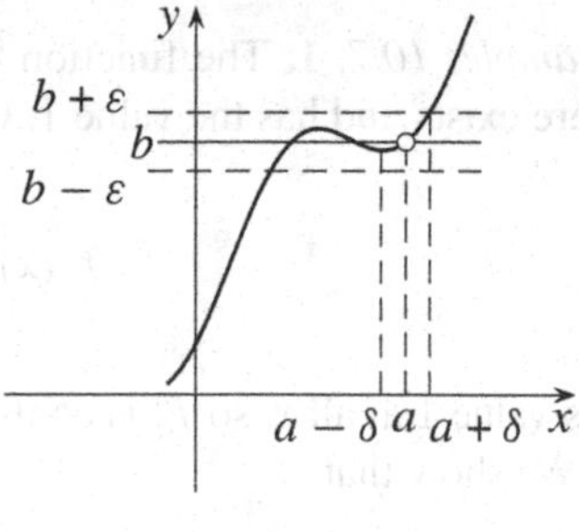

Definition 10.4. Let f be defined on an open interval containing a, excluding perhaps a itself. The *limit* of f at a exists and has the value b if for all $\varepsilon > 0$, there exists a $\delta > 0$ such that

Fig. 10.6

$$|f(x) - b| < \varepsilon, \quad \text{if } 0 < |x - a| < \delta. \quad (10.5)$$

(See Figure 10.6.) Noting the definition of continuity, we see that the following is clearly equivalent:

Definition 10.5. Let f be defined on an open interval containing a, excluding perhaps a itself. The limit of f at a exists and has the value b if the function

$$f^*(x) = \begin{cases} f(x), & \text{if } x \neq a \\ b, & \text{if } x = a \end{cases}$$

is continuous at a.

We denote that f has the limit b at a by

$$\lim_{x \to a} f(x) = b, \quad \text{or} \quad f(x) \to b, \quad \text{if } x \to a;$$

$f(x) \not\to b$ denotes that f does not tend to b.

If f is continuous at a, then the conditions of Definition 10.4 are satisfied with $b = f(a)$. Thus, the connection between continuity and limits can also be expressed in the following way:

Let f be defined on an open interval containing a. The function f is continuous at a if and only if $\lim_{x \to a} f(x)$ *exists and has the value $f(a)$.*

The statement $\lim_{x \to a} f(x) = b$ holds the following interpretation in the graph of f: given an arbitrarily "narrow" strip $\{(x,y) : b - \varepsilon < y < b + \varepsilon\}$, there exists a $\delta > 0$ such that the part of the graph over the set $(a - \delta, a + \delta) \setminus \{a\}$ lies inside the strip.

The following theorem states that limits are unique.

Theorem 10.6. *If* $\lim_{x \to a} f(x) = b$ *and* $\lim_{x \to a} f(x) = b'$, *then $b = b'$.*

Proof. Suppose that $b' \neq b$. Let $0 < \varepsilon < |b' - b|/2$. Then the inequalities $|f(x) - b| < \varepsilon$ and $|f(x) - b'| < \varepsilon$ can never both hold at once (since if they did, then

$$|b - b'| \leq |b - f(x)| + |f(x) - b'| < \varepsilon + \varepsilon = 2\varepsilon$$

would follow), which is impossible.   □

Examples 10.7. **1.** The function $f(x) = \text{sgn}^2 x$ is not continuous at 0, but its limit there exists and has the value 1. Clearly, the function

$$f^*(x) = \begin{cases} \text{sgn}^2 x, & \text{if } x \neq 0 \\ 1, & \text{if } x = 0 \end{cases}$$

has value 1 at all x, so f^* is continuous at 0.
2. We show that

$$\lim_{x \to 2} \frac{x - 2}{x^2 - 3x + 2} = 1.$$

Since $x^2 - 3x + 2 = (x - 1)(x - 2)$, then whenever $x \neq 2$,

$$\frac{x - 2}{x^2 - 3x + 2} = \frac{1}{x - 1}.$$

From this, we have that

$$\left| \frac{x - 2}{x^2 - 3x + 2} - 1 \right| = \left| \frac{1}{x - 1} - 1 \right| = \left| \frac{2 - x}{x - 1} \right|.$$

Since whenever $|x - 2| < 1/2$, $|x - 1| > 1/2$, it follows that $|(2 - x)/(x - 1)| < \varepsilon$ as long as $0 < |2 - x| < \min(\varepsilon/2, 1/2)$.

3. Let

$$f(x) = \begin{cases} 0, & \text{if } x \text{ irrational} \\ \frac{1}{q}, & \text{if } x = \frac{p}{q}, \text{ where } p, q \text{ integers, } q > 0 \text{ and } (p,q) = 1. \end{cases}$$

This function has the following strange properties from the point of view of continuity and limits:

a) *The function f has a limit at every value a, which is 0* (even though f is not identically 0).

b) The function f is continuous at every irrational point.

c) The function f is not continuous at any rational point.

To prove statement a), we need to show that if a is an arbitrary value, then for all $\varepsilon > 0$, there exists a $\delta > 0$ such that

$$|f(x) - 0| < \varepsilon, \qquad \text{if} \qquad 0 < |x - a| < \delta. \tag{10.6}$$

For simplicity, let us restrict ourselves to the interval $(-1, 1)$. Let $\varepsilon > 0$ be given, and choose an integer $n > 1/\varepsilon$. By the definition of f, $|f(x)| < 1/n$ for all irrational numbers x and for all rational numbers $x = p/q$ for which $(p, q) = 1$ and $q > n$. That is, $|f(x) - 0| \geq 1/n$ only on the following points of $(-1, 1)$:

$$0, \pm 1, \pm\frac{1}{2}, \pm\frac{1}{3}, \pm\frac{2}{3}, \dots, \pm\frac{1}{n}, \pm\frac{2}{n}, \dots, \pm\frac{n-1}{n}. \tag{10.7}$$

Now if a is an arbitrary point of the interval $(-1, 1)$, then out of the finitely many numbers in (10.7), there is one that is different from a, and among those, the closest to a. Let this one be p_1/q_1, and let $\delta = |p_1/q_1 - a|$. Then inside the interval $(a - \delta, a + \delta)$, there is no number in (10.7) other than a, so $|f(x)| < 1/n < \varepsilon$ if $0 < |x - a| < \delta = |p_1/q_1 - a|$, that is, for ε, $\delta = |p_1/q_1 - a|$ is a good choice. Since $\varepsilon > 0$ was arbitrary, we have shown that $\lim_{x \to a} f(x) = 0$.

b) Since for an irrational number a we have $f(a) = 0$, $\lim_{x \to a} f(x) = f(a)$ follows, so the function is continuous at every irrational point.

c) Since for a rational number a we have $f(a) \neq 0$, $\lim_{x \to a} f(x) \neq f(a)$ follows, so the function is not continuous at the rational points.

Remark 10.8. The function defined in Example 3 is named—after its discoverer—the *Riemann function.*[1] This function is then continuous at all irrational points, and discontinuous at every rational point. It can be proven, however, that there *does not exist* a function which is continuous at each rational point but discontinuous at all irrational points (see Exercise 10.17).

Similar to the idea of one-sided continuity, there is also the notion of one-sided limits.

---

[1] Georg Friedrich Bernhard Riemann (1826–1866), German mathematician.

Definition 10.9. Let f be defined on an open interval (a,c). The right-hand limit of f at a is b if for every $\varepsilon > 0$, there exists a $\delta > 0$ such that $|f(x) - b| < \varepsilon$ if $0 < x - a < \delta$.

Notation: $\lim_{x \to a+0} f(x) = b$, $\lim_{x \to a^+} f(x) = b$ or $f(x) \to b$, if $x \to a + 0$, or even shorter, $f(a+0) = b$.

Left-hand limits can be defined and denoted in a similar way.

Remark 10.10. In the above notation, $a + 0$ and $a - 0$ are not numbers, simply symbols that allow a more abbreviated notation for expressing the property given in the definition.

The following theorem is clear from the definitions.

Theorem 10.11. $\lim_{x \to a} f(x) = b$ *if and only if both* $f(a+0)$ *and* $f(a-0)$ *exist, and* $f(a+0) = f(a-0) = b$.

Similarly to when we talked about limits of sequences, we will need to define the concept of a function tending to positive and negative infinity.

Definition 10.12. Let f be defined on an open interval containing a, except possibly a itself. The limit of the function f at a is ∞ if for all numbers P, there exists a $\delta > 0$ such that $f(x) > P$ whenever $0 < |x - a| < \delta$.

Notation: $\lim_{x \to a} f(x) = \infty$, or $f(x) \to \infty$ as $x \to a$.

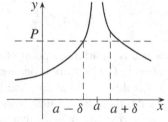

The statement $\lim_{x \to a} f(x) = \infty$ expresses the following property of the graph of f: for arbitrary P, there exists a $\delta > 0$ such that the graph over the set $(a - \delta, a + \delta) \setminus \{a\}$ lies above the horizontal line $y = P$ (Figure 10.7).

We define the limit of f as $-\infty$ at a point a similarly. We will also need the one-sided equivalents for functions tending to ∞.

Fig. 10.7

Definition 10.13. Let f be defined on an open interval (a,c). The right-hand limit of f at a is ∞ if for every number P, there exists a $\delta > 0$ such that $f(x) > P$ if $0 < x - a < \delta$.

Notation: $\lim_{x \to a+0} f(x) = \infty$; $f(x) \to \infty$, if $x \to a + 0$; or $f(a+0) = \infty$.
We define $\lim_{x \to a+0} f(x) = -\infty$ and $\lim_{x \to a-0} f(x) = \pm\infty$ similarly.

But we are not yet done with the different variations of definitions of limits.

Definition 10.14. Let f be defined on the half-line (a, ∞). We say that the limit of f at ∞ is b if for all $\varepsilon > 0$, there exists a K such that $|f(x) - b| < \varepsilon$ if $x > K$ (Figure 10.8).

Notation: $\lim_{x \to \infty} f(x) = b$, or $f(x) \to b$, if $x \to \infty$.

We can similarly define the limit of f at $-\infty$ being b.

Finally, we have one more type, in which both the "place" and the "value" are infinite.

Definition 10.15. Let f be defined on the half-line (a, ∞). We say that the limit of f at ∞ is ∞ if for all P, there exists a K such that $f(x) > P$ if $x > K$.

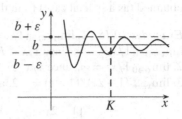

Fig. 10.8

We get three more variations of the last definition if we define the limit at ∞ as $-\infty$, and the limit at $-\infty$ as ∞ or $-\infty$.

To summarize, we have defined the following variants of limits:

$$\lim_{x \to a} f(x) = \begin{cases} b \text{ is finite} \\ \infty \\ -\infty \end{cases} ; \quad \lim_{x \to a \pm 0} f(x) = \begin{cases} b \text{ is finite} \\ \infty \\ -\infty \end{cases} ; \quad \lim_{x \to \pm \infty} f(x) = \begin{cases} b \text{ is finite} \\ \infty \\ -\infty \end{cases} .$$

These are 15 variations of a concept that we can feel is based on a unifying idea. To make this unifying concept clear, we introduce the notion of a **neighborhood**.

Definition 10.16. We define a *neighborhood* of a real number a as an interval of the form $(a - \delta, a + \delta)$, where δ is an arbitrary positive number. Sets of the form $[a, a + \delta)$ and $(a - \delta, a]$ are called the *right-hand*, and respectively the *left-hand neighborhoods* of the real number a.

A *punctured neighborhood* of the real number a is a set of the form $(a - \delta, a + \delta) \setminus \{a\}$, where δ is an arbitrary positive number. Sets of the form $(a, a + \delta)$ and $(a - \delta, a)$ are called the *right-hand* and respectively the *left-hand punctured neighborhoods* of the real number a.

Finally, a *neighborhood* of ∞ is a half-line of the form (K, ∞), where K is an arbitrary real number, while a *neighborhood* of $-\infty$ is a half-line of the form $(-\infty, K)$, where K is an arbitrary real number.

In the above definition, the prefix "punctured" indicates that we leave out the point itself from its neighborhood, or that we "puncture" the set at that point. Now with the help of the concept of neighborhoods, we can give a unified definition for the 15 different types of limits, which also demonstrates the meaning of a limit better.

Definition 10.17. Let α denote the real number a or one of the symbols $a + 0$, $a - 0$, ∞, or $-\infty$. In each case, by the punctured neighborhood of α we mean the punctured neighborhood of a, the right-hand punctured neighborhood of a, the left-hand punctured neighborhood of a, the neighborhood of ∞, or the neighborhood of $-\infty$ respectively. Let β denote the real number b, the symbol ∞, or $-\infty$.

Let f be defined on a punctured neighborhood of α. We say that $\lim_{x \to \alpha} f(x) = \beta$ if for each neighborhood V of β, there exists a punctured neighborhood $\dot{U}$ of α such that $f(x) \in V$ if $x \in \dot{U}$.

We let the reader verify that each listed case of the definition of the limit can be obtained (as a special case) from this definition.

Examples 10.18. **1.** $\lim_{x\to 0+0} 1/x = \infty$, since $1/x > P$, if $0 < x < 1/P$.
$\lim_{x\to 0-0} 1/x = -\infty$, since $1/x < P$, if $1/P < x < 0$.
2. $\lim_{x\to 0} 1/x^2 = \infty$, since $1/x^2 > P$, if $0 < |x| < 1/\sqrt{P}$.
3. $\lim_{x\to\infty}(1-2x)/(1+x) = -2$, as for arbitrary $\varepsilon > 0$,

$$\left|\frac{1-2x}{1+x} - (-2)\right| = \frac{3}{|1+x|} < \varepsilon, \text{ if } x > \frac{3}{\varepsilon} - 1.$$

4. $\lim_{x\to -\infty} 10x/(2x^2 + 3) = 0$, since for arbitrary $\varepsilon > 0$,

$$\left|\frac{10x}{2x^2+3}\right| < \frac{5}{|x|} < \varepsilon, \text{ if } |x| > \frac{5}{\varepsilon},$$

and so the same holds if $x < -5/\varepsilon$.
5. $\lim_{x\to -\infty} x^2 = \infty$, since $x^2 > P$ if $x < -\sqrt{|P|}$.
6. We show that $\lim_{x\to\infty} a^x = \infty$ for all $a > 1$. Let P be given. If $n = [P/(a-1)]+1$ and $x > n$, then using the monotonicity of a^x (Theorem 3.27) and Bernoulli's inequality, we get that

$$a^x > a^n = (1+(a-1))^n \geq 1 + n\cdot(a-1) > \frac{P}{a-1}\cdot(a-1) = P.$$

7. Let $f(x) = x\cdot[1/x]$. Then $\lim_{x\to 0} f(x) = 1$ (Figure 10.9).

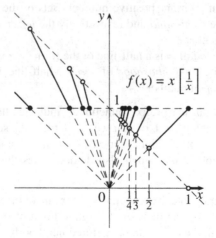

Fig. 10.9

Since

$$f(x) = \begin{cases} 0, & \text{if } x > 1, \\ nx, & \text{if } \frac{1}{n+1} < x \le \frac{1}{n}, \\ -(n+1)x, & \text{if } -\frac{1}{n} < x \le -\frac{1}{n+1}, \\ -x, & \text{if } x \le -1, \end{cases}$$

we can see that if $1/(n+1) < x \le 1/n$ holds, then $n/(n+1) < f(x) \le 1$, that is, $|f(x) - 1| < 1/(n+1)$. Similarly, if $-1/n < x \le -1/(n+1)$, then $1 \le f(x) = -(n+1)x < (n+1)/n$, that is, $|f(x) - 1| < 1/n$. It then follows that

$$|f(x) - 1| < \frac{1}{n}, \text{ ha } 0 < |x| < \frac{1}{n}.$$

That is, for every $\varepsilon > 0$, the choice $\delta = 1/n$ is good if $n > 1/\varepsilon$.

8. Let $f(x) = [x]/x$ (Figure 10.10), that is

$$f(x) = \begin{cases} 0, & \text{if } 0 < x < 1, \\ n/x, & \text{if } n \le x < n+1, \\ -(n+1)/x, & \text{if } -(n+1) \le x < -n. \end{cases}$$

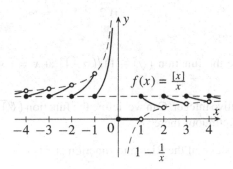

Fig. 10.10

Clearly $\lim_{x \to 0+0} f(x) = 0$; moreover, $\lim_{x \to 0-0} f(x) = \infty$, since $f(x) = -1/x$, if $-1 < x < 0$.

Finally, we show that

$$\lim_{x \to \infty} f(x) = \lim_{x \to -\infty} f(x) = 1.$$

Clearly, $n/(n+1) = 1 - 1/(n+1) < f(x) = n/x \le 1$ if $n \le x < n+1$, that is, $|f(x) - 1| < 1/(n+1)$, if $x \ge n$, so

$$\lim_{x \to \infty} f(x) = 1.$$

Similarly,

$$1 = \frac{-(n+1)}{-(n+1)} \le f(x) = \frac{-(n+1)}{x} < \frac{-(n+1)}{-n} = 1 + \frac{1}{n}$$

if $-(n+1) \le x < -n$; therefore, $|f(x) - 1| < 1/n$ if $x < -n$. Thus we see that $\lim_{x \to -\infty} f(x) = 1$.

Exercises

10.9. In the following functions, the limit β at α exists. Find β, and for each neighborhood V of β, give a neighborhood $\dot{U}$ of α such that $x \in \dot{U}$ implies $f(x) \in V$.

(a) $f(x) = [x]$, $\alpha = 2 + 0$;

(b) $f(x) = \{x\}$, $\alpha = 2 + 0$;

(c) $f(x) = \frac{x}{2x-1}$, $\alpha = \infty$;

(d) $f(x) = \frac{x}{2x-1}$, $\alpha = \frac{1}{2} + 0$;

(e) $f(x) = \frac{x}{x^2-1}$, $\alpha = \infty$;

(f) $f(x) = \frac{x}{x^2-1}$, $\alpha = 1 - 0$.

(g) $f(x) = \sqrt{x+1} - \sqrt{x}$, $\alpha = \infty$;

(h) $\dfrac{\sqrt{x} + \sqrt[3]{x}}{x - \sqrt{x}}$, $\alpha = \infty$;

(i) $\dfrac{x^2 + 5x + 6}{x^2 + 6x + 5}$, $\alpha = \infty$;

(j) $2^{-[1/x]}$, $\alpha = \infty$;

(k) $\sqrt[3]{x^3 + 1} - x$, $\alpha = \infty$.

10.10. Can we define the function $(\sqrt{x} - 1)/(x - 1)$ at $x = 1$ such that it will be continuous there?

10.11. Let n be a positive integer. Can we define the function $(\sqrt[n]{1+x} - 1)/x$ at $x = 0$ such that it will be continuous there?

10.12. Prove that the value of the Riemann function at x is

$$1 - \lim_{n \to \infty} \max(\{x\}, \{2x\}, \ldots, \{nx\}). \tag{H}$$

10.13. Suppose that the function $f \colon \mathbb{R} \to \mathbb{R}$ has a finite limit $b(x)$ at every point $x \in \mathbb{R}$. Show that the function b is continuous everywhere.

10.14. Prove that if $f \colon \mathbb{R} \to \mathbb{R}$ is periodic and $\lim_{x \to \infty} f(x) = 0$, then f is identically 0.

10.15. Prove that there is no function $f \colon \mathbb{R} \to \mathbb{R}$ whose limit at every point x is infinite. (H)

10.16. Prove that if at each point x the limit of the function $f \colon \mathbb{R} \to \mathbb{R}$ equals zero, then there exists a point x at which $f(x) = 0$. (H)

10.17. Prove that if the function $f \colon \mathbb{R} \to \mathbb{R}$ is continuous at every rational point, then there is an irrational point at which it is also continuous. (*S)

10.2 The Transference Principle

The concept of a limit of a function is closely related to the concept of a limit of a sequence. This is expressed by the following theorem. (In the theorem, the meanings of α and β are the same as in Definition 10.17.)

Theorem 10.19. *Let f be defined on a punctured neighborhood $\dot{U}$ of α. We have $\lim_{x \to \alpha} f(x) = \beta$ if and only if whenever a sequence (x_n) satisfies*

$$\{x_n\} \subset \dot{U} \quad \text{and} \quad x_n \to \alpha, \tag{10.8}$$

then $\lim_{n \to \infty} f(x_n) = \beta$.

Proof. Suppose that $\alpha = a$ and $\beta = b$ are finite real numbers. We first prove that if $\lim_{x \to a} f(x) = b$, then $f(x_n) \to b$ whenever $\{x_n\} \subset \dot{U}$ and $x_n \to a$.

Let $\varepsilon > 0$ be given. We know that there exists a $\delta > 0$ such that $|f(x) - b| < \varepsilon$ if $0 < |x - a| < \delta$. If $x_n \to a$, then there exists an n_0 for $\delta > 0$ such that $|x_n - a| < \delta$ for all $n > n_0$. Since by (10.8), x_n is in a punctured neighborhood of a, $x_n \neq a$ for all n. Thus if $n > n_0$, then $0 < |x_n - a| < \delta$, and so $|f(x_n) - b| < \varepsilon$. This shows that $f(x_n) \to b$.

Now we show that if for all sequences satisfying (10.8) we have $f(x_n) \to b$, then $\lim_{x \to a} f(x) = b$. We prove this by contradiction.

Suppose that $\lim_{x \to a} f(x) = b$ does not hold. This means that there exists an $\varepsilon > 0$ for which there does not exist a good $\delta > 0$, that is, in all $(a - \delta, a + \delta) \cap \dot{U}$, there exists an x for which $|f(x) - b| \geq \varepsilon$. This is true for all $\delta = 1/n$ specifically, so for all $n \in \mathbb{N}^+$, there is an $x_n \in \dot{U}$ for which $0 < |x_n - a| < 1/n$ and $|f(x_n) - b| \geq \varepsilon$. The sequence (x_n) we get in this way has the properties that $x_n \to a$ and $x_n \in \dot{U}$, but $f(x_n) \not\to b$. This contradicts our assumptions.

Let us now consider the case that $\alpha = a + 0$ and also $\beta = \infty$. Suppose that $\lim_{x \to a+0} f(x) = \infty$, and let (x_n) be a sequence approaching a from the right.[2] We want to show that $f(x_n) \to \infty$. Let K be fixed. Then there exists a $\delta > 0$ such that $f(x) > K$ for all $a < x < a + \delta$. Since $x_n \to a$ and $x_n > a$, there exists an n_0 such that $a < x_n < a + \delta$ holds for all $n > n_0$. Then $f(x_n) > K$ if $n > n_0$, which shows that $f(x_n) \to \infty$.

Now suppose that $f(x_n) \to \infty$ for all sequences for which $x_n \to a$ and $x_n > a$. We want to show that $\lim_{x \to a+0} f(x) = \infty$. We prove this by contradiction. If the statement is false, then there exists a K for which there is no good δ, that is, for all $\delta > 0$, there exists an $x \in (a, a + \delta)$ such that $f(x) \leq K$. This is also true for each $\delta = 1/n$, so for each $n \in \mathbb{N}^+$, there is an $a < x_n < a + 1/n$ such that $f(x_n) \leq K$. The sequence (x_n) we then get tends to a from the right, and $f(x_n) \not\to \infty$, which contradicts our initial assumption.

The statements for the rest of the cases can be shown similarly. $\qquad\square$

Remark 10.20. We see that a necessary and sufficient condition for the existence of a limit is that for each sequence $x_n \to \alpha$, $\{x_n\} \subset \dot{U}$ (i) $(f(x_n))$ has a limit, and (ii) the value of $\lim_{n \to \infty} f(x_n)$ is independent of the choice of the sequence (x_n).

---

[2] This naturally means that $x_n > a$ for all n, and $x_n \to a$.

Here condition (ii) can actually be dropped, since it follows from (i). We can show this using contradiction: suppose that (i) holds, but (ii) is false. This would mean that there exist a sequence

$$x_n' \to \alpha, \ \{x_n'\} \subset \dot{U}$$

and a sequence

$$x_n'' \to \alpha, \ \{x_n''\} \subset \dot{U}$$

such that

$$\lim_{n \to \infty} f(x_n') \neq \lim_{n \to \infty} f(x_n'').$$

But then the sequence $(x_1', x_1'', x_2', x_2'', \ldots, x_n', x_n'', \ldots)$, which also tends to α, provides us with a sequence

$$(f(x_1'), f(x_1''), f(x_2'), f(x_2''), \ldots, f(x_n'), f(x_n''), \ldots),$$

which would oscillate at infinity, since it would have two subsequences that tend to two different limits. This, however, is impossible by (i).

We call Theorem 10.19 the **transference principle**, since it "transfers" the concept (and value) of limits of functions to limits of sequences.

The significance of the theorem lies in its ability to convert results pertaining to limits of sequences into results about limits of functions. We will also want to use a theorem connecting continuity to limits, but stating this will be much simpler than Theorem 10.19, and we can actually reduce our new theorem to it.

Theorem 10.21. *The function f is continuous at a point a if and only if f is defined in a neighborhood of a and for each sequence (x_n), $x_n \to a$ implies $f(x_n) \to f(a)$.*

Proof. Suppose that f is continuous at a, and let (x_n) be a sequence that tends to a. For a fixed ε, there exists a δ such that $|f(x) - f(a)| < \varepsilon$ for all $x \in (a - \delta, a + \delta)$. Since $x_n \to a$, $x_n \in (a - \delta, a + \delta)$ for all sufficiently large n. Thus $|f(x_n) - f(a)| < \varepsilon$ for all sufficiently large n, which shows that $f(x_n) \to f(a)$.

Now suppose that $f(x_n) \to f(a)$ if $x_n \to a$. By Theorem 10.19, it then follows that $\lim_{x \to a} f(x) = f(a)$, that is, f is continuous at a. $\qquad \square$

For applications in the future, it is worth stating the following theorem.

Theorem 10.22. *The finite limit $\lim_{x \to a-0} f(x)$ exists if and only if for for each sequence $x_n \nearrow a$, $(f(x_n))$ is convergent. That is, in the case of a left-hand limit, it suffices to consider monotone increasing sequences (x_n). A similar statement holds for right-hand limits.*

Proof. To prove the theorem, it is enough to show that if for every sequence $x_n \nearrow a$, $(f(x_n))$ is convergent, then it follows that for all sequences $x_n \to a$ such that $x_n < a$, the sequence $(f(x_n))$ is convergent.

But this is a simple corollary of the fact that every sequence $x_n < a$, $x_n \to a$, can be rearranged into a monotone increasing sequence (x_{k_n}) (See Theorem 6.7 and the remark following it), and if the rearranged sequence $(f(x_{k_n}))$ is convergent, then the original sequence $(f(x_n))$ is also convergent (see Theorem 5.5). □

Another—albeit not as deep—connection between limits of functions and limits of sequences is as follows. An infinite sequence is actually just a function defined on the positive integers. Then the limit of the sequence $a_1 = f(1)$, $a_2 = f(2)$, ..., $a_n = f(n)$, ... is the limit of f at ∞, at least restricted to the set of positive integers. To make this precise, we define the notion of a limit restricted to a set.

Definition 10.23. Let α denote a real number, the symbol ∞, or $-\infty$. We say that α is a *limit point* or an *accumulation point* of the set A if every neighborhood of α contains infinitely many points of A.

Definition 10.24. Let α be a limit point of $A = b \in \mathbb{R}$. The *limit of f at α restricted to A is β* if for each neighborhood V of β there exists a punctured neighborhood $\dot{U}$ of α such that

$$f(x) \in V, \text{ if } x \in \dot{U} \cap A. \tag{10.9}$$

Notation: $\lim\limits_{\substack{x \to \alpha \\ x \in A}} f(x) = \beta$.

Example 10.25. Let f be the Dirichlet function (as in (9) of Examples 9.7). Then for an arbitrary real number c, we have $\lim_{x \to c, x \in \mathbb{Q}} f(x) = 1$ and $\lim_{x \to c, x \in \mathbb{R} \setminus \mathbb{Q}} f(x) = 0$, since $f(x) = 1$ if x is rational, and $f(x) = 0$ if x is irrational. It is clear that every real number c is a limit point of the set of rational numbers as well as the set of irrational numbers, so the above limits have meaning.

Remarks 10.26. **1.** By this definition, the limit $\lim_{x \to a+0} f(x)$ is simply the limit of f at a restricted to the set (a, ∞).
2. The limit of the sequence $a_n = f(n)$ (in the case $n \to \infty$) is simply the limit of f as $x \to \infty$ restricted to the set $\mathbb{N}^+$.
3. If α is not a limit point of the set A, then α has a punctured neighborhood $\dot{U}$ such that $\dot{U} \cap A = \emptyset$. In this case, the conditions of the definition automatically hold (since the statement $x \in \dot{U} \cap A$ is then vacuous). Thus (10.9) is true for *all* neighborhoods V of β. This means that we get a meaningful definition only if α is a limit point of A.

With regard to the above, the following is the most natural definition.

Definition 10.27. Let $a \in A \subset D(f)$. The function f is *continuous at the point a restricted to the set A* if for every $\varepsilon > 0$, there exists a $\delta > 0$ such that $|f(x) - f(a)| < \varepsilon$ if $x \in (a - \delta, a + \delta) \cap A$. If $A = D(f)$, then instead of saying that f is continuous at the point a restricted to $D(f)$, we say that f is *continuous at a for short*.

Remarks 10.28. **1.** By this definition, f being continuous from the right at a can also be defined as f being continuous at a restricted to the interval $[a, \infty)$.
2. In contrast to the definition of defining limits, we have to assume that f is defined at a when we define continuity. However, we *do not* need to assume that a is a limit point of the set A. If $a \in A$, but a is not a limit point of A, then we say that a is an **isolated** point of A. It is easy to see that a is an isolated point of A if and only if there exists a $\delta > 0$ such that $(a - \delta, a + \delta) \cap A = \{a\}$. It then follows that *if a is an isolated point of A, then every function $f: A \to \mathbb{R}$ is continuous at a restricted to A.* Clearly, for every $\varepsilon > 0$, the δ above has the property that $|f(x) - f(a)| < \varepsilon$ if $x \in (a - \delta, a + \delta) \cap A$, since the last condition is satisfied only by $x = a$, and $|f(a) - f(a)| = 0 < \varepsilon$.

The following statements are often used; they are the equivalents to Theorems 5.7, 5.8, and 5.10.

Theorem 10.29 (Squeeze Theorem). *If the inequality $f(x) \leq g(x) \leq h(x)$ is satisfied in a punctured neighborhood of α and $\lim_{x \to \alpha} f(x) = \lim_{x \to \alpha} h(x) = \beta$, then $\lim_{x \to \alpha} g(x) = \beta$.*

Proof. The statement follows from Theorems 5.7 and 10.19. $\qquad\qquad\square$

Theorem 10.30. *If*

$$\lim_{x \to \alpha,\ x \in A} f(x) = b < c = \lim_{x \to \alpha,\ x \in A} g(x),$$

then there exists a punctured neighborhood $\dot{U}$ of α such that $f(x) < g(x)$ for all $x \in \dot{U} \cap A$.

Proof. By the definition of the limit, for $\varepsilon = (c - b)/2$, there exists a punctured neighborhood $\dot{U}_1$ of α such that $|f(x) - b| < (c - b)/2$ whenever $x \in A \cap \dot{U}_1$. Similarly, there is a $\dot{U}_2$ such that $|g(x) - c| < (c - b)/2$ if $x \in A \cap \dot{U}_2$. Let $\dot{U} = \dot{U}_1 \cap \dot{U}_2$. Then $\dot{U}$ is also a punctured neighborhood of α, and if $x \in A \cap \dot{U}$, then $f(x) < b + (c - b)/2 = c - (c - b)/2 < g(x)$. $\qquad\qquad\square$

Theorem 10.31. *If the limits $\lim_{x \to \alpha} f(x) = b$ and $\lim_{x \to \alpha} g(x) = c$ exist, and if $f(x) \leq g(x)$ holds on a punctured neighborhood of α, then $b \leq c$.*

Proof. Let $\dot{U}$ be a punctured neighborhood of α in which $f(x) \leq g(x)$. Suppose that $b > c$. Then by the previous theorem, there exists a punctured neighborhood $\dot{V}$ of α such that $f(x) > g(x)$ if $x \in \dot{V}$. This, however, is impossible, since the set $\dot{U} \cap \dot{V}$ is nonempty, and for each of its elements x, $f(x) \leq g(x)$. $\qquad\qquad\square$

Corollary 10.32. *If f is continuous in a and $f(a) > 0$, then there exists a $\delta > 0$ such that $f(x) > 0$ for all $x \in (a - \delta, a + \delta)$. If $f \geq 0$ holds in a punctured neighborhood of a and f is continuous at a, then $f(a) \geq 0$.* $\square$

Remark 10.33. The converse of Theorem 10.30 is not true: if $f(x) < g(x)$ holds on a punctured neighborhood of α, then we cannot conclude that $\lim_{x \to \alpha} f(x) < \lim_{x \to \alpha} g(x)$. If, for example, $f(x) = 0$ and $g(x) = |x|$, then $f(x) < g(x)$ for all $x \neq 0$, but $\lim_{x \to 0} f(x) = \lim_{x \to 0} g(x) = 0$.

The converse of Theorem 10.31 is also not true: if $\lim_{x \to \alpha} f(x) \leq \lim_{x \to \alpha} g(x)$, then we cannot conclude that $f(x) \leq g(x)$ holds in a punctured neighborhood of α. If, for example, $f(x) = |x|$ and $g(x) = 0$, then $\lim_{x \to 0} f(x) \leq \lim_{x \to 0} g(x) = 0$, but $f(x) > g(x)$ for all $x \neq 0$.

The following theorem for function limits is the analogue to Cauchy's criterion.

Theorem 10.34. *Let f be defined on a punctured neighborhood of α. The limit $\lim_{x \to \alpha} f(x)$ exists and is finite if and only if for all $\varepsilon > 0$, there exists a punctured neighborhood $\dot{U}$ of α such that*

$$|f(x_1) - f(x_2)| < \varepsilon \tag{10.10}$$

if $x_1, x_2 \in \dot{U}$.

Proof. Suppose that $\lim_{x \to \alpha} f(x) = b \in \mathbb{R}$, and let $\varepsilon > 0$ be fixed. Then there exists a punctured neighborhood $\dot{U}$ of α such that $|f(x) - b| < \varepsilon/2$ if $x \in \dot{U}$. It is clear that (10.10) holds for all $x_1, x_2 \in \dot{U}$.

Now suppose that the condition formulated in the theorem holds. If $x_n \to \alpha$ and $x_n \neq \alpha$ for all n, then the sequence $f(x_n)$ satisfies the Cauchy criterion. Indeed, for a given ε, choose a punctured neighborhood $\dot{U}$ such that (10.10) holds for all $x_1, x_2 \in \dot{U}$. Since $x_n \to \alpha$ and $x_n \neq \alpha$ for all n, there exists an N such that $x_n \in \dot{U}$ for all $n \geq N$. If $n, m \geq N$, then by (10.10), we have $|f(x_n) - f(x_m)| < \varepsilon$. Then by Theorem 6.13, the sequence $(f(x_n))$ is convergent.

Fix a sequence $x_n \to \alpha$ that satisfies $x_n \neq \alpha$ for all n, and let $\lim_{n \to \infty} f(x_n) = b$. If $y_n \to \alpha$ is another sequence satisfying $y_n \neq \alpha$ for all n, then the combined sequence $(x_1, y_1, x_2, y_2, \ldots)$ also satisfies this assumption, and so the sequence of function values $s = (f(x_1), f(y_1), f(x_2), f(y_2), \ldots)$ is also convergent. Since $(f(x_n))$ is a subsequence of this, the limit of s can only be b. On the other hand, $(f(y_n))$ is also a subsequence of s, so $f(y_n) \to b$. This holds for all sequences $y_n \to \alpha$ for which $y_n \neq \alpha$ for all n, so by the transference principle, $\lim_{x \to \alpha} f(x) = b$. $\square$

Exercises

10.18. Show that for all functions $f : \mathbb{R} \to \mathbb{R}$, there exists a sequence $x_n \to \infty$ such that $(f(x_n))$ has a limit.

10.19. Let $f : \mathbb{R} \to \mathbb{R}$ be arbitrary. Prove that the limit $\lim_{x \to \infty} f(x)$ exists if and only if whenever the sequences (x_n) and (y_n) tend to ∞, and the limits $\lim_{n \to \infty} f(x_n)$, $\lim_{n \to \infty} f(y_n)$ exist, then they are equal.

10.20. Construct a function $f : \mathbb{R} \to \mathbb{R}$ such that $f(a \cdot n) \to 0$ $(n \to \infty)$ for all $a > 0$, but the limit $\lim_{x \to \infty} f(x)$ does not exist. (H)

10.21. Prove that if $f \colon \mathbb{R} \to \mathbb{R}$ is continuous and $f(a \cdot n) \to 0$ $(n \to \infty)$ for all $a > 0$, then $\lim_{x \to \infty} f(x) = 0$. (∗S)

10.22. Let $f \colon \mathbb{R} \to \mathbb{R}$ be a function such that the sequence $(f(x_n))$ has a limit for all sequences $x_n \to \infty$ for which $x_{n+1}/x_n \to \infty$. Show that the limit $\lim_{x \to \infty} f(x)$ exists.

10.23. Prove that if the points $1/n$ for all $n \in \mathbb{N}^+$ are limit points of H, then 0 is also a limit point of H.

10.24. Show that every point $x \in \mathbb{R}$ is a limit point of each of the sets $\mathbb{Q}$ and $\mathbb{R} \setminus \mathbb{Q}$.

10.25. Prove that (a) every bounded infinite set has a finite limit point; and (b) every infinite set has a limit point.

10.26. Prove that if the set H has only one limit point, then H is countable, and it can be listed as a sequence (x_n) such that $\lim_{n \to \infty} x_n$ exists and is equal to the limit point of H.

10.27. What are the sets that have exactly two limit points?

10.28. Let $f(x) = x$ if x is rational, and $f(x) = -x$ if x is irrational. What can we say about the limits $\lim_{x \to c, x \in \mathbb{Q}} f(x)$, and $\lim_{x \to c, x \in \mathbb{R} \setminus \mathbb{Q}} f(x)$?

10.3 Limits and Operations

In examples up until now, we deduced continuity and values of limits directly from the definitions. The following theorems—which follow from the transference principle and theorems analogous to theorems on limits of sequences—let us deduce the continuity or limits of more complex functions using what we already know about continuity or limits of simple functions.

Theorem 10.35. *Let α denote a number a or one of the symbols $a - 0$, $a + 0$, ∞, $-\infty$. If the finite limits $\lim_{x \to \alpha} f(x) = b$ and $\lim_{x \to \alpha} g(x) = c$ exist, then*

(i) $\lim_{x \to \alpha}(f(x) + g(x))$ *exists and is equal to $b + c$;*
(ii) $\lim_{x \to \alpha}(f(x) \cdot g(x))$ *exists and is equal to $b \cdot c$;*
(iii) *If $c \neq 0$, then $\lim_{x \to \alpha}(f(x)/g(x))$ exists and is equal to b/c.*

Proof. We will go into detail only with (i). Let f and g be defined on a punctured neighborhood $\dot{U}$ of α. By the transference principle, we know that for every sequence $x_n \to \alpha$, $x_n \in \dot{U}$, $\lim_{n \to \infty} f(x_n) = b$ and $\lim_{n \to \infty} g(x_n) = c$. Then by Theorem 5.12,

$$\lim_{n \to \infty} (f(x_n) + g(x_n)) = b + c,$$

which, by the transference principle again, gives us (i). Statements (ii) and (iii) can be shown similarly. $\square$

Remarks 10.36. **1.** In the first half of the proof, we used that the condition for sequences is necessary, while in the second half, we used that it is sufficient for the limit to exist.

2. In statement (iii), we did not suppose that $g(x) \neq 0$. To ensure that the limit $\lim_{x \to a}(f(x)/g(x))$ has meaning, we see that if $c \neq 0$, then there exists a punctured neighborhood $\dot{U}$ of α in which $g(x) \neq 0$. Theorem 10.30 states that if $c < 0$, then in a suitable punctured neighborhood, $g(x) < 0$, and if $c > 0$, then in a suitable $\dot{U}$, $g(x) > 0$.

Examples 10.37. **1.** A simple application of Theorem 10.35 gives

$$\lim_{x \to 1} \frac{x^n - 1}{x^m - 1} = \frac{n}{m}$$

for all $n, m \in \mathbb{N}^+$, since if $x \neq 1$, then

$$\frac{x^n - 1}{x^m - 1} = \frac{x^{n-1} + x^{n-2} + \cdots + 1}{x^{m-1} + x^{m-2} + \cdots + 1}.$$

Here the numerator has n summands, while the denominator has m, each of which tends to 1 if $x \to 1$.

2. Consider the following problem. Find the values of a and b such that

$$\lim_{x \to \infty} \left(\sqrt{x^2 - x + 1} - (ax + b) \right) = 0 \tag{10.11}$$

holds. It is clear that only positive values of a can come into play, and so we have to find limits of type "$\infty - \infty$." The following argument will lead us to an answer:

$$
\begin{aligned}
&\sqrt{x^2 - x + 1} - (ax + b) = \\
&= \frac{(\sqrt{x^2 - x + 1} - (ax + b)) \cdot (\sqrt{x^2 - x + 1} + (ax + b))}{\sqrt{x^2 - x + 1} + (ax + b)} = \\
&= \frac{x^2 - x + 1 - (ax + b)^2}{\sqrt{x^2 - x + 1} + (ax + b)} = \frac{(1 - a^2)x^2 - (2ab + 1)x + (1 - b^2)}{\sqrt{x^2 - x + 1} + (ax + b)} = \\
&= \frac{(1 - a^2)x - (2ab + 1) + \frac{1 - b^2}{x}}{\sqrt{1 - \frac{1}{x} + \frac{1}{x^2}} + a + \frac{b}{x}}.
\end{aligned}
$$

Since

$$\sqrt{1 - \frac{1}{x} + \frac{1}{x^2}} + a + \frac{b}{x} \to a + 1 \quad \text{and} \quad 2ab + 1 + \frac{b^2 - 1}{x} \to 2ab + 1,$$

if $x \to \infty$, the quotient can tend to 0 only if $1 - a^2 = 0$, that is, considering $a > 0$, only if $a = 1$. In this case,

$$\lim_{x \to \infty} \left(\sqrt{x^2 - x + 1} - (x + b) \right) = -\frac{2b + 1}{2}.$$

This is 0 if $b = -1/2$. Thus (10.11) holds if and only if $a = 1$ and $b = -1/2$.

Definition 10.38. If

$$\lim_{x \to \infty} (f(x) - (ax + b)) = 0,$$

then we say that the *asymptote* of $f(x)$ at ∞ is the linear function $ax + b$. (Or from a geometric viewpoint: the asymptote of the curve $y = f(x)$ at ∞ is the line $y = ax + b$.) The asymptote of $f(x)$ at $-\infty$ is defined similarly.

Remark 10.39. In Theorems 5.12, 5.14 and 5.16, we saw that we can interchange the order of taking limits of sequences and performing basic operations on them. We also proved this in numerous cases in which the limit of one (or both) of the sequences considered is infinity. These cases apply to the corresponding statements for functions (with identical proofs), just as with the cases with finite limits. So for example:

If $\lim_{x \to a} f(x) = b$ is finite and $\lim_{x \to a} g(x) = \infty$, then $\lim_{x \to a}(f(x) + g(x)) = \infty$.

Or if $\lim_{x \to a} f(x) = a \neq 0$, $\lim_{x \to a} g(x) = 0$ and $g \neq 0$ in a punctured neighborhood of a, then $\lim_{x \to a} |f(x)/g(x)| = \infty$. If we also know that f/g has unchanging sign, then it follows that $\lim_{x \to a} f(x)/g(x) = \infty$ or $\lim_{x \to a} f(x)/g(x) = -\infty$ depending on the sign.

Example 10.40. Let x_1 and x_2 be the roots of the equation $ax^2 + bx + c = 0$. Find the limits of x_1 and x_2 if b and c are fixed, $b \neq 0$, and $a \to 0$.

Let

$$x_1 = \frac{-b + \sqrt{b^2 - 4ac}}{2a} \quad \text{and} \quad x_2 = \frac{-b - \sqrt{b^2 - 4ac}}{2a}.$$

We can suppose $b > 0$ without sacrificing generality. We see that then $\lim_{a \to 0} x_1$ is a limit of type $0/0$. A simple rearrangement yields

$$x_1 = \frac{\sqrt{b^2 - 4ac} - b}{2a} = \frac{-4ac}{2a(\sqrt{b^2 - 4ac} + b)} = \frac{-4c}{2(\sqrt{b^2 - 4ac} + b)} \to -\frac{c}{b}.$$

For the other root, $\lim_{a \to 0}(-b - \sqrt{b^2 - 4ac}) = -2b < 0$ implies

$$\lim_{a \to 0+0} x_2 = -\infty, \quad \text{and} \quad \lim_{a \to 0-0} x_2 = \infty.$$

We note that critical limits can arise in the case of functions as well—in fact, exactly in the same situations as for sequences. Critical limits occur when the limits of f and g on their own do not determine the limits of $f + g$, $f \cdot g$, or f/g. So for example, if $\lim_{x \to 0} f(x) = \lim_{x \to 0} g(x) = 0$, then the limit $\lim_{x \to 0} f(x)/g(x)$ can be finite or infinite, or it might not exist. Clearly, $\lim_{x \to 0} x/x = 1$, $\lim_{x \to 0} x/x^3 = \infty$, $\lim_{x \to 0} -x/x^3 = -\infty$, and if f is the Riemann function, then the limit $\lim_{x \to 0} f(x)/x$ does not exist (show this). The examples that illustrate critical limits for sequences can usually be transformed into examples involving functions without much difficulty.

The following theorem addresses limits of compositions of functions.

Theorem 10.41. *Suppose that* $\lim_{x \to \alpha} g(x) = \gamma$ *and* $\lim_{t \to \gamma} f(t) = \beta$. *If* $g(x) \neq \gamma$ *on a punctured neighborhood of* α, *or* γ *is finite and* f *continuous at* γ, *then* $\lim_{x \to \alpha} f(g(x)) = \beta$.

Proof. For brevity, we denote the punctured neighborhoods of α by $\dot{U}(\alpha)$.

We have to show that for every neighborhood V of β, we can find a $\dot{U}(\alpha)$ such that if $x \in \dot{U}(\alpha)$, then $f(g(x)) \in V$. Since $\lim_{t \to \gamma} f(t) = \beta$, there exists a $\dot{W}(\gamma)$ such that $f(t) \in V$ whenever $t \in \dot{W}(\gamma)$. Let $W(\gamma) = \dot{W}(\gamma)$ if $\gamma = \infty$ or $-\infty$, and let $W(\gamma) = \dot{W}(\gamma) \cup \{\gamma\}$ if γ is finite. Then $W(\gamma)$ is a neighborhood of γ, so by $\lim_{x \to \alpha} g(x) = \gamma$, there exists a $U_1(\alpha)$ such that if $x \in U_1(\alpha)$, then $g(x) \in W(\gamma)$.

If we know that $g(x) \neq \gamma$ in a punctured neighborhood $\dot{U}_2(\alpha)$ of α, then $\dot{U}(\alpha) = U_1(\alpha) \cap \dot{U}_2(\alpha)$ is a punctured neighborhood of α such that whenever $x \in \dot{U}(\alpha)$, $g(x) \in W(\gamma)$ and $g(x) \neq \gamma$, that is, $g(x) \in \dot{W}(\gamma)$, and thus $f(g(x)) \in V$.

Now let f be continuous at γ. Then $f(t) \in V$ for all $t \in W(\gamma)$, since $f(\gamma) = \beta \in V$. Thus if $x \in U_1(\alpha)$, then $g(x) \in W(\gamma)$ and $f(g(x)) \in V$.   □

Remark 10.42. The crucial condition in the theorem is that either $g(x) \neq \gamma$ in $\dot{U}(\alpha)$, or that f is continuous at γ. If neither of these holds, then the statement of the theorem is not always true, which is illustrated by the following example. Let g be the Riemann function (that is, function 3 in Example 10.7), and let

$$f(t) = \begin{cases} 1, & \text{if } t \neq 0, \\ 0, & \text{if } t = 0. \end{cases}$$

It is easy to see that then $(f \circ g)(x) = f(g(x))$ is exactly the Dirichlet function. In Example 10.7, we saw that $\lim_{x \to 0} g(x) = 0$. On the other hand, it is clear that $\lim_{t \to 0} f(t) = 1$, while the limit $\lim_{x \to 0} f(g(x))$ does not exist.

If $\gamma = \infty$, then the condition $g(x) \neq \gamma$ ($x \in \dot{U}(\alpha)$) automatically holds. This means that if $\lim_{x \to \alpha} g(x) = \infty$ and $\lim_{t \to \infty} f(t) = \beta$, then $\lim_{x \to \alpha} f(g(x)) = \beta$ holds without further assumptions. A special case of the converse of this is also true.

Theorem 10.43. *Let* f *be defined in a neighborhood of* ∞. *Then*

$$\lim_{x \to 0+0} f\left(\frac{1}{x}\right) = \lim_{x \to \infty} f(x) \tag{10.12}$$

in the sense that if one limit exists, then so does the other, and they are equal.

Proof. Since $\lim_{x \to 0+0} 1/x = \infty$, as we saw before, the statement is true whenever the right-hand side exists. If the left-hand side exists, then we apply Theorem 10.41 with the choices $\alpha = \infty$, $g(x) = 1/x$, and $\gamma = 0$. Since $g(x)$ is nowhere zero,

$$\lim_{x \to \infty} h(1/x) = \lim_{x \to 0+0} h(x)$$

whenever the right-hand side exists. If we apply this to the function $h(x) = f(1/x)$, then we get (10.12).   □

As an application of Theorems 10.35 and 10.41, we immediately get the following theorem.

Theorem 10.44.

(i) *If f and g are continuous at a, then $f + g$ and $f \cdot g$ are also continuous at a, and if $g(a) \neq 0$, then f/g is continuous at a as well.*
(ii) *If $g(x)$ is continuous at a and $f(t)$ is continuous at $g(a)$, then $f \circ g$ is also continuous at a.*

We prove the following regarding continuity of the inverse of a function. (We would like to draw the reader's attention to the fact that in the theorem, we do not suppose that the function itself is continuous.)

Theorem 10.45. *Let f be strictly monotone increasing (decreasing) on the interval I. Then*

(i) *The inverse of f, f^{-1}, is strictly monotone increasing (decreasing) on the set $f(I)$, and moreover,*
(ii) *f^{-1} is continuous at every point of $f(I)$ restricted to the set $f(I)$.*

Proof. We can suppose that the interval I is nondegenerate, since otherwise, the statement is trivial: on a set of one element, every function is strictly increasing, decreasing, and continuous (restricted to the set). We can also assume that f is strictly increasing, since the decreasing case can be proved the same way.

Since if $u_1, u_2 \in I$, $u_1 < u_2$, then $f(u_1) < f(u_2)$, f is one-to-one, so it has an inverse. Let $x_1, x_2 \in f(I)$, and suppose that $x_1 < x_2$, but $f^{-1}(x_1) \geq f^{-1}(x_2)$. Since f is monotone increasing, we would have

$$x_1 = f\left(f^{-1}(x_1)\right) \geq f\left(f^{-1}(x_2)\right) = x_2,$$

which is impossible. Thus we see that $f^{-1}(x_1) < f^{-1}(x_2)$ if $x_1 < x_2$, that is, f^{-1} is monotone increasing.

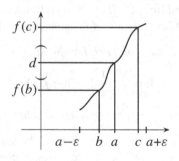

Fig. 10.11

Let $d \in f(I)$ be arbitrary. Then $d = f(a)$ for a suitable $a \in I$. We show that f^{-1} is continuous at d restricted to $f(I)$. Let $\varepsilon > 0$ be fixed. Since $f^{-1}(d) = a$, we have to show that there exists a $\delta > 0$ such that

$$f^{-1}(x) \in (a - \varepsilon, a + \varepsilon) \quad \text{if } x \in (d - \delta, d + \delta) \cap f(I). \tag{10.13}$$

Consider first the case that a is an interior point of I (that is, not an endpoint). Then we can choose points $b, c \in I$ such that $a - \varepsilon < b < a < c < a + \varepsilon$ (see Figure 10.11). Since f is strictly monotone increasing, $f(b) < f(a) < f(c)$, that is, $f(b) < d < f(c)$. Choose a positive δ such that $f(b) < d - \delta < d + \delta < f(c)$. If $x \in (d - \delta, d + \delta) \cap f(I)$, then by the strict monotonicity of f^{-1},

$$b = f^{-1}(f(b)) < f^{-1}(x) < f^{-1}(f(c)) = c.$$

Thus $f^{-1}(x) \in (b, c) \subset (a - \varepsilon, a + \varepsilon)$, which proves (10.13).

If a is the left endpoint of a, then choose a point $c \in I$ such that $a < c < a + \varepsilon$. Then $d = f(a) < f(c)$, so for a suitable $\delta > 0$, $d + \delta < f(c)$. If $x \in (d - \delta, d + \delta) \cap f(I)$, then

$$a \leq f^{-1}(x) < f^{-1}(f(c)) = c.$$

Thus $f^{-1}(x) \in [a, c) \subset (a - \varepsilon, a + \varepsilon)$, so (10.13) is again true. The argument for a being the right endpoint of I is the same. $\square$

Just as with sequences, we define order of magnitude and asymptotic equivalence for functions as well.

Definition 10.46. Suppose that $\lim_{x \to \alpha} f(x) = \lim_{x \to \alpha} g(x) = \infty$. If

$$\lim_{x \to \alpha} \frac{f(x)}{g(x)} = \infty,$$

then we say that f *tends to infinity faster than g* (or g *tends to infinity more slowly than f*). We can also say that the *order of magnitude of f is greater than the order of magnitude of g*.

Similarly, if $\lim_{x \to \alpha} f(x) = \lim_{x \to \alpha} g(x) = 0$ and $\lim_{x \to \alpha} f(x)/g(x) = 0$, then we say that f *tends to zero faster than g* (or g *tends to zero more slowly than f*).

The statement $\lim_{x \to \alpha} f(x)/g(x) = 0$ is sometimes denoted by $f(x) = o(g(x))$ (read: $f(x)$ is little-o of $g(x)$).

If we know only that $f(x)/g(x)$ is bounded in $\dot{U}(\alpha)$, then we denote this by $f(x) = O(g(x))$ (read: $f(x)$ is big-O of $g(x)$).

Example 10.47. We show that if $x \to \infty$, then the function a^x tends to infinity faster than x^k for every $a > 1$ and $k > 0$. To see this, we have to show that $\lim_{x \to \infty} a^x / x^k = \infty$. Let P be a fixed real number. Since the sequence (a^n / n^k) tends to infinity, there exists an n_0 such that $a^n / n^k > 2^k \cdot P$ if $n \geq n_0$. Now if $x \geq 1$, then $a^x \geq a^{[x]}$ and $x^k \leq (2 \cdot [x])^k = 2^k \cdot [x]^k$. Thus if $x \geq n_0$, then $[x] \geq n_0$, and so

$$\frac{a^x}{x^k} \geq \frac{a^{[x]}}{2^k \cdot [x]^k} > P,$$

which concludes the argument.

Consider the following functions:

$$\ldots, \sqrt[n]{x}, \ldots, \sqrt[3]{x}, \sqrt{x}, x, x^2, x^3, \ldots, x^n, \ldots, 2^x, 3^x, \ldots, n^x, \ldots, x^x. \qquad (10.14)$$

It is easy to see that if $x \to \infty$, then in the above arrangement, each function tends to infinity faster than the functions to the left of it. (For the function 2^x, this follows from the previous example.)

Definition 10.48. Suppose that

$$\lim_{x \to \alpha} f(x) = \lim_{x \to \alpha} g(x) = 0 \qquad \text{or} \qquad \lim_{x \to \alpha} f(x) = \lim_{x \to \alpha} g(x) = \pm\infty.$$

If

$$\lim_{x \to \alpha} \frac{f(x)}{g(x)} = 1,$$

then we say that f and g are *asymptotically equal*. We denote this by $f \sim g$ if $x \to \alpha$.

Example 10.49. We show that if $x \to 0$, then $\sqrt{1+x}-1 \sim x/2$. Indeed,

$$\frac{\sqrt{1+x}-1}{(x/2)} = \frac{(\sqrt{1+x}-1)(\sqrt{1+x}+1)}{(x/2)\cdot(\sqrt{1+x}+1)} = \frac{2}{\sqrt{1+x}+1} \to 1$$

if $x \to 0$.

Exercises

10.29. $\lim_{x \to 7} \frac{\sqrt{x+2}-\sqrt[3]{x+20}}{\sqrt[4]{x+9}-2} = ?$

10.30. $\lim_{x \to 1} \frac{\sqrt[359]{x}-1}{\sqrt[5]{x}-1} = ?$

10.31. $\lim_{x \to \infty} x \cdot \left[\sqrt{x^2+2x} - 2\sqrt{x^2+x} + x \right] = ?$

10.32. $\lim_{x \to \infty} x^{3/2} \cdot \left[\sqrt{x+2} + \sqrt{x} - 2\sqrt{x+1} \right] = ?$

10.33. $\lim_{x \to 1} \frac{(1-x)(1-\sqrt{x})(1-\sqrt[3]{x})\cdot\ldots\cdot(1-\sqrt[n]{x})}{(1-x)^n} = ?$

10.34. Prove that

$$\lim_{x \to -d/c+0} \frac{ax+b}{cx+d} = \begin{cases} \infty, & \text{if } bc - ad > 0 \\ -\infty, & \text{if } bc - ad < 0, \end{cases}$$

$$\lim_{x \to -d/c-0} \frac{ax+b}{cx+d} = \begin{cases} -\infty, & \text{if } bc - ad > 0 \\ \infty, & \text{if } bc - ad < 0, \end{cases}$$

and

$$\lim_{x \to \pm\infty} \frac{ax+b}{cx+d} = \frac{a}{c} \qquad (c \neq 0).$$

10.35. Let $p(x)$ be a polynomial of degree at most n; that is, let

$$p(x) = a_n x^n + a_{n-1} x^{n-1} + \cdots + a_1 x + a_0.$$

Prove that if

$$\lim_{x \to 0} \frac{p(x)}{x^n} = 0,$$

then $p(x) = 0$ for all x.

10.36. Construct functions f and g such that $\lim_{x \to 0} f(x) = \infty$, $\lim_{x \to 0} g(x) = -\infty$, and moreover,

(a) $\lim_{x \to 0}(f(x) + g(x))$ exists and is finite;
(b) $\lim_{x \to 0}(f(x) + g(x)) = \infty$;
(c) $\lim_{x \to 0}(f(x) + g(x)) = -\infty$;
(d) $\lim_{x \to 0}(f(x) + g(x))$ does not exist.

10.37. (a) If at $x = a$, f is continuous while g is not, then can $f + g$ be continuous here?
(b) If at $x = a$, neither f nor g is continuous, then can $f + g$ be continuous here?

10.38. Suppose that $\varphi \colon \mathbb{R} \to \mathbb{R}$ is strictly monotone, and let $R(\varphi) = \mathbb{R}$. Prove that if $f \colon \mathbb{R} \to \mathbb{R}$ and if $f \circ \varphi$ is continuous, then f is continuous.

10.39. What can we say about the continuity of $f \circ g$ if in $\mathbb{R}$,

(i) both f and g are continuous,
(ii) f is continuous, g is not continuous,
(iii) neither f nor g is continuous.

10.40. Give an example of functions $f, g \colon \mathbb{R} \to \mathbb{R}$ such that

$$\lim_{x \to 0} f(x) = \lim_{x \to 0} g(x) = 0, \quad \text{but} \quad \lim_{x \to 0} f(g(x)) = 1. \tag{S}$$

10.41. Prove that if f is the Riemann function, then the limit $\lim_{x \to 0} f(x)/x$ does not exist.

10.42. Is the following statement true? If $f_1, f_2, \ldots$ is an infinite sequence of continuous functions and $F(x) = \inf_k\{f_k(x)\}$, then $F(x)$ is also continuous.

10.43. Is it true that if f is strictly monotone on the set $A \subset \mathbb{R}$, then its inverse is continuous on the set $f(A)$?

10.44. Let a and b be positive numbers. Prove that (a) if $x \to \infty$, then the order of magnitude of x^a is greater than the order of magnitude of x^b if and only if $a > b$; and (b) if $x \to 0 + 0$, then the order of magnitude of x^{-a} is greater than the order of magnitude of x^{-b} if and only if $a > b$.

10.45. Prove that in the ordering (10.14), each function tends to infinity faster than the functions to the left of it.

10.46. Let $a > 1$ and $k > 0$. Prove that if $x \to \infty$, then the order of magnitude of $a^{\sqrt{x}}$ is larger than the order of magnitude of x^k.

10.47. Suppose that all of the functions $f_1, f_2, \ldots$ tend to infinity as $x \to \infty$. Prove that there is a function f whose order of magnitude is greater than the order of magnitude of each f_n.

10.4 Continuous Functions in Closed and Bounded Intervals

The following theorems show that if f is continuous on a closed and bounded interval, then it automatically follows that f possesses numerous other important properties.

Definition 10.50. Let $a < b$. The function f *is continuous in the interval* $[a,b]$ if it is continuous at all $x \in (a,b)$, is continuous from the right at a, and continuous from the left at b.

More generally:

Definition 10.51. Let $A \subset D(f)$. The function f is *continuous on the set A* if at each $x \in A$, it is continuous when restricted to the set A.

From now on, we denote the set of all continuous functions on the closed and bounded interval $[a,b]$ by $C[a,b]$.

Theorem 10.52. *If $f \in C[a,b]$, then f is bounded on $[a,b]$.*

Proof. We prove the statement by contradiction. Suppose that f is not bounded on $[a,b]$. Then for no number K does $|f(x)| \leq K$ for all $x \in [a,b]$ hold. Namely, for every n, there exists an $x_n \in [a,b]$ such that $|f(x_n)| > n$.

Consider the sequence (x_n). This is bounded, since each of its terms falls inside $[a,b]$, so it has a convergent subsequence (x_{n_k}). Let $\lim_{k \to \infty} x_{n_k} = \alpha$. Since $\{x_n\} \subset [a,b]$, α is also in $[a,b]$. However, f is continuous at α, so by the transference principle, the sequence $(f(x_{n_k}))$ is convergent (and tends to $f(\alpha)$). It follows that the sequence $(f(x_{n_k}))$ is bounded. This, however, contradicts that $|f(x_{n_k})| > n_k$ of all k. $\qquad \square$

Remark 10.53. In the theorem, it is necessary to suppose that the function f is continuous on a *closed and bounded* interval. Dropping either assumption leads to the statement of the theorem not being true. So for example, the function $f(x) = 1/x$ is continuous on the bounded interval $(0,1]$, but f is not bounded there. The function $f(x) = x^2$ is continuous on $[0, \infty)$, but is also not bounded there.

Definition 10.54. Let f be defined on the set A. If the image of A, $f(A)$, has a greatest element, then we call it the (global) *maximum* of the function f over A, and denote it by $\max f(A)$ or $\max_{x \in A} f(x)$. If $a \in A$ and $f(a) = \max f(A)$, then we say that a is an *global maximum point* of f over A.

If $f(A)$ has a smallest element, then we call this the (global) *minimum* of the function f over A, and denote it by $\min f(A)$ or $\min_{x \in A} f(x)$. If $b \in A$ and $f(b) = \min f(A)$, then we say that b is a *global minimum point* of f over A.

We collectively call the global maximum and minimum points *global extrema*.

Naturally, on a set A, a function can have numerous maximum (or minimum) points.

A set of numbers can have a maximum (or minimum) only if it is bounded from above (below). However, as we have already seen, not every bounded set has a maximum or minimum element. If the image of a set A under a function f is bounded, that alone does not guarantee that there will be a largest or smallest value of f.

For example, the fractional part $f(x) = \{x\}$ is bounded on $[0, 2]$. The least upper bound of the image of this set is 1, but the function is never equal to 1. Thus this function does not have a maximum point in $[0, 2]$.

The following theorem shows that this cannot happen with a continuous function on a closed and bounded interval.

Theorem 10.55 (Weierstrass's Theorem). *If $f \in C[a, b]$, then there exist $\alpha \in [a, b]$ and $\beta \in [a, b]$, such that $f(\alpha) \leq f(x) \leq f(\beta)$ if $x \in [a, b]$. In other words, a continuous function always has an absolute maximum and minimum point over a closed and bounded interval.*

We give two proofs of the theorem.

Proof I. By Theorem 10.52, $f([a, b])$ is bounded. Let the least upper bound of $f([a, b])$ be M. If $M \in f([a, b])$, then this means that $M = \max_{x \in [a, b]} f(x)$. Thus we have only to show that $M \notin f([a, b])$ is impossible. We prove this by contradiction. If $M \notin f([a, b])$, then the values of the function $F(x) = M - f(x)$ are positive for all $x \in [a, b]$. Thus the function $1/F$ is also continuous on $[a, b]$ (see Theorem 10.44), so it is bounded there (by Theorem 10.52 again). Then there exists a $K > 0$ such that

$$\frac{1}{M - f(x)} \leq K$$

for all $x \in [a, b]$. Taking reciprocals of both sides and rearranging (and using that $M - f(x) > 0$ everywhere) yields the inequality

$$f(x) \leq M - \frac{1}{K}$$

if $x \in [a, b]$. However, this contradicts M being the least upper bound of $f([a, b])$.

The existence of $\min f[a, b]$ can be proven similarly. (Or we can reduce it to the statement regarding the maximum if we apply that to $-f$ instead of f.) $\square$

Proof II. Once again, let $M = \sup f([a,b])$; we will show that $M \in f([a,b])$. If n is a positive integer, then $M - (1/n)$ is not an upper bound of $f([a,b])$, since M was the least upper bound. Thus there exists a point $x_n \in [a,b]$ such that $f(x_n) > M - (1/n)$. The sequence (x_n) is bounded (since each of its terms falls in $[a,b]$), so it has a convergent subsequence (x_{n_k}). Let $\lim_{k\to\infty} x_{n_k} = \alpha$. Since $\{x_n\} \subset [a,b]$, we have $\alpha \in [a,b]$. Now f is continuous at α, so by the transference principle, $f(x_{n_k}) \to f(\alpha)$. Since

$$M - \frac{1}{n_k} < f(x_{n_k}) \le M$$

for all k, by the squeeze theorem, $M \le f(\alpha) \le M$, that is, $f(\alpha) = M$. This shows that $M \in f([a,b])$.

The existence of $\min f[a,b]$ can be proven similarly.   □

Remark 10.56. Looking at the conditions of the theorem, it is again essential that we are talking about continuous functions in *closed, bounded* intervals. As we saw in Remark 10.53, if f is continuous on an open interval (a,b), then it might happen that $f((a,b))$ is not bounded from above, and then $\max f((a,b))$ does not exist. But this can occur even if f is bounded. For example, the function $f(x) = x$ is continuous and bounded on the open interval $(0,1)$, but it does not have a greatest value there.

It is equally important for the interval to be bounded. This is illustrated by the function $f(x) = -1/(1+x^2)$, which is bounded in $[0,\infty)$, but does not have a greatest value.

Another important property of continuous functions over closed and bounded intervals is given by the following theorem.

Theorem 10.57 (Bolzano–Darboux[3] Theorem). *If $f \in C[a,b]$, then f takes on every value between $f(a)$ and $f(b)$ on the interval $[a,b]$.*

We give two proofs of this theorem again, since both proofs embody ideas that are frequently used in analysis.

Proof I. Without loss of generality, we may assume that $f(a) < c < f(b)$. We will prove the existence of a point $\alpha \in [a,b]$ such that the function takes on values not larger than c and not smaller than c in every neighborhood of the point. Then by the continuity of f at α, it follows that $f(\alpha) = c$.

We will define α as the intersection of a sequence of nested closed intervals $I_0 \supset I_1 \supset \ldots$. Let $I_0 = [a,b]$.

$$\text{If} \quad f\left(\frac{a+b}{2}\right) \le c, \text{ then let } I_1 = [a_1,b_1] = \left[\frac{a+b}{2}, b\right],$$

$$\text{but if } f\left(\frac{a+b}{2}\right) > c, \text{ then let } I_1 = [a_1,b_1] = \left[a, \frac{a+b}{2}\right].$$

We continue this process. If $I_n = [a_n, b_n]$ is already defined

---

[3] Jean Gaston Darboux (1842–1917), French mathematician.

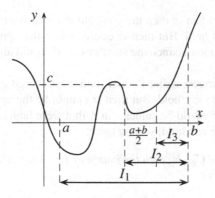

Fig. 10.12

and $f\left(\dfrac{a_n+b_n}{2}\right) \le c$, then let $I_{n+1} = [a_{n+1},b_{n+1}] = \left[\dfrac{a_n+b_n}{2}, b_n\right]$,

but if $f\left(\dfrac{a_n+b_n}{2}\right) > c$, then let $I_{n+1} = [a_{n+1},b_{n+1}] = \left[a_n, \dfrac{a_n+b_n}{2}\right]$.

The interval sequence $I_0 \supset I_1 \supset \dots$ is defined such that

$$f(a_n) \le c < f(b_n) \tag{10.15}$$

holds for all n (Figure 10.12). Since $|I_n| = (b-a)/2^n \to 0$, the interval sequence (I_n) has exactly one shared point. Let this be α. Clearly,

$$\alpha = \lim_{n\to\infty} a_n = \lim_{n\to\infty} b_n,$$

and since f is continuous in α,

$$\lim_{n\to\infty} f(a_n) = \lim_{n\to\infty} f(b_n) = f(\alpha). \tag{10.16}$$

But by (10.15),

$$\lim_{n\to\infty} f(a_n) \le c \le \lim_{n\to\infty} f(b_n),$$

that is, (10.16) can hold only if $f(\alpha) = c$.   □

Proof II. Let us suppose once again that $f(a) < c < f(b)$, and let

$$A = \{x \in [a,b] : f(x) < c\}.$$

The set A is then bounded and nonempty, since $a \in A$. Thus $\alpha = \sup A$ exists, and since $A \subset [a,b]$, we have $\alpha \in [a,b]$. Since f is continuous at a and $f(a) < c$, $f(x) < c$ holds over a suitable interval $[a, a+\delta)$, and so $a < \alpha$. Moreover, since f is continuous at b and $f(b) > c$, $f(x) > c$ holds on a suitable interval $(b-\delta, b]$, and so $\alpha < b$. We will show that $f(\alpha) = c$.

If $f(\alpha)$ were larger than c, then there would exist an interval $(\alpha - \delta, \alpha + \delta)$ in which $f(x) > c$ would hold. But then α could not be the upper limit of the set A, that is, its *least* upper bound, since the smaller $\alpha - \delta$ would also be an upper bound of A.

If, however, $f(\alpha)$ were smaller than c, then there would exist an interval $(\alpha - \delta, \alpha + \delta)$ in which $f(x) < c$ held. But then α cannot be the upper limit of the set A once again, since there would be values x in A that were larger than α. Thus neither $f(\alpha) > c$ nor $f(\alpha) < c$ can hold, so $f(\alpha) = c$. $\square$

Corollary 10.58. *If $f \in C[a,b]$, then the image of f (the set $f([a,b])$) is a closed and bounded interval; in fact,*

$$f([a,b]) = \left[\min_{x \in [a,b]} f(x), \ \max_{x \in [a,b]} f(x) \right].$$

Proof. It follows from Weierstrass's theorem that $M = \max f([a,b])$ and $m = \min f([a,b])$ exist. It is clear that $f([a,b]) \subset [m,M]$. By Theorem 10.57, we also know that the function takes on every value of $[m,M]$ in $[a,b]$, so $f([a,b]) = [m,M]$. $\square$

It is easy to see from the above theorems that if I is any kind of interval and f is continuous on I, then $f(I)$ is also an interval (see Exercise 10.61).

Using the Bolzano–Darboux theorem, we can give a simple proof of the existence of the kth roots of nonnegative numbers (Theorem (3.6)).

Corollary 10.59. *If $a \geq 0$ and k is a positive integer, then there exists a nonnegative real number b such that $b^k = a$.*

Proof. The function x^k is continuous on the interval $[0, a+1]$ (why?). Since we have $f(0) = 0 \leq a$ and $f(a+1) = (a+1)^k \geq a+1 > a$, by the Bolzano–Darboux theorem, there exists a $b \in [0, a+1]$ such that $b^k = f(b) = a$. $\square$

Exercises

10.48. Give an example of a function $f : [a,b] \to \mathbb{R}$ that is continuous on all of $[a,b]$ except for one point, and that is (a) not bounded, (b) bounded, but does not have a greatest value.

10.49. Show that if $f : \mathbb{R} \to \mathbb{R}$ is continuous and $\lim_{x \to \infty} f(x) = \lim_{x \to -\infty} f(x)$, then f has either a largest or smallest value (not necessarily both).

10.50. Which are the functions $f\colon [a,b] \to \mathbb{R}$ which have smallest and largest values on every nonempty set $A \subset [a,b]$?

10.51. Suppose that the function $f\colon [a,b] \to \mathbb{R}$ satisfies the following properties: (a) If $a \le c < d \le b$, then f takes on every value between $f(c)$ and $f(d)$ on $[c,d]$; moreover, (b) whenever $x_n \in [a,b]$ for all n and $x_n \to c$, then on the set $\{x_n : n = 1,2,\ldots\} \cup \{c\}$, the function f has largest and smallest values. Prove that f is continuous.

10.52. Let $f\colon [a,b] \to (0,\infty)$ be continuous. Prove that for suitable $\delta > 0$, $f(x) > \delta$ for all $x \in [a,b]$. Give a counterexample if we write (a,b) instead of $[a,b]$.

10.53. Let f, $g\colon [a,b] \to \mathbb{R}$ be continuous and suppose that $f(x) < g(x)$ for all $x \in [a,b]$. Prove that for suitable $\delta > 0$, $f(x) + \delta < g(x)$ for all $x \in [a,b]$. Give a counterexample if we write (a,b) instead of $[a,b]$.

10.54. Prove that if $f\colon [a,b] \to \mathbb{R}$ is continuous and one-to-one, then it is strictly monotone.

10.55. Show that if $f\colon [a,b] \to \mathbb{R}$ is monotone increasing and the image contains $[f(a),f(b)] \cap \mathbb{Q}$, then f is continuous.

10.56. Prove that if $f\colon [a,b] \to \mathbb{R}$ is continuous, then for every $x_1,\ldots,x_n \in [a,b]$, there exists a $c \in [a,b]$ such that $(f(x_1) + \cdots + f(x_n))/n = f(c)$.

10.57. Prove that every cubic polynomial has a real root. Is it true that every fourth-degree polynomial has a real root? (H)

10.58. Prove that if $f\colon [a,b] \to [a,b]$ is continuous, then there exists $c \in [a,b]$ for which $f(c) = c$. Give a counterexample if we take any other type of interval than $[a,b]$. (H)

10.59. Prove that if $f\colon [0,1] \to \mathbb{R}$ is continuous and $f(0) = f(1)$, then there exists an $x \in [0,1/2]$ such that $f(x) = f(x+(1/2))$. In fact, for every $n \in \mathbb{N}^+$, there exists a $0 \le x \le 1 - (1/n)$ such that $f(x) = f(x+(1/n))$.

10.60. Does there exist a continuous function $f\colon \mathbb{R} \to \mathbb{R}$ for which $f(f(x)) = -x$ for all x? (H)

10.61. Prove that if I is an interval (closed or not, bounded or not, degenerate or not) and $f\colon I \to \mathbb{R}$ is continuous, then $f(I)$ is also an interval. (S)

10.5 Uniform Continuity

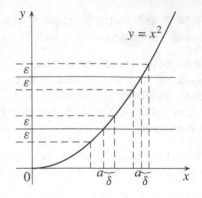

Fig. 10.13

Let the function f be continuous on the interval I. This means that for all $a \in I$ and arbitrary $\varepsilon > 0$, there exists a $\delta > 0$ such that

$$|f(x) - f(a)| < \varepsilon, \quad \text{if } x \in (a - \delta, a + \delta) \cap I.$$
$$(10.17)$$

In many cases, we can determine the largest possible δ for a given a such that (10.17) holds. Let us denote this by $\delta(a)$. If $\varepsilon > 0$ is fixed, then for different a, usually different $\delta(a)$ correspond. It is easy to see, for example, that for the function $f(x) = x^2$, the larger $|a|$ is, the smaller the corresponding $\delta(a)$ is (Figure 10.13). Thus in the interval $[0,1]$, the $\delta(a)$ corresponding to $a = 1$ is the smallest, so for each $a \in [0,1]$, we can choose δ to be $\delta(1)$. In other words, this means that for all $a \in [0,1]$

$$|f(x) - f(a)| < \varepsilon, \quad \text{if} \quad |x - a| < \delta(1).$$

This argument, of course, does not usually work. Since there is not always a smallest number out of infinitely many, we cannot always—at least using the above method—find a δ that is good for all $a \in I$ if we have a continuous function $f : I \to \mathbb{R}$. But such a δ does not always exist. In the case of $f(x) = 1/x$, we found the value of $\delta(a)$ in Example 10.2.4 (see equation (10.2)). We can see that in this case, $\delta(a) \to 0$ if $a \to 0$, that is, there does not exist a δ that would be good at every point of the interval $(0,1)$. As we will soon show, this phenomenon cannot occur in functions that are continuous on a closed and bounded interval: in this case, there must exist a shared, good δ for the whole interval. We call this property uniform continuity. The following is the precise definition:

Definition 10.60. The function f is *uniformly continuous* on the interval I if for every $\varepsilon > 0$, there exists a (shared, independent of the position) $\delta > 0$ such that

$$\text{if} \quad x_0, x_1 \in I \quad \text{and} \quad |x_1 - x_0| < \delta, \quad \text{then} \quad |f(x_1) - f(x_0)| < \varepsilon. \quad (10.18)$$

We can define uniform continuity on an arbitrary set $A \subset \mathbb{R}$ similarly: in the above definition, write A in place of I everywhere.

Theorem 10.61 (Heine's Theorem[4]). *If $f \in C[a,b]$, then f is uniformly continuous in $[a,b]$.*

Proof. We prove the statement by contradiction. Suppose that f is not uniformly continuous in $[a,b]$. This means that there exists an $\varepsilon_0 > 0$ for which there does not exist a $\delta > 0$ such that (10.18) holds. Then (10.18) does not hold with the choice

---

[4] Heinrich Eduard Heine (1821–1881), German mathematician.

$\delta = 1/n$ either, that is, for every n, there exist $\alpha_n \in [a,b]$ and $\beta_n \in [a,b]$ for which

$$|\alpha_n - \beta_n| < \frac{1}{n}, \tag{10.19}$$

but at the same time,

$$|f(\alpha_n) - f(\beta_n)| \geq \varepsilon_0. \tag{10.20}$$

Since $\alpha_n \in [a,b]$, there exists a convergent subsequence (α_{n_k}) whose limit, α, is also in $[a,b]$. Now by (10.19),

$$\beta_{n_k} = (\beta_{n_k} - \alpha_{n_k}) + \alpha_{n_k} \to 0 + \alpha = \alpha.$$

Since f is continuous on $[a,b]$, it is continuous at α (restricted to $[a,b]$). Thus by the transference principle, $f(\alpha_{n_k}) \to f(\alpha)$ and $f(\beta_{n_k}) \to f(\alpha)$, so

$$\lim_{k \to \infty} \left(f(\alpha_{n_k}) - f(\beta_{n_k}) \right) = 0.$$

This, however, contradicts (10.20).   □

Remark 10.62. In Theorem 10.61, both the boundedness and the closedness of the interval $[a,b]$ are necessary. For example, the function $f(x) = 1/x$ is continuous on $(0,1)$, but it is not uniformly continuous there. This shows that the closedness assumption cannot be dropped. The function $f(x) = x^2$ is continuous on $(-\infty, \infty)$, but it is not uniformly continuous there. This shows that the boundedness assumption cannot be dropped either.

Later, we will see that uniform continuity is a very useful property, and we often need to determine whether a function is uniformly continuous on a set A. If A is a closed and bounded interval, then our job is easy: by Heine's theorem, the function is uniformly continuous on A if and only if it is continuous at every point of A. If, however, A is an interval that is neither bounded nor closed (perhaps A is not even an interval), then Heine's theorem does not help. This is why it is important for us to know that there is a simple property that is easy to check that implies uniform continuity.

Definition 10.63. The function f is said to have the *Lipschitz*[5] property (is Lipschitz, for short) on the set A if there exists a constant $K \geq 0$ such that

$$|f(x_1) - f(x_0)| \leq K \cdot |x_1 - x_0| \tag{10.21}$$

for all $x_0, x_1 \in A$.

Theorem 10.64. *If f is Lipschitz on the set A, then f is uniformly continuous on A.*

---

[5] Rudolph Otto Sigismund Lipschitz (1832–1903), German mathematician.

Proof. If (10.21) holds for all $x_0, x_1 \in A$, then $x_0, x_1 \in A$ and $|x_1 - x_0| < \varepsilon/K$ imply

$$|f(x_1) - f(x_0)| \leq K \cdot |x_1 - x_0| < K \cdot \frac{\varepsilon}{K} = \varepsilon.$$

$\square$

Remark 10.65. The converse is generally not true: uniform continuity does not generally imply the Lipschitz property. (That is, the Lipschitz property is stronger than uniform continuity.) So for example, the function $\sqrt{x}$ is *not* Lipschitz on the interval $[0, 1]$. Indeed, for every constant $K > 0$, if $x_0 = 0$ and $0 < x_1 < \min(1, 1/K^2)$, then $x_1 > K^2 \cdot x_1^2$, and so

$$\left| \sqrt{x_1} - \sqrt{x_0} \right| = \sqrt{x_1} > K \cdot x_1 = K \cdot |x_1 - x_0| .$$

On the other hand, $\sqrt{x}$ is uniformly continuous on $[0, 1]$, since it is continuous there.

Exercises

10.62. The functions given below are uniformly continuous on the given intervals by Heine's theorem. For all $\varepsilon > 0$, give a δ that satisfies the definition of uniform continuity.

  (a) x^2 on $[0, 1]$;   (b) x^3 on $[-2, 2]$;   (c) $\sqrt{x}$ on $[0, 1]$.

10.63. Prove that (a) $f(x) = x^3$ is not uniformly continuous on $\mathbb{R}$; and (b) $f(x) = 1/x^2$ is not uniformly continuous on $(0, 1)$, but is uniformly continuous on $[1, +\infty)$.

10.64. Prove that if f is continuous on $\mathbb{R}$ and

$$\lim_{x \to \infty} f(x) = \lim_{x \to -\infty} f(x) = 0,$$

then f is uniformly continuous on $\mathbb{R}$.

10.65. Prove that if f is uniformly continuous on a bounded set A, then f is bounded on A. Does this statement still hold if we do not assume that A is bounded?

10.66. Prove that if $f \colon \mathbb{R} \to \mathbb{R}$ and $g \colon \mathbb{R} \to \mathbb{R}$ are uniformly continuous on $\mathbb{R}$, then $f + g$ is also uniformly continuous on $\mathbb{R}$.

10.67. Is it true that if $f \colon \mathbb{R} \to \mathbb{R}$ and $g \colon \mathbb{R} \to \mathbb{R}$ are uniformly continuous on $\mathbb{R}$, then $f \cdot g$ is also uniformly continuous on $\mathbb{R}$?

10.68. Prove that if f is continuous on $[a, b]$, then for every $\varepsilon > 0$, we can find a piecewise linear function $\ell(x)$ in $[a, b]$ such that $|f(x) - \ell(x)| < \varepsilon$ for all $x \in [a, b]$ (that is, the graph of f can be approximated to within less than ε by a piecewise linear function).

(The function $\ell(x)$ is a piecewise linear function on $[a,b]$ if the interval $[a,b]$ can be subdivided with points $a_0 = a < a_1 < \cdots < a_{n-1} < a_n = b$ into subintervals $[a_{k-1}, a_k]$ on which $\ell(x)$ is linear, that is, $\ell(x) = c_k x + d_k$ if $x \in [a_{k-1}, a_k]$ and $k = 1, \ldots, n$.)

10.69. Prove that the function x^k is Lipschitz on every bounded set (where k is an arbitrary positive integer).

10.70. Prove that the function $\sqrt{x}$ is Lipschitz on the interval $[a,b]$ for all $0 < a < b$.

10.71. Suppose that f and g are Lipschitz on A. Prove that then

(i) $f + g$ and $c \cdot f$ are Lipschitz on the set A for all $c \in \mathbb{R}$; and
(ii) if the set A is bounded, then $f \cdot g$ is also Lipschitz on A. (H)

10.72. Give an example for Lipschitz functions $f, g: \mathbb{R} \to \mathbb{R}$ for which $f \cdot g$ is not Lipschitz.

10.73. Suppose that f is Lipschitz on the closed and bounded interval $[a,b]$. Prove that if f is nowhere zero, then $1/f$ is also Lipschitz on $[a,b]$.

10.74. Suppose that the function $f: \mathbb{R} \to \mathbb{R}$ satisfies $|f(x_1) - f(x_2)| \leq |x_1 - x_2|^2$ for all $x_1, x_2 \in \mathbb{R}$. Prove that then f is constant.

10.6 Monotonicity and Continuity

Let f be defined on a punctured neighborhood of a. The function f is continuous at a if and only if all of the following conditions hold:

(i) $\lim_{x \to a} f(x)$ exists,
(ii) $a \in D(f)$,
(iii) $\lim_{x \to a} f(x) = f(a)$.

If any one of these three conditions does not hold, then the function is not continuous at a; we then say that f has a **point of discontinuity** at a. We classify points of discontinuity as follows.

Definition 10.66. Let f be defined on a punctured neighborhood of a, and suppose that f is not continuous at a. If $\lim_{x \to a} f(x)$ exists and is finite, but $a \notin D(f)$ or $f(a) \neq \lim_{x \to a} f(x)$, then we say that a is a *removable discontinuity* of f.[6]
    If $\lim_{x \to a} f(x)$ does not exist, but the finite limits

$$\lim_{x \to a+0} f(x) = f(a+0) \quad \text{and} \quad \lim_{x \to a-0} f(x) = f(a-0)$$

both exist (and then are necessarily different), then we say that f has a *jump discontinuity* at a. We call removable discontinuities and jump discontinuities *discontinuities of the first type* collectively.
    In all other cases, we say that f has a *discontinuity of the second type at a*.

---

[6] Since then, by setting $f(a) = \lim_{x \to a} f(x)$, f can be made continuous at a.

Examples 10.67. **1.** The functions $\{x\}$ and $[x]$ have jump discontinuities at every positive integer value. Similarly, the function $\operatorname{sgn} x$ has a jump discontinuity at 0.
2. The Riemann function (function 3. in Example 10.7) has a removable discontinuity at every rational point.
3. The Dirichlet function (function (9) in Example 9.7) has discontinuities of the second type at every point.

Below, we will show that the points of discontinuity of a monotone function are of the first type, and these can only be jump discontinuities. This is equivalent to saying that a monotone function possesses both one-sided limits at every point.

Theorem 10.68. *Let f be monotone increasing on the finite or infinite open interval (a,b). Then*

(i) *for every $x_0 \in (a,b)$, the finite limits $f(x_0 - 0)$ and $f(x_0 + 0)$ exist, and*

$$f(x_0 - 0) \le f(x_0) \le f(x_0 + 0).$$

(ii) *If f is bounded from above on (a,b), then the finite limit $f(b-0)$ exists. If f is bounded from below on (a,b), then the finite limit $f(a+0)$ exists.*

(iii) *If f is not bounded from above on (a,b), then $f(b-0) = \infty$; if f is not bounded from below on (a,b), then $f(a+0) = -\infty$.*

A similar statement can be formulated for monotone decreasing functions, as well as for intervals that are unbounded. We give two proofs of the theorem.

Proof I. (i) Since $f(x) \le f(x_0)$ for all $x \in (a,x_0)$, the set $f((a,x_0))$ is bounded from above, and $f(x_0)$ is an upper bound. Let $\alpha = \sup f((a,x_0))$; then $\alpha \le f(x_0)$.

Let $\varepsilon > 0$ be fixed. Since α is the least upper bound of the set $f((a,x_0))$, $\alpha - \varepsilon$ cannot be an upper bound. Thus there exists $x_\varepsilon \in (a,x_0)$ for which $\alpha - \varepsilon < f(x_\varepsilon)$. Now by the monotonicity of f and the definition of α,

$$\alpha - \varepsilon < f(x_\varepsilon) \le f(x) \le \alpha$$

if $a < x_\varepsilon < x < x_0$, which clearly shows that $\lim_{x \to x_0 - 0} f(x) = \alpha$. Thus we saw that $f(x_0 - 0)$ exists and is finite, as well as that $f(x_0 - 0) \le f(x_0)$. The argument is similar for $f(x_0 + 0) \ge f(x_0)$.

Statements (ii) and (iii) can be proven similarly; in the first statement of (ii), $\sup f((a,b))$ takes on the role of $f(x_0)$. □

Proof II. We will go into detail only in proving (i). By Theorem 10.22, it suffices to show that for every sequence $x_n \nearrow x_0$, $(f(x_n))$ is convergent, and its limit is at most $f(x_0)$. By the monotonicity of f, if $x_n \nearrow x_0$, then $(f(x_n))$ is monotone increasing; it has a (finite or infinite) limit. Since also $f(x_n) \le f(x_0)$ for all n, $\lim_{n \to \infty} f(x_n) \le f(x_0)$. □

Corollary 10.69. *If f is monotone on (a,b), then at every point $x_0 \in (a,b)$, it either is continuous or has a jump discontinuity: a monotone function on (a,b) can have only jump discontinuities.*

We now show that discontinuities of a monotone function are limited not only by type, but
by quantity.

Theorem 10.70. *If f is monotone on the open
interval I, then it can have at most countably
many discontinuities on I.*

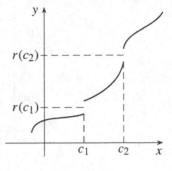

Fig. 10.14

Proof. Without loss of generality, we may assume that f is monotone increasing on I. If f
is not continuous at a $c \in I$, then $f(c-0) <$
$f(c+0)$. Let $r(c)$ be a rational number for
which $f(c-0) < r(c) < f(c+0)$. If $c_1 < c_2$,
then by the monotonicity of f, $f(c_1+0) \leq$
$f(c_2-0)$. Thus if f has both c_1 and c_2 as points of discontinuity, then $r(c_1) < r(c_2)$
(Figure 10.14).

This means that we have created a one-to-one correspondence between the points
of discontinuity and a subset of the rational numbers. Since the set of rational numbers is countable, f can have only a countable number of discontinuities.   □

Remark 10.71. Given an arbitrary countable set of numbers A, we can construct a
function f that is monotone on $(-\infty,\infty)$ and whose set of points of discontinuity
is exactly A (see Exercise 10.76). So for example, we can construct a monotone
increasing function on $(-\infty,\infty)$ that is continuous at every irrational point and discontinuous at every rational point.

In Theorem 10.45, we saw that if f is strictly monotone in the interval I, then its
inverse is continuous on the interval $f(I)$. If the function f is also continuous, then
we can expand on this in the following way.

Theorem 10.72. *Let f be strictly increasing and continuous on the interval I. Then*

(i) *$f(I)$ is also an interval; namely,*
   if $I = [a,b]$, then $f(I) = [f(a), f(b)]$;
   if $I = [a,b)$, where b is finite or infinite, then $f(I) = [f(a), \sup f(I))$;
   if $I = (a,b]$, where a is finite or infinite, then $f(I) = (\inf f(I), f(b)]$;
   *if $I = (a,b)$, where each of a and b is either finite or infinite, then $f(I) =$
   $(\inf f(I), \sup f(I))$.*
(ii) *The inverse of f, f^{-1}, is strictly monotone increasing and continuous on the
   interval $f(I)$ restricted to $f(I)$.*

*A similar statement can be made for strictly monotone decreasing and continuous
functions.*

Proof. We need only prove (i). If $I = [a,b]$, then $f(I) = [f(a), f(b)]$ is clear from
the Bolzano–Darboux theorem (Figure 10.15).

Next suppose that $I = [a,b)$. It is clear that then, $f(I) \subset [f(a), \sup f(I)]$. If
$f(a) \leq c < \sup f(I)$, then let us choose a point $u \in I$ for which $c < f(u)$. By the

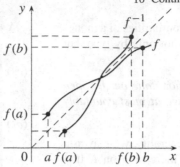

Fig. 10.15

Bolzano–Darboux theorem, f takes on every value between $f(a)$ and $f(u)$ on the interval $[a,u]$, so $c \in f([a,u]) \subset f(I)$.

This shows that $[f(a), \sup f(I)) \subset f(I)$. To prove that $f(I) = [f(a), \sup f(I))$, we just need to show that $\sup f(I) \notin f(I)$. Indeed, if $c \in f(I)$ and $c = f(u)$, where $u \in I$, then $u < v \in I$ implies $c = f(u) < f(v) \leq \sup f(I)$, so $c \neq \sup f(I)$.

The rest of the statements can be proved similarly.   □

Remark 10.73. By the previous theorem, the inverse of a function f defined on an interval $[a,b]$ exists and is also defined on a closed and bounded interval if the function f is strictly monotone and continuous. This condition is far from necessary, as the following example illustrates (Figure 10.16).

Let f be defined for $x \in [0,1]$ as

$$f(x) = \begin{cases} x, & \text{if } x \text{ is rational}, \\ 1-x, & \text{if } x \text{ is irrational}. \end{cases}$$

It is easy to see that in $[0,1]$, f

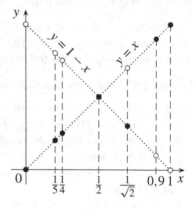

a) is not monotone on any subinterval,
b) is nowhere continuous except for the
    point $x = 1/2$; yet
c) the inverse of f exists.

Moreover, $f([0,1]) = [0,1]$, so f is a one-to-one correspondence from $[0,1]$ to itself that is nowhere monotone and is continuous nowhere except at one point.

Fig. 10.16

We see, however, that if f is continuous on an interval, then strict monotonicity of f is necessary and sufficient for the inverse function to exist (see Exercise 10.54).

Exercises

10.75. Give a function $f: [0,1] \to [0,1]$ that is monotone and has infinitely many jump discontinuities.

10.76. Prove that for every countable set $A \subset \mathbb{R}$, there exists a monotone increasing function $f : \mathbb{R} \to \mathbb{R}$ that jumps at every point of A but is continuous at every point of $\mathbb{R} \setminus A$. (H)

10.77. Let f be defined on a neighborhood of a, and let

$$m(h) = \inf\{f(x) : x \in [a-h, a+h]\}, \quad M(h) = \sup\{f(x) : x \in [a-h, a+h]\}$$

for all $h > 0$. Prove that the limits $\lim_{h \to 0+0} M(h) = M$ and $\lim_{h \to 0+0} m(h) = m$ exist, and moreover, that f is continuous at a if and only if $m = M$.

10.78. Can f have an inverse function in $[-1,1]$ if $f([-1,1]) = [-1,1]$ and f has exactly two points of discontinuity in $[-1,1]$?

10.79. Construct a function $f: \mathbb{R} \to \mathbb{R}$ that is continuous at every point different from zero and has a discontinuity of the second type at zero.

10.80. Let $f: \mathbb{R} \to \mathbb{R}$ be a function such that $f(x-0) \le f(x) \le f(x+0)$ for all x. Is it true that f is monotone increasing? (H)

10.81. Prove that the set of discontinuities of the first type of every function $f: \mathbb{R} \to \mathbb{R}$ is countable. (H)

10.82. Prove that if there is a discontinuity of the first type at every rational point of the function $f: \mathbb{R} \to \mathbb{R}$, then there is an irrational point where it is continuous. (H)

10.7 Convexity and Continuity

Our first goal is to prove that a convex function in an open interval is necessarily continuous. As we will see, this follows from the fact that if f is convex, then every point c has a neighborhood in which f can be surrounded by two continuous (linear) functions that share the value $f(c)$ at c. To see this, we first prove a helping theorem. We recall that (for a given f) the linear function that agrees with f at a and b is $h_{a,b}$, that is,

$$h_{a,b}(x) = \frac{f(b) - f(a)}{b-a} \cdot (x-a) + f(a).$$

Lemma 10.74. *Let f be convex on the interval I. If a, $b \in I$, $a < b$, and $x \in I \setminus [a,b]$, then*

$$f(x) \ge h_{a,b}(x). \tag{10.22}$$

If f is strictly convex on I, then strict inequality holds in (10.22). (That is, outside the interval [a,b], the points of the graph of f lie above the line connecting the points $(a, f(a))$ and $(b, f(b))$; see Figure 9.2.)

Proof. Suppose that $a < b < x$. By the definition of convexity, $f(b) \leq h_{a,x}(b)$, that is,

$$f(b) \leq \frac{f(x) - f(a)}{x - a} \cdot (b - a) + f(a),$$

which yields (10.22) after a simple rearrangement. If instead $x < a < b$, then $f(a) \leq h_{x,b}(a)$, that is,

$$f(a) \leq \frac{f(b) - f(x)}{b - x} \cdot (a - x) + f(x),$$

which yields (10.22) after a simple rearrangement.

If f is strictly convex, then we can repeat the above argument using strict inequalities.   □

Now we can easily prove the continuity of convex functions.

Theorem 10.75. *If f is convex on the open interval I, then f is continuous on I.*

Proof. Let $c \in I$ be fixed, and choose points $a, b \in I$ such that $a < c < b$. If $x \in (c, b)$, then by the above lemma and the convexity of f,

$$h_{a,c}(x) \leq f(x) \leq h_{c,b}(x).$$

Since $\lim_{x \to c} h_{a,c}(x) = \lim_{x \to c} h_{c,b}(x) = f(c)$, by the squeeze theorem we have $\lim_{x \to c+0} f(x) = f(c)$. We can similarly get that $\lim_{x \to c-0} f(x) = f(c)$.   □

If f is convex on the interval I, then for arbitrary $a, b \in I$,

$$f\left(\frac{a+b}{2}\right) \leq \frac{f(a) + f(b)}{2}. \tag{10.23}$$

Indeed, if $a = b$, then (10.23) is clear, while if $a < b$, (10.23) follows from the inequality $f(x) \leq h_{a,b}(x)$ applied to $x = (a+b)/2$. The functions that satisfy (10.23) for all $a, b \in I$ are called **weakly convex** functions.[7] The function f is **weakly concave** if $f((a+b)/2) \geq (f(a) + f(b))/2$ for all $a, b \in I$.

The condition for weak convexity—true to its name—is a weaker condition than convexity, that is, there exist functions that are weakly convex but not convex. One can show that there exists a function $f \colon \mathbb{R} \to \mathbb{R}$ that is **additive** in the sense that $f(x + y) = f(x) + f(y)$ holds for all $x, y \in \mathbb{R}$, but is not continuous. (The proof of this fact, however, is beyond the scope of this book.) Now it is easy to see that such a function is weakly convex, and it actually satisfies the stronger condition $f((a+b)/2) = (f(a) + f(b))/2$ as well for all a, b. On the other hand, f is not convex, since it is not continuous.

---

[7] Weakly convex functions are often called **Jensen-convex** functions as well.

In the following theorem, we prove that if f is continuous, then the weak convexity of f is equivalent to the convexity of f. This means that in talking about continuous functions, to determine convexity it is enough to check the conditions for weak convexity, which is usually easier to do.

Theorem 10.76. *Suppose that f is continuous and is weakly convex on the interval I. Then f is convex on I.*

Proof. We have to show that if $a, x_0, b \in I$ and $a < x_0 < b$, then $f(x_0) \le h_{a,b}(x_0)$. Suppose that this is not true, that is, $f(x_0) > h_{a,b}(x_0)$. This means that the function $g(x) = f(x) - h_{a,b}(x)$ is positive at x_0. Since $g(a) = 0$, there exists a last point before x_0 where g vanishes. To see this, let $A = \{x \in [a, x_0] : g(x) = 0\}$, and let $\alpha = \sup A$. Then $a \le \alpha \le x_0$. We show that $g(\alpha) = 0$. We can choose a sequence $x_n \in A$ that tends to α, and so by the continuity of g, we have $g(x_n) \to g(\alpha)$. Since $g(x_n) = 0$ for all n, we must have $g(\alpha) = 0$. It follows that $\alpha < x_0$, and the function g is positive on the interval $(\alpha, x_0]$: if there were a point $\alpha < x_1 \le x_0$ such that $g(x_1) \le 0$, then by the Bolzano–Darboux theorem, g would have a root in $[x_1, x_0]$, which contradicts the fact that α is the supremum of the set A.

By the exact same argument, there is a first point β after x_0 where g vanishes, and so the function g is positive in the interval $[x_0, \beta)$. Then $g(\alpha) = g(\beta) = 0$, and $g(x) > 0$ for all $x \in (\alpha, \beta)$. Now we got g by subtracting a linear function ℓ from f. It follows that g is also weakly convex; since ℓ is linear, $\ell((a+b)/2) = (\ell(a) + \ell(b))/2$ for all a, b, so if f satisfies inequality (10.23), then subtracting ℓ does not change this. However, $g(\alpha) = g(\beta) = 0$ and $g((\alpha + \beta)/2) > 0$, so (10.23) is not satisfied with the choices $a = \alpha$ and $b = \beta$. This is a contradiction, which shows that f is convex. $\qquad\square$

Remark 10.77. If the function $f : I \to \mathbb{R}$ satisfies the condition

$$f\left(\frac{a+b}{2}\right) < \frac{f(a) + f(b)}{2} \tag{10.24}$$

for all $a, b \in I$, $a \ne b$, then we call f **strictly weakly convex**. We can similarly define **strictly weakly concave** functions. By the previous theorem, it follows that *if f is continuous and strictly weakly convex on the interval I, then f is strictly convex on I.* Indeed, it is easy to see that if f is convex but not strictly convex on the interval I, then there is a subinterval J on which f is linear (see Exercise 10.83). Then, however, (10.24) does not hold for the points of J, since if $a, b \in J$, then equality holds in (10.24).

We can similarly see that every continuous and strictly weakly concave function is strictly concave.

We mention that the conditions of Theorem 10.76 can be greatly weakened: instead of assuming the continuity of f, it suffices to assume that I has a subinterval on which f is bounded from above (see Exercises 10.99–10.102).

Exercises

10.83. Prove that if f is convex but not strictly convex on the interval I, then I has a subinterval on which f is linear.

10.84. Let us call a function $f: I \to \mathbb{R}$ **barely convex** if whenever $a, b, c \in I$ and $a < b < c$, then $f(b) \le \max(f(a), f(c))$. Prove that if $f: I \to \mathbb{R}$ is convex on the interval I, then f is barely convex.

10.85. Let f be barely convex on the interval (a, b), and suppose that $a < c < d < b$ and $f(c) > f(d)$. Show that f is monotone decreasing on $(a, c]$. Similarly, show that if $a < c < d < b$ and $f(c) < f(d)$, then f is monotone increasing on $[d, b)$.

10.86. Prove that the function $f: I \to \mathbb{R}$ is barely convex on the interval (a, b) if and only if one of the following cases applies:

(a) f is monotone decreasing on (a, b).
(b) f is monotone increasing on (a, b).
(c) There exists a point $c \in (a, b)$ such that f is monotone decreasing on (a, c), monotone increasing on (c, b), and $f(c) \le \max(f(c-0), f(c+0))$.

10.87. Prove that if $f: I \to \mathbb{R}$ is convex on the interval (a, b), then one of the following cases applies:

(a) f is monotone decreasing on (a, b).
(b) f is monotone increasing on (a, b).
(c) There exists a point $c \in (a, b)$ such that f is monotone decreasing on $(a, c]$ and monotone increasing on $[c, b)$.

10.88. Let f be convex on $(-\infty, \infty)$, and suppose that $\lim_{x \to -\infty} f(x) = \infty$. Is it possible that $\lim_{x \to \infty} f(x) = -\infty$? (S)

10.89. Let f be convex on $(-\infty, \infty)$, and suppose that $\lim_{x \to -\infty} f(x) = 0$. Is it possible that $\lim_{x \to \infty} f(x) = -\infty$? (H)

10.90. Let f be convex on $(0, 1)$. Is it possible that $\lim_{x \to 1-0} f(x) = -\infty$? (H)

10.91. Let f be weakly convex on the interval I. Prove that

$$f\left(\frac{x_1 + \cdots + x_n}{n}\right) \le \frac{f(x_1) + \cdots + f(x_n)}{n}$$

for all $x_1, \ldots, x_n \in I$. (S)

10.92. Let $f: \mathbb{R} \to \mathbb{R}$ be an additive function (that is, suppose that for all x, y, $f(x + y) = f(x) + f(y)$). Prove that $f(rx) = r \cdot f(x)$ for every real number x and rational number r.

10.93. Prove that if $f: \mathbb{R} \to \mathbb{R}$ is additive, then the function $g(x) = f(x) - f(1) \cdot x$ is also additive and periodic, namely that every rational number is a period.

10.94. Let $f: \mathbb{R} \to \mathbb{R}$ be an additive function. Prove that if f is bounded from above on an interval, then $f(x) = f(1) \cdot x$ for all x. (H)

10.95. Let $f: \mathbb{R} \to \mathbb{R}$ be an additive function. Prove that f^2 is weakly convex. (If f is not a linear function, then f^2 is a weakly convex function that is bounded from below, but is not convex.)

10.96. Let f be continuous on the interval I, and suppose that for all $a, b \in I$, $a < b$, there exists a point $a < x < b$ such that $f(x) \le h_{a,b}(x)$. Prove that f is convex. (H)

10.97. Let f be bounded on the interval I, and suppose that for all $a, b \in I$, $a < b$, there exists a point $a < x < b$ such that $f(x) \le h_{a,b}(x)$. Does it then follow that f is convex?

10.98. Let f be convex on the open interval I. Prove that f is Lipschitz on every closed and bounded subinterval of I.

The following four questions will take us through the proof that if f is weakly convex on an open interval I, and I has a subinterval in which f is bounded from above, then f is convex.

10.99. Let f be weakly convex on the open interval I, and let $x_0 \in I$. Prove that if f is bounded from above on $(x_0 - \delta, x_0 + \delta)$, then f is bounded on $(x_0 - \delta, x_0 + \delta)$. (S)

10.100. Let f be weakly convex on the open interval I. Let $n \ge 1$ be an integer, and let x and h be numbers such that $x \in I$ and $x + 2^n h \in I$. Prove that

$$f(x+h) - f(x) \le \frac{1}{2^n} \cdot [f(x + 2^n h) - f(x)]. \tag{S}$$

10.101. Let f be weakly convex on the open interval I, and let $x_0 \in I$. Prove that if f is bounded from above on $(x_0 - \delta, x_0 + \delta)$, then f is continuous at x_0. (S)

10.102. Let f be weakly convex on the interval I, and suppose that I contains a nondegenerate subinterval on which f is bounded from above. Prove that f is continuous (and so by Theorem 10.76, convex) on I. (H)

10.8 Arc Lengths of Graphs of Functions

One of the key objectives of analysis is the measurement of lengths, areas, and volumes. Our next goal is to deal with a special case: the notion of the arc length of the graph of a function[8].

We denote the line segment connecting the points $p, q \in \mathbb{R}^2$ by $[p,q]$, that is, $[p,q] = \{p + t(q-p) : t \in [0,1]\}$. The length of the line segment $[p,q]$ (by definition) is the distance between its endpoints, which is $|q - p|$. We call a set that is

---

[8] We will have need of this in defining trigonometric functions. We return to dealing with arc lengths of more general curves in Chapter 16.

a union of connected line segments a **broken line** (or a polygonal path). Thus a broken line is of the form $[p_0,p_1] \cup [p_1,p_2] \cup \ldots \cup [p_{n-1},p_n]$, where $p_0,\ldots,p_n$ are points of the plane. The length of the broken line is the sum of the lengths of the lines that it comprises, that is, $|p_1 - p_0| + |p_2 - p_1| + \cdots + |p_n - p_{n-1}|$.

Since "the shortest distance between two points is a straight line", the length of a curve (no matter how we define it) should not be smaller than the distance between its endpoints. If we inscribe a broken line $[p_0,p_1] \cup [p_1,p_2] \cup \ldots \cup [p_{n-1},p_n]$ "on top of" a curve, then the part of the arc connecting p_{i-1} and p_i has length at least $|p_i - p_{i-1}|$, and so the length of the whole curve needs to be at least $|p_1 - p_0| + |p_2 - p_1| + \cdots + |p_n - p_{n-1}|$. On the other hand—again just using intuition—we can expect a "very fine" broken line inscribed on the curve to "approximate" it well enough so that the two lengths will be close. What we can take away from this is that the arc length should be equal to the supremum of the lengths of the broken lines on the curve. This finding is what we will accept as the definition. We remind ourselves that we denote the graph of the function $f\colon [a,b] \to \mathbb{R}$ by graph f.

Definition 10.78. Let $f\colon [a,b] \to \mathbb{R}$ be an arbitrary function and let $a = x_0 < x_1 < \cdots < x_n = b$ be a partition F of the interval $[a,b]$. The *inscribed polygonal path* over the partition F of f is the broken line over the points

$$(x_0, f(x_0)), \ldots, (x_n, f(x_n)).$$

The *arc length* of graph f is the least upper bound of the set of lengths of all inscribed polygonal paths on f. (The arc length can be infinite). We denote the arc length of the graph of f by $s(f;[a,b])$. Thus

$$s(f;[a,b]) = \sup\left\{ \sum_{i=1}^n |p_i - p_{i-1}| : a = x_0 < x_1 < \cdots < x_n = b, \right.$$

$$\left. p_i = (x_i, f(x_i)) \ (i = 0,\ldots,n) \right\}.$$

We say that graph f is *rectifiable* if $s(f;[a,b])$ is finite.

Let us note that if $a = b$, then $s(f;[a,b]) = 0$ for all functions f.

Theorem 10.79.

(i) *For an arbitrary function* $f\colon [a,b] \to \mathbb{R}$,

$$\sqrt{(b-a)^2 + (f(b) - f(a))^2} \le s(f;[a,b]), \qquad (10.25)$$

and so if $a < b$, *then* $s(f;[a,b]) > 0$.
(ii) *If* $f\colon [a,b] \to \mathbb{R}$ *is monotone, then* graph f *is rectifiable, and*

$$s(f;[a,b]) \le (b-a) + |f(b) - f(a)|. \qquad (10.26)$$

Proof. It is clear that $s(f;[a,b])$ is not smaller than any of its inscribed polygonal paths. Now the segment connecting $(a,f(a))$ and $(b,f(b))$ is such a path that corresponds to the partition $a = x_0 < x_1 = b$. Since the length of this is $\sqrt{(b-a)^2 + (f(a) - f(b))^2}$, (10.25) holds.

Now suppose that f is monotone increasing, let $F: a = x_0 < x_1 < \cdots < x_n = b$ be a partition of the interval $[a,b]$, and denote the point $(x_i, f(x_i))$ by p_i for all $i = 0, \ldots, n$. Then, using the monotonicity of f,

$$|p_i - p_{i-1}| = \sqrt{(x_i - x_{i-1})^2 + (f(x_i) - f(x_{i-1}))^2} \le$$
$$\le (x_i - x_{i-1}) + (f(x_i) - f(x_{i-1}))$$

for all i, so

$$\sum_{i=1}^{n} |p_i - p_{i-1}| \le \left[\sum_{i=1}^{n}(x_i - x_{i-1})\right] + \left[\sum_{i=1}^{n}(f(x_i) - f(x_{i-1}))\right] =$$
$$= (x_n - x_0) + (f(x_n) - f(x_0)) = (b - a) + (f(b) - f(a)).$$

Since the partition was arbitrary, we have established (10.26). If f is monotone decreasing, then we can argue similarly, or we can reduce the statement to one about monotone increasing functions by considering the function $-f$. □

Remark 10.80. Since not every monotone function is continuous, by (ii) of the previous theorem, there exist functions that are not everywhere continuous but whose graphs are rectifiable. Thus rectifiability is a more general concept then what the words "arc length" intuitively suggest.

The statement of the following theorem can also be expressed by saying that *arc lengths are additive.*

Theorem 10.81. *Let $a < b < c$ and $f: [a,c] \to \mathbb{R}$. If graph f is rectifiable, then*

$$s(f; [a,c]) = s(f; [a,b]) + s(f; [b,c]). \tag{10.27}$$

We give a proof for this theorem in the appendix of the chapter.
We will need the following simple geometric fact soon.

Lemma 10.82. *If A, B are convex polygons, and $A \subset B$, then the perimeter of A is not larger than the perimeter of B.*

Proof. If we cut off part of the polygon B at a line given by an edge of the polygon A, then we get a polygon B_1 with perimeter not larger than B but still containing A. Repeating the process, we get the sequence $B, B_1, \ldots, B_n = A$, in which the perimeter of each polygon is at most as big as the one before it. □

Arc length of a circle. Let C denote the unit circle centered at the origin. Let the part of C falling into the upper half of the plane $\{(x,y) : y \ge 0\}$ be denoted by C^+. It is clear that C^+ agrees with the graph of the function $c(x) = \sqrt{1 - x^2}$ defined on the interval $[-1,1]$. Since c is monotone on both the intervals $[-1,0]$ and $[0,1]$, by the above theorems, the graph of c is rectifiable.

We denote the arc length of graph c (that is, of the half-circle C^+) by π.

By the previous two theorems, we can extract the approximation $2\sqrt{2} \leq \pi \leq 4$, where the value $2\sqrt{2}$ is the length of the inscribed broken path corresponding to the partition $-1 = x_0 < 0 = x_1 < 1 = x_2$. Inscribing different broken lines into C^+ gives us different lower bounds for π, and with the help of these, we can approximate π with arbitrary precision (at least in theory).

If we inscribe C into an arbitrary convex polygon P, then by Lemma 10.82, every polygon inscribed in C will have smaller or equal perimeter than the perimeter of P. Thus the supremum of the perimeters of the inscribed polygons, 2π, cannot be larger than the perimeter of P.

The lower and upper approximations that we get this way can help us show that $\pi = 3.14159265\ldots$. The number π, like e, is irrational. One can also show that π, again like e, is transcendental, but the proof of that is beyond the scope of this book.

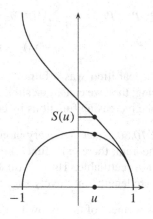

Fig. 10.17

Remark 10.83. To define trigonometric functions, we will need the (seemingly trivial) fact that starting from the point $(0,1)$, we can measure arcs of any length on C. Consider the case $x \in [0, \pi]$. We have to show that there is a number $u \in [-1,1]$ such that $s(c; [u,1]) = x$. With the notation $S(u) = s(c; [u,1])$, this means that the function $S(u)$ takes on every value between 0 and π on the interval $[-1,1]$ (Figure 10.17).

Theorem 10.84. *The function S is strictly monotone decreasing and continuous on $[-1,1]$.*

Proof. If $-1 \leq u < v \leq 1$, then by Theorem 10.81,

$$S(u) = s(c; [u,1]) = s(c; [u,v]) + s(c; [v,1]) = S(v) + s(c; [u,v]).$$

Since $s(c; [u,v]) > 0$, we know that S is strictly monotone decreasing on $[-1,1]$.
Since the function c is monotone both on $[-1,0]$ and on $[0,1]$, by (10.26),

$$s(c; [u,v]) \leq (v-u) + |c(v) - c(u)|$$

if $-1 \leq u < v \leq 0$ or $0 \leq u < v \leq 1$.
Thus

$$|S(u) - S(v)| \leq |v - u| + |c(v) - c(u)| \tag{10.28}$$

whenever $u, v \in [-1, 0]$ or $u, v \in [0, 1]$. Since the function $c(u) = \sqrt{1 - u^2}$ is continuous on $[-1, 1]$, we have that

$$\lim_{v \to u} (|v - u| + |c(v) - c(u)|) = 0$$

for all $u \in [-1, 1]$, so by (10.28) we immediately have that S is continuous on $[-1, 1]$.
□

By the previous theorem and by the Bolzano–Darboux theorem, the function $S(u)$ takes on every value between $S(-1)$ and $S(1)$, moreover exactly once. Since $S(-1) = \pi$ (since this was the definition of π) and $S(1) = 0$, we have seen that if $0 \leq x \leq \pi$, then we can measure out an arc of length x onto the circle C. What about other lengths? Since the arc length of the semicircle is π, if we can measure one of length x, then we can measure one of length $x + \pi$ (or $x - \pi$) as well, in which case we just jump to the antipodal point.

Exercises

10.103. Let $f: [a, b] \to \mathbb{R}$ be a function for which $s(f; [a, b]) = b - a$. Prove that f is constant.

10.104. Prove that the function $f: [a, b] \to \mathbb{R}$ is linear (that is, it is of the form $mx + b$ with suitable constants m and b) if and only if

$$s(f; [a, b]) = \sqrt{(b - a)^2 + (f(b) - f(a))^2}.$$

10.105. Prove that if the graph of $f: [a, b] \to \mathbb{R}$ is rectifiable, then f is bounded on $[a, b]$.

10.106. Prove that if the graph of $f: [a, b] \to \mathbb{R}$ is rectifiable, then at every point $x \in [a, b)$, the right-hand limit of f exists, and at every point $x \in (a, b]$, the left-hand limit exists.

10.107. Prove that neither the graph of the Dirichlet function nor the graph of the Riemann function over the interval $[0, 1]$ is rectifiable.

10.108. Let the function $f: [0, 1] \to \mathbb{R}$ be defined as follows: $f(x) = x$ if $x = 1/2^n$ ($n = 1, 2, \ldots$), and $f(x) = 0$ otherwise. Prove that the graph of f is rectifiable. What is its arc length?

10.109. Prove that if $f: [a, b] \to \mathbb{R}$ is Lipschitz, then its graph is rectifiable.

10.9 Appendix: Proof of Theorem 10.81

Proof. Let us denote by S_1, S_2, and S_3 the sets of the lengths of the inscribed polygonal paths of the intervals $[a,b]$, $[b,c]$, and $[a,c]$, respectively. Then $s(f;[a,b]) = \sup S_1$, $s(f;[b,c]) = \sup S_2$, and $s(f;[a,c]) = \sup S$ by the definition of arc length.

Since one partition of $[a,b]$ and one of $[b,c]$ together yield a partition of the interval $[a,c]$, the sum of any number in S_1 with any number in S_2 is in S. This means that $S \supset S_1 + S_2$. By Theorem 3.20, $\sup(S_1 + S_2) = \sup S_1 + \sup S_2$, which implies

$$s(f;[a,c]) = \sup S \geq \sup(S_1 + S_2) = \sup S_1 + \sup S_2 = s(f;[a,b]) + s(f;[b,c]).$$

Now we show that

$$s(f;[a,c]) \leq s(f;[a,b]) + s(f;[b,c]). \tag{10.29}$$

Let $F: a = x_0 < x_1 < \cdots < x_n = c$ be a partition of the interval $[a,c]$, and denote the point $(x_i, f(x_i))$ by p_i. Then the length of the inscribed polygonal path on F is $h_F = \sum_{i=1}^{n} |p_i - p_{i-1}|$. If the point b is equal to one of the points x_i, say to x_k, then $F_1: a = x_0 < x_1 < \cdots < x_k = b$, and $F_2: b = x_k < x_{k+1} < \cdots < x_n = c$ are partitions of the intervals $[a,b]$ and $[b,c]$ respectively, so

$$h_{F_1} = \sum_{i=1}^{k} |p_i - p_{i-1}| \leq s(f;[a,b]) \quad \text{and} \quad h_{F_2} = \sum_{i=k+1}^{n} |p_i - p_{i-1}| \leq s(f;[b,c]).$$

Since $h_F = h_{F_1} + h_{F_2}$, then $h_F \leq s(f;[a,b]) + s(f;[b,c])$. If the point b is not equal to any of the points x_i and $x_{k-1} < b < x_k$, then let $F_1: a = x_0 < x_1 < \cdots < x_{k-1} < b$ and $F_2: b < x_k < x_{k+1} < \cdots < x_n = c$. Let q denote the point $(b, f(b))$. The lengths of the inscribed polygonal paths corresponding to partitions F_1 and F_2 are

$$h_{F_1} = \sum_{i=1}^{k-1} |p_i - p_{i-1}| + |q - p_{k-1}| \leq s(f;[a,b])$$

and

$$h_{F_2} = |p_k - q| + \sum_{i=k+1}^{n} |p_i - p_{i-1}| \leq s(f;[b,c]).$$

Now by the triangle inequality,

$$|p_k - p_{k-1}| \leq |q - p_{k-1}| + |p_k - q|,$$

so it follows that $h_F \leq h_{F_1} + h_{F_2} \leq s(f;[a,b]) + s(f;[b,c])$. Thus $h_F \leq s(f;[a,b]) + s(f;[b,c])$ for all partitions F, which makes (10.29) clear. $\qquad\square$

Chapter 11
Various Important Classes of Functions (Elementary Functions)

In this chapter, we will familiarize ourselves with the most commonly occurring functions in mathematics and in applications of mathematics to the sciences. These are the polynomials, rational functions, exponential, power, and logarithm functions, trigonometric functions, hyperbolic functions, and their inverses. We call the functions that we can get from the above functions using basic operations and composition **elementary functions**.

11.1 Polynomials and Rational Functions

We call the function $p : \mathbb{R} \to \mathbb{R}$ a **polynomial function** (a polynomial, for short) if there exist real numbers $a_0, a_1, \ldots, a_n$ such that

$$p(x) = a_n x^n + a_{n-1} x^{n-1} + \cdots + a_1 x + a_0 \qquad (11.1)$$

for all x. Suppose that in the above description, $a_n \neq 0$. If x_1 is a root of p (that is, if $p(x_1) = 0$), then

$$p(x) = p(x) - p(x_1) = a_n(x^n - x_1^n) + \cdots + a_1(x - x_1).$$

Here using the equality

$$x^k - x_1^k = (x - x_1)(x^{k-1} + x_1 x^{k-2} + \cdots + x_1^{k-2} x + x_1^{k-1}),$$

and then taking out the common factor $x - x_1$ we get that $p(x) = (x - x_1) \cdot q(x)$, where $q(x) = b_{n-1} x^{n-1} + \cdots + b_1 x + b_0$ and $b_{n-1} = a_n \neq 0$.

If x_2 is a root of q, then by repeating this process with q, we obtain that $p(x) = (x - x_1)(x - x_2) \cdot r(x)$, where $r(x) = c_{n-2} x^{n-2} + \cdots + c_1 x + c_0$ and $c_{n-2} = a_n \neq 0$.

It is clear that this process ends in at most n steps, and in the last step, we get the following.

© Springer New York 2015
M. Laczkovich, V.T. Sós, *Real Analysis*, Undergraduate Texts in Mathematics, DOI 10.1007/978-1-4939-2766-1_11

Lemma 11.1. *Suppose that in* (11.1), $a_n \neq 0$. *If p has a root, then there exist not necessarily distinct real numbers $x_1, \ldots, x_k$ and a polynomial p_1 such that $k \leq n$, the polynomial p_1 has no roots, and*

$$p(x) = (x - x_1) \cdot \ldots \cdot (x - x_k) \cdot p_1(x) \tag{11.2}$$

for all x. It then follows that p can have at most n roots.

The above lemma has several important consequences.

1. If a polynomial is not identically zero, then it has only finitely many roots. Clearly, in the expression (11.1), not all coefficients are zero. If a_m is the nonzero coefficient with largest index, then we can omit the terms with larger indices. Then by the lemma, p can have at most m roots.
2. If two polynomials agree in infinitely many points, then they are equal everywhere. (Apply the previous point to the difference of the two polynomials.)
3. The identically zero function can be expressed as (11.1) only if $a_0 = \ldots = a_n = 0$ (since the identically zero function has infinitely many roots).
4. If in an expression (11.1), $a_n \neq 0$ and the polynomial p defined by (11.1) has an expression of the form

$$p(x) = b_k x^k + b_{k-1} x^{k-1} \cdots + b_1 x + b_0,$$

where $b_k \neq 0$, then necessarily $k = n$ and $b_i = a_i$ for all $i = 0, \ldots, n$. We see this by noting that the difference is the identically zero function, so this statement follows from the previous one.

The final corollary means that a not identically zero polynomial has a unique expression of the form (11.1) in which a_n is nonzero.

In this presentation of a polynomial, we call the coefficient a_n the **leading coefficient** of p, and the number n the **degree** of the polynomial. We denote the degree of p by $\operatorname{gr} p$.[1] The zero-degree polynomials are thus the constant functions different from zero. The identically zero function does not have a degree.

If a polynomial p is not identically zero, then its presentation of the form (11.2) is unique. Clearly, if $p(x) = (x - y_1) \cdot \ldots \cdot (x - y_m) \cdot p_2(x)$ is another presentation, then x_1 is also a root of this, whence one of $y_1, \ldots, y_m$ must be equal to x_1 (since p_2 has no roots). We can suppose that $y_1 = x_1$. Then

$$(x - x_2) \cdot \ldots \cdot (x - x_k) \cdot p_1(x) = (x - y_2) \cdot \ldots \cdot (x - y_m) \cdot p_2(x)$$

for all $x \neq x_1$. Then the two sides agree at infinitely many points, so they are equal everywhere. Since x_2 is a root of the right-hand side, it must be equal to one of $y_2, \ldots, y_m$. We can assume that $y_2 = x_2$. Repeating this argument, we run out of $x - x_i$ terms on the left-hand side, and at the kth step, we get that

$$p_1(x) = (x - y_{k+1}) \cdot \ldots \cdot (x - y_m) \cdot p_2(x).$$

---

[1] The notation is based on the Latin *gradus* = degree.

Since p_1 has no roots, necessarily $m = k$ and $p_1 = p_2$.

If in the presentation (11.2), an $x - \alpha$ term appears ℓ times, then we say that α is a **root of multiplicity** ℓ. So, for example, the polynomial $p(x) = x^5 - x^4 - x + 1$ has 1 as a root of multiplicity two, and -1 is a root of multiplicity one (often called a simple root), since $p(x) = (x-1)^2(x+1)(x^2+1)$, and $x^2 + 1$ has no roots.[2]

As for the analytic properties of polynomials, first of all, we should note that *a polynomial is continuous everywhere*. This follows from Theorem 10.44, taking into account the fact that constant functions and the function x are continuous everywhere. We now show that if in the presentation (11.1), $n > 0$ and $a_n \neq 0$, then

$$\lim_{x \to \infty} p(x) = \begin{cases} \infty, & \text{if } a_n > 0, \\ -\infty, & \text{if } a_n < 0. \end{cases} \qquad (11.3)$$

This is clear from the rearrangement

$$p(x) = x^n \left(a_n + \frac{a_{n-1}}{x} + \cdots + \frac{a_0}{x^n} \right),$$

using that $\lim_{x \to \infty} x^n = \infty$ and

$$\lim_{x \to \infty} \left(a_n + \frac{a_{n-1}}{x} + \cdots + \frac{a_0}{x^n} \right) = a_n.$$

Rational functions are functions of the form p/q, where p and q are polynomials, and q is not identically zero. The rational function p/q is defined where the denominator is nonzero, so everywhere except for a finite number of points. By 10.44, it again follows that a rational function is continuous at every point where it is defined.

The following theorem is analogous to the limit relation (11.3).

Theorem 11.2. *Let*

$$p(x) = a_n x^n + a_{n-1} x^{n-1} + \cdots + a_1 x + a_0$$

and

$$q(x) = b_k x^k + b_{k-1} x^{k-1} \cdots + b_1 x + b_0,$$

where $a_n \neq 0$ and $b_k \neq 0$. Then

$$\lim_{x \to \infty} \frac{p(x)}{q(x)} = \begin{cases} \infty, & \text{ha } a_n/b_k > 0 \text{ and } n > k, \\ -\infty, & \text{ha } a_n/b_k < 0 \text{ and } n > k, \\ a_n/b_k, & \text{if } n = k, \\ 0, & \text{if } n < k. \end{cases}$$

---

[2] Since we defined polynomials on $\mathbb{R}$, we have been talking about only real roots the entire time. Among complex numbers, every nonconstant polynomial has a root; see the second appendix of the chapter.

Exercises

11.1. Show that if p and q are polynomials, then so are $p+q$, $p \cdot q$, and $p \circ q$.

11.2. Let p and q be polynomials. Prove that

(a) if none of p, q, $p+q$ are identically zero, then $\mathrm{gr}(p+q) \leq \max(\mathrm{gr}\, p, \mathrm{gr}\, q)$;
(b) if neither p nor q is identically zero, then $\mathrm{gr}(p \cdot q) = (\mathrm{gr}\, p) + (\mathrm{gr}\, q)$;
(c) if none of p, q, and $p \circ q$ are identically zero then $\mathrm{gr}(p \circ q) = (\mathrm{gr}\, p) \cdot (\mathrm{gr}\, q)$.
   Does it suffice to assume that only p and q are not identically zero?

11.3. Let $p(x) = a_n x^n + a_{n-1} x^{n-1} + \cdots + a_1 x + a_0$, where $a_n > 0$. Prove that p is monotone increasing on the half-line (K, ∞) if K is sufficiently large.

11.4. Prove that if the polynomial p is not constant, then p takes on each of its values at most k times, where $k = \mathrm{gr}\, p$.

11.5. Prove that if the rational function p/q is not constant, then it takes on each of its values at most k times, where $k = \max(\mathrm{gr}\, p, \mathrm{gr}\, q)$.

11.6. Prove that every polynomial is Lipschitz on every bounded interval.

11.7. Prove that every rational function is Lipschitz on every closed and bounded interval on which it is defined.

11.2 Exponential and Power Functions

Before we define the two important classes of the exponential and power functions, we fulfill our old promise (which we made after Theorem 3.27) and show that the identities regarding taking powers still apply when we take arbitrary real powers.

The simple proof is made possible by our newly gained knowledge of limits of sequences and their properties. We will use the following lemma in the proof of all three identities.

Lemma 11.3. *If $a > 0$ and $x_n \to x$, then $a^{x_n} \to a^x$.*

Proof. Suppose first that $a > 1$. In Theorem 3.25, we saw that

$$\sup\{a^r : r \in \mathbb{Q},\ r < x\} = \inf\{a^s : s \in \mathbb{Q},\ s > x\},$$

and by definition, the shared value is a^x. Let $\varepsilon > 0$ be given. Then there exist rational numbers $r < x$ and $s > x$ such that

$$a^x - \varepsilon < a^r \quad \text{and} \quad a^s < a^x + \varepsilon.$$

Since $x_n \to x$, for suitable n_0 we have $r < x_n < s$ if $n > n_0$. Now according to Theorem 3.27, for every $u < v$, $a^u < a^v$. Thus for every $n > n_0$,

$$a^x - \varepsilon < a^r < a^{x_n} < a^s < a^x + \varepsilon.$$

Since ε was arbitrary, we have shown that $a^{x_n} \to a^x$.

The statement can be proved similarly if $0 < a < 1$, while the case $a = 1$ is trivial. $\qquad \square$

Theorem 11.4. *For arbitrary* $a, b > 0$ *and real exponents* x, y,

$$(ab)^x = a^x \cdot b^x, \qquad a^{x+y} = a^x \cdot a^y \qquad and \qquad (a^x)^y = a^{xy}. \qquad (11.4)$$

Proof. We have already seen these inequalities for *rational* exponents in Theorem 3.23.

We begin by showing that the first two equalities in (11.4) hold for all positive a, b and real numbers x, y. Choose two sequences (r_n) and (s_n) of rational numbers that tend to x and y respectively. (For example, if $r_n \in \left(x - (1/n), x + (1/n)\right) \cap \mathbb{Q}$ and $s_n \in \left(y - (1/n), y + (1/n)\right) \cap \mathbb{Q}$, then these sequences work.) Then by Lemma 11.3,

$$(ab)^x = \lim_{n \to \infty} (ab)^{r_n} = \lim_{n \to \infty} a^{r_n} \cdot b^{r_n} = a^x \cdot b^x$$

and

$$a^{x+y} = \lim_{n \to \infty} a^{r_n + s_n} = \lim_{n \to \infty} a^{r_n} \cdot a^{s_n} = a^x \cdot a^y.$$

We only outline the proof of the third identity for the case $a > 1$ and $x, y > 0$. (The remaining cases cant be proven similarly, or can be reduced to our case by considering reciprocals.)

Let now $r_n \to x$ and $s_n \to y$ be sequences consisting of rational numbers that satisfy $0 < r_n < x$ and $0 < s_n < y$ for all n. Then

$$a^{r_n s_n} = (a^{r_n})^{s_n} < (a^x)^{s_n} < (a^x)^y. \qquad (11.5)$$

Here, other than using Theorem 3.27, in the middle inequality we used the fact that if $0 < u < v$ and $s > 0$ is rational, then $u^s < v^s$. This follows from $v^s / u^s = (v/u)^s > (v/u)^0 = 1$, since $v/u > 1$, and then we can apply Theorem 3.27 again. Now from (11.5), we get that

$$a^{xy} = \lim_{n \to \infty} a^{r_n s_n} \le (a^x)^y.$$

The inequality $a^{xy} \ge (a^x)^y$ can be proven similarly if we take sequences $r_n \to x$ and $s_n \to y$ consisting of rational numbers such that $r_n > x$ and $s_n > y$ for all n. $\qquad \square$

We note that by the second identity of (11.4),

$$a^x \cdot a^{-x} = a^{x+(-x)} = a^0 = 1,$$

and so $a^{-x} = 1/a^x$ holds for all $a > 0$ and real numbers x.

Now we can continue and define exponential and power functions. If in the power a^b we consider the base to be fixed and let the exponent vary, then we get the exponential functions; if we consider the exponent to be fixed and let the base be a variable, then we get the power functions. The precise definition is the following.

Definition 11.5. Given arbitrary $a > 0$, the function $x \mapsto a^x$ $(x \in \mathbb{R})$ is called the *exponential function with base a*.

Given arbitrary $b \in \mathbb{R}$, the function $x \mapsto x^b$ $(x > 0)$ is called the *power function with exponent b*.

Since $1^x = 1$ for all x, the function that is identically 1 is one of the exponential functions. Similarly, by $x^0 = 1$ $(x > 0)$ and $x^1 = x$, the functions 1 and x are power functions over the interval $(0, \infty)$.

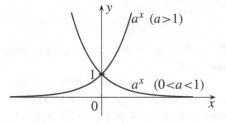

Fig. 11.1

The most important properties of exponential functions are summarized by the following theorem.

Theorem 11.6.

(i) *If $a > 1$, then the exponential function a^x is positive everywhere, strictly monotone increasing, and continuous on $\mathbb{R}$. Moreover,*

$$\lim_{x \to \infty} a^x = \infty \;\; and \;\; \lim_{x \to -\infty} a^x = 0. \tag{11.6}$$

(ii) *If $0 < a < 1$, then the exponential function a^x is positive everywhere, strictly monotone decreasing, and continuous on $\mathbb{R}$. Moreover,*

$$\lim_{x \to \infty} a^x = 0 \;\; and \;\; \lim_{x \to -\infty} a^x = \infty.$$

(iii) *Given arbitrary $a > 0$, the function a^x is convex on $\mathbb{R}$ (Figure 11.1).*

Proof. We have already seen in Theorem 3.27 that if $a > 1$, then the function a^x is positive and strictly monotone increasing. Thus by Theorem 10.68, the limits $\lim_{x \to \infty} a^x$ and $\lim_{x \to -\infty} a^x$ exist. And since $a^n \to \infty$ and $a^{-n} \to 0$ if $n \to \infty$, (11.6) holds. The analogous statements when $0 < a < 1$ can be proven similarly.

The continuity of exponential functions is clear by Lemma 11.3, using Theorem 10.19. Thus we have proved statements (i) and (ii).

Let $a > 0$ and $x, y \in \mathbb{R}$. If we apply the inequality between the arithmetic and geometric means to the numbers a^x and a^y, then we get that

$$a^{(x+y)/2} = \sqrt{a^x \cdot a^y} \le \frac{a^x + a^y}{2}.$$

This means that the function a^x is weakly convex. Since it is continuous, it is convex by Theorem 10.76. $\square$

The corresponding properties of power functions are given by the following theorem (Figure 11.2).

Theorem 11.7.

(i) *If $b > 0$, then the power function x^b is positive, strictly monotone increasing, and continuous on the interval $(0, \infty)$, and moreover,*

$$\lim_{x \to 0+0} x^b = 0 \quad and \quad \lim_{x \to \infty} x^b = \infty.$$

(ii) *If $b < 0$, then the power function x^b is positive, strictly monotone decreasing, and continuous on the interval $(0, \infty)$, and moreover,*

$$\lim_{x \to 0+0} x^b = \infty \quad and \quad \lim_{x \to \infty} x^b = 0.$$

(iii) *If $b \geq 1$ or $b \leq 0$, then the function x^b is convex on $(0, \infty)$. If $0 \leq b \leq 1$, then x^b is concave on $(0, \infty)$.*

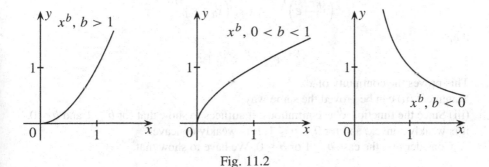

Fig. 11.2

To prove this theorem, we will require a generalization of Bernoulli's inequality (Theorem 2.5).

Theorem 11.8. *Let $x > -1$.*

(i) *If $b \geq 1$ or $b \leq 0$, then $(1+x)^b \geq 1+bx$.*
(ii) *If $0 \leq b \leq 1$, then $(1+x)^b \leq 1+bx$.*

Proof. We already proved the statement for nonnegative exponents in Exercise 3.33. The following simple proof is based on the convexity of the exponential function.

Let us change our notation: write a instead of x, and x instead of b. We have to show that if $a > -1$, then $x \in [0, 1]$ implies $(1+a)^x \leq ax+1$, while $x \notin (0, 1)$ implies $(1+a)^x \geq ax+1$. Both statements follow from the fact that the function $(1+a)^x$ is convex. We see this by noting that the chord connecting the points 0 and 1 is exactly $y = ax + 1$; in other words, $h_{0,1}(x) = ax + 1$. Thus if $x \in [0, 1]$, then the inequality $(1+a)^x \leq h_{0,1}(x)$ follows from the definition of convexity, while for $x \notin (0, 1)$, we have $(1+a)^x \geq h_{0,1}(x)$ by Lemma 10.74. $\qquad \square$

Proof (Theorem 11.7). (i) Let $b > 0$. If $t > 1$, then $t^b > t^0 = 1$ by Theorem 3.27. Now if $0 < x < y$, then

$$y^b = \left(\frac{y}{x} \cdot x\right)^b = \left(\frac{y}{x}\right)^b \cdot x^b > 1 \cdot x^b = x^b,$$

which shows that x^b is strictly monotone increasing. Since for arbitrary $K > 0$, we have $x^b > K$ if $x > K^{1/b}$, it follows that $\lim_{x \to \infty} x^b = \infty$. Similarly, for arbitrary $\varepsilon > 0$, we have $x^b < \varepsilon$ if $x < \varepsilon^{1/b}$, and so $\lim_{x \to 0+0} x^b = 0$. (In both arguments, we used that $\left(a^{1/b}\right)^b = a^1 = a$ for all $a > 0$.)

Let $x_0 > 0$ be given; we will see that the function x^b is continuous at x_0. If $0 < \varepsilon < x_0^b$, then by the monotonicity of the power function with exponent $1/b$, we have

$$\left(x_0^b - \varepsilon\right)^{1/b} < x_0 < \left(x_0^b + \varepsilon\right)^{1/b}.$$

Now if

$$\left(x_0^b - \varepsilon\right)^{1/b} < x < \left(x_0^b + \varepsilon\right)^{1/b},$$

then

$$x_0^b - \varepsilon < x^b < x_0^b + \varepsilon.$$

This proves the continuity of x^b.

Statement (ii) can be proved the same way.

(iii) Since the function x^b is continuous, it suffices to show that for $b \geq 1$ and $b \leq 0$, it is weakly convex, and for $0 \leq b \leq 1$, it is weakly concave.

Consider first the case $b \geq 1$ or $b \leq 0$. We have to show that

$$\left(\frac{x+y}{2}\right)^b \leq \frac{x^b + y^b}{2} \tag{11.7}$$

for all $x, y > 0$. Let us introduce the notation $(x+y)/2 = t$, $x/t = u$, $y/t = v$. Then $u + v = 2$. By statement (i) of Theorem 11.8, $u^b \geq 1 + b \cdot (u - 1)$ and $v^b \geq 1 + b \cdot (v - 1)$. Thus

$$\frac{u^b + v^b}{2} \geq 1 + b \cdot \frac{u + v - 2}{2} = 1.$$

If we multiply this inequality through by t^b, then we get (11.7).

Now suppose that $0 \leq b \leq 1$; we have to show that $\left((x+y)/2\right)^b \geq (x^b + y^b)/2$. We get this result by repeating the argument we used above, except that we apply statement (ii) of Theorem 11.8 instead of (i). □

As another application of Theorem 11.8, we inspect the function $\left(1 + 1/x\right)^x$.

Theorem 11.9. *The function* $f(x) = \left(1 + 1/x\right)^x$ *is monotone increasing on the intervals* $(-\infty, -1)$ *and* $(0, \infty)$, *and*

$$\lim_{x \to -\infty} \left(1 + \frac{1}{x}\right)^x = \lim_{x \to \infty} \left(1 + \frac{1}{x}\right)^x = e. \tag{11.8}$$

Proof. If $0 < x < y$, then $y/x > 1$, so by Theorem 11.8,

$$\left(1+\frac{1}{y}\right)^{y/x} \geq 1+\frac{y}{x}\cdot\frac{1}{y} = 1+\frac{1}{x},$$

and so by the monotonicity of the power function,

$$\left(1+\frac{1}{y}\right)^{y} \geq \left(1+\frac{1}{x}\right)^{x}. \qquad (11.9)$$

If, however, $x < y < -1$, then $0 < y/x < 1$, so again by Theorem 11.8,

$$\left(1+\frac{1}{y}\right)^{y/x} \leq 1+\frac{y}{x}\cdot\frac{1}{y} = 1+\frac{1}{x},$$

and then by the monotone decreasing property of the power function with exponent x, we get (11.9) again. Thus we have shown that f is monotone increasing on the given intervals.

By Theorem 10.68, it then follows that f has limits at infinity in both directions. Since $(1+1/n)^n \to e$ (since this was the definition of the number e), the limit at infinity can only be e. On the other hand,

$$\left(1-\frac{1}{n}\right)^{n} = \left(\frac{n-1}{n}\right)^{-n} = \left(\frac{n}{n-1}\right)^{n} = \left(1+\frac{1}{n-1}\right)^{n-1}\cdot\left(1+\frac{1}{n-1}\right),$$

so

$$\lim_{n\to-\infty}\left(1+\frac{1}{n}\right)^{n} = \lim_{n\to\infty}\left(1-\frac{1}{n}\right)^{-n} = e.$$

This gives us the first equality of (11.8). □

We can generalize Theorem 11.9 in the following way.

Theorem 11.10. *For an arbitrary real number b,*

$$\lim_{x\to-\infty}\left(1+\frac{b}{x}\right)^{x} = \lim_{x\to\infty}\left(1+\frac{b}{x}\right)^{x} = e^{b}. \qquad (11.10)$$

Proof. The statement is clear for $b = 0$. If $b > 0$, then using Theorem 10.41 for limits of compositions yields

$$\lim_{x\to\infty}\left(1+\frac{b}{x}\right)^{x/b} = \lim_{x\to\infty}\left(1+\frac{1}{x}\right)^{x} = e,$$

and so by the continuity of power functions with exponent b,

$$\lim_{x\to\infty}\left(1+\frac{b}{x}\right)^{x} = \lim_{x\to\infty}\left[\left(1+\frac{b}{x}\right)^{x/b}\right]^{b} = e^{b}.$$

We can find the limit at $-\infty$ similarly, as well as in the $b < 0$ case. □

Corollary 11.11. *For an arbitrary real number b, $\lim_{h \to 0}(1 + bh)^{1/h} = e^b$.*

Proof. Applying Theorem 10.41 twice, we get

$$\lim_{h \to 0 \pm 0} (1 + bh)^{1/h} = \lim_{x \to \pm\infty} \left(1 + \frac{b}{x}\right)^x = e^b.$$

$\square$

By Theorem 11.10,

$$\lim_{n \to \infty} \left(1 + \frac{b}{n}\right)^n = e^b \tag{11.11}$$

for all real numbers b. This fact has several important applications.

Examples 11.12. **1.** Suppose that a bank pays p percent yearly interest on a bond. A 1-dollar bond would then grow to $a + q$ dollars after a year, where $q = p/100$. If, however, after half a year, the bank pays interest of $p/2$ percent, and from this point on, the new interest applies to this increased value, then at the end of the year, our bond will be worth $1 + (q/2) + [1 + (q/2)] \cdot (q/2) = [1 + (q/2)]^2$ dollars. If now we divide the year into n equal parts, and the interest is added to our bond after every $1/n$ year, which is then included in calculating interest from that point on, then by the end of the year, the bond will be worth $[1 + (q/n)]^n$ dollars. This sequence is monotone increasing (why?), and as we have seen, its limit is $e^q = e^{p/100}$. This means that no matter how often we add the interest to our bond during the year, its value cannot exceed $e^{p/100}$, but it can get arbitrarily close to it.
2. In this application, we inspect how much a certain material (say a window pane) absorbs a certain radiation (say a fixed wavelength of light). The amount of absorption is a function of the thickness of the material. This function is not linear, but experience shows that for thin slices of the material, a linear approximation works well. This means that there exists a positive constant α (called the absorption coefficient) such that a slice of the material with thickness h absorbs about $\alpha \cdot h$ of entering radiation if h is sufficiently small.

Consider a slab of the material with thickness h, where h is now arbitrary. Subdivide this slab into n equal slices. If n is sufficiently large, then each slice with thickness h/n will absorb $\alpha \cdot (h/n)$ of the radiation that has made it that far. Thus after the ith slice, $(1 - (\alpha h/n))^i$ of the radiation remains, so after the light leaves the whole slab, $1 - (1 - (\alpha h/n))^n$ of it is absorbed. If we let n go to infinity, we get that a slab of thickness h will absorb $1 - e^{-\alpha h}$ of the radiation.

The following theorem gives an interesting characterization of exponential functions.

Theorem 11.13. *The function $f: \mathbb{R} \to \mathbb{R}$ is an exponential function if and only if it is continuous, not identically zero, and satisfies the identity*

$$f(x_1 + x_2) = f(x_1) \cdot f(x_2) \tag{11.12}$$

for all $x_1, x_2 \in \mathbb{R}$.

Proof. We already know that the conditions are satisfied for an exponential function. Suppose now that f satisfies these conditions, and let $a = f(1)$. We will show that $a > 0$ and $f(x) = a^x$ for all x.

Since $f(x) = f((x/2) + (x/2)) = f(x/2)^2$ for all x, f is nonnegative everywhere. If f vanishes at a point x_0, then by the identity $f(x) = f((x - x_0) + x_0) = f(x - x_0) \cdot f(x_0)$, it would follow that f is identically zero, which we have excluded. Thus f is positive everywhere, so specifically $a = f(1)$ is positive. Since $f(0) = f(0 + 0) = f(0)^2$ and $f(0) > 0$, we have $f(0) = 1$. Thus we get that $1 = f(0) = f(x + (-x)) = f(x) \cdot f(-x)$, so $f(-x) = 1/f(x)$ for all x.

From assumption (11.12), using induction we know that

$$f(x_1 + \cdots + x_n) = f(x_1) \cdot \ldots \cdot f(x_n)$$

holds for all n and $x_1, \ldots, x_n$. Then by the choice $x_1 = \cdots = x_n = x$, we get $f(nx) = f(x)^n$. Thus if p and q are positive integers, then

$$\left(f\left(\frac{p}{q} \right) \right)^q = f\left(\frac{p}{q} \cdot q \right) = f(p \cdot 1) = f(1)^p = a^p,$$

that is, $f(p/q) = a^{p/q}$. Since $f(-p/q) = 1/f(p/q) = 1/a^{p/q} = a^{-p/q}$, we have shown that $f(r) = a^r$ for all rational numbers r.

If x is an arbitrary real number, then let (r_n) be a sequence of rational numbers that tends to x. Since f is continuous on x,

$$f(x) = \lim_{n \to \infty} f(r_n) = \lim_{n \to \infty} a^{r_n} = a^x.$$

$\square$

Remark 11.14. Theorem 11.13 characterizes exponential functions with the help of a **functional equation**. We ran into a similar functional equation in Chapter 8, when we mentioned functions that satisfy

$$f(x_1 + x_2) = f(x_1) + f(x_2) \tag{11.13}$$

while talking about weakly convex functions; this functional equation is called **Cauchy's functional equation**. We also mentioned that there exist solutions of (11.13) that are not continuous. These solutions cannot be bounded from above on any interval, by Exercise 10.94. If f is such a function, then the function $e^{f(x)}$ satisfies the functional equation (11.12), and it is not bounded in any interval. This remark shows that in Theorem 11.13, the continuity condition cannot be dropped (although it can be weakened).

We will run into two relatives of the functional equations above, whose continuous solutions are exactly the power functions and logarithmic functions (see Exercises 11.15 and 11.26). Equally noteworthy is **d'Alembert's**[3] **functional equation:**

$$f(x_1 + x_2) + f(x_1 - x_2) = 2f(x_1)f(x_2).$$

See Exercises 11.35 and 11.44 regarding the continuous solutions of this functional equation.

Generalized mean. If $a > 0$ and $b \neq 0$, then we will also denote the power $a^{1/b}$ by $\sqrt[b]{a}$ (which is strongly connected to the fact that $a^{1/k} = \sqrt[k]{a}$ for positive integers k by definition). Let $a_1, \ldots, a_n$ be positive numbers, and let $b \neq 0$. The quantity

$$G(b; a_1, \ldots, a_n) = \sqrt[b]{\frac{a_1^b + \cdots + a_n^b}{n}}$$

is called the **generalized mean with exponent** b of the numbers a_i. It is clear that $G(-1; a_1, \ldots, a_n)$, $G(1; a_1, \ldots, a_n)$, and $G(2; a_1, \ldots, a_n)$ are exactly the harmonic, arithmetic, and quadratic means of the numbers a_i. We also consider the geometric mean as a generalized mean, since we define the generalized mean with exponent 0 by

$$G(0; a_1, \ldots, a_n) = \sqrt[n]{a_1 \cdot \ldots \cdot a_n}.$$

(To see the motivation behind this definition, see Exercise 12.35.)

Let $b \geq 1$. Apply Jensen's inequality (Theorem 9.19) to x^b. We get that

$$\left(\frac{a_1 + \cdots + a_n}{n}\right)^b \leq \frac{a_1^b + \cdots + a_n^b}{n}.$$

Raising both sides to the power $1/b$, we get the inequality

$$\frac{a_1 + \cdots + a_n}{n} \leq G(b; a_1, \ldots, a_n) = \sqrt[b]{\frac{a_1^b + \cdots + a_n^b}{n}}.$$

This is called the **generalized mean inequality** (which implies the inequality of arithmetic and quadratic means as a special case). However, this inequality is also just a special case of the following theorem.

Theorem 11.15. *Let* $a_1, \ldots, a_n$ *be fixed positive numbers. Then the function*

$$b \mapsto G(b; a_1, \ldots, a_n) \quad (b \in \mathbb{R})$$

is monotone increasing on $\mathbb{R}$.

Proof. Suppose first that $0 < b < c$. Then $c/b > 1$. Apply Jensen's inequality to the function $x^{c/b}$ and the numbers a_i^b $(i = 1, \ldots, n)$. We get that

---

[3] Jean le Rond d'Alembert (1717–1783), French mathematician.

$$\left(\frac{a_1^b + \cdots + a_n^b}{n}\right)^{c/b} \le \frac{a_1^c + \cdots + a_n^c}{n}.$$

If we raise both sides to the power $1/c$, then we get that

$$G(b; a_1, \ldots, a_n) \le H(c; a_1, \ldots, a_n). \tag{11.14}$$

Now let $b = 0$ and $c > 0$. Apply the inequality of arithmetic and geometric means for the numbers a_i^c $(i = 1, \ldots, n)$. We get that

$$G(0; a_1, \ldots, a_n)^c = \sqrt[n]{a_1^c \cdot \ldots \cdot a_n^c} \le \frac{a_1^c + \cdots + a_n^c}{n},$$

and if here we raise both sides to the power $1/c$, then we get (11.14) again.

It is easy to check that

$$G(-b; a_1, \ldots, a_n) = \frac{1}{G\left(b; \frac{1}{a_1}, \ldots, \frac{1}{a_n}\right)}$$

for all b. So using the inequalities we just proved, we get that for $b < c \le 0$, we have

$$G(b; a_1, \ldots, a_n) = \frac{1}{G\left(-b; \frac{1}{a_1}, \ldots, \frac{1}{a_n}\right)} \le \frac{1}{G\left(-c; \frac{1}{a_1}, \ldots, \frac{1}{a_n}\right)} =$$

$$= G(c; a_1, \ldots, a_n),$$

which proves the theorem. □

As we saw in Theorem 11.7, $\lim_{x \to 0+0} x^b = 0$ if $b > 0$; then it is in our best interest for all positive powers of zero to be zero. Thus we define $0^b = 0$ for all $b > 0$ (but we still do not define the nonpositive powers of zero). Regarding this change, for $b > 0$ we can extend the power function x^b to be defined at zero as well, where its value is zero. The new extended function x^b is then continuous from the right at zero if $b > 0$.

Exercises

11.8. Prove that for the numbers $0 < a < b$, $a^b = b^a$ holds if and only if there exists a positive number x such that

$$a = \left(1 + \frac{1}{x}\right)^x \quad \text{and} \quad b = \left(1 + \frac{1}{x}\right)^{x+1}.$$

11.9. Prove that if $a > 0$ and $a \ne 1$, then the function a^x is strictly convex.

11.10. Prove that if $b > 1$ or $b < 0$, then the function x^b is strictly convex.

11.11. Prove that if $0 < b < 1$, then the function x^b is strictly concave.

11.12. Let $x > -1$ and $b \in \mathbb{R}$. Prove that $(1+x)^b = 1 + bx$ holds if and only if at least one of $x = 0$, $b = 0$, $b = 1$ holds.

11.13. $\lim_{x \to -1-0} \left(1 + \frac{1}{x}\right)^x = ?$

11.14. Let $G(b; a_1, \ldots, a_n)$ be the generalized mean with exponent b of the positive numbers $a_1, \ldots, a_n$. Prove that

$$\lim_{b \to \infty} G(b; a_1, \ldots, a_n) = \max(a_1, \ldots, a_n)$$

and

$$\lim_{b \to -\infty} G(b; a_1, \ldots, a_n) = \min(a_1, \ldots, a_n).$$

11.15. Prove that the function $f : (0, \infty) \to \mathbb{R}$ is a power function if and only if it is continuous, not identically zero, and satisfies the identity

$$f(x_1 \cdot x_2) = f(x_1) \cdot f(x_2)$$

for all positive x_1, x_2.

11.3 Logarithmic Functions

If $a > 0$ and $a \neq 1$, then the function a^x is strictly monotone and continuous on $\mathbb{R}$ by Theorem 11.6. Thus if $a > 0$ and $a \neq 1$, then a^x has an inverse function, which we call the **logarithm to base** a, and we denote it by $\log_a x$. Since the image of a^x is $(0, \infty)$, the function $\log_a x$ is defined on the open interval $(0, \infty)$, but its image is $\mathbb{R}$. Having the definition of the inverse function in mind, we come to the conclusion that if $a > 0$, $a \neq 1$, and $x > 0$, then $\log_a x$ is the only real number for which $a^{\log_a x} = x$ holds. Specifically, $\log_a 1 = 0$ and $\log_a a = 1$ (Figure 11.3).

Theorem 11.16.

(i) *If $a > 1$, then the function $\log_a x$ is strictly monotone increasing, continuous, and strictly concave on $(0, \infty)$. Moreover,*

$$\lim_{x \to 0+0} \log_a x = -\infty \quad \text{and} \quad \lim_{x \to \infty} \log_a x = \infty. \quad (11.15)$$

(ii) *If $0 < a < 1$, then the function $\log_a x$ is strictly monotone decreasing, continuous, and strictly convex on $(0, \infty)$. Moreover,*

$$\lim_{x \to 0+0} \log_a x = \infty \quad \text{and} \quad \lim_{x \to \infty} \log_a x = -\infty. \quad (11.16)$$

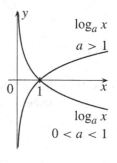

Fig. 11.3

(iii) *For all $a > 0$, $a \neq 1$, and $x, y > 0$, the identities*

$$\log_a(xy) = \log_a x + \log_a y,$$
$$\log_a(x/y) = \log_a x - \log_a y, \tag{11.17}$$
$$\log_a(1/y) = -\log_a y,$$

as well as

$$\log_a(x^y) = y \cdot \log_a x \quad and \quad \log_a \sqrt[y]{x} = \frac{1}{y} \cdot \log_a x \tag{11.18}$$

hold.

Proof. As an application of Theorems 11.6 and 10.72, we get that the function $\log_a x$ is continuous everywhere; if $a > 1$, it is strictly monotone increasing; and if $0 < a < 1$, it is strictly monotone decreasing. Since its image is $\mathbb{R}$, the limit relations (11.15) and (11.16) follow from this. Thus we have proved (i) and (ii) except for the statements about convexity and concavity.

Let $a > 0$, $a \neq 1$. By the second identity in (11.4), we obtain that for arbitrary $x, y > 0$ we have

$$a^{\log_a x + \log_a y} = a^{\log_a x} \cdot a^{\log_a x} = x \cdot y = a^{\log_a(xy)}.$$

Since the function a^x is strictly monotone, we get the first identity (11.17) from this. Thus

$$\log_a x = \log_a \left(\frac{x}{y} \cdot y \right) = \log_a \left(\frac{x}{y} \right) + \log_a y,$$

which is the second identity in (11.17). Applying this to $x = 1$ gives us the third one. If we now apply the third identity of (11.4), then we get that

$$a^{\log_a(x^y)} = x^y = \left(a^{\log_a x} \right)^y = a^{y \cdot \log_a x},$$

which implies the first identity of (11.18). Applying this to $1/y$ instead of y will yield the second identity of (11.18), where we note that $\sqrt[y]{x} = x^{1/y}$.

If $0 < x < y$, then by the inequality of arithmetic and geometric means, $\sqrt{xy} < (x+y)/2$. Applying (11.17) and (11.18), we get that if $a > 1$, then

$$\frac{\log_a x + \log_a y}{2} < \log_a \left(\frac{x+y}{2} \right),$$

where we note also that the function $\log_a$ is strictly monotone increasing for $a > 1$. This means that if $a > 1$, then $\log_a$ is strictly weakly concave. Similarly, we get that if $0 < a < 1$, then the function $\log_a$ is strictly weakly convex. Since we are talking about continuous functions, when $a > 1$, the function $\log_a x$ is strictly concave, while if $0 < a < 1$, then it is strictly convex. This concludes the proof of the theorem. $\square$

Remark 11.17. Let a and b be positive numbers different from 1. Then by applying (11.18), we get

$$\log_a x = \log_a \left(b^{\log_b x} \right) = \log_b x \cdot \log_a b,$$

which gives us

$$\log_b x = \frac{\log_a x}{\log_a b} \tag{11.19}$$

for all $x > 0$. This means that *two logarithmic functions differ by only a constant multiple.* Thus it is useful to choose one logarithmic function and write the rest of them as multiples of this one. But which one should we choose among the infinitely many logarithmic functions? This is determined by necessity; clearly, it is most useful to choose the one that we use the most. Often in engineering, the base-10 logarithm is used, while in computer science, it is base 2. We will use the logarithmic function at base e, since this will make the equations that arise in differentiation the simplest. *From now on, we will write* $\log x$ *instead of* $\log_e x$. (Sometimes, the logarithm at base e is denoted by $\ln x$, which stands for *logaritmus naturalis*, the natural logarithm.)

If we apply (11.19) to $a = e$, we get that

$$\log_b x = \frac{\log x}{\log b} \tag{11.20}$$

whenever $b > 0$, $b \neq 1$, and $x > 0$. Just as every logarithmic function can be expressed using the natural logarithm, we can also express every exponential function using e^x. In fact, if $a > 0$, then by the definition of $\log a$ and by the third identity of (11.4),

$$a^x = \left(e^{\log a}\right)^x = e^{x \cdot \log a} = (e^x)^{\log a},$$

that is, a^x is a power of the function e^x.

Considering that $e > 1$, the function $\log x$ is concave. This fact makes proving an important inequality possible.

Theorem 11.18 (Hölder's Inequality[4]). *Let p and q be positive numbers such that $1/p + 1/q = 1$. Then for arbitrary real numbers $a_1, \ldots, a_n$ and $b_1, \ldots, b_n$,*

$$|a_1 b_1 + \cdots + a_n b_n| \leq \sqrt[p]{|a_1|^p + \cdots + |a_n|^p} \cdot \sqrt[q]{|b_1|^q + \cdots + |b_n|^q}. \tag{11.21}$$

Proof. First we show that

$$ab \leq \frac{a^p}{p} + \frac{b^q}{q} \tag{11.22}$$

for all $a, b \geq 0$. This is clear if $a = 0$ or $b = 0$, so we can assume that $a > 0$ and $b > 0$. Since $\log x$ is concave, by Lemma 9.17,

$$\log (ta^p + (1-t)b^q) \geq t \log a^p + (1-t) \log b^q \tag{11.23}$$

for all $0 < t < 1$. Now we apply this inequality with $t = 1/p$. Then $1 - t = 1/q$, and so the right-hand side of (11.23) becomes $\log a + \log b = \log(ab)$. Since $\log x$ is monotone increasing, we get that $(1/p)a^p + (1/q)b^q \geq ab$, which is exactly (11.22).

---

[4] Otto Ludwig Hölder(1859–1937), German mathematician.

Now to continue the proof of the theorem, let $A = \sqrt[p]{|a_1|^p + \cdots + |a_n|^p}$ and $B = \sqrt[q]{|b_1|^q + \cdots + |b_n|^q}$. If $A = 0$, then $a_1 = \ldots = a_n = 0$, and so (11.21) holds, since both sides are zero. The same is the case if $B = 0$, so we can assume that $A > 0$ and $B > 0$.

Let $\alpha_i = |a_i|/A$ and $\beta_i = |b_i|/B$ for all $i = 1, \ldots, n$. Then

$$\alpha_1^p + \cdots + \alpha_n^p = \beta_1^q + \cdots + \beta_n^q = 1. \tag{11.24}$$

Now by (11.22),

$$\alpha_i \beta_i \leq \frac{1}{p}\alpha_i^p + \frac{1}{q}\beta_i^q$$

for all i. Summing these inequalities, we get that

$$\alpha_1 \beta_1 + \cdots + \alpha_n \beta_n \leq \frac{1}{p} \cdot (\alpha_1^p + \cdots + \alpha_n^p) + \frac{1}{q} \cdot (\beta_1^q + \cdots + \beta_n^q) = \frac{1}{p} \cdot 1 + \frac{1}{q} \cdot 1 = 1,$$

using (11.24) and the assumption on the numbers p and q. If now we write $|a_i|/A$ and $|b_i|/B$ in place of α_i and multiply both sides by AB, then we get that

$$|a_1 b_1| + \cdots + |a_n b_n| \leq AB,$$

from which we immediately get (11.21) by the triangle inequality. □

Hölder's inequality for the special case $p = q = 2$ gives the following, also very notable, inequality.

Theorem 11.19 (Cauchy–Schwarz[5]–Bunyakovsky[6]Inequality). *For arbitrary real numbers $a_1, \ldots, a_n$ and $b_1, \ldots, b_n$,*

$$|a_1 b_1 + \cdots + a_n b_n| \leq \sqrt{a_1^2 + \cdots + a_n^2} \cdot \sqrt{b_1^2 + \cdots + b_n^2}.$$

We also give a direct proof.

Proof. For arbitrary $i, j = 1, \ldots, n$, the number

$$A_{i,j} = a_i^2 b_j^2 + a_j^2 b_i^2 - 2a_i a_j b_i b_j$$

is nonnegative, since $A_{i,j} = (a_i b_j - a_j b_i)^2$. If we add all the numbers $A_{i,j}$ for all $1 \leq i < j \leq n$, then we get the difference

$$(a_1^2 + \cdots + a_n^2) \cdot (b_1^2 + \cdots + b_n^2) - (a_1 b_1 + \cdots + a_n b_n)^2,$$

which is thus also nonnegative. □

---

[5] Hermann Amandus Schwarz (1843–1921), German mathematician.

[6] ViktorYakovlevich Bunyakovsky (1804–1889), Russian mathematician.

Exercises

11.16. Prove that $1 + 1/2 + \cdots + 1/n > \log n$ for all n. (H)

11.17. Prove that the sequence $1 + 1/2 + \cdots + 1/n - \log n$ is monotone decreasing and convergent.[7]

11.18. Let f be a strictly monotone increasing and convex (concave) function on the open interval I. Prove that the inverse of f is concave (convex).

11.19. Let f be strictly monotone decreasing and convex (concave) on the open interval I. Prove that the inverse of f is convex (concave). Check these statements for the cases that f is an exponential, power, or logarithmic function.

11.20. Prove that $\lim_{x \to 0+0} x^{\varepsilon} \cdot \log x = 0$ for all $\varepsilon > 0$.

11.21. Prove that $\lim_{x \to \infty} x^{-\varepsilon} \cdot \log x = 0$ for all $\varepsilon > 0$.

11.22. $\lim_{x \to 0+0} x^x = ?$ $\lim_{x \to \infty} \sqrt[x]{x} = ?$ $\lim_{x \to 0+0} (1 + 1/x)^x = ?$

11.23. Let $\lim_{n \to \infty} a_n = a$, $\lim_{n \to \infty} b_n = b$. When does $\lim_{n \to \infty} a_n^{b_n} = a^b$ hold?

11.24. Prove that equality holds in (11.22) only if $a^p = b^q$.

11.25. Suppose that the numbers $a_1, \ldots, a_n, b_1, \ldots, b_n$ are nonnegative. Prove that equality holds in (11.21) if and only if $a_1 = \ldots = a_n = 0$ or if there exists a number t such that $b_i^q = t \cdot a_i^p$ for all $i = 1, \ldots, n$.

11.26. Prove that the function $f \colon (0, \infty) \to \mathbb{R}$ is a logarithmic function if and only if it is continuous, not identically zero, and satisfies the identity

$$f(x_1 \cdot x_2) = f(x_1) + f(x_2)$$

for all positive x_1, x_2.

11.4 Trigonometric Functions

Defining trigonometric functions. Since trigonometric functions are functions that deal with angles, or more precisely, map angles to real numbers, we first need to clarify how angles are measured.

Let O be a point in the plane, let h and k be rays starting at point O, and let $(h,k) \lessdot$ denote the section of the plane (one of them, that is) determined by h and k. Let D be a disk centered at O, and C a circle centered at O. It is clear by inspection that the angle determined by h and k is proportional to the area of the sector $D \cap (h,k) \lessdot$, as well as the length of the arc $C \cap (h,k) \lessdot$. It would be a good choice, then, to measure

---

[7] The limit of this sequence is called **Euler's constant**. *Whether this number is rational has been an open question for a long time.*

the angle by the area of the sector $D \cap (h,k) \lhd$ or the length of the arc $C \cap (h,k) \lhd$ (or any other quantity directly proportional to these). We will choose the length of the arc. We agree that an angle at the point O is measured by the length of the arc of a unit circle centered at O that subtends the angle. We call this number the **angular measure**, and the units for it are **radians**. Thus the angular measure of a straight angle is the length of the unit semicircle, that is, π radians; the angular measure of a right angle is half of this, that is, $\pi/2$ radians. From now on, we omit the word "radians," so measurements of angles (unless otherwise specified) will be given in radians.

In trigonometry, we define the co-sine (or sine) of an angle x smaller than $\pi/2$ by considering the right tri-angle with acute angle x and taking the quotient of the length of the side adjacent (or opposite) to the angle x and the hypotenuse. Let $0 < u < 1$ and $v = c(u) = \sqrt{1 - u^2}$. Then the points $O = (0,0)$, $P = (u,0)$, and $Q = (u,v)$ define a right triangle whose angle at O is defined by the rays $\overrightarrow{OP}$ and $\overrightarrow{OQ}$. The arc of the circle C that falls into this

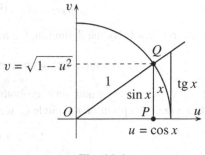

Fig. 11.4

section agrees with the graph of c over the interval $[u, 1]$, whose length is $s(c; [u, 1])$ (see Definition 10.78). If $s(c; [u, 1]) = x$, then the angular measure of the angle $POQ \lhd$ is x, and so $\cos x = \overline{OP}/\overline{OQ} = u/1 = u$, and $\sin x = \overline{PQ}/\overline{OQ} = v/1 = v$. We can also formulate this by saying that *if starting from the point* $(1,0)$, *we measure out an arc of length x onto the circle C in the positive direction, then the point we end up at has coordinates* $(\cos x, \sin x)$. Here the "positive direction" means that we move in a counterclockwise direction, that is, we start measuring the arc in the upper half-plane.[8]

We will accept the statement above as the definition of the functions $\cos x$ and $\sin x$.

Definition 11.20. Starting at the point $(1,0)$ on the unit circle centered at the origin, measure an arc of length x in the positive or negative direction according to whether $x > 0$ or $x < 0$. (If $|x| \geq 2\pi$, then we will circle around more than once as necessary.) The first coordinate of the point we end up at is denoted by $\cos x$, while the second coordinate is $\sin x$. Thus we have defined the functions $\cos$ and $\sin$ on the set of all real numbers.

Remark 11.21. In Remark 10.83, we saw that for $0 \leq x \leq \pi$, there exists a $u \in [-1, 1]$ such that $S(u) = s(c; [u, 1]) = x$. Thus the endpoint of an arc of length x measured on C is the point $(u, c(u)) = (u, \sqrt{1 - u^2})$. It then follows that $\cos x = u$ and $\sin x = \sqrt{1 - u^2} = \sqrt{1 - \cos x^2}$. The relation $\cos x = u$ means that on the interval $[0, \pi]$, the function $\cos x$ is exactly the inverse of the function S.

---

[8] That is, in the part of the plane given by $\{(x,y) : y > 0\}$.

We often need to use the expressions $\sin x/\cos x$ and $\cos x/\sin x$, for which we use the notation $\operatorname{tg} x$ and $\operatorname{ctg} x$ respectively.

Properties of Trigonometric Functions. As we noted, measuring an arc of length π from any point $(\cos x, \sin x)$ of the circle C gets us into the antipodal point, whose coordinates are $(-\cos x, -\sin x)$. Thus

$$\cos(x+\pi) = -\cos x \text{ and } \sin(x+\pi) = -\sin x \tag{11.25}$$

for all x. Since $(\cos 0, \sin 0) = (1,0)$, we have $\cos 0 = 1$ and $\sin 0 = 0$. Thus by (11.25),

$$\cos(k\pi) = (-1)^k \text{ and } \sin(k\pi) = 0 \tag{11.26}$$

for all integers k. By the definition, we immediately get that

$$\cos(x+2\pi) = \cos x \text{ and } \sin(x+2\pi) = \sin x$$

for all x, that is, $\cos x$ and $\sin x$ are both periodic functions with period 2π. Since $(\cos x, \sin x)$ is a point on the circle C, we have

$$\sin^2 x + \cos^2 x = 1 \tag{11.27}$$

for all x. The circle C is symmetric with respect to the horizontal axis. So if we measure a segment of length $|x|$ in the positive or negative direction starting at the point $(1,0)$, then we arrive at a point that is symmetric with respect to the horizontal axis. This means that $(\cos(-x), \sin(-x)) = (\cos x, -\sin x)$, that is,

$$\cos(-x) = \cos x \text{ and } \sin(-x) = -\sin x \tag{11.28}$$

for all x. In other words, the function $\cos x$ is even, while the function $\sin x$ is odd. Connecting the identities (11.28) and (11.25), we obtain that

$$\cos(\pi - x) = -\cos x \text{ and } \sin(\pi - x) = \sin x \tag{11.29}$$

for all x. Substituting $x = \pi/2$ into this, we get $\cos(\pi/2) = 0$, so by (11.25), we see that

$$\cos\left(\frac{\pi}{2} + k\pi\right) = 0 \quad (k \in \mathbb{Z}). \tag{11.30}$$

Since $\sin(\pi/2) = \sqrt{1 - \cos^2(\pi/2)} = 1$, then again by (11.25), we have

$$\sin\left(\frac{\pi}{2} + k\pi\right) = (-1)^k \quad (k \in \mathbb{Z}). \tag{11.31}$$

The circle C is also symmetric with respect to the $45°$ line passing through the origin. If we measure x in the positive direction from the point $(1,0)$ and then reflect over this line, we get the point that we would get if we measured x in the negative direction starting at $(0,1)$. Since $(0,1) = (\cos(\pi/2), \sin(\pi/2))$, the endpoint of the arc we mirrored is the same as the endpoint of the arc of length $x - (\pi/2)$ measured from $(1,0)$ in the negative direction, which is the same as $(\pi/2) - x$ in the positive direction. Thus we see that the point $(\sin x, \cos x)$—the reflection of $(\cos x, \sin x)$ about the $45°$ line passing through the origin—agrees with the point $(\cos((\pi/2) - x), \sin((\pi/2) - x))$, so

$$\cos\left(\frac{\pi}{2} - x\right) = \sin x \quad \text{and} \quad \sin\left(\frac{\pi}{2} - x\right) = \cos x \qquad (11.32)$$

for all x. The following identities are the addition formulas for sin and cos.

$$\begin{aligned}
\sin(x+y) &= \sin x \cos y + \cos x \sin y, \\
\sin(x-y) &= \sin x \cos y - \cos x \sin y, \\
\cos(x+y) &= \cos x \cos y - \sin x \sin y, \\
\cos(x-y) &= \cos x \cos y + \sin x \sin y.
\end{aligned} \qquad (11.33)$$

The proofs of the addition formulas are outlined in the first appendix of the chapter. The proof there is based on rotations about the origin. Later, using differentiation, we will give a proof that does not use geometric concepts and does not rely on geometric inspections (see the second appendix of Chapter 13).

The following identities are simple consequences of the addition formulas.

$$\begin{aligned}
\sin 2x &= 2\sin x \cos x, \\
\cos 2x &= \cos^2 x - \sin^2 x = 1 - 2\sin^2 x = 2\cos^2 x - 1,
\end{aligned} \qquad (11.34)$$

$$\cos^2 x = \frac{1 + \cos 2x}{2}, \quad \sin^2 x = \frac{1 - \cos 2x}{2}, \qquad (11.35)$$

$$\begin{aligned}
\cos x \cos y &= \tfrac{1}{2}\left(\cos(x+y) + \cos(x-y)\right), \\
\sin x \sin y &= \tfrac{1}{2}\left(\cos(x-y) - \cos(x+y)\right), \\
\sin x \cos y &= \tfrac{1}{2}\left(\sin(x+y) + \sin(x-y)\right),
\end{aligned} \qquad (11.36)$$

$$\begin{aligned}
\sin x + \sin y &= 2\sin\frac{x+y}{2}\cos\frac{x-y}{2}, \\
\sin x - \sin y &= 2\sin\frac{x-y}{2}\cos\frac{x+y}{2}, \\
\cos x + \cos y &= 2\cos\frac{x+y}{2}\cos\frac{x-y}{2}, \\
\cos x - \cos y &= -2\sin\frac{x-y}{2}\sin\frac{x+y}{2}.
\end{aligned} \qquad (11.37)$$

Now we turn to the analytic properties of the functions sin and cos (Figure 11.5).

Theorem 11.22.

(i) *The function $\cos x$ is strictly monotone decreasing on the intervals $[2k\pi, (2k+1)\pi]$ and strictly monotone increasing on the intervals $[(2k-1)\pi, 2k\pi]$, $(k \in \mathbb{Z})$. The only roots of the function $\cos x$ are the points $(\pi/2) + k\pi$.*

(ii) *The function $\sin x$ is strictly monotone increasing on the intervals $[2k\pi - (\pi/2), 2k\pi + (\pi/2)]$ and strictly monotone decreasing on the intervals $[2k\pi + (\pi/2), 2k\pi + (3\pi/2)]$, $(k \in \mathbb{Z})$. The only roots of the function $\sin x$ are the points $k\pi$.*

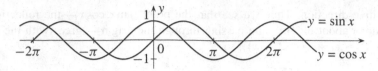

Fig. 11.5

Proof. (i) By remark 11.21, on the interval $[0, \pi]$ the function cos is none other than the inverse of the function S. Since S is strictly monotone decreasing on $[-1, 1]$, the inverse function $\cos x$ is also strictly monotone decreasing on $[0, \pi]$. Thus by (11.25), it is clear that if $k \in \mathbb{Z}$ is even, then $\cos x$ is strictly monotone decreasing on $[2k\pi, (2k+1)\pi]$ and is strictly monotone increasing on $[(2k-1)\pi, 2k\pi]$. Now the statement about the roots of $\cos x$ is clear from this.

Statement (ii) follows from (i) via the identity (11.32).   □

The following inequalities are especially important for applications.

Theorem 11.23. *For all x, the equations*

$$|\sin x| \leq |x| \tag{11.38}$$

and

$$0 \leq 1 - \cos x \leq x^2 \tag{11.39}$$

hold.

Proof. It suffices to prove the inequality (11.38) for nonnegative x, since both sides are even. If $x > \pi/2$, then $|\sin x| \leq 1 < \pi/2 < x$, and the statement holds. We can suppose that $0 \leq x \leq \pi/2$. Let $u = \cos x$ and $v = \sin x$. Then by the definition of $\cos x$ and $\sin x$, in the graph of the function $c(t) = \sqrt{1 - t^2}$, the arc length over the interval $[u, 1]$ is exactly x, since $(\cos x, \sin x) = (u, v)$ are the coordinates of the points that we get by measuring an arc of length x onto the circle C; see Figure 11.4. Then by Theorem 10.79,

$$0 \leq \sin x = v \leq \sqrt{(1-u)^2 + (v-0)^2} \leq s(c; [u, 1]) = x,$$

which proves (11.38).

For inequality (11.39), it again suffices to consider only the nonnegative x, since cos is an even function. If $x > \pi/2$, then

$$1 - \cos x \leq 2 < \left(\frac{3}{2}\right)^2 < \left(\frac{\pi}{2}\right)^2 < x^2,$$

and so (11.39) is true. If, however, $0 \le x \le \pi/2$, then $\cos x \ge 0$, and so

$$1 - \cos x = \frac{1 - \cos^2 x}{1 + \cos x} = \frac{\sin^2 x}{1 + \cos x} \le \sin^2 x \le x^2,$$

which gives (11.39). □

Theorem 11.24. *For all $x, y \in \mathbb{R}$, the inequalities*

$$|\cos x - \cos y| \le |x - y| \tag{11.40}$$

and

$$|\sin x - \sin y| \le |x - y| \tag{11.41}$$

hold.

Proof. By Theorem 11.23 and the identity (11.37), we get that

$$|\cos x - \cos y| = 2 \cdot \left| \sin \frac{x - y}{2} \right| \cdot \left| \sin \frac{x + y}{2} \right| \le 2 \cdot \left| \frac{x - y}{2} \right| \cdot 1 = |x - y|$$

and

$$|\sin x - \sin y| = 2 \cdot \left| \sin \frac{x - y}{2} \right| \cdot \left| \cos \frac{x + y}{2} \right| \le 2 \cdot \left| \frac{x - y}{2} \right| \cdot 1 = |x - y|.$$

□

Theorem 11.25.

(i) *The functions* sin *and* cos *are continuous everywhere, and in fact, they are Lipschitz.*

(ii) *The function* $\sin x$ *is strictly concave on the intervals* $[2k\pi, (2k + 1)\pi]$ *and strictly convex on the intervals* $[(2k - 1)\pi, 2k\pi]$, $(k \in \mathbb{Z})$.

(iii) *The function* $\cos x$ *is strictly concave on the intervals* $[2k\pi - (\pi/2), 2k\pi + (\pi/2)]$ *and strictly convex on the intervals* $[2k\pi + (\pi/2), 2k\pi + (2\pi/2)]$, $(k \in \mathbb{Z})$.

Proof. Statement (i) is clear by the previous theorem.

If $0 \le x < y \le \pi$, then applying the first identity in (11.37) yields

$$\frac{\sin x + \sin y}{2} = \sin \frac{x + y}{2} \cos \frac{x - y}{2} < \sin \frac{x + y}{2}.$$

This shows that $\sin x$ is strictly weakly concave on the interval $[0, \pi]$. Since it is continuous, it is strictly concave there. Then statement (ii) follows immediately by the identity $\sin(x + \pi) = -\sin x$.

Finally, statement (iii) follows from (ii) using the identity (11.32). □

Theorem 11.26. *If* $|x| < \pi/2$ *and* $x \neq 0$, *then*

$$\cos x \leq \frac{\sin x}{x} \leq 1. \tag{11.42}$$

Proof. Since the function $(\sin x)/x$ is even, it suffices to consider the case $x > 0$. The inequality $(\sin x)/x \leq 1$ is clear by (11.38).

So all we need to show is that if $0 < x < \pi/2$, then

$$x \leq \operatorname{tg} x. \tag{11.43}$$

In Figure 11.6, we have $A = (\cos x, 0)$, $B = (\cos x, \sin x)$. For the circle K, the tangent at point B intersects the horizontal axis at point C. The reflection of the point B over the horizontal axis is D. Since OAB and OBC are similar triangles, $\overline{BC} = \sin x / \cos x = \operatorname{tg} x$.

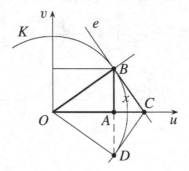

Fig. 11.6

Let us inscribe an arbitrary polygonal path in the arc DB. Extending this with segments OD and OB gives us a convex polygon T, which is part of the quadrilateral $ODCB$. By Lemma 10.82 it then follows that the perimeter of T is at most the perimeter of $ODCB$, which is $2 + 2 \operatorname{tg} x$. Since the supremum of polygonal paths inscribed into DB is $2x$, we get that $2 + 2x \leq 2 + 2 \operatorname{tg} x$, which proves (11.43). $\square$

Theorem 11.27. *The limit relations*

$$\lim_{x \to 0} \frac{1 - \cos x}{x} = 0 \tag{11.44}$$

and

$$\lim_{x \to 0} \frac{\sin x}{x} = 1 \tag{11.45}$$

hold.

Proof. The two statements follow from inequalities (11.39) and (11.42) by applying the squeeze theorem. $\square$

Now we summarize the properties of the functions $\mathrm{tg}\,x$ and $\mathrm{ctg}\,x$. The function $\mathrm{tg}\,x = \sin x/\cos x$ is defined where the denominator is nonzero, that is, at the points $x \neq (\pi/2) + k\pi$, where k is an arbitrary integer. By the addition formulas of sin and cos, it is easy to deduce the following identities:

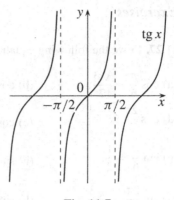

Fig. 11.7

$$\mathrm{tg}(x+y) = \frac{\mathrm{tg}\,x + \mathrm{tg}\,y}{1 - \mathrm{tg}\,x \cdot \mathrm{tg}\,y},$$

$$\mathrm{tg}(x-y) = \frac{\mathrm{tg}\,x - \mathrm{tg}\,y}{1 + \mathrm{tg}\,x \cdot \mathrm{tg}\,y}, \qquad (11.46)$$

$$\mathrm{tg}\,2x = \frac{2\,\mathrm{tg}\,x}{1 - \mathrm{tg}^2 x}.$$

The function $\mathrm{tg}\,x$ is continuous on its domain, since it is the quotient of two continuous functions. The function $\mathrm{tg}\,x$ is odd, and it is periodic with period π, since

$$\mathrm{tg}(x+\pi) = \frac{\sin(x+\pi)}{\cos(x+\pi)} = \frac{-\sin x}{-\cos x} = \frac{\sin x}{\cos x} = \mathrm{tg}\,x$$

for all $x \neq (\pi/2) + k\pi$. Since on the interval $[0, \pi/2)$, the function $\sin x$ is strictly monotone increasing, $\cos x$ is strictly monotone decreasing, and both are positive, so there $\mathrm{tg}\,x$ is strictly monotone increasing. Since $\mathrm{tg}\,0 = 0$ and $\mathrm{tg}\,x$ is odd, $\mathrm{tg}\,x$ is strictly increasing on the whole interval $(-\pi/2, \pi/2)$.

The limit relations

$$\lim_{x \to -(\pi/2)+0} \mathrm{tg}\,x = -\infty \qquad \text{and} \qquad \lim_{x \to (\pi/2)-0} \mathrm{tg}\,x = \infty \qquad (11.47)$$

hold, following from the facts

$$\lim_{x \to \pm\pi/2} \sin x = \sin(\pm\pi/2) = \pm 1, \qquad \lim_{x \to \pm\pi/2} \cos x = \cos(\pm\pi/2) = 0,$$

and that $\cos x$ is positive on $(-\pi/2, \pi/2)$ (Figure 11.7).

The function $\mathrm{ctg}\,x = \cos x/\sin x$ is defined where the denominator is not zero, that is, at the points $x \neq k\pi$, where k is an arbitrary integer. The function $\mathrm{ctg}\,x$ is continuous on its domain, is odd, and is periodic with period π. By (11.32) we get that

$$\mathrm{ctg}\,x = \mathrm{tg}\left(\frac{\pi}{2} - x\right) \qquad (11.48)$$

for all $x \neq k\pi$. It then follows that the function $\mathrm{ctg}\,x$ is strictly monotone decreasing on the interval $(0, \pi)$, and that

$$\lim_{x \to 0+0} \mathrm{ctg}\,x = \infty \qquad \text{and} \qquad \lim_{x \to \pi-0} \mathrm{ctg}\,x = -\infty. \qquad (11.49)$$

Exercises

11.27. Prove the following equalities:

(a) $\cos\dfrac{\pi}{6} = \dfrac{\sqrt{3}}{2}$,   (b) $\cos\dfrac{\pi}{4} = \dfrac{\sqrt{2}}{2}$,   (c) $\cos\dfrac{\pi}{3} = \dfrac{1}{2}$,

(d) $\cos\dfrac{2\pi}{3} = -\dfrac{1}{2}$,   (e) $\cos\dfrac{3\pi}{4} = -\dfrac{\sqrt{2}}{2}$,   (f) $\cos\dfrac{5\pi}{6} = -\dfrac{\sqrt{3}}{2}$,

(g) $\sin\dfrac{\pi}{6} = \dfrac{1}{2}$,   (h) $\sin\dfrac{\pi}{4} = \dfrac{\sqrt{2}}{2}$,   (i) $\sin\dfrac{\pi}{3} = \dfrac{\sqrt{3}}{2}$,

(j) $\sin\dfrac{2\pi}{3} = \dfrac{\sqrt{3}}{2}$,   (k) $\sin\dfrac{3\pi}{4} = \dfrac{\sqrt{2}}{2}$,   (l) $\sin\dfrac{5\pi}{6} = \dfrac{1}{2}$.

11.28. Prove that

$$\sin 3x = 4 \cdot \sin x \cdot \sin\left(x + \frac{\pi}{3}\right) \cdot \sin\left(x + \frac{2\pi}{3}\right)$$

for all x.

11.29. Prove that

$$\sin 4x = 8 \cdot \sin x \cdot \sin\left(x + \frac{\pi}{4}\right) \cdot \sin\left(x + \frac{2\pi}{4}\right) \cdot \sin\left(x + \frac{3\pi}{4}\right)$$

for all x. How can the statement be generalized?

11.30. Let (a_n) denote sequence (15) in Example 4.1, that is, let $a_1 = 0$ and $a_{n+1} = \sqrt{2 + a_n}$ $(n \geq 1)$. Prove that $a_n = 2 \cdot \cos(\pi/2^n)$.

11.31. Prove that if n is a positive integer, then $\cos nx$ can be written as an nth-degree polynomial in $\cos x$, that is, there exists a polynomial T_n of degree n such that $\cos nx = T_n(\cos x)$ for all x. (H)

11.32. Prove that if n is a positive integer, then $\sin nx/\sin x$ can be written as an nth-degree polynomial of $\cos x$, that is, there exists a polynomial U_n of degree n such that $\sin nx = \sin x \cdot U_n(\cos x)$ for all x.[9]

11.33. Can $\sin nx$ be written as a polynomial of $\sin x$ for all $n \in \mathbb{N}^+$?

11.34. Prove that the equation $x \cdot \sin x = 100$ has infinitely many solutions.

11.35. Let $f : \mathbb{R} \to \mathbb{R}$ be continuous, not identically zero, and suppose that $|f(x)| \leq 1$ and $f(x+y) + f(x-y) = 2f(x)f(y)$ for all x, y. Prove that for a suitable constant c, $f(x) = \cos cx$ for all x. (∗H)

---

[9] The polynomials T_n and U_n defined as such are called the **Chebyshev polynomials**.

11.36. (a) Let $A_n = \sum_{k=1}^{n} \sin^{-2}(k\pi/2n)$. Show that $A_1 = 1$ and $A_{2n} = 4 \cdot A_n - 1$ for all $n = 1, 2, \ldots$.

(b) Prove that $A_{2^n} = (2/3) \cdot 4^n + (1/3)$ ($n = 0, 1, \ldots$).

(c) Prove that $(\sin^{-2}x) - 1 < x^{-2} < \sin^{-2}x$ for all $0 < x < \pi/2$, and deduce from this that

$$A_n - n < (2n/\pi)^2 \cdot \sum_{k=1}^{n} 1/k^2 < A_n$$

for all n.

(d) Prove that $\sum_{k=1}^{\infty} 1/k^2 = \pi^2/6$. (H S)

11.5 The Inverse Trigonometric Functions

Since the function $\cos x$ is strictly monotone on the interval $[0, \pi]$, it has an inverse there, which we call the **arccosine** and denote by $\arccos x$. We are already familiar with this function (Figures 11.8 and 11.9). In Remark 11.21, we noted that the function $\cos x$ on the interval $[0, \pi]$ agrees with the inverse of $S(u) = s(c; [u, 1])$. Thus the function arccos is none other than the function S. The function $\arccos x$ is thus defined on the interval $[-1, 1]$, and there it is strictly monotone decreasing and continuous.

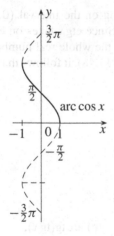

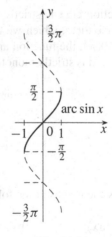

Fig. 11.8   Fig. 11.9

The function $\sin x$ is strictly monotone increasing on the interval $[-\pi/2, \pi/2]$, so it has an inverse there, which we call the **arcsine** function, and we denote it by $\arcsin x$. This function is defined on the interval $[-1, 1]$, and is strictly monotone increasing and continuous there. From the identities (11.32), we see that for all $x \in [-1, 1]$,

$$\arccos x = \frac{\pi}{2} - \arcsin x. \tag{11.50}$$

The function $\operatorname{tg} x$ is strictly monotone increasing on the interval $(-\pi/2, \pi/2)$, so it has an inverse there, which we call the **arctangent**, and we denote it by $\operatorname{arctg} x$. By the limit relation (11.47) and the continuity of $\operatorname{tg} x$, it follows that the function $\operatorname{tg} x$ takes on every real value over $(-\pi/2, \pi/2)$. Thus the function $\operatorname{arctg} x$ is defined on all of the real number line, is continuous, and is strictly monotone increasing (Figure 11.10). It is also clear that

$$\lim_{x \to -\infty} \operatorname{arctg} x = -\frac{\pi}{2} \qquad \text{and} \qquad \lim_{x \to \infty} \operatorname{arctg} x = \frac{\pi}{2}. \tag{11.51}$$

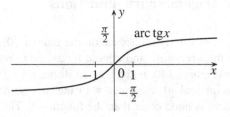

Fig. 11.10

The function $\operatorname{ctg} x$ is strictly monotone decreasing on the interval $(0, \pi)$, so it has an inverse there, which we denote by $\operatorname{arcctg} x$. Since $\operatorname{ctg} x$ takes on every real value over $(0, \pi)$, the function $\operatorname{arcctg} x$ is defined on the whole real number line, is continuous, and is strictly monotone decreasing. By (11.48), it follows that for all x,

$$\operatorname{arcctg} x = \frac{\pi}{2} - \operatorname{arctg} x. \tag{11.52}$$

Exercises

11.37. Draw the graphs of the following functions:

(a) $\arcsin(\sin x)$,   (b) $\arccos(\cos x)$,   (c) $\operatorname{arctg}(\operatorname{tg} x)$,

(d) $\operatorname{arcctg}(\operatorname{ctg} x)$,   (e) $\operatorname{arctg} x - \operatorname{arcctg}(1/x)$.

11.38. Prove the following identities:

(a) $\arcsin x = \arccos \sqrt{1 - x^2}$ $(x \in [0, 1])$.
(b) $\operatorname{arctg} x = \arcsin \dfrac{x}{\sqrt{1+x^2}}$ $(x \in \mathbb{R})$.

(c) $\arcsin x = \operatorname{arctg} \dfrac{x}{\sqrt{1-x^2}}$ $(x \in (-1,1))$.

(d) $\operatorname{arctg} x + \operatorname{arctg}(1/x) = \pi/2$ $(x > 0)$.

11.39. Solve the following equation: $\sin(2\operatorname{arctg} x) = 1/x$.

11.6 Hyperbolic Functions and Their Inverses

The so-called hyperbolic functions defined below share many properties with their trigonometric analogues. We call the functions

$$\operatorname{sh} x = \frac{e^x - e^{-x}}{2}, \quad \text{and} \quad \operatorname{ch} x = \frac{e^x + e^{-x}}{2}$$

hyperbolic sine and **hyperbolic cosine** respectively. By their definition, it is clear that $\operatorname{sh} x$ and $\operatorname{ch} x$ are defined everywhere and are continuous, and moreover, that $\operatorname{ch} x$ is even, while $\operatorname{sh} x$ is odd.

Since the function e^x is strictly monotone increasing, and the function e^{-x} is strictly monotone decreasing, *the function* $\operatorname{sh} x$ *is strictly monotone increasing on* $\mathbb{R}$ (Figure 11.11). By the limit relation (11.6), it is clear that

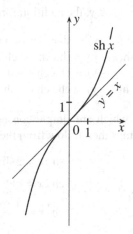

Fig. 11.11

$$\lim_{x \to -\infty} \operatorname{sh} x = -\infty \quad \text{and} \quad \lim_{x \to \infty} \operatorname{sh} x = \infty. \tag{11.53}$$

Since e^x and e^{-x} are strictly convex, it is easy to see that *the function* $\operatorname{ch} x$ *is strictly convex on* $\mathbb{R}$.

The following properties, which follow easily from the definitions, show some of the similarities between the hyperbolic and trigonometric functions. First of all,

$$\operatorname{ch}^2 x - \operatorname{sh}^2 x = 1$$

for all x, that is, the point $(\operatorname{ch} u, \operatorname{sh} u)$ lies on the hyperbola with equation $x^2 - y^2 = 1$ (whence the name "hyperbolic"; how this is analogous to $\cos^2 x + \sin^2 x = 1$ is clear)

Considering that $\mathrm{ch}\,x$ is positive everywhere,

$$\mathrm{ch}\,x = \sqrt{1 + \mathrm{sh}^2\,x} \qquad (11.54)$$

for all x. It immediately follows that *the small-est value of* $\mathrm{ch}\,x$ *is 1*, and moreover, that $\mathrm{ch}\,x$ *is strictly monotone increasing on* $[0,\infty)$ *and strictly monotone decreasing on* $(-\infty,0]$ (Figure 11.12).

For all x,y, the addition formulas

$$\begin{aligned}
\mathrm{sh}(x+y) &= \mathrm{sh}\,x\,\mathrm{ch}\,y + \mathrm{ch}\,x\,\mathrm{sh}\,y,\\
\mathrm{sh}(x-y) &= \mathrm{sh}\,x\,\mathrm{ch}\,y - \mathrm{ch}\,x\,\mathrm{sh}\,y,\\
\mathrm{ch}(x+y) &= \mathrm{ch}\,x\,\mathrm{ch}\,y + \mathrm{sh}\,x\,\mathrm{sh}\,y,\\
\mathrm{ch}(x-y) &= \mathrm{ch}\,x\,\mathrm{ch}\,y - \mathrm{sh}\,x\,\mathrm{sh}\,y
\end{aligned} \qquad (11.55)$$

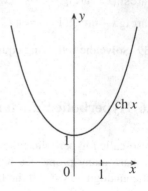

Fig. 11.12

hold, which are simple consequences of the identity $e^{x+y} = e^x \cdot e^y$. The following identities then follow from the addition formulas:

$$\begin{aligned}
\mathrm{sh}\,2x &= 2\,\mathrm{sh}\,x\,\mathrm{ch}\,x,\\
\mathrm{ch}\,2x &= \mathrm{ch}^2\,x + \mathrm{sh}^2\,x = 1 + 2\,\mathrm{sh}^2\,x = 2\,\mathrm{ch}^2\,x - 1,\\
\mathrm{ch}^2\,x &= \frac{1 + \mathrm{ch}\,2x}{2},\\
\mathrm{sh}^2\,x &= \frac{-1 + \mathrm{ch}\,2x}{2}.
\end{aligned} \qquad (11.56)$$

Remark 11.28. The similarities to trigonometric functions we see above are surprising, and what makes the analogy even more puzzling is how differently the two families of functions were defined. The answer lies in the fact that—contrary to how it seems—exponential functions very much have a connection with trigonometric functions. This connection, however, can be seen only through the complex numbers, and since dealing with complex numbers is not our goal here, we only outline this connection in the second appendix of the chapter (as well as Remarks 13.18 and 11.29).

Much like the functions tg and ctg, we introduce $\mathrm{th}\,x = \mathrm{sh}\,x/\mathrm{ch}\,x$ and $\mathrm{cth}\,x = \mathrm{ch}\,x/\mathrm{sh}\,x$ (Figures 11.13 and 11.14). We leave it to the reader to check that the function $\mathrm{th}\,x$ (**hyperbolic tangent**) is defined on the real numbers, continuous everywhere, odd, and strictly monotone increasing, and moreover, that

$$\lim_{x\to-\infty} \mathrm{th}\,x = -1 \qquad \text{and} \qquad \lim_{x\to+\infty} \mathrm{th}\,x = 1. \qquad (11.57)$$

The function $\mathrm{cth}\,x$ (**hyperbolic cotangent**) is defined on the set $\mathbb{R} \setminus \{0\}$, and it is continuous there; it is strictly monotone decreasing on the intervals $(-\infty,0)$ and $(0,\infty)$, and moreover,

$$\lim_{x\to\pm\infty} \mathrm{cth}\,x = \pm 1 \qquad \text{and} \qquad \lim_{x\to 0\pm 0} \mathrm{cth}\,x = \pm\infty. \qquad (11.58)$$

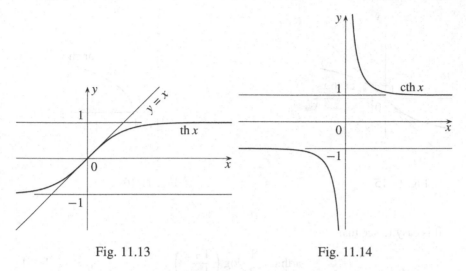

Fig. 11.13   Fig. 11.14

The inverse of the function $\operatorname{sh} x$ is called the **area hyperbolic sine** function, denoted by $\operatorname{arsh} x$ (Figure 11.15). Since $\operatorname{sh} x$ is strictly monotone increasing, continuous, and by (11.53) it takes on every value, $\operatorname{arsh} x$ is defined on all real numbers, is continuous everywhere, and is strictly monotone increasing. We can actually express arsh with the help of power and logarithmic functions. Notice that $\operatorname{arsh} x = y$ is equivalent to $\operatorname{sh} y = x$. If we write the definition of $\operatorname{sh} x$ into this and multiply both sides by $2e^y$, we get the equation $e^{2y} - 1 = 2xe^y$. This is a quadratic equation in e^y, from which we get $e^y = x \pm \sqrt{x^2 + 1}$. Since $e^y > 0$, only the positive sign is considered. Finally, taking the logarithm of both sides, we get that

$$\operatorname{arsh} x = \log\left(x + \sqrt{x^2 + 1}\right) \tag{11.59}$$

for all x.

The function $\operatorname{ch} x$ is strictly monotone increasing and continuous on the interval $[0, \infty)$, and by (11.54), its image there is $[1, \infty)$. The inverse of the function $\operatorname{ch} x$ restricted to the interval $[0, \infty)$ is called the **area hyperbolic cosine**, and we denote it by $\operatorname{arch} x$ (Figure 11.16). By the above, $\operatorname{arch} x$ is defined on the interval $[1, \infty)$, and it is continuous and strictly monotone increasing. It is easy to see that

$$\operatorname{arch} x = \log\left(x + \sqrt{x^2 - 1}\right) \tag{11.60}$$

for all $x \geq 1$.

The inverse of the function $\operatorname{th} x$ is called the **area hyperbolic tangent**, and we denote it by $\operatorname{arth} x$ (Figure 11.17). By the properties of $\operatorname{th} x$, we see that $\operatorname{arth} x$ is defined on the interval $(-1, 1)$, is continuous, and is strictly monotone increasing.

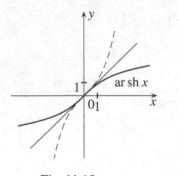

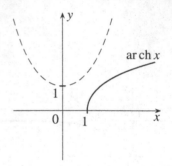

Fig. 11.15   Fig. 11.16

It is easy to see that

$$\operatorname{arth} x = \frac{1}{2} \cdot \log\left(\frac{1+x}{1-x}\right)$$ (11.61)

for all $x \in (-1, 1)$. We leave the definition of arcth and the verification of its most important properties to the reader.

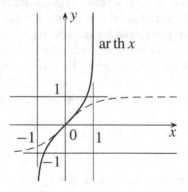

Fig. 11.17

Remark 11.29. The names of trigonometric and hyperbolic functions come from Latin words. The notation $\sin x$ comes from the word *sinus* meaning *bend, fold.*

The notation $\operatorname{tg} x$ comes from the word *tangens*, meaning *tangent*. The name is justified by the fact that if $0 < x < \pi/2$, then $\operatorname{tg} x$ is the length of the segment tangent to the circle at $(1, 0)$, starting there and ending where the line going through the origin and the point $(\cos x, \sin x)$ crosses it (see Figure 11.4).

The inverse trigonometric functions all have the *arc* prefix, which implies that $\arccos x$ corresponds to the length of some arc. For inverse hyperbolic functions, instead of *arc*, the word used is *area*. This is justified by the following observation. Let $u \geq 1$ and $v = \sqrt{u^2 - 1}$. The line segments connecting the origin to (u, v) and

$(u, -v)$, and the hyperbola $x^2 - y^2 = 1$ between the points (u, v) and $(u, -v)$ define a section A_u of the plane. One can show that the area of A_u is equal to $\mathrm{arch}\, u$ (see Exercise 16.9).

Exercises

11.40. Check the addition formulas for $\mathrm{sh}\, x$ and $\mathrm{ch}\, x$.

11.41. Find and prove formulas analogous to (11.46) about $\mathrm{th}\, x$.

11.42. Prove that $\log\left(\sqrt{x^2 + 1} - x\right) = -\,\mathrm{arsh}\, x$ for all x.

11.43. Prove that $\log\left(x - \sqrt{x^2 - 1}\right) = -\,\mathrm{arch}\, x$ for all $x \geq 1$.

11.44. Let $f : \mathbb{R} \to \mathbb{R}$ be continuous, and suppose that $f(x + y) + f(x - y) = 2f(x)f(y)$ for all x, y. Prove that one of the following holds.

(a) f is the function that is identically zero.
(b) There exists a constant c such that $f(x) = \cos cx$ for all x.
(c) There exists a constant c such that $f(x) = \mathrm{ch}\, cx$ for all x.

11.45. We say that the function f is **algebraic** if there exist polynomials $p_0(x)$, $p_1(x), \ldots, p_n(x)$ such that p_n is not identically zero, and

$$p_0(x) + p_1(x)f(x) + \cdots + p_n(x)f^n(x) = 0$$

for all $x \in D(f)$. We call a function f **transcendental** if it is not algebraic.

(a) Prove that every polynomial and every rational function are algebraic.
(b) Prove that $\sqrt{1+x}$, $\sqrt[3]{\frac{x^2-1}{x+2}}$ and $|x|$ are algebraic functions.
(c) Prove that e^x, $\log x$, $\sin x$, $\cos x$ are transcendental functions.
(d) Can a (nonconstant) periodic function be algebraic?

11.7 First Appendix: Proof of the Addition Formulas

Let O_α denote the positive rotation around the origin by α degrees. Then for arbitrary $\alpha \in \mathbb{R}$ and $x \in \mathbb{R}^2$, $O_\alpha(x)$ is the point that we get by rotating x around the origin by α in the positive direction. We will need the following properties of the rotations O_α.

(i) For arbitrary $\alpha, \beta \in \mathbb{R}$ and $x \in \mathbb{R}^2$, $O_{\alpha+\beta}(x) = O_\alpha\left(O_\beta(x)\right)$.

(ii) The map O_α is linear, that is,

$$O_\alpha(px + qy) = p \cdot O_\alpha(x) + q \cdot O_\alpha(y)$$

for all $x, y \in \mathbb{R}^2$ and $p, q \in \mathbb{R}$.

We will use these properties without proof. (Both properties seem clear; we can convince ourselves of (ii) if we think about what the geometric meaning is of the sum of two vectors and of multiplying a vector by a real number.) We now show that

$$O_\alpha((1,0)) = (\cos \alpha, \sin \alpha). \tag{11.62}$$

Let $O_\alpha((1,0)) = P$, and let h denote the ray starting at the origin and crossing P. Then the angle formed by h and the positive half of the x-axis is α, that is, h intersects the circle C at the point $(\cos \alpha, \sin \alpha)$. Since rotations preserve distances (another geometric fact that we accept), the distance of P from the origin is 1. Thus P is on the circle C, that is, it agrees with the intersection of h and C, which is $(\cos \alpha, \sin \alpha)$.

By the above, $(0,1) = O_{\pi/2}((1,0))$, so by property (i),

$$O_\alpha((0,1)) = O_{\alpha+(\pi/2)}((1,0)) =$$
$$= \left(\cos\left(\alpha + \frac{\pi}{2}\right), \sin\left(\alpha + \frac{\pi}{2}\right)\right) =$$
$$= (-\sin \alpha, \cos \alpha), \tag{11.63}$$

where we used (11.28) and (11.32).

Let the coordinates of x be (x_1, x_2). Then by (11.62), (11.63), and (ii), we get that

$$O_\alpha(x) = x_1 \cdot O_\alpha((1,0)) + x_2 \cdot O_\alpha((0,1)) =$$
$$= x_1 \cdot (\cos \alpha, \sin \alpha) + x_2 \cdot (-\sin \alpha, \cos \alpha) =$$
$$= (x_1 \cdot \cos \alpha - x_2 \cdot \sin \alpha, x_1 \cdot \sin \alpha + x_2 \cdot \cos \alpha).$$

By all these,

$$(\cos(\alpha + \beta), \sin(\alpha + \beta)) = O_{\alpha+\beta}((1,0)) = O_\alpha\left(O_\beta((1,0))\right) =$$
$$= O_\alpha((\cos \beta, \sin \beta)) =$$
$$= (\cos \beta \cdot \cos \alpha - \sin \beta \cdot \sin \alpha, \cos \beta \cdot \sin \alpha + \sin \beta \cdot \cos \alpha),$$

so after comparing coordinates, we get the first and third identities of (11.33). If we apply these to $-y$ instead of y, then also using that $\cos x$ is even and $\sin x$ is odd, we get the other two identities.

11.8 Second Appendix: A Few Words on Complex Numbers

The introduction of complex numbers is motivated by the need to include a number whose square is -1 into our investigations. We denote this number by i.[10] We call the formal expressions $a + bi$, where a and b are real numbers, **complex numbers**. We consider the real numbers complex numbers as well by identifying the real number a with the complex number $a + 0 \cdot i$. Addition and multiplication are defined on the complex numbers in a way to preserve the usual rules, and so that $i^2 = -1$ also holds. Thus the sum and product of the complex numbers $z_1 = a_1 + b_1 i$ and $z_2 = a_2 + b_2 i$ are defined by the formulas

$$z_1 + z_2 = (a_1 + a_2) + (b_1 + b_2)i$$

and

$$z_1 \cdot z_2 = (a_1 a_2 - b_1 b_2) + (a_1 b_2 + a_2 b_1)i.$$

One can show that the complex numbers form a field with these operations if we take the zero element to be $0 = 0 + 0 \cdot i$ and the identity element to be $1 = 1 + 0 \cdot i$. Thus *the complex numbers form a field that contains the real numbers.*

A remarkable fact is that among the complex numbers, every nonconstant polynomial (with complex coefficients) has a root (so for example, the polynomial $x^2 + 1$ has the roots i and $-i$). One can prove that *a degree-n polynomial has n roots counting multiplicity.* This statement is called the **fundamental theorem of algebra.**

To define complex powers, we will use the help of (11.11). Since $(1 + z/n)^n$ is defined for all complex numbers z, it is reasonable to define e^z as the limit of the sequence $(1 + z/n)^n$ as $n \to \infty$. (We say that the sequence of complex numbers $z_n = a_n + b_n i$ tends to $z = a + bi$ if $a_n \to a$ and $b_n \to b$.) One can show that this is well defined, that is, the limit exists for all complex numbers z, and that the powers defined in this way satisfy $e^{z+w} = e^z \cdot e^w$ for all complex z and w. Moreover—and this is important to us—if $z = ix$, where x is real, then the limit of $(1 + z/n)^n$ is $\cos x + i \cdot \sin x$, so

$$e^{ix} = \cos x + i \cdot \sin x \tag{11.64}$$

for all real numbers x.[11] If we apply (11.64) to $-x$ instead of x, then we get that

$$e^{-ix} = \cos x - i \cdot \sin x.$$

We can express both $\cos x$ and $\sin x$ from this expression, and we get the identities

$$\cos x = \frac{e^{ix} + e^{-ix}}{2}, \qquad \sin x = \frac{e^{ix} - e^{-ix}}{2i}. \tag{11.65}$$

---

[10] As the first letter of the word "imaginary."

[11] It is worth checking that $e^{i(x+y)} = e^{ix} \cdot e^{iy}$ holds for all $x, y \in \mathbb{R}$. By (11.64), this is equivalent to the addition formulas for the functions cos and sin.

These are called **Euler's formulas.** These two identities help make the link between trigonometric and hyperbolic functions clear. If we extend the definitions of ch and sh to the complex numbers, we get that

$$\cos x = \text{ch}(ix) \quad \text{and} \quad \sin x = \text{sh}(ix)/i \tag{11.66}$$

for all real x.

Chapter 12
Differentiation

12.1 The Definition of Differentiability

Consider a point that is moving on a line, and let $s(t)$ denote the location of the point on the line at time t. Back when we talked about real-life problems that could lead to the definition of limits (see Chapter 9, p. 121), we saw that the definition of instantaneous velocity required taking the limit of the fraction $(s(t) - s(t_0))/(t - t_0)$ in t_0. Having precisely defined what a limit is, we can now *define the instantaneous velocity of the point at a t_0 to be the limit*

$$\lim_{t \to t_0} \frac{s(t) - s(t_0)}{t - t_0}$$

(assuming, of course, that this limit exists and is finite).

We also saw that if we want to define the tangent line to the graph of a function f at $(a, f(a))$, then the slope of that line is exactly the limit of $(f(x) - f(a))/(x - a)$ in a. *We agree to define the tangent line as the line that contains the point $(a, f(a))$ and has slope*

$$\lim_{x \to a} \frac{f(x) - f(a)}{x - a},$$

again assuming that this limit exists and is finite.

Other than the two examples above, many problems in mathematics, physics, and other fields can be grasped in the same form as above. This is the case when we have to find the rate of some change (not necessarily happening in space). If, for example, the temperature of an object at time t is $H(t)$, then we can ask how fast the temperature is changing at time t_0. The average change over the interval $[t_0, t]$ is $(H(t) - H(t_0))/(t - t_0)$. Clearly, the instantaneous change of temperature will be defined as the limit

$$\lim_{t \to t_0} \frac{H(t) - H(t_0)}{t - t_0}$$

(assuming that it exists and is finite).

© Springer New York 2015
M. Laczkovich, V.T. Sós, *Real Analysis*, Undergraduate Texts
in Mathematics, DOI 10.1007/978-1-4939-2766-1_12

We use the following names for the quotients appearing above. If f is defined at the points a and b, then the quotient $(f(b) - f(a))/(b - a)$ is called the **difference quotient** of f between a and b. It is clear that the difference quotient $(f(b) - f(a))/(b - a)$ agrees with the slope of the line passing through the points $(a, f(a))$ and $(b, f(b))$.

In many cases, using the notation $b - a = h$, the difference quotient between a and $x = a + h$ is written as

$$\frac{f(a+h) - f(a)}{h}.$$

Definition 12.1. Let f be defined on a neighborhood of the point a. We say that the function f is *differentiable at the point a* if the finite limit

$$\lim_{x \to a} \frac{f(x) - f(a)}{x - a} \tag{12.1}$$

exists. The limit (12.1) is called the *derivative* of f at a.

The derivative of f at a is most often denoted by $f'(a)$. Sometimes, other notations,

$$\dot{f}(a), \qquad \left.\frac{df}{dx}\right|_{x=a}, \qquad \left.\frac{df(x)}{dx}\right|_{x=a}, \qquad y'(a), \qquad \left.\frac{dy}{dx}\right|_{x=a}$$

are used (the last two in accordance with the notation $y = f(x)$).

Equipped with the above definition, we can say that if the function $s(t)$ describes the location of a moving point, then the instantaneous velocity at time t_0 is $s'(t_0)$. Similarly, if the temperature of an object at time t is given by $H(t)$, then the instantaneous temperature change at time t_0 is $H'(t_0)$. The definition of a tangent should also be updated with our new definition. The tangent of f at a is the line that passes through the point $(a, f(a))$ and has slope

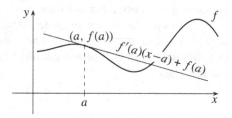

Fig. 12.1

$$\lim_{x \to a} \frac{f(x) - f(a)}{x - a} = f'(a).$$

Since the equation of this line is $y = f'(a) \cdot (x - a) + f(a)$, we accept the following definition.

Definition 12.2. Let f be differentiable at the point a. The *tangent line of f at* $(a, f(a))$ is the line with equation $y = f'(a) \cdot (x - a) + f(a)$.

The visual meaning of the derivative $f'(a)$ is then the slope of the tangent of graph f at $(a, f(a))$ (Figure 12.1).

Examples 12.3. **1.** The constant function $f(x) = c$ is differentiable at all values a, and its derivative is zero. This is because

$$\frac{f(x) - f(a)}{x - a} = \frac{c - c}{x - a} = 0$$

for all $x \neq a$.

2. The function $f(x) = x$ is differentiable at all values a, and $f'(a) = 1$. This is because

$$\frac{f(x) - f(a)}{x - a} = \frac{x - a}{x - a} = 1$$

for all $x \neq a$.

3. The function $f(x) = x^2$ is differentiable at all values a, and $f'(a) = 2a$. This is because

$$\frac{f(x) - f(a)}{x - a} = \frac{x^2 - a^2}{x - a} = x + a,$$

and so

$$f'(a) = \lim_{x \to a} \frac{f(x) - f(a)}{x - a} = 2a.$$

Thus by the definition of tangent lines, the tangent of the parabola $y = x^2$ at the point (a, a^2) is the line with equation $y = 2a(x - a) + a^2 = 2ax - a^2$. Since this passes through the point $(a/2, 0)$, we can construct the tangent by drawing a line connecting the point $(a/2, 0)$ to the point (a, a^2) (Figure 12.2).[1]

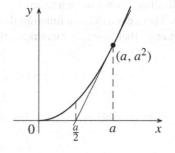

Fig. 12.2

Differentiability is a stronger condition than continuity. This is shown by the theorem below and the remarks following it.

Theorem 12.4. *If f is differentiable at a, then f is continuous at a.*

Proof. If f is differentiable at a, then

$$\lim_{x \to a}(f(x) - f(a)) = \lim_{x \to a}\left[\frac{f(x) - f(a)}{x - a} \cdot (x - a)\right] = f'(a) \cdot 0 = 0.$$

This means exactly that f is continuous at a.   □

Remarks 12.5. **1.** Continuity is a necessary but not sufficient condition for differentiability. There exist functions that are continuous at a point a but are not differentiable there. An easy example is the function $f(x) = |x|$ at $a = 0$. Clearly,

---

[1] So the calculus was correct; see page 5.

$$f(x) = \begin{cases} x, & \text{if } x \geq 0, \\ -x, & \text{if } x < 0, \end{cases}$$

so

$$\frac{f(x) - f(0)}{x - 0} = \begin{cases} 1, & \text{if } x > 0, \\ -1, & \text{if } x < 0. \end{cases}$$

Then

$$\lim_{x \to 0+0} \frac{f(x) - f(0)}{x - 0} = 1 \text{ and } \lim_{x \to 0-0} \frac{f(x) - f(0)}{x - 0} = -1,$$

and so f is not differentiable at 0.

2. There even exist functions that are continuous everywhere but not differentiable anywhere. Using the theory of series of functions, one can show that for suitable $a, b > 0$, the function

$$f(x) = \sum_{n=1}^{\infty} a^n \cdot \sin(b^n x)$$

has this property (for example with the choice $a = 1/2$, $b = 10$). Similarly, one can show that the function

$$g(x) = \sum_{n=1}^{\infty} \frac{\langle 2^n x \rangle}{2^n}$$

is everywhere continuous but nowhere differentiable, where $\langle x \rangle$ denotes the smallest distance from x to an integer.

3. There even exists a function that is differentiable at a point a but is not continuous at any other point. For example, the function

$$f(x) = \begin{cases} x^2, & \text{if } x \text{ is rational}, \\ -x^2, & \text{if } x \text{ is irrational} \end{cases}$$

is differentiable at 0. This is because

$$\left| \frac{f(x) - f(0)}{x - 0} \right| = \left| \frac{x^2}{x} \right| = |x| \to 0, \qquad \text{if } x \to 0,$$

so

$$f'(0) = \lim_{x \to 0} \frac{f(x) - f(0)}{x - 0} = 0.$$

At the same time, it can easily be seen—see function 6 in Example 10.2—that the function $f(x)$ is not continuous at any $x \neq 0$.

As we saw, the function $|x|$ is not differentiable at 0, and accordingly, the graph does not have a tangent at 0 (here the graph has a cusp). If, however, we look only on the right-hand side of the point, then the difference quotient has a limit from that side. Accordingly, in the graph of $|x|$, the "right-hand chords" of $(0,0)$ have a limit (which is none other than the line $y = x$). The situation is similar on the left-hand side of 0.

As the above example illustrates, it is reasonable to introduce the one-sided variants of derivatives (and hence one-sided differentiability).

Definition 12.6. If the finite limit

$$\lim_{x \to a+0} \frac{f(x) - f(a)}{x - a}$$

exists, then we call this limit the *right-hand derivative of f at a*. We can similarly define the left-hand derivative as well.

We denote the right-hand derivative at a by $f'_+(a)$, and the left-hand derivative at a by $f'_-(a)$.

Remark 12.7. It is clear that f is differentiable at a if and only if both the right- and left-hand derivatives of f exist at a, and $f'_+(a) = f'_-(a)$. (Then this shared value is $f'(a)$.)

As we have seen, differentiability does not follow from continuity. We will now show that convexity—which is a stronger property than continuity—implies one-sided differentiability.

Theorem 12.8. *If f is convex on the interval (a,b), then f is differentiable from the left and from the right at all points $c \in (a,b)$.*

Proof. By Theorem 9.20, the function $x \mapsto (f(x) - f(c))/(x - c)$ is monotone increasing on the set $(a,b) \setminus \{c\}$. Fix $d \in (a,c)$. Then

$$\frac{f(d) - f(c)}{d - c} \le \frac{f(x) - f(c)}{x - c}$$

for all $x \in (c,b)$, so the function $(f(x) - f(c))/(x - c)$ is monotone increasing and bounded from below on the interval (c,b).

By statement (ii) of Theorem 10.68, it then follows that the limit

$$\lim_{x \to c+0} \frac{f(x) - f(c)}{x - c}$$

exists and is finite, which means that f is differentiable from the right at c. It can be shown similarly that f has a left-hand derivative at c.   □

Linear Approximation. In working with a function that arises in a problem, we can frequently get a simpler and more intuitive result if instead of the function, we use another simpler one that "approximates the original one well." One of the simplest classes of functions comprises the linear functions $(y = mx + b)$. We show that the differentiability of a function f at a point a means exactly that the function can be "well approximated" by a linear function. As we will soon see, the best linear approximation for f at a point a is the function $y = f'(a)(x - a) + f(a)$.

If f is continuous at a, then for all c,

$$f(x) - [c \cdot (x - a) + f(a)] \to 0, \quad \text{if} \quad x \to a.$$

Thus every linear function $\ell(x)$ such that $\ell(a) = f(a)$ "approximates f well" in the sense that $f(x) - \ell(x) \to 0$ as $x \to a$. The differentiability of f at a, by the theorem below, means exactly that the function

$$t(x) = f'(a) \cdot (x-a) + f(a) \tag{12.2}$$

approximate significantly better: not only does the difference $f - t$ tend to zero as $x \to a$, but it tends to zero faster than $(x-a)$.

Theorem 12.9. *The function f is differentiable at a if and only if at a, it can be "well approximated" locally by a linear polynomial in the following sense: there exists a number α (independent of x) such that*

$$f(x) = \alpha \cdot (x-a) + f(a) + \varepsilon(x) \cdot (x-a),$$

where $\varepsilon(x) \to 0$ as $x \to a$. The number α is the derivative of f at a.

Proof. Suppose first that f is differentiable at a, and let

$$\varepsilon(x) = \frac{f(x) - f(a)}{x - a} - f'(a).$$

Since

$$f'(a) = \lim_{x \to a} \frac{f(x) - f(a)}{x - a},$$

we have $\varepsilon(x) \to 0$ as $x \to a$. Thus

$$f(x) = f(a) + f'(a)(x-a) + \varepsilon(x)(x-a),$$

where $\varepsilon(x) \to 0$ as $x \to a$.
    Now suppose that

$$f(x) = \alpha \cdot (x-a) + f(a) + \varepsilon(x)(x-a),$$

where $\varepsilon(x) \to 0$ as $x \to a$. Then

$$\frac{f(x) - f(a)}{x - a} = \alpha + \varepsilon(x) \to \alpha, \quad \text{if} \quad x \to a.$$

Thus f is differentiable at a, and $f'(a) = \alpha$. $\square$

The following theorem expresses that $t(x) = f'(a) \cdot (x-a) + f(a)$ is the "best" linear function approximating f.

Theorem 12.10. *If the function f is differentiable at a, then for all $c \neq f'(a)$,*

$$\lim_{x \to a} \frac{f(x) - [f'(a)(x-a) + f(a)]}{f(x) - [c(x-a) + f(a)]} = 0.$$

Proof. As $x \to a$,

$$\frac{f(x) - [f'(a)(x-a) + f(a)]}{f(x) - [c(x-a) + f(a)]} = \frac{\frac{f(x)-f(a)}{x-a} - f'(a)}{\frac{f(x)-f(a)}{x-a} - c} \to \frac{f'(a) - f'(a)}{f'(a) - c} = 0.$$

□

Theorem 12.9 gives a necessary and sufficient condition for f to be differentiable at a. With the help of this, we can give another (equivalent to (12.1)) definition of differentiability.

Definition 12.11. The function f is *differentiable* at a if there exists a number α (independent of x) such that

$$f(x) = \alpha \cdot (x - a) + f(a) + \varepsilon(x)(x - a),$$

where $\varepsilon(x) \to 0$ if $x \to a$.

The significance of this equivalent definition lies in the fact that if we want to extend the concept of differentiability to other functions—not necessarily real-valued or of a real variable—then we cannot always find a definition analogous to Definition 12.1, but the generalization of Definition 12.11 is generally feasible.

Derivative Function. We will see that the derivative is the most useful tool in investigating the properties of a function. This is true locally and globally. The existence of the derivative $f'(a)$ and its value describes local properties of f: from the value of $f'(a)$, we can deduce the behavior of f at a.[2]

If, however, f is differentiable at every point of an interval, then from the values of $f'(x)$ we get global properties of f. In applications, we mostly come upon functions that are differentiable in some interval. The definition of this is the following.

Definition 12.12. Let $a < b$. We say that f is *differentiable on the interval* (a,b) if it is differentiable at every point of (a,b). We say that f is *differentiable on* $[a,b]$ if it is differentiable on (a,b) and differentiable from the right at a and from the left at b.

Generally, we can think of differentiation as an operation that maps functions to functions.

Definition 12.13. The *derivative function* of the function f is the function that is defined at every x where f is differentiable and has the value $f'(x)$ there. It is denoted by f'.

The basic task of differentiation is to find relations between functions and their derivatives, and to apply them. For the applications, however, first we need to decide where a function is differentiable, and we have to find its derivative. We begin with this latter problem, and only thereafter do we inspect what derivatives tell us about functions and how to use that information.

---

[2] One such link was already outlined when we saw that differentiability implies continuity.

Exercises

12.1. Where is the function $(\{x\} - 1/2)^2$ differentiable?

12.2. Let $f(x) = x^2$ if $x \le 1$, and $f(x) = ax + b$ if $x > 1$. For what values of a and b will f be differentiable everywhere?

12.3. Let $f(x) = |x|^\alpha \cdot \sin|x|^\beta$ if $x \ne 0$, and let $f(0) = 0$. For what values of α and β will f be continuous at 0? When will it be differentiable at 0?

12.4. Prove that the graph of the function x^2 has the tangent line $y = mx + b$ if the line and the graph of the function intersect at exactly one point.

12.5. Where are the tangent lines to the function $2x^3 - 3x^2 + 8$ horizontal?

12.6. When is the x-axis tangent to the graph of $x^3 + px + q$?

12.7. Let $f(2^{-n}) = 3^{-n}$ for all positive integers n, and let $f(x) = 0$ otherwise. Where is f differentiable?

12.8. Are there any points where the Riemann function is differentiable?

12.9. Are there any points where the square of the Riemann function is differentiable? (H)

12.10. At what angle does the graph of x^2 intersect the line $y = 2x$? (That is, what is the angle between the tangent of the function and the line?)

12.11. Prove that the function $f(x) = \sqrt{x}$ is differentiable for all $a > 0$, and that $f'(a) = 1/(2\sqrt{a})$.

12.12. Prove that the function $1/x$ is differentiable at all points $a > 0$, and find its derivative. Show that every tangent of the function $1/x$ forms a triangle with the two axes whose area does not depend on which point the tangent is taken at.

12.13. Prove that if f is differentiable at 0, then the function $f(|x|)$ is differentiable at 0 if and only if $f'(0) = 0$.

12.14. Prove that if f is differentiable at a, then

$$\lim_{h \to 0} \frac{f(a+h) - f(a-h)}{2h} = f'(a).$$

Show that the converse of this statement does not hold.

12.15. Let f be differentiable at the point a. Prove that if $x_n < a < y_n$ for all n and if $y_n - x_n \to 0$, then

$$\lim_{n \to \infty} \frac{f(y_n) - f(x_n)}{y_n - x_n} = f'(a). \text{ (H S)}$$

12.16. Suppose that f is differentiable everywhere on $(-\infty,\infty)$. Prove that if f is even (odd), then f' is odd (even).

12.17. Let

$B = \{f: f \text{ is bounded on } [a,b]\}$,
$C = \{f: f \text{ is continuous on } [a,b]\}$,
$M = \{f: f \text{ is monotone on } [a,b]\}$,
$X = \{f: f \text{ is convex on } [a,b]\}$,
$D = \{f: f \text{ is differentiable on } [a,b]\}$,
$I = \{f: f \text{ has an inverse on } [a,b]\}$.

From the point of containment, how are the sets B, C, M, X, D, and I related?

12.2 Differentiation Rules and Derivatives of the Elementary Functions

The differentiability of some basic functions can be deduced from the properties we have seen. These include the polynomials, trigonometric functions, and logarithmic functions. On the other hand, it is easy to see that all of the elementary functions can be expressed in terms of polynomials, trigonometric functions, and logarithmic functions using the four basic arithmetic operations, taking inverses, and composition. Thus to determine the differentiability of the remaining elementary functions, we need theorems that help us deduce their differentiability and to calculate their derivatives based on the differentiability and derivatives of the component functions in terms of which they can be expressed. These are called the **differentiation rules.** Below, we will determine the derivatives of the power functions with integer exponent, trigonometric functions, and logarithmic functions; we will introduce the differentiation rules, and using all this information, we will determine the derivatives of the rest of the elementary functions.

Theorem 12.14. *For an arbitrary positive integer n, the function x^n is differentiable everywhere on $(-\infty,\infty)$, and $(x^n)' = n \cdot x^{n-1}$ for all x.*

Proof. For all a,

$$\lim_{x \to a} \frac{x^n - a^n}{x - a} = \lim_{x \to a} \left(x^{n-1} + x^{n-2} \cdot a + \cdots + x \cdot a^{n-2} + a^{n-1} \right) = na^{n-1}$$

by the continuity of x^k. $\qquad\square$

Theorem 12.15.

(i) *The functions $\sin x$ and $\cos x$ are differentiable everywhere on $(-\infty,\infty)$. Moreover,*

$$(\sin x)' = \cos x \text{ and } (\cos x)' = -\sin x$$

for all x.

(ii) *The function* tg x *is differentiable at all points* $x \neq \frac{\pi}{2} + k\pi$ $(k \in \mathbb{Z})$, *and at those points,*

$$(\text{tg}\, x)' = \frac{1}{\cos^2 x}.$$

(iii) *The function* ctg x *is differentiable for all* $x \neq k\pi$ $(k \in \mathbb{Z})$, *and at those points,*

$$(\text{ctg}\, x)' = -\frac{1}{\sin^2 x}.$$

Proof. (i) For arbitrary $a \in \mathbb{R}$ and $x \neq a$,

$$\frac{\sin x - \sin a}{x - a} = \frac{2\sin(x-a)/2 \cdot \cos(x+a)/2}{x-a} = \frac{\sin((x-a)/2)}{(x-a)/2} \cdot \cos\frac{x+a}{2}$$

by the second identity of (11.37). Now $\lim_{x \to a} \sin((x-a)/2)/((x-a)/2) = 1$ and $\lim_{x \to a} \cos(x+a)/2 = \cos a$ by (11.45), by the continuity of the cos function and the theorem about limits of compositions of functions. Therefore,

$$\lim_{x \to a} \frac{\sin x - \sin a}{x - a} = \cos a.$$

Similarly, for arbitrary $a \in \mathbb{R}$ and $x \neq a$,

$$\frac{\cos x - \cos a}{x - a} = -\frac{2\sin(x-a)/2 \cdot \sin(x+a)/2}{x-a} = -\frac{\sin((x-a)/2)}{(x-a)/2} \cdot \sin\frac{x+a}{2}$$

by the fourth identity of (11.37). Then using (11.45) and the continuity of the sin function, we get that

$$\lim_{x \to a} \frac{\cos x - \cos a}{x - a} = -\sin a.$$

(ii) For arbitrary $a \neq (\pi/2) + k\pi$ and $x \neq a$,

$$\frac{\text{tg}\, x - \text{tg}\, a}{x - a} = \left(\frac{\sin x}{\cos x} - \frac{\sin a}{\cos a}\right) \cdot \frac{1}{x-a} = \frac{\sin x \cos a - \sin a \cos x}{\cos x \cos a} \cdot \frac{1}{x-a} =$$

$$= \frac{\sin(x-a)}{x-a} \cdot \frac{1}{\cos x \cos a}.$$

Then using (11.45) and the continuity of cos, we get that

$$\lim_{x \to a} \frac{\text{tg}\, x - \text{tg}\, a}{x - a} = \frac{1}{\cos^2 a}.$$

(iii) For arbitrary $a \neq k\pi$ and $x \neq a$,

$$\frac{\operatorname{ctg} x - \operatorname{ctg} a}{x - a} = \left(\frac{\cos x}{\sin x} - \frac{\cos a}{\sin a} \right) \cdot \frac{1}{x - a} = \frac{\cos x \sin a - \cos a \sin x}{\sin x \sin a} \cdot \frac{1}{x - a} =$$

$$= -\frac{\sin(x - a)}{x - a} \cdot \frac{1}{\sin x \sin a},$$

which gives us

$$\lim_{x \to a} \frac{\operatorname{ctg} x - \operatorname{ctg} a}{x - a} = -\frac{1}{\sin^2 a}.$$

$\square$

Theorem 12.16. *If $a > 0$ and $a \neq 1$, then the function $\log_a x$ is differentiable at every point $x > 0$, and*

$$(\log_a x)' = \frac{1}{\log a} \cdot \frac{1}{x}. \tag{12.3}$$

Proof. By Corollary 11.11,

$$\lim_{h \to 0} \left(1 + \frac{h}{x} \right)^{1/h} = e^{1/x}$$

for all $x > 0$. Thus

$$\lim_{h \to 0} \frac{\log_a(x + h) - \log_a x}{h} = \lim_{h \to 0} \log_a \left(1 + \frac{h}{x} \right)^{1/h} = \log_a e^{1/x} =$$

$$= \frac{1}{x} \cdot \log_a e = \frac{1}{\log a} \cdot \frac{1}{x}.$$

$\square$

It is clear that if $a > 0$ and $a \neq 1$, then the function $\log_a |x|$ is differentiable on the set $\mathbb{R} \setminus \{0\}$, and $(\log_a |x|)' = 1/(x \cdot \log a)$ for all $x \neq 0$.

We now turn our attention to introducing the **differentiation rules.** As we will see, a large portion of these are consequences of the definition of the derivative and theorems about limits.

Theorem 12.17. *If the functions f and g are differentiable at a, then cf $(c \in \mathbb{R})$, $f + g$, and $f \cdot g$ are also differentiable at a, and*

(i) $(cf)'(a) = cf'(a),$

(ii) $(f + g)'(a) = f'(a) + g'(a),$

(iii) $(fg)'(a) = f'(a)g(a) + f(a)g'(a).$

If $g(a) \neq 0$, then $1/g$ and f/g are differentiable at a. Moreover,

(iv) $\left(\dfrac{1}{g} \right)'(a) = -\dfrac{g'(a)}{g^2(a)},$

(v) $\left(\dfrac{f}{g} \right)'(a) = \dfrac{f'(a)g(a) - f(a)g'(a)}{g^2(a)}.$

Proof. The shared idea of the proofs of these is to express the difference quotients of each function with the help of the difference quotients $(f(x) - f(a))/(x-a)$ and $(g(x) - g(a))/(x-a)$:

(i) The difference quotient of the function $F = cf$ is

$$\frac{F(x) - F(a)}{x - a} = \frac{cf(x) - cf(a)}{x - a} = c \cdot \frac{f(x) - f(a)}{x - a} \to c \cdot f'(a).$$

(ii) The difference quotient of $F = f + g$ is

$$\frac{F(x) - F(a)}{x - a} = \frac{(f(x) + g(x)) - (f(a) + g(a))}{x - a} = \frac{f(x) - f(a)}{x - a} + \frac{g(x) - g(a)}{x - a}.$$

Thus

$$\lim_{x \to a} \frac{F(x) - F(a)}{x - a} = f'(a) + g'(a).$$

(iii) The difference quotient of $F = f \cdot g$ is

$$\frac{F(x) - F(a)}{x - a} = \frac{f(x) \cdot g(x) - f(a) \cdot g(a)}{x - a} =$$

$$= \frac{f(x) - f(a)}{x - a} \cdot g(x) + f(a) \cdot \frac{g(x) - g(a)}{x - a}.$$

Since $g(x)$ is differentiable at a, it is continuous there (by Theorem 12.4), and so

$$\lim_{x \to a} \frac{F(x) - F(a)}{x - a} = \lim_{x \to a} \frac{f(x) - f(a)}{x - a} \cdot \lim_{x \to a} g(x) + f(a) \cdot \lim_{x \to a} \frac{g(x) - g(a)}{x - a} =$$
$$= f'(a)g(a) + f(a)g'(a).$$

If $g(a) \neq 0$, then by the continuity of g, it follows that $g(x) \neq 0$ on a neighborhood of a; that is, the functions $1/g(x)$ and $f(x)/g(x)$ are defined here.

(iv) The difference quotient of $F = 1/g$ is

$$\frac{F(x) - F(a)}{x - a} = \frac{1/g(x) - 1/g(a)}{x - a} = \frac{g(a) - g(x)}{g(a)g(x)} \cdot \frac{1}{x - a} =$$

$$= -\frac{1}{g(x)g(a)} \cdot \frac{g(x) - g(a)}{x - a} \to -\frac{1}{g^2(a)} \cdot g'(a).$$

(v) The difference quotient of $F = f/g$ is

$$\frac{F(x) - F(a)}{x - a} = \frac{f(x)/g(x) - f(a)/g(a)}{x - a} = \frac{1}{g(a)g(x)} \frac{f(x)g(a) - f(a)g(x)}{x - a} =$$

$$= \frac{1}{g(x)g(a)} \left(\frac{f(x) - f(a)}{x - a} \cdot g(x) - \frac{g(x) - g(a)}{x - a} \cdot f(x) \right),$$

which implies

$$\lim_{x \to a} \frac{F(x) - F(a)}{x - a} = \frac{f'(a)g(a) - f(a)g'(a)}{g^2(a)}.$$

$\square$

Remarks 12.18. **1.** The statements of the theorem hold for right- and left-hand derivatives as well.

2. Let I be an interval, and suppose that the functions $f: I \to \mathbb{R}$ and $g: I \to \mathbb{R}$ are differentiable on I as stated in Definition 12.12. By the above theorem, it follows that cf, $f + g$, and $f \cdot g$ are also differentiable on I, and the equalities

$$(cf)' = cf', \ (f + g)' = f' + g', \ (fg)' = f'g + fg'$$

hold. If we also suppose that $g \neq 0$ on I, then $1/g$ and f/g are also differentiable on I, and

$$\left(\frac{1}{g} \right)' = -\frac{g'}{g^2}, \ \text{moreover} \ \left(\frac{f}{g} \right)' = \frac{f'g - fg'}{g^2}.$$

We emphasize that here we are talking about equality of *functions*, not simply numbers.

3. From the above theorem, we can use induction to easily prove the following statements.

If the functions $f_1, \dots, f_n$ are differentiable at a, then

(i) $f_1 + \cdots + f_n$ is differentiable at a, and

$$(f_1 + \cdots + f_n)'(a) = f_1'(a) + \cdots + f_n'(a);$$

moreover,

(ii) $f_1 \cdot \ldots \cdot f_n$ is differentiable at a, and

$$(f_1 \cdot \ldots \cdot f_n)'(a) = \left(f_1' \cdot f_2 \cdots f_n + f_1 \cdot f_2' \cdot f_3 \cdots f_n + \cdots + f_1 \cdots f_{n-1} \cdot f_n' \right)(a).$$

4. By (ii) above, it follows that if $f_1(a) \cdot \ldots \cdot f_n(a) \neq 0$, then

$$\left(\frac{(f_1 \cdot \ldots \cdot f_n)'}{f_1 \cdot \ldots \cdot f_n} \right)(a) = \left(\frac{f_1'}{f_1} + \cdots + \frac{f_n'}{f_n} \right)(a).$$

Thus if $f_1, \dots, f_n$ are defined on the interval I, are nowhere zero, and are differentiable on the interval I, then

$$\frac{(f_1 \cdot \ldots \cdot f_n)'}{f_1 \cdot \ldots \cdot f_n} = \frac{f_1'}{f_1} + \cdots + \frac{f_n'}{f_n}. \tag{12.4}$$

The next theorem is known as the **chain rule.**

Theorem 12.19. *If the function g is differentiable at a and the function f is differentiable at $g(a)$, then the function $h = f \circ g$ is differentiable at a, and*

$$h'(a) = f'(g(a)) \cdot g'(a).$$

With the notation $y = g(x)$ and $z = f(y)$, the statement of the theorem can easily be remembered in the form

$$\frac{dz}{dx} = \frac{dz}{dy} \cdot \frac{dy}{dx}.$$

This formula is where the name "chain rule" comes from.

Proof. By the assumptions, it follows that the function h is defined on a neighborhood of the point a. Indeed, f is defined on a neighborhood V of $g(a)$. Since g is differentiable at a, it is continuous at a, so there exists a neighborhood U of a such that $g(x) \in V$ for all $x \in U$. Thus h is defined on U. After these preparations, we give two proofs of the theorem.

I. Following the proof of the previous theorem, let us express the difference quotient of h using the difference quotients of f and g. Suppose first that $g(x) \neq g(a)$ on a punctured neighborhood of a. Then

$$\frac{h(x) - h(a)}{x - a} = \frac{f(g(x)) - f(g(a))}{x - a} = \frac{f(g(x)) - f(g(a))}{g(x) - g(a)} \cdot \frac{g(x) - g(a)}{x - a}. \tag{12.5}$$

Since g is continuous at a, if $x \to a$, then $g(x) \to g(a)$. Thus by the theorem on the limits of compositions of functions,

$$\lim_{x \to a} \frac{h(x) - h(a)}{x - a} = \lim_{t \to g(a)} \frac{f(t) - f(g(a))}{t - g(a)} \cdot \lim_{x \to a} \frac{g(x) - g(a)}{x - a} = f'(g(a)) \cdot g'(a).$$

The proof of the special case above used twice the fact that $g(x) \neq g(a)$ on a punctured neighborhood of a: first, when we divided by $g(x) - g(a)$, and second, when we applied Theorem 10.41 for the limit of the composition. Recall that the assumption of Theorem 10.41 requires that the inner function shouldn't take on its limit in a punctured neighborhood of the place, unless the outer function is continuous. This, however, does not hold for us, since the outer function is the difference quotient $(f(t) - f(g(a)))/(t - g(a))$, which is not even defined at $g(a)$.

Exactly this circumstance gives an idea for the proof in the general case. Define the function $F(t)$ as follows: let

$$F(t) = \frac{f(t) - f(g(a))}{t - g(a)}$$

if $t \in V$ and $t \neq g(a)$, and let $F(t) = f'(g(a))$ if $t = g(a)$. Then F is continuous at $g(a)$, so by Theorem 10.41, $\lim_{x \to a} F(g(x)) = f'(g(a))$. To finish the proof, it suffices to show that

$$\frac{h(x) - h(a)}{x - a} = F(g(x)) \cdot \frac{g(x) - g(a)}{x - a} \tag{12.6}$$

for all $x \in U$. We distinguish two cases. If $g(x) \neq g(a)$, then (12.6) is clear from (12.5). If, however, $g(x) = g(a)$, then $h(x) = f(g(x)) = f(g(a)) = h(a)$, so both sides of (12.6) are zero. Thus the proof is complete.

II. This proof is based on the definition of differentiability given in 12.11. According to this, the differentiability of f at $g(a)$ means that

$$f(t) - f(g(a)) = f'(g(a))(t - g(a)) + \varepsilon_1(t)(t - g(a)) \tag{12.7}$$

for all $t \in V$, where $\varepsilon_1(t) \to 0$ if $t \to g(a)$.

Let $\varepsilon_1(g(a)) = 0$. Similarly, by the differentiability of the function g, it follows that

$$g(x) - g(a) = g'(a)(x - a) + \varepsilon_2(x)(x - a) \tag{12.8}$$

for all $x \in U$, where $\varepsilon_2(x) \to 0$ as $x \to a$. If we substitute $t = g(x)$ in (12.7), and then apply (12.8), we get that

$$
\begin{aligned}
h(x) - h(a) = f(g(x)) - f(g(a)) = \\
= f'(g(a))(g(x) - g(a)) + \varepsilon_1(g(x))(g(x) - g(a)) = \\
= f'(g(a))g'(a)(x - a) + \varepsilon(x)(x - a),
\end{aligned}
$$

where

$$\varepsilon(x) = f'(g(a))\varepsilon_2(x) + \varepsilon_1(g(x))(g'(a) + \varepsilon_2(x)).$$

Since $g(x) \to g(a)$ as $x \to a$, we have $\varepsilon_1(g(x)) \to 0$ as $x \to a$ (since by $\varepsilon_1(g(a)) = 0$, ε_1 is continuous at $g(a)$). Then from $\varepsilon_2(x) \to 0$, we get that $\varepsilon(x) \to 0$ if $x \to a$. Now this, by Definition 12.11, means exactly that the function h is differentiable at a, and $h'(a) = f'(g(a))g'(a)$. $\square$

The following theorem gives the differentiation rule for inverse functions.

Theorem 12.20. *Let f be strictly monotone and continuous on the interval (a, b), and let φ denote the inverse of f. If f is differentiable at the point $c \in (a, b)$ and $f'(c) \neq 0$, then φ is differentiable at $f(c)$, and*

$$\varphi'(f(c)) = \frac{1}{f'(c)}.$$

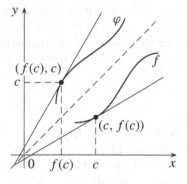

Fig. 12.3

Proof. The function φ is defined on the interval $J = f((a, b))$. By the definition of the inverse function, $\varphi(f(c)) = c$ and $f(\varphi(y)) = y$ for all $y \in J$. Let $F(x)$ denote the difference quotient $(f(x) - f(c))/(x - c)$. If $y \neq f(c)$, then

$$\frac{\varphi(y) - \varphi(f(c))}{y - f(c)} = \frac{\varphi(y) - c}{f(\varphi(y)) - f(c)} =$$

$$= \frac{1}{F(\varphi(y))}. \tag{12.9}$$

Since φ is strictly monotone, if $y \neq f(c)$, then $\varphi(y) \neq c$. Thus we can apply Theorem 10.41 on the limits of compositions of functions. We get that

$$\varphi'(f(c)) = \lim_{y \to f(c)} \frac{\varphi(y) - \varphi(f(c))}{y - f(c)} = \lim_{y \to f(c)} \frac{1}{F(\varphi(y))} = \lim_{x \to c} \frac{1}{F(x)} = \frac{1}{f'(c)}.$$

□

Remarks 12.21. **1.** The statement of the theorem can be illustrated with the following geometric argument. The graphs of the functions f and φ are the mirror images of each other in the line $y = x$. The mirror image of the tangent to graph f at the point $(c, f(c))$ gives us the tangent to graph φ at $(f(c), c)$. The slopes of these are the reciprocals of each other, that is, $\varphi'(f(c)) = 1/f'(c)$ (Figure 12.3).
2. Let f be strictly monotone and continuous on the interval (a, b), and suppose that f is differentiable everywhere on (a, b). If f' is nonvanishing everywhere, then by the above theorem, φ is everywhere differentiable on the interval $J = f((a, b))$, and $\varphi'(f(x)) = 1/f'(x)$ for all $x \in (a, b)$. If $y \in J$, then $\varphi(y) \in I$ and $f(\varphi(y)) = y$. Thus $\varphi'(y) = 1/f'(\varphi(y))$. Since this holds for all $y \in J$, we have that

$$\varphi' = \frac{1}{f' \circ \varphi}. \tag{12.10}$$

3. If $f'(c) = 0$ (that is, if the tangent line to graph f is parallel to the x-axis at the point $(c, f(c))$), then by (12.9), it is easy to see that the difference quotient $(\varphi(y) - \varphi(f(c)))/(y - f(c))$ does not have a finite limit at $f(c)$. Indeed, in this case, the limit of the difference quotient $F(x) = (f(x) - f(c))/(x - c)$ at c is zero, so $\lim_{y \to f(c)} F(\varphi(y)) = 0$. If, however, f is strictly monotone increasing, then the difference quotient $F(x)$ is positive everywhere, and so $\lim_{y \to f(c)} 1/F(\varphi(y)) = \infty$ (see Remark 10.39). Then

$$\lim_{y \to f(c)} \frac{\varphi(y) - \varphi(f(c))}{y - f(c)} = \infty,$$

and we can similarly get that if f is strictly monotone decreasing, then the value of the above limit is $-\infty$. (These observations agree with the fact that if $f'(c) = 0$, then the tangent line to graph φ at $(f(c), c)$ is parallel to the y-axis.) This remark motivates the following extension of the definition of the derivative.

Definition 12.22. Let f be defined on a neighborhood of a point a. We say that the *derivative of f at a is infinite* if

$$\lim_{x \to a} \frac{f(x) - f(a)}{x - a} = \infty,$$

and we denote this by $f'(a) = \infty$. We define $f'(a) = -\infty$ similarly.

Remark 12.23. Definitions 12.1 and 12.22 can be stated jointly as follows: if the limit (12.1) exists and has the value β (which can be finite or infinite), then we say that **the derivative of** f **at** a **exists,** and we use the notation $f'(a) = \beta$. We emphasize that the "differentiable" property is reserved for the cases in which the derivative is finite. Thus a function f is differentiable at a if and only if its derivative there exists and is *finite*.

We extend the definition of the one-sided derivatives (Definition 12.6) for the cases in which the one-sided limits of the difference quotient are infinite. We use the notation $f'_+(a) = \infty$, $f'_-(a) = \infty$, $f'_+(a) = -\infty$, and $f'_-(a) = -\infty$; their meaning is straightforward.

Using the concepts above, we can extend Theorem 12.20 as follows.

Theorem 12.24. *Let* f *be strictly monotone and continuous on the interval* (a,b). *Let* φ *denote the inverse function of* f. *If* $f'(c) = 0$, *then* φ *has a derivative at* $f(c)$, *and in fact,* $\varphi'(f(c)) = \infty$ *if* f *is strictly monotone increasing, and* $\varphi'(f(c)) = -\infty$ *if* f *is strictly monotone decreasing.*

Now let us return to the elementary functions. With the help of the differentiation rules, we can now determine the derivatives of all of them.

Theorem 12.25. *If* $a > 0$, *then the function* a^x *is differentiable everywhere, and*

$$(a^x)' = \log a \cdot a^x \tag{12.11}$$

for all x.

Proof. The statement is clear for $a = 1$, so we can suppose that $a \neq 1$. Since $(\log_a x)' = 1/(x \cdot \log a)$, by the differentiation rule for inverse functions, we get that

$$(a^x)' = a^x \cdot \log a.$$

$\square$

We note that the differentiability of the function a^x can also be deduced easily from Theorem 12.8.

Applying Theorems 12.16 and 12.25 for $a = e$, we get the following.

Theorem 12.26.

(i) *For all x,*

$$(e^x)' = e^x. \tag{12.12}$$

(ii) *For all* $x > 0$,

$$(\log x)' = \frac{1}{x}. \tag{12.13}$$

According to (12.11), the function e^x *is the only exponential function that is the derivative of itself.* This fact motivates us to consider e to be one of the most important constants[3] of analysis (and of mathematics, more generally). By equality (12.13), out of all the logarithmic functions, the derivative of the logarithm with base e is the simplest. This is why we chose the logarithm with base e from among all the other logarithm functions (see Remark 11.17).

Using the derivatives of the exponential and logarithmic functions, we can easily find the derivatives of the power functions.

Theorem 12.27. *For arbitrary $b \in \mathbb{R}$, the function x^b is differentiable at all points $x > 0$, and*

$$(x^b)' = b \cdot x^{b-1}. \tag{12.14}$$

Proof. Since $x^b = e^{b \log x}$ for all $x > 0$, we can apply the differentiation rule for compositions of functions, Theorem 12.19. □

The derivatives of the inverse trigonometric functions can easily be found by the differentiation rule for inverse functions.

Theorem 12.28.

(i) *The function $\arcsin x$ is differentiable on the interval $(-1,1)$, and*

$$(\arcsin x)' = \frac{1}{\sqrt{1-x^2}} \tag{12.15}$$

for all $x \in (-1,1)$.

(ii) *The function $\arccos x$ is differentiable on the interval $(-1,1)$, and*

$$(\arccos x)' = -\frac{1}{\sqrt{1-x^2}} \tag{12.16}$$

for all $x \in (-1,1)$.

(iii) *The function $\operatorname{arctg} x$ is differentiable everywhere, and*

$$(\operatorname{arctg} x)' = \frac{1}{1+x^2} \tag{12.17}$$

for all x.

(iv) *The function $\operatorname{arcctg} x$ is differentiable everywhere, and*

$$(\operatorname{arcctg} x)' = -\frac{1}{1+x^2} \tag{12.18}$$

for all x.

Proof. The function $\sin x$ is strictly increasing and differentiable on $\left[-\frac{\pi}{2}, \frac{\pi}{2}\right]$, and its derivative is $\cos x$. Since $\cos x \neq 0$ if $x \in \left(-\frac{\pi}{2}, \frac{\pi}{2}\right)$, by Theorem 12.20, if $x \in (-1,1)$, then

$$(\arcsin x)' = \frac{1}{\cos(\arcsin x)} = \frac{1}{\sqrt{1 - \sin^2(\arcsin x)}} = \frac{1}{\sqrt{1-x^2}},$$

which proves (i). Statement (ii) follows quite simply from (i) and (11.50).

---

[3] The other central constant is π. The relation between these two constants is given by $e^{i\pi} = -1$, which is a special case of the identity (11.64).

The function $\mathrm{tg}\,x$ is strictly increasing on the interval $\left(-\frac{\pi}{2},\frac{\pi}{2}\right)$, and its derivative is $1/\cos^2 x \neq 0$ there. Thus

$$(\mathrm{arc\,tg}\,x)' = \cos^2(\mathrm{arc\,tg}\,x).$$

However,

$$\cos^2 x = \frac{1}{1+\mathrm{tg}^2 x},$$

so

$$\cos^2(\mathrm{arc\,tg}\,x) = \frac{1}{1+x^2},$$

which establishes (iii). Statement (iv) is clear from (iii) and (11.52).   □

By the definition of the hyperbolic functions and (12.12), the assertions of the following theorem, which strengthen their link to the trigonometric functions, are clear once again.

Theorem 12.29.

(i) *The functions* $\mathrm{sh}\,x$ *and* $\mathrm{ch}\,x$ *are differentiable everywhere on* $(-\infty,\infty)$, *and*

$$(\mathrm{sh}\,x)' = \mathrm{ch}\,x \quad and \quad (\mathrm{ch}\,x)' = \mathrm{sh}\,x$$

   for all x.
(ii) *The function* $\mathrm{th}\,x$ *is differentiable everywhere on* $(-\infty,\infty)$. *Moreover,*

$$(\mathrm{th}\,x)' = \frac{1}{\mathrm{ch}^2 x}$$

   for all x.
(iii) *The function* $\mathrm{cth}\,x$ *is differentiable at all points* $x \neq 0$, *and there,*

$$(\mathrm{cth}\,x)' = -\frac{1}{\mathrm{sh}^2 x}.$$

Finally, consider the inverse hyperbolic functions.

Theorem 12.30.

(i) *The function* $\mathrm{arsh}\,x$ *is differentiable everywhere, and*

$$(\mathrm{arsh}\,x)' = \frac{1}{\sqrt{x^2+1}} \qquad\qquad (12.19)$$

   for all x.
(ii) *The function* $\mathrm{arch}\,x$ *is differentiable on the interval* $(1,\infty)$, *and*

$$(\mathrm{arch}\,x)' = \frac{1}{\sqrt{x^2-1}} \qquad\qquad (12.20)$$

   for all x > 1.

(iii) *The function* $\text{arth}\,x$ *is differentiable everywhere on* $(-1, 1)$, *and*

$$(\text{arth}\,x)' = \frac{1}{1 - x^2} \tag{12.21}$$

for all $x \in (-1, 1)$.

Proof. (i) Since the function $\text{sh}\,x$ is strictly increasing on $\mathbb{R}$ and its derivative there is $\text{ch}\,x \neq 0$, the function $\text{arsh}\,x$ is differentiable everywhere. The derivative can be computed either from Theorem 12.20 or by the identity (11.59).
(ii) Since the function $\text{ch}\,x$ is strictly increasing on $(0, \infty)$ and its derivative there is $\text{sh}\,x \neq 0$, the function $\text{arch}\,x$ is differentiable on the interval $(1, \infty)$.

We leave the proofs of (12.20) and statement (iii) to the reader.   □

Remark 12.31. It is worth noting that the derivatives of the functions $\log x$, $\text{arc\,tg}\,x$, and $\text{arth}\,x$ are rational functions, and the derivatives of $\text{arc\,sin}\,x$, $\text{arccos}\,x$, $\text{arsh}\,x$, and $\text{arch}\,x$ are algebraic functions. (The definition of an algebraic function can be found in Exercise 11.45.)

Exercises

12.18. Suppose $f + g$ is differentiable at a, and g is not differentiable at a. Can f be differentiable at a?

12.19. Let $f(x) = x^2 \cdot \sin(1/x)$, $f(0) = 0$. Prove that f is differentiable everywhere. (S)

12.20. Prove that if $0 < c < 1$, then the right-hand derivative of x^c at 0 is infinity.

12.21. Prove that if n is a positive odd integer, then the derivative of $\sqrt[n]{x}$ at 0 is infinity.

12.22. Where is the tangent line of the function $\sqrt[3]{\sin x}$ vertical?

12.23. Prove that the graphs of the functions $\sqrt{4a(a - x)}$ and $\sqrt{4b(b + x)}$ cross each other at right angles, that is, the tangent lines at the intersection point are perpendicular. (S)

12.24. Prove that the curves $x^2 - y^2 = a$ and $xy = b$ cross each other at right angles. That is, the graphs of the functions $\pm\sqrt{x^2 - a}$ and b/x cross each other perpendicularly.

12.25. At what angle do the graphs of the functions 2^x and $(\pi - e)^x$ cross each other? (S)

12.26. Give a closed form for $x + 2x^2 + \cdots + nx^n$. (Hint: differentiate the function $1 + x + \cdots + x^n$.) Use this to compute the sums

$$\frac{1}{2} + \frac{2}{4} + \frac{3}{8} + \cdots + \frac{n}{2^n} \quad \text{and} \quad \frac{1}{3} + \frac{2}{9} + \frac{3}{27} + \cdots + \frac{n}{3^n}.$$

12.27. Let $f(x) = x \cdot (x+1) \cdot \ldots \cdot (x+100)$, and let $g = f \circ f \circ f$. Compute the value of $g'(0)$.

12.28. Prove that the function x^x is differentiable for all $x > 0$, and compute its derivative.

12.29. The function x^x is strictly monotone on $[1, \infty)$. What is the value of the derivative of its inverse at the point 27?

12.30. The function $x^5 + x^2$ is strictly monotone on $[0, \infty)$. What is the value of the derivative of its inverse at the point 2?

12.31. Prove that the function $x + \sin x$ is strictly monotone increasing. What is the value of the derivative of its inverse at the point $1 + (\pi/2)$?

12.32. Let $f(x) = \log_x 3$ $(x > 0, x \neq 1)$. Compute the derivatives of f and f^{-1}.

12.33. Let us apply differentiation to find limits. The method consists in changing the function being considered into a difference quotient and finding its limit through differentiation. For example, instead of

$$\lim_{x \to 0} (x + e^x)^{1/x},$$

we can take its logarithm to get the quotient

$$\frac{\log (x + e^x)}{x},$$

which is the difference quotient of the numerator at 0. The limit of the quotient is thus the derivative of the numerator at 0. If this limit is A, then the original limit is e^A. Finish this computation.

12.34. Apply the method above, or one of its variants, to find the following limits:

(a) $\lim_{x \to 0} (\cos x)^{1/\sin x}$,

(b) $\lim_{x \to 0} \left(\frac{e^x + 1}{2} \right)^{1/\operatorname{sh} x}$,

(c) $\lim_{x \to 0} \frac{\operatorname{sh}^2 x}{\log \cos 3x}$,

(d) $\lim_{x \to 1} (2 - x)^{1/\cos(\pi/(2x))}$,

(e) $\lim_{x \to \infty} (x^{1/x} - 1) \cdot \frac{x}{\log x}$.

12.35. Prove, using the method above, that if $a_1, \ldots, a_n > 0$, then

$$\lim_{x \to 0} \sqrt[x]{\frac{a_1^x + \cdots + a_n^x}{n}} = \sqrt[n]{a_1 \cdot \ldots \cdot a_n}.$$

12.36. Let T_n denote the nth Chebyshev polynomial (see Exercise 11.32). Prove that if $T_n(a) = 0$, then $|T_n'(a)| = n/\sqrt{1 - a^2}$.

12.37. Let f be convex on the open interval I.

(a) Prove that the function $f_+'(x)$ is monotone increasing on I.
(b) Prove that if the function $f_+'(x)$ is continuous at a point x_0, then f is differentiable at x_0.
(c) Prove that the set $\{x \in I : f \text{ is not differentiable at } x\}$ is countable.

12.3 Higher-Order Derivatives

Definition 12.32. Let the function f be differentiable in a neighborhood of a point a. If the derivative function f' has a derivative at a, then we call the derivative of f' at a the *second derivative* of f. We denote this by $f''(a)$. Thus

$$f''(a) = \lim_{x \to a} \frac{f'(x) - f'(a)}{x - a}.$$

If $f''(a)$ exists and is finite, then we say that f is twice differentiable at a. The *second derivative function* of f, denoted by f'', is the function defined for the points x at which f is twice differentiable, and its value there is $f''(x)$.

We can define the kth-order derivatives by induction:

Definition 12.33. Let the function f be $k - 1$ times differentiable in a neighborhood of the point a. Let the $(k - 1)$th derivative function of f be denoted by $f^{(k-1)}$. The derivative of $f^{(k-1)}$ at a, if it exists, is called the *kth (order) derivative* of f. The kth derivative function is denoted by $f^{(k)}$; this is defined where f is k times differentiable.

The kth derivative at a can be denoted by the symbols

$$\left. \frac{d^k f}{dx^k} \right|_{x=a}, \quad \left. \frac{d^k f(x)}{dx^k} \right|_{x=a}, \quad y^{(k)}(a), \quad \left. \frac{d^k y}{dx^k} \right|_{x=a}$$

as well. To keep our notation consistent, we will sometimes use the notation

$$f^{(0)} = f, \quad f^{(1)} = f', \quad f^{(2)} = f''$$

too.

If $f^{(k)}$ exists for all $k \in \mathbb{N}^+$ at a, then we say that f is *infinitely differentiable at a*.

It is easy to see that if p is an nth-degree polynomial, then its kth derivative is an $(n-k)$th-degree polynomial for all $k \leq n$. Thus the nth derivative of the polynomial p is constant, and the kth derivative is identically zero for $k > n$. It follows that *every polynomial is infinitely differentiable.* With the help of higher-order derivatives, we can easily determine the multiplicity of a root of a polynomial.

Theorem 12.34. *The number a is a root of the polynomial p with multiplicity k if and only if*

$$p(a) = p'(a) = \ldots = p^{(k-1)}(a) = 0 \text{ and } p^{(k)}(a) \neq 0. \qquad (12.22)$$

Proof. Clearly, it is enough to show that if a is a root of multiplicity k, then (12.22) holds (since, for different k's the statements (12.22) exclude each other). We prove this by induction. If $k = 1$, then $p(x) = (x-a) \cdot q(x)$, where $q(a) \neq 0$. Then $p'(x) = q(x) + (x-a) \cdot q'(x)$, which gives $p'(a) = q(a) \neq 0$, so (12.22) holds for $k = 1$.

Let $k > 1$, and suppose that the statement holds for $k-1$. Since $p(x) = (x-a)^k \cdot q(x)$, where $q(a) \neq 0$, we have

$$p'(x) = k \cdot (x-a)^{k-1} \cdot q(x) + (x-a)^k \cdot q'(x) = (x-a)^{k-1} \cdot r(x),$$

where $r(a) \neq 0$. Then the number a is a root of the polynomial p' with multiplicity $k-1$, so by the induction hypothesis,

$$p'(a) = p''(a) = \ldots = p^{(k-1)}(a) = 0 \text{ and } p^{(k)}(a) \neq 0.$$

Since $p(a) = 0$ is also true, we have proved (12.22). □

Some of the differentiation rules apply for higher-order derivatives as well. Out of these, we will give those for addition and multiplication. To define the rule for multiplication, we need to introduce **binomial coefficients.**

Definition 12.35. If $0 \leq k \leq n$ are integers, then the number $\dfrac{n!}{k!(n-k)!}$ is denoted by $\binom{n}{k}$, where $0!$ is defined to be 1.

By the definition, it is clear that $\binom{n}{0} = \binom{n}{n} = 1$ for all n. It is also easy to check that

$$\binom{n}{k} = \binom{n-1}{k-1} + \binom{n-1}{k} \qquad (12.23)$$

for all $n \geq 2$ and $k = 1, \ldots, n-1$.

Theorem 12.36 (Binomial Theorem). *The following identity holds:*

$$(a+b)^n = \sum_{k=0}^{n} \binom{n}{k} a^{n-k} b^k. \qquad (12.24)$$

The name of the theorem comes from the fact that a *binomial* (that is, a polynomial with two terms) appears on the left-hand side of (12.24).

Proof. We prove this by induction. The statement is clear for $n = 1$. If it holds for n, then

$$(a+b)^{n+1} = (a+b)^n \cdot (a+b) =$$

$$= \left[a^n + \binom{n}{1} a^{n-1} \cdot b + \cdots + \binom{n}{n-1} a \cdot b^{n-1} + b^n \right] \cdot (a+b).$$

If we multiply this out, then in the resulting sum, the terms a^{n+1} and b^{n+1} appear, and moreover, for all $1 \le k \le n$, the terms $\binom{n}{k-1} a^{n-k+1} \cdot b^k$ and $\binom{n}{k} a^{n-k+1} \cdot b^k$ also appear. The sum of these two terms, according to (12.23), is exactly $\binom{n+1}{k} a^{n-k+1} \cdot b^k$. Thus we get the identity (12.24) (with $n+1$ replacing n). $\square$

Theorem 12.37. *If f and g are n times differentiable at a, then $f + g$ and $f \cdot g$ are also n times differentiable there, and*

$$(f+g)^{(n)}(a) = f^{(n)}(a) + g^{(n)}(a), \tag{12.25}$$

as well as

$$(f \cdot g)^{(n)}(a) = \sum_{k=0}^{n} \binom{n}{k} f^{(n-k)}(a) \cdot g^{(k)}(a). \tag{12.26}$$

The identity (12.26) is called the **Leibniz rule.**

Proof. (12.25) is straightforward by induction. We also use induction to prove (12.26). If $n = 1$, then the statement is clear by the differentiation rule for products.

Suppose that the statement holds for n. If f and g are $n+1$ times differentiable at a, then they are n times differentiable in a neighborhood U of a, so by the induction hypothesis,

$$(f \cdot g)^{(n)} = f^{(n)} \cdot g + \binom{n}{1} f^{(n-1)} \cdot g' + \cdots + \binom{n}{n-1} f' \cdot g^{(n-1)} + f \cdot g^{(n)} \tag{12.27}$$

in U. Using the differentiation rules for sums and products, we get that $(f \cdot g)^{(n+1)}(a)$ is the sum of the terms $f^{(n+1)}(a) \cdot g(a)$ and $f(a) \cdot g^{(n+1)}(a)$, as well as terms of the form $\binom{n}{k-1} f^{(n-k+1)}(a) \cdot g^{(k)}(a)$ and $\binom{n}{k} f^{(n-k+1)}(a) \cdot g^{(k)}(a)$ for all $k = 1, \ldots, n$. These sum to $\binom{n+1}{k} f^{(n-k+1)}(a) \cdot g^{(k)}(a)$, which shows that (12.26) holds for $n+1$. $\square$

The higher-order derivatives of some elementary functions are easy to compute.

Examples 12.38. **1.** It is easy to see that the exponential function a^x is infinitely differentiable, and that

$$(a^x)^{(n)} = (\log a)^n \cdot a^x \tag{12.28}$$

for all n.

2. It is also easy to check that the power function x^b is infinitely differentiable on the interval $(0, \infty)$, and that

$$(x^b)^{(n)} = b(b-1) \cdot \ldots \cdot (b-n+1) \cdot x^{b-n} \tag{12.29}$$

for all n and $x > 0$.

3. The functions $\sin x$ and $\cos x$ are also infinitely differentiable, and their higher-order derivatives are

$$(\sin x)^{(2n)} = (-1)^n \cdot \sin x, \qquad (\sin x)^{(2n+1)} = (-1)^n \cdot \cos x,$$
$$(\cos x)^{(2n)} = (-1)^n \cdot \cos x, \qquad (\cos x)^{(2n+1)} = (-1)^{n+1} \cdot \sin x \qquad (12.30)$$

for all n and x.

Remark 12.39. The equalities $(\sin x)'' = -\sin x$ and $(\cos x)'' = -\cos x$ can be expressed by saying that the functions $\sin x$ and $\cos x$ satisfy the relation

$$y'' + y = 0;$$

this means that if we write $\sin x$ or $\cos x$ in place of y, then we get an equality. Such a relation that links the derivatives of a function to the function itself (possibly using other known functions) is called a **differential equation.** A differential equation is said to have order n if the highest-order derivative that appears in the differential equation is the nth. So we can say that the functions $\sin x$ and $\cos x$ satisfy the second-order differential equation $y'' + y = 0$. More specifically, we will say that this differential equation is an **algebraic differential equation**, since only the basic operations are applied to the function and its derivatives.

It is clear that the exponential function a^x satisfies the first-order differential equation $y' - \log a \cdot y = 0$. But it is not immediately clear that every exponential function satisfies the *same* differential equation. Indeed, if $y = a^x$, then $y'/y = \log a$, which is constant. Thus $(y'/y)' = 0$, that is, a^x satisfies the second-order algebraic differential equation

$$y'' \cdot y - (y')^2 = 0.$$

We can similarly show that every power function satisfies the same second-order algebraic differential equation. The function x will appear in this differential equation, but can be removed by increasing the order (see Exercise 12.43).

The function $\log_a x$ satisfies the equation $y' - (\log a \cdot x)^{-1} = 0$. From this, we get that $x \cdot y'$ is a constant, that is,

$$x \cdot y'' + y' = 0.$$

It is easy to see that the logarithmic functions satisfy a single third-order algebraic differential equation, in which x does not appear. The inverse trigonometric and hyperbolic functions satisfy similar equations.

One can show that if two functions both satisfy an algebraic differential equation, then their sum, product, quotient, and composition also satisfy an algebraic differential equation (a different one, generally more complicated). It follows that *every elementary function satisfies an algebraic differential equation* (which, of course, depends on the function).

In the next chapter, we will discuss differential equations in more detail.

Exercises

12.38. Prove that if f is twice differentiable at a, then f is continuous in a neighborhood of a.

12.39. How many times is the function $|x|^3$ differentiable at 0?

12.40. Give a function that is k times differentiable at 0 but is not $k+1$ times differentiable there.

12.41. Prove that for the Chebyshev polynomial $T_n(x)$ (see Exercise 11.31),

$$(1 - x^2)T_n''(x) - x\, T_n'(x) + n^2 T_n(x) = 0$$

for all x.

12.42. Prove that the **Legendre polynomial**

$$P_n(x) = \frac{1}{2^n\, n!} \left((x^2 - 1)^n\right)^{(n)}$$

satisfies

$$(1 - x^2)P_n''(x) - 2x\, P_n'(x) + n(n+1)P_n(x) = 0$$

for all x.

12.43. (a) Prove that every power function satisfies a second-order algebraic differential equation.
(b) Prove that every power function satisfies a single third-order algebraic differential equation that does not contain x.

12.44. Prove that the logarithmic functions satisfy a single third-order algebraic differential equation that does not contain x. (S)

12.45. Prove that each of the functions $\arcsin x$, $\arccos x$, $\operatorname{arctg} x$, $\operatorname{arsh} x$, $\operatorname{arch} x$, and $\operatorname{arth} x$ (individually) satisfies a third-degree algebraic differential equation that does not contain x.

12.46. Prove that the function $e^x + \log x$ satisfies an algebraic differential equation.

12.47. Prove that the function $e^x \cdot \sin x$ satisfies an algebraic differential equation.

12.4 Linking the Derivative and Local Properties

Definition 12.40. Let the function f be defined on a neighborhood of the point a. We say that f is *locally increasing at* a if there exists a $\delta > 0$ such that $f(x) \leq f(a)$ for all $a - \delta < x < a$, and $f(x) \geq f(a)$ for all $a < x < a + \delta$.

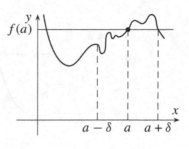

Fig. 12.4

Let f be defined on a right-hand neighborhood of a. We say that f is *locally increasing on the right at* a if there exists a $\delta > 0$ such that $f(x) \geq f(a)$ for all $a < x < a + \delta$ (Figure 12.4).

We similarly define the concepts of *strictly locally increasing, locally decreasing,* and *strictly locally decreasing at* a, as well as *(strictly) locally increasing and decreasing from the left.*

Remark 12.41. We have to take care in distinguishing the concepts of local and monotone increasing (or decreasing) functions. The precise link between the two is the following.

On the one hand, it is clear that if f is monotone increasing on (a,b), then f is locally increasing for all points in (a,b).

On the other hand, it can be shown that if f is locally increasing at every point in (a,b), then it is monotone increasing in (a,b) (but since we will not need this fact, we leave the proof of it as an exercise; see Exercise 12.54).

It is possible, however, for a function f to be locally increasing at a point a but not be monotone increasing on any neighborhood $U(a)$. Consider the following examples.

1. The function

$$f(x) = \begin{cases} x \cdot \sin^2(1/x), & \text{if } x \neq 0, \\ 0, & \text{if } x = 0 \end{cases}$$

   is locally increasing at 0, but it does not have a neighborhood of 0 in which f is monotone (Figure 12.5).
2. The function

$$f(x) = \begin{cases} 1/x, & \text{if } x \neq 0, \\ 0, & \text{if } x = 0 \end{cases}$$

   is strictly locally increasing at 0, but it is not monotone increasing in any interval. In fact, in the *right-hand* punctured neighborhoods of 0, f is strictly monotone decreasing.

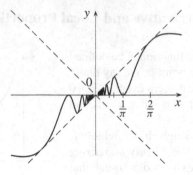

Fig. 12.5

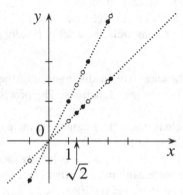

Fig. 12.6

3. Similarly, the function

$$f(x) = \begin{cases} \operatorname{tg} x, & \text{if } x \in (0, \pi) \setminus \{\frac{\pi}{2}\}, \\ 0, & \text{if } x = \frac{\pi}{2} \end{cases}$$

is strictly locally decreasing at $\frac{\pi}{2}$ but is strictly monotone increasing on the intervals $(0, \pi/2)$ and $(\pi/2, \pi)$.

4. The function

$$f(x) = \begin{cases} x, & \text{if } x \text{ is irrational}, \\ 2x, & \text{if } x \text{ is rational} \end{cases}$$

is strictly locally increasing at 0, but there is no interval on which it is monotone (Figure 12.6).

Definition 12.42. We say that the function f has a *local maximum* (or *minimum*) at a if a has a neighborhood U in which f is defined and for all $x \in U$, $f(x) \le f(a)$ (or $f(x) \ge f(a)$). We often refer to the point a itself as the *local maximum (or minimum) of the function*.

If for all $x \in U \setminus \{a\}$, $f(x) < f(a)$ (or $f(x) > f(a)$), then we say that a is a *strict local maximum (or minimum)*.

Local maxima and local minima are collectively called *local extrema*.

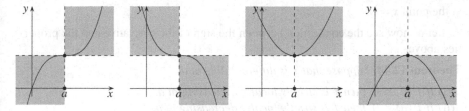

Fig. 12.7

Remark 12.43. We defined absolute (global) extrema in Definition 10.54. The following connections exist between absolute and local extrema.

An absolute extremum is not necessarily a local extremum, since a condition for a point to be a local extremum is that the function be defined in a neighborhood of the point. So for example, the function x on the interval $[0,1]$ has an absolute minimum at 0, but this is not a local minimum. However, if the function $f : A \to \mathbb{R}$ has an absolute extremum at $a \in A$, and A contains a neighborhood of a, then a is a local extremum.

A local extremum is not necessarily an absolute extremum, since the fact that f does not have a value larger than $f(a)$ in a neighborhood of a does not prevent it from having a larger value outside the neighborhood.

Consider the following three properties:

I. The function f is locally increasing at a.
II. The function f is locally decreasing at a.
III. a is a local extremum of the function f.

The function f satisfies one of the properties I, II, and III if and only if there exists a $\delta > 0$ such that the graph of f over the interval $(a - \delta, a)$ lies entirely in one section of the plane separated by the horizontal line $y = f(a)$ and the graph of f over $(a, a + \delta)$ lies in one of the sections separated by $y = f(a)$. The four possibilities correspond to the function at a being locally increasing, locally decreasing, a local maximum, and a local minimum (Figure 12.7).

It is clear that properties I and II can hold simultaneously only if f is constant in a neighborhood of a.

In the strict variants of the properties above, however, only one can hold at one time.

Of course, it is possible that none of I, II, and III holds. This is the case, for example, with the function

$$f(x) = \begin{cases} x\sin 1/x, & \text{if } x \neq 0, \\ 0, & \text{if } x = 0 \end{cases}$$

at the point $x = 0$.

Let us now see the connection between the sign of the derivative and the properties above.

Theorem 12.44. *Suppose that f is differentiable at a.*

(i) *If $f'(a) > 0$, then f is strictly locally increasing at a.*
(ii) *If $f'(a) < 0$, then f is strictly locally decreasing at a.*
(iii) *If f is locally increasing at a, then $f'(a) \geq 0$.*
(iv) *If f is locally decreasing at a, then $f'(a) \leq 0$.*
(v) *If f has a local extremum at a, then $f'(a) = 0$.*

Proof. (i) If

$$f'(a) = \lim_{x \to a} \frac{f(x) - f(a)}{x - a} > 0,$$

then by Theorem 10.30, there exists a $\delta > 0$ such that

$$\frac{f(x) - f(a)}{x - a} > 0$$

for all $0 < |x - a| < \delta$. Thus $f(x) > f(a)$ if $a < x < a + \delta$, and $f(x) < f(a)$ if $a - \delta < x < a$. But this means precisely that the function f is strictly locally increasing at a (Figure 12.8). Statement (ii) can be proved similarly.
(iii) If f is locally increasing at a, then there exists a $\delta > 0$ such that

$$\frac{f(x) - f(a)}{x - a} \geq 0$$

if $0 < |x - a| < \delta$. But then

$$f'(a) = \lim_{x \to a} \frac{f(x) - f(a)}{x - a} \geq 0.$$

Statement (iv) can be proved similarly.
(v) If $f'(a) \neq 0$, then by (i) and (ii), f is either locally strictly increasing or locally strictly decreasing at a, so a cannot be a local extremum. Thus if f has a local extremum at a, then necessarily $f'(a) = 0$. $\square$

Remarks 12.45. **1.** The one-sided variants of the statements (i)–(iv) above also hold (and can be proved the same way). That is, if $f'_+(a) > 0$, then f is strictly locally increasing on the right at a; if f is locally increasing on the right at a, then $f'_+(a) \geq 0$, assuming that the right-hand derivative exists.

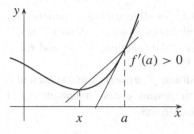

Fig. 12.8

2. *None of the converses of the statements (i)–(v) is true.* If we know only that f is strictly locally increasing at a, cannot deduce that $f'(a) > 0$. For example, the function $f(x) = x^3$ is strictly locally increasing at 0 (and in fact strictly monotone increasing on the whole real line), but $f'(0) = 0$.

If we know only that $f'(a) \geq 0$, we cannot deduce that the function f is locally increasing at a. For example, for the function $f(x) = -x^3$, $f'(0) = 0 \geq 0$, but f is not locally increasing at 0 (and in fact, it is strictly locally decreasing there).

Similarly, if we know only that $f'(a) = 0$, cannot deduce that the function f has a local extremum at a. For example, for the function $f(x) = x^3$, $f'(0) = 0$, but f does not have a local extremum at 0 (since x^3 is strictly monotone increasing on the entire real line). We can also express this by saying that if f is differentiable at a, then the assumption $f'(a) = 0$ is a necessary *but not sufficient* condition for f to have a local extremum at a.

3. If in statement (iii), we assume f to be strictly locally increasing instead, then we still cannot say more than $f'(a) \geq 0$ generally (since the converse of statement (i) does not hold).

4. From $f'(a) > 0$, we can deduce only that f is locally increasing at a, and not that it is monotone increasing. Consider the following example. Let f be a function such that $x - x^2 \leq f(x) \leq x + x^2$ for all x (Figure 12.9). Then $f(0) = 0$, and so if $x > 0$, then

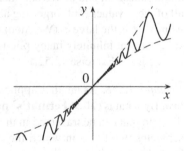

Fig. 12.9

$$1 - x \leq \frac{f(x)}{x} = \frac{f(x) - f(0)}{x - 0} \leq 1 + x,$$

while if $x < 0$, the reverse inequalities hold. Thus by the squeeze theorem,

$$f'(0) = \lim_{x \to 0} \frac{f(x) - f(0)}{x - 0} = 1 > 0.$$

On the other hand, it is clear that we can choose the function f such that it is not monotone increasing in any neighborhood of 0. For this, if we choose $\delta > 0$, we need $-\delta < x < y < \delta$ such that $f(x) > f(y)$. If, for example, $f(x) = x - x^2$ for all rational x and $f(x) = x + x^2$ for all irrational x, then this holds for sure. We can even construct f to be differentiable everywhere: draw a "smooth" (that is, differentiable) wave between the graphs of the functions $x - x^2$ and $x + x^2$ (one such function can be seen in Exercise 12.53).

Even though the condition $f'(a) = 0$ is not sufficient for f to have a local extremum at a, in certain important cases, statement (v) of Theorem 12.44 is still applicable for finding the extrema.

Example 12.46. Find the (absolute) maximum of the function $f(x) = x \cdot (1 - x)$ in the interval $[0, 1]$. Since the function is continuous, by Weierstrass's theorem (Theorem 10.55) f has an absolute maximum in $[0, 1]$. Suppose that f takes on its largest value at the point $a \in [0, 1]$. Then either $a = 0$, $a = 1$, or $a \in (0, 1)$. In the last case f has a local maximum at a, and since f is everywhere differentiable, by statement (v) of Theorem 12.44 we have $f'(a) = 0$. Now $f'(x) = 1 - 2x$, so the condition $f'(a) = 0$ is satisfied only by $a = 1/2$. We get that the function attains its maximum at one of the points 0, 1, 1/2. However, $f(0) = f(1) = 0$ and $f(1/2) = 1/4 > 0$, so 0 and 1 cannot be maxima of the function. Thus only $a = 1/2$ is possible. Thus we have shown that the function $f(x) = x \cdot (1 - x)$ over the interval $[0, 1]$ attains its maximum at the point $1/2$; that is, $a = 1/2$ is its absolute maximum.

Remark 12.47. This argument can be applied in the cases in which we are dealing with a function f that is continuous on a closed and bounded interval $[a, b]$ and is differentiable inside, on (a, b). Then f has a largest value by Weierstrass's theorem. If this is attained at a point c, then either $c = a$, $c = b$, or $c \in (a, b)$. In this last case, we are talking about a local extremum as well, so $f'(c) = 0$. Thus if we find all points $c \in (a, b)$ where f' vanishes, then the absolute maximum points of f must be among these, a, and b. We can then locate the absolute maxima by computing f at all of these values (not forgetting about a and b), and determining those at which the value of f is the largest. (We should note that in some cases, we have to compute the value of f at infinitely many points. It can happen that f' has infinitely many roots in (a, b); see Exercise 12.52.)

Example 12.48. As another application of the argument above, we deduce **Snell's**[4] **law.** By what is called **Fermat's**[5] **principle,** light traveling between two points takes the path that can be traversed in the least amount of time. In Figure 12.10, the x-axis separates two fluids in which the speed of light is respectively v_1 and v_2. Looking from the point P_1, we will see point P_2 in the direction that a light ray arrives at P_1 if it starts at P_2. The light ray—by Fermat's principle—"chooses" the path that takes the shortest time to traverse. To determine the path of the light ray, we thus need to solve the following problem.

---

[4] Willebrord Snellius (1580–1626), Dutch mathematician.

[5] Pierre de Fermat (1601–1665), French mathematician.

Let a line e be given in the plane, and in the two half-planes determined by this line, let there be given the points P_1 and P_2. If a moving point travels with velocity v_1 in the half-plane in which P_1 is located, and with velocity v_2 if it is in the half-plane of P_2, what path must it take to get from P_1 to P_2 in the shortest amount of time?

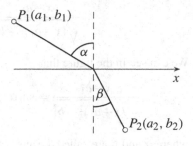

Fig. 12.10

Let the line e be the x-axis, let the coordinates of P_1 be (a_1, b_1), and let the coordinates of P_2 be (a_2, b_2). We may assume that $a_1 < a_2$ (Figure 12.10). Clearly, the point needs to travel in a straight line in both half-planes, so the problem is simply to find where the point crosses the x-axis, that is, where the path bends (is refracted).

If the path intersects the x-axis at the point x, the time necessary for the point to traverse the entire path is

$$f(x) = \frac{1}{v_1} \cdot \sqrt{(x-a_1)^2 + b_1^2} + \frac{1}{v_2} \cdot \sqrt{(x-a_2)^2 + b_2^2},$$

and so

$$f'(x) = \frac{1}{v_1} \cdot \frac{x-a_1}{\sqrt{(x-a_1)^2 + b_1^2}} + \frac{1}{v_2} \cdot \frac{x-a_2}{\sqrt{(x-a_2)^2 + b_2^2}}. \qquad (12.31)$$

Our task is to find the absolute minimum of f. Since if $x < a_1$, then $f(x) > f(a_1)$, and if $x > a_2$, then $f(x) > f(a_2)$, it suffices to find the minimum of f in the interval $[a_1, a_2]$. Since f is continuous, Weierstrass's theorem applies, and f attains its minimum on $[a_1, a_2]$. Since f is also differentiable, the minima can be only at the endpoints of the interval and at the points where the derivative is zero.

Now by (12.31),

$$f'(a_1) = \frac{(a_1 - a_2)}{v_2 \cdot \sqrt{(a_1 - a_2)^2 + b_2^2}} < 0,$$

and so by Theorem 12.44, f is strictly locally decreasing at a_1. Thus in a suitable right-hand neighborhood of a_1, every value of f is smaller that $f(a_1)$, so a_1 cannot be a minimum. Similarly,

$$f'(a_2) = \frac{(a_2 - a_1)}{v_1 \cdot \sqrt{(a_2 - a_1)^2 + b_1^2}} > 0,$$

and so by Theorem 12.44, f is strictly locally increasing at a_2. Thus in a suitable left-hand neighborhood of a_2, every value of f is smaller that $f(a_2)$, so a_2 cannot be a minimum either. Thus the minimum of the function f is at a point $x \in (a_1, a_2)$ where $f'(x) = 0$. By (12.31), this is equivalent to saying that

$$\frac{x-a_1}{\sqrt{(x-a_1)^2+b_1^2}} : \frac{a_2-x}{\sqrt{(x-a_2)^2+b_2^2}} = \frac{v_1}{v_2}.$$

We can see in the figure that

$$\frac{x-a_1}{\sqrt{(x-a_1)^2+b_1^2}} = \sin\alpha \quad \text{and} \quad \frac{a_2-x}{\sqrt{(x-a_2)^2+b_2^2}} = \sin\beta,$$

where α and β are called the angle of incidence and angle of refraction respectively. Thus the path taking the least time will intersect the line separating the two fluids at the point where

$$\frac{\sin\alpha}{\sin\beta} = \frac{v_1}{v_2}.$$

This is **Snell's law**.

Exercises

12.48. We want to create a box with no lid out of a rectangle with sides a and b by cutting out a square of size x at each corner of the rectangle. How should we chose x to maximize the volume of the box? (S)

12.49. Which cylinder inscribed into a given sphere has the largest volume?

12.50. Which right circular cone inscribed into a given sphere has the largest volume?

12.51. Which right circular cone inscribed into a given sphere has the largest surface area? (The surface area of the cone includes the base circle.)

12.52. Let

$$f(x) = \begin{cases} x^2 \cdot \sin 1/x, & \text{if } x \neq 0, \\ 0, & \text{if } x = 0. \end{cases}$$

Prove that the derivative of f has infinitely many roots in $(0,1)$.

12.53. Let

$$f(x) = \begin{cases} x + 2x^2 \cdot \sin 1/x, & \text{if } x \neq 0, \\ 0, & \text{if } x = 0. \end{cases}$$

Show that $f'(0) > 0$, but f is not monotone increasing in any neighborhood of 0. (S)

12.54. Prove that if f is locally increasing at all points in (a,b), then it is monotone increasing in (a,b). (H)

12.55. Determine the absolute extrema of the functions below in the given intervals.

(a) $x^2 - x^4$, $[-2,2]$;

(b) $x - \arctan x$, $[-1,1]$;

(c) $x + e^{-x}$, $[-1,1]$;

(d) $x + x^{-2}$, $[1/10, 10]$;

(e) $\arctan(1/x)$, $[1/10, 10]$;

(f) $\cos x^2$, $[0, \pi]$;

(g) $\sin(\sin x)$, $[-\pi/2, \pi/2]$;

(h) $x \cdot e^{-x}$, $[-2, 2]$;

(i) $x^n \cdot e^{-x}$, $[-2n, 2n]$;

(j) $x - \log x$, $[1/2, 2]$;

(k) $1/(1 + \sin^2 x)$, $(0, \pi)$;

(l) $\sqrt{1 - e^{-x^2}}$, $[-2,2]$;

(m) $x \cdot \sin(\log x)$, $[1, 100]$;

(n) x^x, $(0, \infty)$;

(o) $\sqrt[x]{x}$, $(0, \infty)$;

(p) $(\log x)/x$, $(0, \infty)$;

(q) $x \cdot \log x$, $(0, \infty)$;

(r) $x^x \cdot (1 - x)^{1-x}$, $(0, 1)$.

12.5 Intermediate Value Theorems

The following three theorems—each a generalization of the one that it follows—are some of the most frequently used theorems in differentiation. When we are looking for a link between properties of a function and its derivative, most often we use one of these intermediate value theorems.

Theorem 12.49 (Rolle's Theorem[6]). *Suppose that the function f is continuous on* $[a,b]$ *and differentiable on* (a,b). *If* $f(a) = f(b)$, *then there exists a* $c \in (a,b)$ *such that* $f'(c) = 0$.

Proof. If $f(x) = f(a)$ for all $x \in (a,b)$, then f is constant in (a,b), so $f'(x) = 0$ for all $x \in (a,b)$. Then we can choose c to be any number in (a,b).

We can thus suppose that there exists an $x_0 \in (a,b)$ for which $f(x_0) \neq f(a)$. Consider first the case $f(x_0) > f(a)$. By Weierstrass's theorem, f has an absolute maximum in $[a,b]$. Since $f(x_0) > f(a) = f(b)$, neither a nor b can be its absolute maximum. Thus if c is an abso-

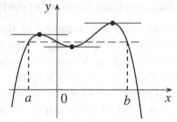

Fig. 12.11

lute maximum, then $c \in (a,b)$, and so c is also a local maximum too. By statement (v) of Theorem 12.44, it then follows that $f'(c) = 0$.

If $f(x_0) < f(a)$, then we argue similarly, considering the absolute minimum of f instead (Figure 12.11). $\square$

An important generalization of Rolle's theorem is the following theorem.

---

[6] Michel Rolle (1652–1719), French mathematician.

Theorem 12.50 (Mean Value Theorem). *If the function f is continuous on* $[a,b]$ *and differentiable on* (a,b)*, then there exists a* $c \in (a,b)$ *such that*

$$f'(c) = \frac{f(b) - f(a)}{b - a}.$$

Proof. The equation for the chord between the points $(a, f(a))$ and $(b, f(b))$ is given by

$$y = h_{a,b}(x) = \frac{f(b) - f(a)}{b - a}(x - a) + f(a).$$

The function

$$F(x) = f(x) - h_{a,b}(x)$$

satisfies the conditions of Rolle's theorem. Indeed, since f and $h_{a,b}$ are both continuous in $[a,b]$ and differentiable on (a,b), their difference also has these properties. Since $F(b) = F(a) = 0$, we can apply Rolle's theorem to F. We get that there exists a $c \in (a,b)$ such that $F'(c) = 0$. But this means that

$$0 = F'(c) = f'(c) - h'_{a,b}(c) = f'(c) - \frac{f(b) - f(a)}{b - a},$$

so

$$f'(c) = \frac{f(b) - f(a)}{b - a},$$

which concludes the proof of the theorem (Figure 12.12). $\square$

The geometric meaning of the mean value theorem is the following: if the function f is continuous on $[a,b]$ and differentiable on (a,b), then the graph of f has a point in (a,b) where the tangent line is parallel to the chord $h_{a,b}$.

The following theorem is a generalization of the previous one.

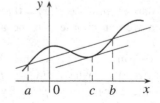

Fig. 12.12

Theorem 12.51 (Cauchy's Mean Value Theorem). *If the functions f and g are continuous on* $[a,b]$*, differentiable on* (a,b)*, and for* $x \in (a,b)$ *we have* $g'(x) \neq 0$*, then there exists a* $c \in (a,b)$ *such that*

$$\frac{f'(c)}{g'(c)} = \frac{f(b) - f(a)}{g(b) - g(a)}.$$

Proof. By Rolle's theorem, we know that $g(a) \neq g(b)$. Indeed, if $g(a) = g(b)$ held, then the derivative of g would be zero at at least one point of the interval (a,b), which we did not allow. Let

$$F(x) = f(x) - f(a) - \frac{f(b) - f(a)}{g(b) - g(a)}(g(x) - g(a)).$$

The function F is continuous on $[a,b]$, differentiable on (a,b), and $F(a) = F(b) = 0$. Thus by Rolle's theorem, there exists a $c \in (a,b)$ such that $F'(c) = 0$. Then

$$0 = F'(c) = f'(c) - \frac{f(b) - f(a)}{g(b) - g(a)} g'(c).$$

Since by the assumptions, $g'(c) \neq 0$, we get that

$$\frac{f'(c)}{g'(c)} = \frac{f(b) - f(a)}{g(b) - g(a)},$$

which concludes the proof.   □

It is clear that the mean value theorem is a special case of Cauchy's mean value theorem if we apply the latter with $g(x) = x$.

A simple but important corollary of the mean value theorem is the following.

Theorem 12.52. *If f is continuous on $[a,b]$, differentiable on (a,b), and $f'(x)=0$ for all $x \in (a,b)$, then the function f is constant on $[a,b]$.*

Proof. By the mean value theorem, for every x in $(a,b]$, there exists a $c \in (a,b)$ such that

$$f'(c) = \frac{f(x) - f(a)}{x - a}.$$

So by $f'(c) = 0$, we have $f(x) = f(a)$.   □

The following corollary is sometimes called the *fundamental theorem of integration;* we will later see why.

Corollary 12.53. *If f and g are continuous on $[a,b]$, differentiable on (a,b), and moreover, $f'(x) = g'(x)$ for all $x \in (a,b)$, then with a suitable constant c, we have $f(x) = g(x) + c$ for all $x \in [a,b]$.*

Proof. Apply Theroem 12.52 to the function $f - g$.   □

Exercises

12.56. Give an example of a differentiable function $f : \mathbb{R} \to \mathbb{R}$ that has a point c such that $f'(c)$ is not equal to the difference quotient $(f(b) - f(a))/(b - a)$ for any $a < b$. Why does this not contradict the mean value theorem?

12.57. Prove that if f is twice differentiable on $[a,b]$, and for a $c \in (a,b)$ we have $f''(c) \neq 0$, then there exist $a \leq x_1 < c < x_2 \leq b$ such that

$$f'(c) = \frac{f(x_1) - f(x_2)}{x_1 - x_2}. \tag{H}$$

12.58. Prove that

$$(\alpha - \beta) \cdot \cos \alpha \le \sin \alpha - \sin \beta \le (\alpha - \beta) \cdot \cos \beta$$

for all $0 < \beta < \alpha < \pi/2$.

12.59. Prove that $|\arctan x - \arctan y| \le |x - y|$ for all x, y.

12.60. Let f be differentiable on the interval I, and suppose that the function f' is bounded on I. Prove that f is Lipschitz on I.

12.61. Prove that if $f'(x) = x^2$ for all x, then there exists a constant c such that $f(x) = (x^3/3) + c$.

12.62. Prove that if $f'(x) = f(x)$ for all x, then there exists a constant c such that $f(x) = c \cdot e^x$ for all x.

12.63. Let $f : \mathbb{R} \to (0, \infty)$ be differentiable and strictly monotone increasing. Suppose that the tangent line of the graph of f at every point $(x, f(x))$ intersects the x-axis at the point $x - a$, where $a > 0$ is a constant. Prove that f is an exponential function.

12.64. Let $f : (0, \infty) \to (0, \infty)$ be differentiable and strictly monotone increasing. Suppose that the tangent line to the graph of f at every point $(x, f(x))$ intersects the x-axis at the point $c \cdot x$, where $c > 0$ is a constant. Prove that f is a power function.

12.65. Prove that if f and g are differentiable everywhere, $f(0) = 0$, $g(0) = 1$, $f' = g$, and $g' = -f$, then $f(x) = \sin x$ and $g(x) = \cos x$ for all x. (H)

12.66. Prove that the function $x^5 - 5x + 2$ has three real roots.

12.67. Prove that the function $x^7 + 8x^2 + 5x - 23$ has at most three real roots.

12.68. At most how many real roots can the function $x^{16} + ax + b$ have?

12.69. For what values of k does the function $x^3 - 6x^2 + 9x + k$ have exactly one real root?

12.70. Prove that if p is an nth-degree polynomial, then the function $e^x - p(x)$ has at most $n + 1$ real roots.

12.71. Let f be n times differentiable on (a, b). Prove that if f has n distinct roots in (a, b), then $f^{(n-k)}$ has at least k roots in (a, b) for all $k = 1, \ldots, n - 1$. (H)

12.72. Prove that if p is an nth-degree polynomial and every root of p is real, then every root of p' is also real.

12.73. Prove that every root of the Legendre polynomial

$$P_n(x) = \frac{1}{2^n n!} \left((x^2 - 1)^n \right)^{(n)}$$

is real.

12.74. Let f and g be n times differentiable functions on $[a,b]$, and suppose that they have n common roots in $[a,b]$. Prove that if the functions $f^{(n)}$ and $g^{(n)}$ have no common roots in $[a,b]$, then for all $x \in [a,b]$ such that $g(x) \neq 0$, there exists a $c \in (a,b)$ such that

$$\frac{f(x)}{g(x)} = \frac{f^{(n)}(c)}{g^{(n)}(c)}.$$

12.75. Let f be continuous on (a,b) and differentiable on $(a,b) \setminus \{c\}$, where $a < c < b$. Prove that if $\lim_{x \to c} f'(x) = A$, where A is finite, then f is differentiable at c and $f'(c) = A$.

12.76. Prove that if f is twice differentiable at a, then

$$\lim_{h \to 0} \frac{f(a+2h) - 2f(a+h) + f(a)}{h^2} = f''(a). \tag{H}$$

12.77. Let f be differentiable on $(0, \infty)$. Prove that if there exists a sequence $x_n \to \infty$ such that $f(x_n) \to 0$, then there also exists a sequence $y_n \to \infty$ such that $f'(y_n) \to 0$.

12.78. Let f be differentiable on $(0, \infty)$. Prove that if $\lim_{x \to \infty} f'(x) = 0$, then $\lim_{x \to \infty} f(x)/x = 0$.

12.6 Investigation of Differentiable Functions

We begin with monotonicity criteria.

Theorem 12.54. *Let f be continuous on $[a,b]$ and differentiable on (a,b).*

(i) *f is monotone increasing (decreasing) on $[a,b]$ if and only if $f'(x) \geq 0$ ($f'(x) \leq 0$) for all $x \in (a,b)$.*

(ii) *f is strictly monotone increasing (decreasing) on $[a,b]$ if and only if $f'(x) \geq 0$ ($f'(x) \leq 0$) for all $x \in (a,b)$, and $[a,b]$ does not have a nondegenerate subinterval on which f' is identically zero.*

Proof. (i) Suppose that $f'(x) \geq 0$ for all $x \in (a,b)$. By the mean value theorem, for arbitrary $a \leq x_1 < x_2 \leq b$ there exists a $c \in (x_1, x_2)$ such that

$$\frac{f(x_1) - f(x_2)}{x_1 - x_2} = f'(c).$$

Since $f'(c) \geq 0$, we have $f(x_1) \leq f(x_2)$, which means exactly that f is monotone increasing on $[a,b]$.

Conversely, if f is monotone increasing on $[a,b]$, then it is locally increasing at every x in (a,b). Thus by statement (iii) of Theorem 12.44, we see that $f'(x) \geq 0$.

The proof is similar for the monotone decreasing case.

(ii) It is easy to see that a function f is strictly monotone on $[a,b]$ if and only if it is monotone in $[a,b]$, and if $[a,b]$ does not have a subinterval on which f is constant. Then the statement can be proved easily by Theorem 12.52 $\square$

As an application of the theorem above, we introduce a simple but useful method for proving inequalities.

Corollary 12.55. *Let f and g be continuous on $[a,b]$ and differentiable on (a,b). If $f(a) = g(a)$ and $a < x \le b$ implies $f'(x) \ge g'(x)$, then $f(x) \ge g(x)$ for all $x \in [a,b]$. Also, if $f(a) = g(a)$ and $a < x \le b$ imply $f'(x) > g'(x)$, then $f(x) > g(x)$ for all $x \in (a,b]$.*

Proof. Let $h = f - g$. If $a < x \le b$ implies $f'(x) \ge g'(x)$, then $h'(x) \ge 0$ for all $x \in (a,b)$, and so by statement (i) of Theorem 12.54, h is monotone increasing on $[a,b]$. If $f(a) = g(a)$, then $h(a) = 0$, and so $h(x) \ge h(a) = 0$, and thus $f(x) \ge g(x)$ for all $x \in [a,b]$. The second statement follows similarly, using statement (ii) of Theorem 12.54. $\square$

Example 12.56. To illustrate the method, let us show that

$$\log(1+x) > \frac{2x}{x+2} \tag{12.32}$$

for all $x > 0$. Since for $x = 0$ we have equality, by Corollary 12.55 it suffices to show that the derivatives of the given functions satisfy the inequality for all $x > 0$. It is easy to check that $(2x/(x+2))' = 4/(x+2)^2$. Thus we have only to show that for $x > 0$, $1/(1+x) > 4/(x+2)^2$, which can be checked by multiplying through.

More applications of Corollary 12.55 can be found among the exercises.

Remark 12.57. With the help of Theorem 12.54, we can find the local and absolute extrema of an arbitrary differentiable function even if it is not defined on a closed and bounded interval. This is because by the sign of the derivative, we can determine on which intervals the function is increasing and on which it is decreasing, and this generally gives us enough information to find the extrema.

Consider the function $f(x) = x \cdot e^{-x}$, for example. Since $f'(x) = e^{-x} - x \cdot e^{-x}$, we have $f'(x) > 0$ if $x < 1$, and $f'(x) < 0$ if $x > 1$. Thus f is strictly monotone increasing on $(-\infty, 1]$, and strictly monotone decreasing on $[1, \infty)$. It follows that f has an absolute maximum at 1 (which is also a local maximum), and that f does not have any local or absolute minima.

In Theorem 12.44 we saw that if f is differentiable at a, then for f to have a local extremum at a, it is necessary (but generally not sufficient) for $f'(a) = 0$ to hold. The following theorems give *sufficient* conditions for the existence of local extrema.

Theorem 12.58. *Let f be differentiable in a neighborhood of the point a.*

(i) *If $f'(a) = 0$ and f' is locally increasing (decreasing) at a,[7] then a is a local minimum (maximum) of f.*

(ii) *If $f'(a) = 0$ and f' is strictly locally increasing (decreasing) at a, then the point a is a strict local minimum (maximum) of f.*

Proof. (i) Consider the case that f' is locally increasing at a. Then there exists a $\delta > 0$ such that $f'(x) \leq 0$ if $a - \delta < x < a$, and $f'(x) \geq 0$ if $a < x < a + \delta$. By Theorem 12.54, it then follows that f is monotone decreasing on $[a - \delta, a]$, and monotone increasing on $[a, a + \delta]$. Thus if $a - \delta < x < a$, then $f(x) \geq f(a)$. Moreover, if $a < x < a + \delta$, then again $f(x) \geq f(a)$. This means exactly that f has a local minimum at a. The statement for the local maximum can be proved similarly.

(ii) If f' is strictly locally increasing at a, then there exists a $\delta > 0$ such that $f'(x) < 0$ if $a - \delta < x < a$, and $f'(x) > 0$ if $a < x < a + \delta$.

It then follows that f is strictly monotone decreasing on $[a - \delta, a]$, and strictly monotone increasing on $[a, a + \delta]$. Thus if $a - \delta < x < a$, then $f(x) > f(a)$. Moreover, if $a < x < a + \delta$, then again $f(x) > f(a)$. This means exactly that f has a strict local minimum at a. One can argue similarly for the case of a strict local maximum. $\square$

Remark 12.59. The sign change of f' at a is not necessary for f to have a local extremum at the point a. Let f be a function such that $x^2 \leq f(x) \leq 2x^2$ for all x. Then the point 0 is a strict local (and absolute) minimum of f. On the other hand, it is possible that f is not differentiable (or even continuous) at any point other than 0; this is the case, for example, if $f(x) = x^2$ for all rational x, and $f(x) = 2x^2$ for all irrational x.

We can also construct such an f to be differentiable; for this we need to place a differentiable function between the graphs of the functions x^2 and $2x^2$ each of whose one-sided neighborhoods of 0 contains a section that is monotone decreasing and a section that is monotone increasing as well. We give such a function in Exercise 12.95.

Theorem 12.60. *Let f be twice differentiable at a. If $f'(a) = 0$ and $f''(a) > 0$, then f has a strict local minimum at a. If $f'(a) = 0$ and $f''(a) < 0$, then f has a strict local maximum at a.*

Proof. Suppose that $f''(a) > 0$. By Theorem 12.44, it follows that f' is strictly locally increasing at a. Now apply the previous theorem (Figure 12.13). The proof for the case $f''(a) < 0$ is similar. $\square$

Remark 12.61. If $f'(a) = 0$ and $f''(a) = 0$, then we cannot deduce whether f has a local extremum at a. The different possibilities are illustrated by the functions

---

[7] That is, if f' changes signs at the point a, meaning that it is nonpositive on a left-sided neighborhood of a and nonnegative on a right-sided neighborhood, or vice versa.

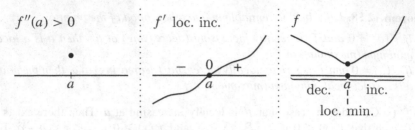

$$f''(a) > 0 \qquad\qquad f' \text{ loc. inc.} \qquad\qquad f$$

dec. a inc.

loc. min.

Fig. 12.13

$f(x) = x^3$, $f(x) = x^4$, and $f(x) = -x^4$ at $a = 0$. In this case, we can get suffi-cient conditions for f to have a local extremum at a by the value of higher-order derivatives.

Theorem 12.62.

(i) *Let the function f be $2k$ times differentiable at the point a, where $k \geq 1$. If*

$$f'(a) = \ldots = f^{(2k-1)}(a) = 0 \ \text{and} \ f^{(2k)}(a) > 0, \qquad (12.33)$$

then f has a strict local minimum at a. If

$$f'(a) = \ldots = f^{(2k-1)}(a) = 0 \ \text{and} \ f^{(2k)}(a) < 0,$$

then f has a strict local maximum at a.

(ii) *Let the function f be $2k+1$ times differentiable at a, where $k \geq 1$. If*

$$f'(a) = \ldots = f^{(2k)}(a) = 0 \ \text{and} \ f^{(2k+1)}(a) \neq 0, \qquad (12.34)$$

then f is strictly monotone in a neighborhood of a, that is, f does not have a local extremum there.

Proof. (i) We prove only the first statement, using induction. We already saw the $k = 1$ case in Theorem 12.60. Let $k > 1$, and suppose that the statement holds for $k - 1$. If (12.33) holds for f, then for the function $g = f''$, we have

$$g'(a) = \ldots = g^{(2k-3)}(a) = 0 \ \text{and} \ g^{(2k-2)}(a) > 0.$$

Thus by the induction hypothesis, f'' has a strict local minimum at a. Since by $k > 1$ we have $f''(a) = 0$, there must exist a $\delta > 0$ such that $f''(x) > 0$ at all points $x \in (a - \delta, a + \delta) \setminus \{a\}$. Then by Theorem 12.54, it follows that f' is strictly mono-tone increasing on $(a - \delta, a + \delta)$. Thus f' is strictly locally increasing at a, so we can apply Theorem 12.58.

(ii) Suppose (12.34) holds. Then by the already proved statement (i), a is a strict local extremum of f'. Since $f'(a) = 0$, either $f'(x) > 0$ for all $x \in (a - \delta, a + \delta) \setminus \{a\}$, or $f'(x) < 0$ for all $x \in (a - \delta, a + \delta) \setminus \{a\}$. By Theorem 12.54, it then follows that f is strictly monotone on $(a - \delta, a + \delta)$, so it does not have a local extremum at the point a (Figure 12.14). $\qquad\square$

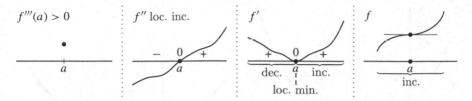

$f'''(a) > 0$

f'' loc. inc.

f'

$-$ 0 $+$
a

$+$ 0 $+$
dec. a inc.
loc. min.

f

a
inc.

Fig. 12.14

We now turn to the conditions for convexity.

Theorem 12.63. *Let f be differentiable on the interval I.*

(i) *The function f is convex (concave) on I if and only if f' is monotone increasing (decreasing) on I.*

(ii) *The function f is strictly convex (concave) on I if and only if f' is strictly monotone increasing (decreasing) on I.*

Proof. (i) Suppose that f' is monotone increasing on I. Let $a, b \in I$, $a < b$, and let $a < x < b$ be arbitrary. By the mean value theorem, there exist points $u \in (a, x)$ and $v \in (x, b)$ such that

$$f'(u) = \frac{f(x) - f(a)}{x - a} \text{ and } f'(v) = \frac{f(b) - f(x)}{b - x}.$$

Since $u < v$ and f' is monotone increasing, $f'(u) \le f'(v)$, so

$$\frac{f(x) - f(a)}{x - a} \le \frac{f(b) - f(x)}{b - x}.$$

Then by a simple rearrangement, we get that

$$f(x) \le \frac{f(b) - f(a)}{b - a} \cdot (x - a) + f(a),$$

which shows that f is convex.

Now suppose that f is convex on I, and let $a, b \in I$, $a < b$, be arbitrary. By Theorem 9.20, the function $F(x) = (f(x) - f(a))/(x - a)$ is monotone increasing on the set $I \setminus \{a\}$. Thus $F(x) \le ((f(b) - f(a))/(b - a)$ for all $x < b$, $x \ne a$. Since $f'(a) = \lim_{x \to a} F(x)$, we have that (Figures 12.15)

$$f'(a) \le \frac{f(b) - f(a)}{b - a}. \tag{12.35}$$

Similarly, the function $G(x) = (f(x) - f(b))/(x - b)$ is monotone increasing on the set $I \setminus \{b\}$, so $G(x) \ge ((f(b) - f(a))/(b - a)$ for all $x > a$, $x \ne b$.

Since $f'(b) = \lim_{x \to b} G(x)$, we have that

$$f'(b) \ge \frac{f(b) - f(a)}{b - a}. \tag{12.36}$$

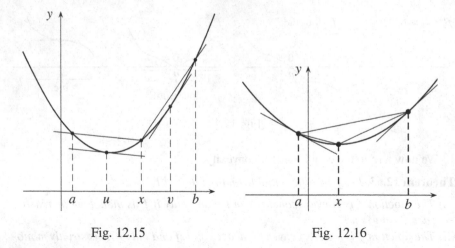

Fig. 12.15   Fig. 12.16

If we combine (12.35) and (12.36), we get that $f'(a) \leq f'(b)$. Since this is true for all $a, b \in I$ if $a < b$, f' is monotone increasing on I (Figure 12.16).

The statement for concavity can be proved in the same way. Statement (ii) follows by a straightforward change of the argument above.   □

Rearranging equation (12.35), we get that if $a < b$, then $f(b) \geq f'(a) \cdot (b-a) + f(a)$. Equation (12.36) states that if $a < b$, then $f(a) \geq f'(b) \cdot (a-b) + f(b)$. This means that for arbitrary $a, x \in I$,

$$f(x) \geq f'(a) \cdot (x-a) + f(a), \tag{12.37}$$

that is, a tangent line drawn at any point on the graph of f lies below the graph itself. Thus we have proved the "only if" part of the following theorem.

Theorem 12.64. *Let f be differentiable on the interval I. The function f is convex on I if and only if for every $a \in I$, the graph of the function f lies above the tangent of the graph at the point a, that is, if and only if (12.37) holds for all $a, x \in I$.*

Proof. We now have to prove only the "if" part of the statement. Suppose that (12.37) holds for all $a, x \in I$. If $a, b \in I$ and $a < b$, then it follows that both (12.35) and (12.36) are true, so $f'(a) \leq f'(b)$. Thus f' is monotone increasing on I, so by Theorem 12.63, f is convex.   □

Theorem 12.65. *Let f be twice differentiable on I. The function f is convex (concave) on I if and only if $f''(x) \geq 0$ ($f''(x) \leq 0$) for all $x \in I$.*

Proof. The statement of the theorem is a simple corollary of Theorems 12.63 and 12.54.   □

Definition 12.66. We say that a point a is an *inflection point* of the function f if f is continuous at a, f has a (finite or infinite) derivative at a, and there exists a $\delta > 0$ such that f is convex on $(a - \delta, a]$ and concave on $[a, a + \delta)$, or vice versa.

So for example, 0 is an inflection point of the functions x^3 and $\sqrt[3]{x}$.

Theorem 12.67. *If f is twice differentiable at a, and f has an inflection point at a, then $f''(a) = 0$.*

Proof. If f is convex on $(a - \delta, a]$, then f' is monotone increasing there; if it is concave on $[a, a + \delta)$, then f' is monotone decreasing there. Thus f' has a local maximum at a, and so $f''(a) = 0$.

The proof is similar in the case that f is concave on $(a - \delta, a]$ and convex on $[a, a + \delta)$.   □

Remark 12.68. Let f be differentiable on a neighborhood of the point a. By Theorem 12.63, a is an inflection point of f if and only if a is a local extremum of f' such that f' is increasing in a left-hand neighborhood of a and is decreasing in a right-hand neighborhood of a, or the other way around. From this observation and by Theorem 12.67, we get the following theorem.

Theorem 12.69. *Let f be twice differentiable on a neighborhood of the point a. Then a is an inflection point of f if and only if f'' changes sign at the point a, that is, if $f''(a) = 0$ and f'' is locally increasing or decreasing at a.*

Corollary 12.70. *Let f be three times differentiable at a. If $f''(a) = 0$ and $f'''(a) \neq 0$, then f has an inflection point at a.*

Remark 12.71. In the case $f''(a) = f'''(a) = 0$, it is possible that f has an inflection point at a, but it is also possible that it does not. The different cases are illustrated by the functions $f(x) = x^4$ and $f(x) = x^5$ at the point $a = 0$. As in the case of extrema, we can refer to the values of higher-order derivatives to help determine when a point of inflection exists.

Theorem 12.72.

(i) *Let the function f be $2k + 1$ times differentiable at a, where $k \geq 1$. If*

$$f''(a) = \ldots = f^{(2k)}(a) = 0 \text{ and } f^{(2k+1)}(a) \neq 0, \qquad (12.38)$$

then f has an inflection point at a.

(ii) *If*

$$f''(a) = \ldots = f^{(2k-1)}(a) = 0 \text{ and } f^{(2k)}(a) \neq 0, \qquad (12.39)$$

then f is strictly convex or concave in a neighborhood of a, and so a is not a point of inflection.

Proof. (i) We already saw the case for $k = 1$ in the previous theorem, so we may assume that $k > 1$. If (12.38) holds, then by statement (ii) of Theorem 12.62, f'' is strictly monotone in a neighborhood of a. Thus f'' is locally increasing or decreasing at the point a, and we can apply Theorem 12.69.

(ii) Assume (12.39). Then necessarily $k > 1$. By statement (i) of Theorem 12.62, f'' has a strict local extremum at a. Since $f''(a) = 0$, this means that for a suitable neighborhood U of a, the sign of f'' does not change in $U \setminus \{a\}$. Thus f' is strictly monotone on U, so f is either strictly convex or strictly concave on U, and so a cannot be a point of inflection. □

Remark 12.73. If the function f is infinitely differentiable at a and $f^{(n)}(a) \neq 0$ for at least one $n \geq 2$, then with the help of Theorems 12.62 and 12.72, we can determine whether f has a local extremum or a point of inflection at a. This is because if $f'(a) \neq 0$, then a cannot be a local extremum. Now suppose that $f'(a) = 0$, and let n be the smallest positive integer for which $f^{(n)}(a) \neq 0$. In this case, the function f has a local extremum at a if and only if n is even.

If $f''(a) \neq 0$, then a is not an inflection point of f. If, however, $f''(a) = 0$, and n is the smallest integer greater than 2 for which $f^{(n)}(a) \neq 0$, then a is a point of inflection of f if and only if n is odd.

It can occur, however, that f is infinitely differentiable at a, and that $f^{(n)}(a) = 0$ for all n, while f is not zero in any neighborhood of a. In the following chapter we will see that among such functions we will find some that have local extrema at a, but we will also find those that do not; we will find those that have points of inflection at a, and those that do not (see Remark 13.17).

A **complete investigation of the function** f is accomplished by finding the following pieces of information about the function:

1. the (one-sided) limits of the accumulation points of the domain;
2. the intervals on which f is monotone increasing or decreasing;
3. the local and absolute extrema of f;
4. the points where f is continuous or differentiable;
5. the intervals on which f is convex or concave;
6. the inflection points of f.

Example 12.74. Carry out a complete investigation of the function $x^{2n}e^{-x^2}$ for all $n \in \mathbb{N}$.

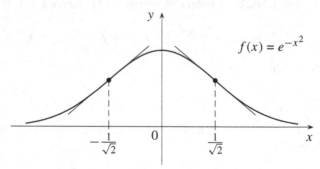

$$f(x) = e^{-x^2}$$

Fig. 12.17

1. Consider first the case $n = 0$. If $f(x) = e^{-x^2}$, then $f'(x) = -2x \cdot e^{-x^2}$, so the sign of $f'(x)$ agrees with the sign of $-x$. Then f is strictly increasing on the interval $(-\infty, 0]$, and strictly decreasing on the interval $[0, \infty)$, so f must have a strict local and global maximum at $x = 0$.

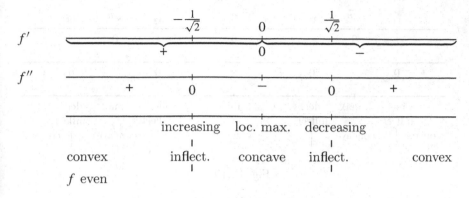

Fig. 12.18

On the other hand, $f''(x) = e^{-x^2}(4x^2 - 2)$, so $f''(x) > 0$ if $|x| > 1/\sqrt{2}$, and $f''(x) < 0$ if $|x| < 1/\sqrt{2}$. So f is strictly concave on $[-1/\sqrt{2}, 1/\sqrt{2}]$, and strictly convex on the intervals $(-\infty, -1/\sqrt{2}]$ and $[1/\sqrt{2}, \infty)$, so the points $\pm 1/\sqrt{2}$ are points of inflection. All these properties are summarized in Figure (12.18).

If we also consider that $\lim_{x \to \pm\infty} f(x) = 0$, then we can sketch the graph of the function as seen in Figure 12.17. Inspired by the shape of its graph, the function e^{-x^2}—which appears many times in probability—is called a **bell curve**.

2. Now let $f(x) = x^{2n} e^{-x^2}$, where $n \in \mathbb{N}^+$. Then

$$f'(x) = \left(2n \cdot x^{2n-1} - 2x^{2n+1}\right) e^{-x^2} = 2 \cdot x^{2n-1} e^{-x^2} \left(n - x^2\right).$$

Thus $f'(x)$ is positive on the intervals $(-\infty, -\sqrt{n})$ and $(0, \sqrt{n})$, and negative on the intervals $(-\sqrt{n}, 0)$ and $(\sqrt{n}, \infty)$. On the other hand,

$$\tfrac{1}{2} \cdot f''(x) = \left[\left(n \cdot x^{2n-1} - x^{2n+1}\right) \cdot e^{-x^2}\right]' =$$
$$= \left(n(2n-1)x^{2n-2} - (2n+1)x^{2n} - 2nx^{2n} + 2x^{2n+2}\right) \cdot e^{-x^2} =$$
$$= x^{2n-2} \cdot \left(2x^4 - (4n+1)x^2 + n(2n-1)\right) \cdot e^{-x^2}.$$

It is easy to see that the roots of f'' are the numbers

$$\pm \sqrt{\frac{4n + 1 \pm \sqrt{16n + 1}}{4}}.$$

If we denote these by x_i ($1 \le i \le 4$), then $x_1 < x_2 < 0 < x_3 < x_4$. It is also easy to check that f'' is positive if $|x| < x_3$ or $|x| > x_4$, and negative if $x_3 < |x| < x_4$ (Figure 12.19). The behavior of the function can be summarized by the Figure 12.18.

If we also consider that $f(0) = 0$ and $\lim_{x \to \pm\infty} f(x) = 0$, then we can sketch the graph as in Figure 12.20.

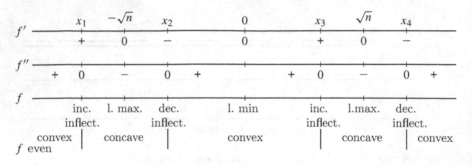

Fig. 12.19

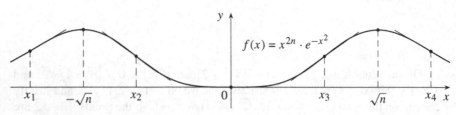

Fig. 12.20

Exercises

12.79. Prove that if $x \in [0,1]$, then $2^x \le 1 + x \le e^x$ and $2x/\pi \le \sin x \le x$.

12.80. Prove that if f is a rational function, then there exists an a such that f is monotone and either concave or convex on $(-\infty, a)$. A similar statement is true on a suitable half-line (b, ∞).

12.81. For which $a > 0$, does the equation $a^x = x$ have roots? (H)

12.82. For which $a > 0$ is the sequence defined by the recurrence $a_1 = a$, $a_{n+1} = a^{a_n}$ convergent? (H)

12.83. Prove that $\left(1 + \frac{1}{x}\right)^{x+(1/2)} > e$ for all $x > 0$. (H)

12.84. Prove that for all $x \ge 0$ and $n \ge 1$,
$$e^x \ge 1 + x + \frac{x^2}{2!} + \cdots + \frac{x^n}{n!}.$$

12.85. Prove that for all $0 \le x \le K$ and $n \ge 1$,
$$e^x \le 1 + x + \frac{x^2}{2!} + \cdots + \frac{x^n}{n!} + \frac{x^{n+1}}{(n+1)!} \cdot e^K.$$

12.86. Prove that for all $x \ge 0$,
$$e^x = 1 + x + \frac{x^2}{2!} + \cdots. \tag{12.40}$$

As a special case,

$$e = 1 + \frac{1}{1!} + \frac{1}{2!} + \cdots.$$

12.87. Prove that e is irrational. (S)

12.88. Prove that for all $x \geq 0$ and $n \geq 1$,

$$1 - x + \frac{x^2}{2!} - \cdots - \frac{x^{2n-1}}{(2n-1)!} \leq e^{-x} \leq 1 - x + \frac{x^2}{2!} - \cdots + \frac{x^{2n}}{(2n)!}.$$

12.89. Prove that (12.40) holds for all x.

12.90. Prove that for all $x \geq 0$ and $n \geq 1$,

$$1 - \frac{x^2}{2!} + \frac{x^4}{4!} - \cdots - \frac{x^{4n-2}}{(4n-2)!} \leq \cos x \leq 1 - \frac{x^2}{2!} + \frac{x^4}{4!} - \cdots + \frac{x^{4n}}{(4n)!}$$

and

$$x - \frac{x^3}{3!} + \frac{x^5}{5!} - \cdots - \frac{x^{4n-1}}{(4n-1)!} \leq \sin x \leq x - \frac{x^3}{3!} + \frac{x^5}{5!} - \cdots + \frac{x^{4n+1}}{(4n+1)!}.$$

12.91. Prove that for all $x \geq 0$ and $n \geq 1$,

$$x - \frac{x^2}{2} + \cdots - \frac{x^{2n}}{2n} \leq \log(1+x) \leq x - \frac{x^2}{2} + \cdots + \frac{x^{2n+1}}{2n+1}.$$

12.92. Prove that for all $x \in [0,1]$,

$$\log(1+x) = x - \frac{x^2}{2} + \frac{x^3}{3} - \cdots.$$

As a special case,

$$\log 2 = 1 - \frac{1}{2} + \frac{1}{3} - \cdots. \tag{12.41}$$

12.93. Prove that

$$x - \frac{x^3}{3} + \frac{x^5}{5} - \cdots - \frac{x^{4n-1}}{4n-1} \leq \arctan x \leq x - \frac{x^3}{3} + \frac{x^5}{5} - \cdots + \frac{x^{4n+1}}{4n+1}$$

for every $x \geq 0$ and $n \geq 1$.

12.94. Prove that

$$\arctan x = x - \frac{x^3}{3} + \frac{x^5}{5} - \cdots.$$

for every $|x| \leq 1$.

As a special case,

$$\frac{\pi}{4} = 1 - \frac{1}{3} + \frac{1}{5} - \dots \tag{12.42}$$

12.95. Let

$$f(x) = \begin{cases} x^4 \cdot (2 + \sin(1/x)), & \text{if } x \neq 0, \\ 0, & \text{if } x = 0. \end{cases}$$

Show that f has a strict local minimum at 0, but that f' does not change sign at 0. (S)

12.96. Let

$$f(x) = \begin{cases} e^{\sin(1/x) - (1/x)}, & \text{if } x > 0, \\ 0, & \text{if } x = 0. \end{cases}$$

Prove that

(a) f is continuous on $[0, \infty)$,
(b) f is differentiable on $[0, \infty)$,
(c) f is strictly monotone increasing on $[0, \infty)$,
(d) $f'(1/(2\pi k)) = 0$ if $k \in \mathbb{N}^+$, that is, f' is 0 at infinitely many places in the interval $[0, 1]$.

12.97. Give a function that is monotone decreasing and differentiable on $(0, \infty)$, satisfies $\lim_{x \to \infty} f(x) = 0$, but $\lim_{x \to \infty} f'(x) \neq 0$.

12.98. Let f be differentiable on a punctured neighborhood of a, and let

$$\lim_{x \to a} \frac{f(x) - f(a)}{x - a} = \infty.$$

Does it then follow that $\lim_{x \to a} f'(x) = \infty$?

12.99. Let f be convex and differentiable on the open interval I. Prove that f has a minimum at the point $a \in I$ if and only if $f'(a) = 0$.

12.100. Let f be convex and differentiable on $(0, 1)$. Prove that if $\lim_{x \to 0+0} f(x) = \infty$, then $\lim_{x \to 0+0} f'(x) = -\infty$. Show that the statement no longer holds if we drop the convexity assumption.

12.101. Carry out a complete investigation of each of the functions below.

$x^3 - 3x$,	$x^2 - x^4$,	$x - \arctan x$,	$x + e^{-x}$,
$x + x^{-2}$,	$\arctan(1/x)$,	$\cos x^2$,	$\sin(\sin x)$,
$\sin(1/x)$,	$x \cdot e^{-x}$,	$x^n \cdot e^{-x}$,	$x - \log x$,
$1/(1 + \sin^2 x)$,	$(1 + \frac{1}{x})^x$,	$(1 + \frac{1}{x})^{x+1}$,	$\sqrt{1 - e^{-x^2}}$,
$x \cdot \sin(\log x)$,	x^x,	$\sqrt[x]{x}$,	$(\log x)/x$,

$x \cdot \log x,$ $\qquad\qquad x^x \cdot (1-x)^{1-x},$ $\qquad\quad \operatorname{arc tg} x - \frac{1}{2}\log(1+x^2),$

$\operatorname{arc tg} x - \frac{x}{x+1},$ $\qquad x^4/(1+x)^3,$ $\qquad\quad e^x/(1+x),$ $\qquad\qquad e^x/\operatorname{sh} x,$

$\sqrt[\log x]{x},$ $\qquad\qquad e^{-x} \cdot \left[\frac{1-x^2}{2}\sin x - \frac{(1+x)^2}{2}\cos x\right],$

$x^{2n+1}e^{-x^2} \quad (n \in \mathbb{N}).$

Chapter 13
Applications of Differentiation

13.1 L'Hôpital's Rule

The following theorem gives a useful method for determining critical limits.

Theorem 13.1 (L'Hôpital's Rule). *Let f and g be differentiable on a punctured neighborhood of* α, *and suppose that* $g \neq 0$ *and* $g' \neq 0$ *there. Moreover, assume that either*

$$\lim_{x \to \alpha} f(x) = \lim_{x \to \alpha} g(x) = 0, \qquad (13.1)$$

or

$$\lim_{x \to \alpha} |g(x)| = \infty. \qquad (13.2)$$

If

$$\lim_{x \to \alpha} \frac{f'(x)}{g'(x)} = \beta,$$

then it follows that

$$\lim_{x \to \alpha} \frac{f(x)}{g(x)} = \beta. \qquad (13.3)$$

Here α *can denote a number* a *or one of the symbols* $a+0$, $a-0$, ∞, $-\infty$, *while* β *can denote a number* b, ∞, *or* $-\infty$.

Proof. We first give the proof for the special case in which $\alpha = a$ is finite, and $\lim_{x \to a} f(x) = \lim_{x \to a} g(x) = f(a) = g(a) = 0$. In this case, by Cauchy's mean value theorem, in a neighborhood of a for all $x \neq a$ there exists a $c \in (x, a)$ such that

$$\frac{f(x)}{g(x)} = \frac{f(x) - f(a)}{g(x) - g(a)} = \frac{f'(c)}{g'(c)}.$$

© Springer New York 2015
M. Laczkovich, V.T. Sós, *Real Analysis*, Undergraduate Texts
in Mathematics, DOI 10.1007/978-1-4939-2766-1_13

Thus if (x_k) is a sequence that tends to a, then there exists a sequence $c_k \to a$ such that

$$\frac{f(x_k)}{g(x_k)} = \frac{f'(c_k)}{g'(c_k)}$$

for all k. Thus

$$\lim_{k \to \infty} \frac{f(x_k)}{g(x_k)} = \lim_{k \to \infty} \frac{f'(c_k)}{g'(c_k)} = \beta, \qquad (13.4)$$

so by the transference principle, (13.3) holds.

In the general case, when $f(\alpha)$ or $g(\alpha)$ is not defined or nonzero, or α is not finite, the above proof is modified by expressing the quotient $f(x)/g(x)$ by the differential quotient $(f(x) - f(y))/(g(x) - g(y))$. First we assume that α is one of the symbols $a+0$, $a-0$, ∞, $-\infty$. For arbitrary $y \neq x$ in a suitable one-sided neighborhood of α, we have $g(y) \neq g(x)$ by Rolle's theorem. Thus

$$\frac{f(x)}{g(x)} = \frac{f(x) - f(y)}{g(x) - g(y)} \cdot \frac{g(x) - g(y)}{g(x)} + \frac{f(y)}{g(x)} = \frac{f(x) - f(y)}{g(x) - g(y)} \left(1 - \frac{g(y)}{g(x)}\right) + \frac{f(y)}{g(x)}.$$

It then follows that with suitable $c \in (x, y)$,

$$\frac{f(x)}{g(x)} = \frac{f'(c)}{g'(c)} \left(1 - \frac{g(y)}{g(x)}\right) + \frac{f(y)}{g(x)}. \qquad (13.5)$$

Thus if we show that for every sequence $x_k \to \alpha$, $x_k \neq \alpha$, there exists another sequence $y_k \to \alpha$, $y_k \neq \alpha$, such that

$$\frac{f(y_k)}{g(x_k)} \to 0 \quad \text{and} \quad \frac{g(y_k)}{g(x_k)} \to 0, \qquad (13.6)$$

then (13.4) follows by (13.5), and so does (13.3) by the transference principle.

Assume first (13.1). Let $x_k \to \alpha$, $x_k \neq \alpha$, be given. For each k, there exists an n_k such that

$$\left|\frac{f(x_{n_k})}{g(x_k)}\right| < \frac{1}{k} \quad \text{and} \quad \left|\frac{g(x_{n_k})}{g(x_k)}\right| < \frac{1}{k},$$

since for a fixed k, $\dfrac{f(x_n)}{g(x_k)} \to 0$ and $\dfrac{g(x_n)}{g(x_k)} \to 0$ if $n \to \infty$. If $y_k = x_{n_k}$ for all k, then (13.6) clearly holds.

Now assume (13.2). Then for all sequences $x_k \to \alpha$ such that $x_k \neq \alpha$, we have $\lim |g(x_k)| = \infty$, so for all i, there exists an N_i such that

$$\left|\frac{f(x_i)}{g(x_m)}\right| < \frac{1}{i} \quad \text{and} \quad \left|\frac{g(x_i)}{g(x_m)}\right| < \frac{1}{i}$$

if $m \geq N_i$. We can assume that $N_1 < N_2 < \dots$. Choose the sequence (y_k) as follows: let $y_k = x_i$ if $N_i < k \leq N_{i+1}$. It is clear that (13.6) holds again.

Finally, if $\alpha = a$ is finite, then by applying the already proved $\alpha = a + 0$ and $\alpha = a - 0$ cases, we get that

$$\lim_{x \to a-0} \frac{f(x)}{g(x)} = \lim_{x \to a+0} \frac{f(x)}{g(x)} = \beta,$$

which implies that

$$\lim_{x \to a} \frac{f(x)}{g(x)} = \beta.$$

$\square$

Examples 13.2. **1.** In Example 10.47, we saw that *if $a > 1$, then as $x \to \infty$, the function a^x tends to infinity faster than any positive power of x.* We can also see this by an n-fold application of L'Hôpital's rule:

$$\lim_{x \to \infty} \frac{a^x}{x^n} = \lim_{x \to \infty} \frac{\log a \cdot a^x}{n \cdot x^{n-1}} = \cdots = \lim_{x \to \infty} \frac{(\log a)^n \cdot a^x}{n(n-1) \cdot \ldots \cdot 1} = \infty.$$

2. Similarly, an n-fold application of L'Hôpital's rule gives

$$\lim_{x \to \infty} \frac{\log^n x}{x} - \lim_{x \to \infty} \frac{n \log^{n-1} x}{x} = \cdots = \lim_{x \to \infty} \frac{n!}{x} = 0.$$

It then simply follows that

$$\lim_{x \to \infty} \frac{\log^\alpha x}{x^\beta} = 0$$

for all positive α and β. Thus *every positive power of x tends to ∞ as $x \to \infty$ faster than any positive power of $\log x$.* (Although this also follows from 1. quite clearly.)

3. $\displaystyle \lim_{x \to 0+0} x \log x = \lim_{x \to 0+0} \frac{\log x}{1/x} = \lim_{x \to 0+0} \frac{1/x}{-1/x^2} = 0.$

We can similarly see that $\lim_{x \to 0+0} x^\alpha \log x = 0$ for all $\alpha > 0$.

4. $\displaystyle \lim_{x \to 0} \frac{\sin x - x}{x^3} = \lim_{x \to 0} \frac{\cos x - 1}{3x^2} = \lim_{x \to 0} \frac{-\sin x}{6x} = \lim_{x \to 0} \frac{-\cos x}{6} = -\frac{1}{6}.$

(Although this also follows easily from Exercise 12.90.)

5. $\displaystyle \lim_{x \to 0+0} \left(\frac{1}{\sin x} - \frac{1}{x} \right) = \lim_{x \to 0+0} \frac{x - \sin x}{x \sin x} = \lim_{x \to 0+0} \frac{1 - \cos x}{\sin x + x \cos x} =$

$$= \lim_{x \to 0+0} \frac{\sin x}{\cos x + \cos x - x \sin x} = 0.$$

Remark 13.3. L'Hôpital's rule is not always applicable. It can occur that the limit $\lim_{x \to a} f(x)/g(x)$ exists, but $\lim_{x \to a} f'(x)/g'(x)$ does not, even though

$$\lim_{x \to a} f(x) = \lim_{x \to a} g(x) = 0$$

and g and g' are nonzero everywhere other than at a.

Let, for example, $a = 0$, $f(x) = x^2 \cdot \sin(1/x)$, and $g(x) = x$. Then

$$\lim_{x \to 0} \frac{f(x)}{g(x)} = \lim_{x \to 0} \frac{x^2 \cdot \sin(1/x)}{x} = \lim_{x \to 0} x \cdot \sin(1/x) = 0.$$

On the other hand,

$$\frac{f'(x)}{g'(x)} = \frac{2x \cdot \sin(1/x) - \cos(1/x)}{1} = 2x \cdot \sin(1/x) - \cos(1/x),$$

so $\lim_{x \to 0} f'(x)/g'(x)$ does not exist.

Exercises

13.1. Determine the following limits.

$$\lim_{x \to 0} \frac{\sin 3x}{\operatorname{tg} 5x}, \qquad \lim_{x \to 0} \frac{\log(\cos ax)}{\log(\cos bx)}, \qquad \lim_{x \to 0} \frac{e^x - e^{-x} - 2x}{x - \sin x},$$

$$\lim_{x \to 0} \left[x^{-2} - (\sin x)^{-2} \right], \qquad \lim_{x \to 0} \frac{\operatorname{tg} x - x}{x - \sin x}, \qquad \lim_{x \to 0} \frac{x \operatorname{ctg} x - 1}{x^2},$$

$$\lim_{x \to 0} \left[\operatorname{ctg} x - \frac{1}{x} \right], \qquad \lim_{x \to 0} \left(\frac{\sin x}{x} \right)^{x^{-2}}, \qquad \lim_{x \to 0} \left(\frac{1 + e^x}{2} \right)^{\operatorname{ctg} x},$$

$$\lim_{x \to 1} (1 - x) \cdot \operatorname{tg}(\pi x/2), \qquad \lim_{x \to 1} x^{1/(1-x)}, \qquad \lim_{x \to 1} (2 - x)^{\operatorname{tg}(\pi x/2)}.$$

13.2 Polynomial Approximation

While dealing with the definition of differentiability, we saw that a function is differentiable at a point if and only if it is well approximated by a linear polynomial there (see Theorem 12.9). We also saw that if f is differentiable at the point a, then out of all the linear functions, the polynomial $f'(a) \cdot (x - a) + f(a)$ is the one that approximates f the best around that point (see Theorem 12.10). Below, we will generalize these results by finding the polynomial of degree at most n that approximates our function f the best locally.[1]

To do this, we first need to make precise what we mean by "best local approximation" (as we did in the case of linear functions). A function can be approximated "better" or "worse" by other functions in many different senses. Our goal might be for the difference

---

[1] In the following, we also include the identically zero polynomial in the polynomials of degree at most n.

$$\max_{x\in[a,b]} |f(x) - p_n(x)|$$

to be the smallest on the interval $[a,b]$. But it could be that our goal is to minimize some sort of average of the difference $f(x) - p_n(x)$. If we want to approximate a function locally, we want the differences very close to a to be "very small."

As we already saw in the case of the linear functions, if we consider a polynomial $p(x)$ that is continuous at a and $p(a) = f(a)$, then it is clear that as long as f is continuous at a,

$$\lim_{x\to a} ((f(x) - p(x)) = f(a) - p(a) = 0.$$

In this case, for x close to a, $p(x)$ will be "close" to $f(x)$. If we want a truly good approximation, then we must require more of it. In Theorem 12.9, we saw that if f is differentiable at a, then the linear function $e(x) = f'(a)(x-a) + f(a)$ approximates our function so well that not only $\lim_{x\to a} ((f(x) - e(x)) = 0$, but even

$$\lim_{x\to a} \frac{f(x) - e(x)}{x - a} = 0$$

holds, that is, $f(x) - e(x)$ tends to 0 faster than $(x - a)$ does. It is easy to see that for linear functions, we cannot expect more—that is, for $f(x) - e(x)$ to tend to zero even faster than $(x - a)^\alpha$ for some number $\alpha > 1$—than this. If, however, f is n times differentiable at a, then we will soon see that using an nth-degree polynomial, we can approximate up to order $(x - a)^n$ in the sense that for a suitable polynomial $t_n(x)$ of degree at most n, we have

$$\lim_{x\to a} \frac{f(x) - t_n(x)}{(x-a)^n} = 0. \tag{13.7}$$

More than this cannot generally be said.

In Theorem 12.10, we saw that the best linear (or polynomial of degree at most 1) approximation in the sense above is unique. This linear polynomial $p_1(x)$ satisfies $p_1(a) = f(a)$ and $p_1'(a) = f'(a)$, and it is easy to see that these two conditions already determine $p_1(x)$. In the following theorem, we show that these statements can be generalized in a straightforward way to approximations using polynomials of degree at most n.

Theorem 13.4. *Let the function f be n times differentiable at the point a, and let*

$$t_n(x) = f(a) + f'(a)(x-a) + \cdots + \frac{f^{(n)}(a)}{n!}(x - a)^n. \tag{13.8}$$

(i) *The polynomial t_n is the only polynomial of degree at most n whose ith derivative at a is equal to $f^{(i)}(a)$ for all $i \le n$. That is,*

$$t_n(a) = f(a), \; t_n'(a) = f'(a), \ldots, t_n^{(n)}(a) = f^{(n)}(a), \tag{13.9}$$

and if a polynomial p of degree at most n satisfies

$$p(a) = f(a), \; p'(a) = f'(a), \ldots, p^{(n)}(a) = f^{(n)}(a), \tag{13.10}$$

then necessarily $p = t_n$.

(ii) *The polynomial t_n satisfies (13.7). If a polynomial p of degree at most n satisfies*

$$\lim_{x \to a} \frac{f(x) - p(x)}{(x-a)^n} = 0, \tag{13.11}$$

then necessarily $p = t_n$. Thus out of the polynomials of degree at most n, t_n is the one that approximates f locally at a the best.

Proof. (i) The equalities (13.9) follow simply from the definition of t_n. Now suppose (13.10) holds, where p is a polynomial of degree at most n. Let $q = p - t_n$. Then

$$q(a) = q'(a) = \cdots = q^{(n)}(a) = 0. \tag{13.12}$$

We show that q is identically zero. Suppose that $q \neq 0$. By Theorem 12.34, the number a is a root of q of multiplicity at least $n + 1$. This, however, is impossible, since q is of degree at most n. Thus q is identically zero, and $p = t_n$.
(ii) Let $g = f - t_n$ and $h(x) = (x - a)^n$. Then

$$g(a) = g'(a) = \ldots = g^{(n)}(a) = 0,$$

and $h^{(i)}(x) = n \cdot (n-1) \cdot \ldots \cdot (n-i+1) \cdot (x-a)^{n-i}$ for all $i < n$, which gives

$$h(a) = h'(a) = \cdots = h^{(n-1)}(a) = 0 \text{ and } h^{(n-1)}(x) = n! \cdot (x-a).$$

Applying L'Hôpital's rule $n - 1$ times, we get that

$$\lim_{x \to a} \frac{g(x)}{h(x)} = \lim_{x \to a} \frac{g'(x)}{h'(x)} = \ldots = \lim_{x \to a} \frac{g^{(n-1)}(x)}{n! \cdot (x-a)}.$$

Since

$$\lim_{x \to a} \frac{g^{(n-1)}(x)}{x - a} = \lim_{x \to a} \frac{g^{(n-1)}(x) - g^{(n-1)}(a)}{x - a} = g^{(n)}(a) = 0,$$

we have (13.7).

Now assume (13.11), where p is a polynomial of degree at most n. If $q = p - t_n$, then by (13.7) and (13.11), we have

$$\lim_{x \to a} \frac{q(x)}{(x - a)^n} = 0. \tag{13.13}$$

Suppose that q is not identically zero. If a is a root of q with multiplicity k, then $q(x) = (x - a)^k \cdot r(x)$, where $r(x)$ is a polynomial such that $r(a) \neq 0$. Since q has degree at most n, we know that $k \leq n$, and so the limit of

$$\frac{q(x)}{(x - a)^n} = r(x) \cdot \frac{1}{(x - a)^{n-k}}$$

at the point a cannot be zero. This contradicts (13.13), so we see that $q = 0$ and $p = t_n$.   □

Remark 13.5. Statement (ii) cannot be strengthened, that is, for every $\alpha > 0$, there exists an n times differentiable function such that

$$\lim_{x \to a} \frac{f(x) - p_n(x)}{(x-a)^{n+\alpha}} \neq 0$$

for every nth-degree polynomial $p_n(x)$. It is easy to check that for example, the function $f(x) = |x|^{n+\alpha}$ has this property at $a = 0$.

Definition 13.6. The polynomial t_n defined in (13.8) is called the *nth Taylor[2] polynomial* of f at the point a.

Thus the 0th Taylor polynomial is the constant function $f(a)$, while the first Taylor polynomial is the linear function $f(a) + f'(a) \cdot (x - a)$.

The difference $f(x) - t_n(x)$ can be obtained in several different ways. We give two of these in the following theorem.

Theorem 13.7 (Taylor's Formula). *Let the function f be $(n + 1)$ times differentiable on the interval $[a, x]$. Then there exists a number $c \in (a, x)$ such that*

$$f(x) = \sum_{k=0}^{n} \frac{f^{(k)}(a)}{k!}(x-a)^k + \frac{f^{(n+1)}(c)}{(n+1)!}(x-a)^{n+1}, \tag{13.14}$$

and there exists a number $d \in (a, x)$ such that

$$f(x) = \sum_{k=0}^{n} \frac{f^{(k)}(a)}{k!}(x-a)^k + \frac{f^{(n+1)}(d)}{n!}(x-d)^n(x-a). \tag{13.15}$$

If f is $(n + 1)$ times differentiable on the interval $[x, a]$, then there exists a $c \in (x, a)$ such that (13.14) holds, and there exists a $d \in (x, a)$ such that (13.15) holds.

Equality (13.14) is called **Taylor's formula with Lagrange remainder,** while (13.15) is **Taylor's formula with Cauchy remainder.** In the case $a = 0$, (13.14) is often called **Maclaurin's[3] formula.**

Proof. We prove only the case $a < x$; the case $x < a$ can be handled in the same way. For arbitrary $t \in [a, x]$, let

$$R(t) = \left[f(t) + \frac{f'(t)}{1!}(x-t) + \cdots + \frac{f^{(n)}(t)}{n!}(x-t)^n \right] - f(x). \tag{13.16}$$

(Here we consider x to be fixed, so R is a function of t.) Then $R(x) = 0$. Our goal is to find a suitable formula for $R(a)$. Since in (13.16), every term is differentiable on $[a, x]$, so is R, and

---

[2] Brook Taylor (1685–1731), English mathematician.
[3] Colin Maclaurin (1698–1746), Scottish mathematician.

$$R'(t) =$$

$$= f'(t) + \left[\frac{f''(t)}{1!}(x-t) - f'(t)\right] + \left[\frac{f'''(t)}{2!}(x-t)^2 - \frac{f''(t)}{1!}(x-t)\right] + \cdots +$$

$$+ \left[\frac{f^{(n+1)}(t)}{n!}(x-t)^n - \frac{f^{(n)}(t)}{(n-1)!}(x-t)^{n-1}\right].$$

We can see that here, everything cancels out except for one term, so

$$R'(t) = \frac{f^{(n+1)}(t)}{n!}(x-t)^n. \tag{13.17}$$

Let $h(t) = (x-t)^{n+1}$. By Cauchy's mean value theorem, there exists a number $c \in (a,x)$ such that

$$\frac{R(a)}{(x-a)^{n+1}} = \frac{R(x) - R(a)}{h(x) - h(a)} = \frac{R'(c)}{h'(c)} = -\frac{R'(c)}{(n+1) \cdot (x-c)^n} =$$

$$= -\frac{\left(f^{(n+1)}(c)/n!\right) \cdot (x-c)^n}{(n+1) \cdot (x-c)^n} = -\frac{f^{(n+1)}(c)}{(n+1)!}.$$

Then $R(a) = -(f^{(n+1)}(c)/(n+1)!) \cdot (x-a)^{n+1}$, so we have obtained (13.14).

If we apply the mean value theorem to R, then we get that for suitable $d \in (a,x)$,

$$\frac{R(a) - R(x)}{a - x} = R'(d) = \frac{f^{(n+1)}(d)}{n!}(x-d)^n.$$

Then $R(a) = -(f^{(n+1)}(d)/n!) \cdot (x-d)^n \cdot (x-a)$, so we get (13.15). $\qquad\square$

In Theorem 13.4, we showed that the nth Taylor polynomial of an n times differentiable function f approximates f well locally in the sense that if $x \to a$ then $f(x) - t_n(x)$ tends to 0 very fast. In other questions, however, we want a function f to be *globally* approximated by a polynomial. In such cases, we want polynomials $p_1(x), \ldots, p_n(x), \ldots$ such that for arbitrary $x \in [a,b]$, $|f(x) - p_n(x)| \to 0$ as $n \to \infty$. As we will see, the Taylor polynomials play an important role in this aspect as well.

An important class of functions have the property that the Taylor polynomials corresponding to a fixed position a will tend to $f(x)$ at every point x as $n \to \infty$. (That is, we do not think about $t_n(x)$ as $x \to a$ for a *fixed* n, but as $n \to \infty$ with a *fixed* x.)

It is easy to see that if the function f is infinitely differentiable at the point a, then the Taylor polynomials of f corresponding to the point a form partial sums of an infinite series. As we will soon see, in many cases these infinite series are convergent, and their value is exactly $f(x)$. It is useful to introduce the following naming conventions.

Definition 13.8. Let f be a function that is infinitely differentiable at the point a. The infinite series

$$\sum_{k=0}^{\infty} \frac{f^{(k)}(a)}{k!}(x-a)^k$$

is called the *Taylor series* of f corresponding to the point a.

If a Taylor series is convergent at a point x and its sum is $f(x)$, then we say that **the Taylor series represents f at x.**

Theorem 13.9. *If f is infinitely differentiable on the interval I and there exists a number K such that $|f^{(n)}(x)| \leq K$ for all $x \in I$ and $n \in \mathbb{N}$, then*

$$f(x) = \sum_{k=0}^{\infty} \frac{f^{(k)}(a)}{k!}(x-a)^k \tag{13.18}$$

for all a, $x \in I$; that is, the Taylor series of f corresponding to every point $a \in I$ represents f in the interval I.

Proof. Equation (13.14) of Theorem 13.7 together with the assumption $|f^{(n+1)}| \leq K$ implies

$$\left| f(x) - \sum_{k=0}^{n} \frac{f^{(k)}(a)}{k!}(x-a)^k \right| \leq \frac{K}{(n+1)!}|x-a|^{n+1}.$$

Now by Theorem 5.26, $|x-a|^n/n! \to 0$, so the partial sums of the infinite series on the right-hand side of (13.18) tend to $f(x)$ as $n \to \infty$. By the definition of convergence of infinite series, this means exactly that (13.18) holds. □

Remark 13.10. The theorem above states that we can obtain the function f in the interval I as the sum of an infinite series consisting of power functions. Thus we have an infinite series whose terms are functions of x. These infinite series are called function series. They are the subject of an important chapter of analysis, but we will not discuss them here.

Remark 13.11. The statement of Theorem 13.9 is actually quite surprising: if for a function f, $|f^{(n)}| \leq K$ holds for all n in an interval I, then the derivatives of f at a *alone* determine the values of the function f at every other point of I. From this, it also follows that for these functions, the values taken on in an arbitrarily small neighborhood of a already determine all other values of the function on I.

Let us see some applications.

Example 13.12. Let $f(x) = \sin x$. By the identities (12.30), it follows that $|f^{(n)}(x)| \leq 1$ for all $x \in \mathbb{R}$ and $n \in \mathbb{N}$. Also by (12.30), the $(4n+1)$th Taylor polynomial of $\sin x$ corresponding to the point 0 is

$$x - \frac{x^3}{3!} + \frac{x^5}{5!} - \ldots + \frac{x^{4n+1}}{(4n+1)!}.$$

Then by Theorem 13.9,

$$\sin x = x - \frac{x^3}{3!} + \frac{x^5}{5!} - \cdots + \frac{x^{4n+1}}{(4n+1)!} - \cdots \qquad (13.19)$$

for all x. We can similarly show that

$$\cos x = 1 - \frac{x^2}{2!} + \frac{x^4}{4!} - \cdots + \frac{x^{4n}}{(4n)!} - \cdots \qquad (13.20)$$

for all x. (These statements can also easily be deduced from Exercise 12.90.)

Remark 13.13. The equalities (13.19) and (13.20) belong to some of the most surprising identities of mathematics. We can express the periodic and bounded functions $\sin x$ and $\cos x$ as the sum of functions that are neither periodic nor bounded (except for the term 1 in the expression of $\cos x$).

Example 13.14. Now let $f(x) = e^x$. Then $|f^{(n)}(x)| \le e^b$ for all $x \in (-\infty, b]$ and $n \in \mathbb{N}$. It is easy to see that the nth Taylor polynomial of the function e^x around 0 is

$$1 + \frac{x}{1!} + \frac{x^2}{2!} + \cdots + \frac{x^n}{n!},$$

so

$$e^x = 1 + x + \frac{x^2}{2!} + \cdots + \frac{x^n}{n!} + \cdots \qquad (13.21)$$

for all x, just as we saw already in Exercises 12.86 and 12.89. If we apply (13.21) to $-x$, then we get that

$$e^{-x} = 1 - x + \frac{x^2}{2!} - \cdots + (-1)^n \frac{x^n}{n!} + \cdots . \qquad (13.22)$$

If we take the sum of (13.21) and (13.22), and their difference, and we then divide both sides by 2, then we get the expressions

$$\operatorname{sh} x = x + \frac{x^3}{3!} + \frac{x^5}{5!} + \cdots + \frac{x^{2n+1}}{(2n+1)!} + \cdots \qquad (13.23)$$

and

$$\operatorname{ch} x = 1 + \frac{x^2}{2!} + \frac{x^4}{4!} + \cdots + \frac{x^{2n}}{(2n)!} + \cdots . \qquad (13.24)$$

Example 13.15. By Exercise 12.92,

$$\log(1 + x) = x - \frac{x^2}{2} + \cdots + (-1)^{n-1} \frac{x^n}{n} + \cdots \qquad (13.25)$$

for all $x \in [0, 1]$. This can easily be seen from (13.14) as well. Let $f(x) = \log(1 + x)$. It can be computed that

$$f^{(n)}(x) = (-1)^{n-1} \cdot \frac{(n-1)!}{(1+x)^n}$$

for all $x > -1$ and $n \in \mathbb{N}^+$. Thus the nth Taylor polynomial of $\log(1+x)$ at 0 is

$$t_n(x) = x - \frac{x^2}{2} + \cdots + (-1)^{n-1} \frac{x^n}{n}.$$

If $x \in (0,1]$, then by (13.14), for all n there exists a number $c \in (0,x)$ such that

$$\log(1+x) - t_n(x) = (-1)^n \cdot \frac{1}{(n+1)(1+c)^{n+1}} x^{n+1}. \qquad (13.26)$$

Since

$$\left| (-1)^n \cdot \frac{1}{(n+1)(1+c)^{n+1}} x^{n+1} \right| \leq \frac{1}{n+1} \to 0 \qquad (13.27)$$

if $n \to \infty$, it follows that (13.25) holds.

This argument cannot be used if $x < 0$, since then $c < 0$, and the bound in (13.27) is not valid. For $-1 < x < 0$, however, we can apply (13.15). According to this, for all n there exists a number $d \in (x,0)$ such that

$$\log(1+x) - t_n(x) = (-1)^n \cdot \frac{1}{(1+d)^{n+1}} \cdot (x-d)^n \cdot x.$$

Here $1+d > 1+x > 0$ and $|(x-d)/(1+d)| < |x|$, so

$$|\log(1+x) - t_n(x)| \leq \frac{1}{1+x} \cdot |x|^{n+1} \to 0 \quad \text{as } n \to \infty,$$

proving our statement. To summarize, equation (13.25) holds for all $x \in (-1,1]$.

Remark 13.16. If we substitute $x = 1$ into the equations (13.21) and (13.25), we get the series representations

$$e = 1 + \frac{1}{1!} + \frac{1}{2!} + \cdots$$

and

$$\log 2 = 1 - \frac{1}{2} + \frac{1}{3} - \cdots.$$

(We have already seen these in Exercises 12.86 and 12.92.)

The examples above can be summarized by saying that the functions $\sin x$, $\cos x$, and e^x are represented by their Taylor series everywhere; the function $\log(1+x)$ is represented by its Taylor series centered at 0 for all $x \in (-1,1]$. By Theorem 13.9, it easily follows that the Taylor series of the functions $\sin x$ and $\cos x$ centered at *any* point represent the function. The same holds for the function e^x (see Exercise 13.4).

Remark 13.17. We cannot omit the condition on the boundedness of the derivatives in Theorem 13.9. Consider the following example.

Let $f(x) = e^{-1/x^2}$ if $x \neq 0$, and let $f(0) = 0$. We show that f is infinitely differentiable everywhere. Using differentiation rules and induction, it is easy to see that f is infinitely differentiable at every point $x \neq 0$ and that for every n, there exists a polynomial p_n such that

$$f^{(n)}(x) = \frac{p_n(x)}{x^{3n}} \cdot e^{-1/x^2}$$

for all $x \neq 0$. Since $\lim_{y \to \infty} y^{2n}/e^y = 0$, it follows from the theorem on the limits of compositions (Theorem 10.41) that

$$\lim_{x \to 0} \frac{1}{x^{4n}} \cdot e^{-1/x^2} = 0,$$

and so

$$\lim_{x \to 0} f^{(n)}(x) = \lim_{x \to 0} p_n(x) \cdot x^n \cdot \frac{1}{x^{4n}} \cdot e^{-1/x^2} = 0.$$

Now we show that f is also infinitely differentiable at the point 0 and that $f^{(n)}(0) = 0$ for all n. The statement $f^{(n)}(0) = 0$ holds for $n = 0$. If we assume it holds for n, then (by L'Hôpital's rule, for example)

$$\lim_{x \to 0} \frac{f^{(n)}(x) - f^{(n)}(0)}{x - 0} = \lim_{x \to 0} \frac{f^{(n)}(x)}{x} = \lim_{x \to 0} \frac{f^{(n+1)}(x)}{1} = 0,$$

so $f^{(n+1)}(0) = 0$.

This shows that f is infinitely differentiable everywhere, and that $f^{(n)}(0) = 0$ for all n. Thus every term of the Taylor series of f centered at 0 will be zero for all x. Since $f(x) > 0$ if $x \neq 0$, the statement of Theorem 13.9 does not hold for the function f and the point $a = 0$.

Moreover, we have constructed a function that is *infinitely differentiable everywhere, but its Taylor series corresponding to the point 0 does not represent it at any point $x \neq 0$.*

We can observe that the function f has an absolute minimum at the origin. It is also easy to see that the function

$$g(x) = \begin{cases} -e^{-1/x^2}, & \text{if } x < 0, \\ e^{-1/x^2}, & \text{if } x > 0, \\ 0, & \text{if } x = 0 \end{cases}$$

is infinitely differentiable, strictly monotone increasing, and all of its derivatives are zero at the origin.

Remark 13.18. In Remark 11.28 and in the second appendix of Chapter 11, we showed that the trigonometric and hyperbolic functions are closely linked, which can be expressed with the help of complex numbers by Euler's formulas (11.64), (11.65), and (11.66).

This connection can be seen from the corresponding Taylor series as well. If we apply the equality (13.21) to ix instead of x, then using (13.19) and (13.20), we get (11.64). If we substitute ix into (13.23) and (13.24), then we get (11.66).

Global Approximation with Polynomials. Approximation with Taylor polynomials can be used only for functions that are many times differentiable. By the theorem below, for a function f to be arbitrarily closely approximated by a polynomial on a closed and bounded interval, it suffices for f to be continuous there. We remind ourselves that the set of continuous functions on the closed and bounded interval $[a,b]$ is denoted by $C[a,b]$ (see Theorems 10.52–10.58).

Theorem 13.19 (Weierstrass Approximation Theorem). *If $f \in C[a,b]$, then for every $\varepsilon > 0$, there exists a polynomial p such that*

$$|f(x) - p(x)| < \varepsilon \tag{13.28}$$

for all $x \in [a,b]$.

We prove the theorem by specifying a polynomial that satisfies (13.28). For the case $[a,b] = [0,1]$, let

$$B_n(x;f) = \sum_{k=0}^{n} f\left(\frac{k}{n}\right) \cdot \binom{n}{k} x^k (1-x)^{n-k}. \tag{13.29}$$

These polynomials are weighted averages of the values $f(k/n)$ $(k = 0, \ldots, n)$, where the weights $\binom{n}{k} x^k (1-x)^{n-k}$ depend on x. The polynomial $B_n(x;f)$ is called the *nth* **Bernstein**[4] **polynomial** *of the function f.* Our goal is to prove the following theorem.

Theorem 13.20. *If $f \in C[0,1]$, then for all $\varepsilon > 0$, there exists an $n_0 = n_0(\varepsilon;f)$ such that $n > n_0$ implies*

$$|f(x) - B_n(x;f)| < \varepsilon$$

for all $x \in [0,1]$.

The general case can be reduced to this. This is because if $f : [a,b] \to \mathbb{R}$, then the function $x \mapsto g(x) = f(a + (b-a) \cdot x)$ is defined on $[0,1]$. Thus let

$$B_n(x;f) = B_n\left(\frac{x-a}{b-a};g\right)$$

for all $x \in [a,b]$. In other words, in the general case, we get the Bernstein polynomial by transforming the function using an inner linear composition into a function defined on $[0,1]$. Then, using the inverse transformation, we transform the Bernstein polynomial of the new function into one defined on $[a,b]$. It is clear that if $|B_n(x;g) - g(x)| < \varepsilon$ for all $x \in [0,1]$, then $|B_n(x;f) - f(x)| < \varepsilon$ for all $x \in [a,b]$. We prove Theorem 13.20 in the first appendix of the chapter.

The Bernstein polynomial—which is a weighted average of the values $f(k/n)$—more or less tries to recreate the function f from the values it takes on at the points k/n $(k = 0, \ldots, n)$. The need to recreate a function from its values at finitely many

---

[4] Sergei Natanovich Bernstein (1880–1968), Russian mathematician.

points appears many times. If, for example, we are performing measurements, then we might deduce other values from the ones we measured—at least approximately. One of the simplest methods for this is to find a polynomial that takes the same values at these points as the given function and has lowest possible degree.

Theorem 13.21. *Let f be defined on the interval $[a,b]$, and let*

$$a \le x_0 < x_1 < \cdots < x_n \le b$$

be fixed points. There exists exactly one polynomial p of degree at most n such that $p(x_i) = f(x_i)$ $(i = 0, \ldots, n)$.

Proof. For every $k = 0, \ldots, n$ let $l_k(x)$ be the product of the linear functions $(x - x_i)/(x_k - x_i)$ for all $0 \le i \le n$, $i \ne k$. Then $l_0, \ldots, l_n$ are nth-degree polynomials such that

$$l_k(x_j) = \begin{cases} 1, & \text{if } k = j, \\ 0, & \text{if } k \ne j. \end{cases}$$

It is then clear that if

$$L_n(x; f) = \sum_{k=0}^{n} f(x_k) \cdot l_k(x), \tag{13.30}$$

then $L_n(x; f)$ is a polynomial of degree at most n that agrees with f at every point x_i.

Now suppose that a polynomial p also satisfies these conditions. Then the polynomial $q = p - L_n(x; f)$ has degree at most n, and has at least $n + 1$ roots, since it vanishes at all the points $x_0, \ldots, x_n$. Thus by Lemma 11.1, q is identically zero, so $p = L_n(x; f)$.   □

The polynomial given in (13.30) is called the **Lagrange interpolation polynomial** of f corresponding to the points $x_0, \ldots, x_n$.

Remark 13.22. It can be proved that taking the points $\{\pm k/n : 0 \le k \le n\}$ yields Lagrange interpolation polynomials for $f(x) = \sin x$ and $g(x) = \sin 2x$ that approximate them well, but do not approximate $h(x) = |x|$ well at all. Namely, one can show that

$$|f(x) - L_n(x; f)| < K \cdot \frac{1}{n!}; \qquad |g(x) - L_n(x; g)| < K \cdot \frac{2^n}{n!};$$

while $|h(x) - L_n(x; h)|$ can be arbitrarily large for any $x \in [-1, 1]$, $x \ne 0, 1, -1$ (see Exercise 13.10).

Generally, the Lagrange interpolation polynomials corresponding to the points $\{\pm k/n : 0 \le k \le n\}$ *do not converge for every continuous function*, even though the polynomial agrees with the function on an ever denser set of points. If, however, the function f is infinitely differentiable and $\max_{x \in [-1,1]} |f^{(n)}(x)| = K_n$, then

$$|f(x) - L_n(x, f)| < 2^n \cdot \frac{K_n}{n!}.$$

Thus the tighter we can bound K_n, the tighter we can bound $|f(x) - L_n(x, f)|$.

Exercises

13.2. Prove that if f is a polynomial, then its Taylor series corresponding to any point f will represent f everywhere. (H)

13.3. Prove that the Taylor series of the function $1/x$ corresponding to any point $a > 0$ represents the function at all $x \in (0, 2a)$.

13.4. Prove that the Taylor series of the function e^x corresponding to any point will represent the function everywhere.

13.5. Check that the nth Bernstein polynomial of the function $f : [-1, 1] \to \mathbb{R}$ is

$$\frac{1}{2^n} \sum_{k=0}^{n} \binom{n}{k} (1+x)^k (1-x)^{n-k} f\left(\frac{2k}{n} - 1\right). \tag{S}$$

13.6. Prove that if the function $f : [-1, 1] \to \mathbb{R}$ is even, then all of its Bernstein polynomials are also even (H).

13.7. Find the nth Bernstein polynomial of the function $f(x) = |x|$ defined on $[-1, 1]$ for $n = 1, 2, 3, 4, 5!$ (S)

13.8. Find the Bernstein polynomials of the function $f(x) = e^{2x}$ in $[a, b]$.

13.9. Determine the Bernstein polynomials of $f(x) = \cos x$ in $\left[-\frac{\pi}{2}, \frac{\pi}{2}\right]$. (H)

13.10. Determine the Lagrange interpolation polynomials of

(a) the functions $f(x) = \sin x$ and $g(x) = \sin 2x$ corresponding to the points $\{\pm\frac{k}{n} \cdot \frac{\pi}{2} : 0 \le k \le n\}$ for $n = 2, 3$, and 4, and of
(b) the function $h(x) = |x|$ corresponding to the points $\{\pm k/n : 0 \le k \le n\}$ for $n = 2, 3, 4$ and $n = 10$.

Check that we get much better approximations in (a) than in (b).

13.11. Show that $\sum_{k=0}^{n} \frac{k}{n} \binom{n}{k} x^k (1-x)^{n-k} = x$ for all x. (S)

13.12. Show that

$$\sum_{k=0}^{n} \frac{k^2}{n^2} \binom{n}{k} x^k (1-x)^{n-k} = x^2 + (x - x^2)/n$$

for all x. (S)

13.3 The Indefinite Integral

It is a common problem to reconstruct a function from its derivative. This is the case when we want to determine the location of a particle based on its speed, or the speed (and thus location) of a particle based on its acceleration. Computing volume and area provide further examples.

Examples 13.23. **1.** Let f be a nonnegative monotone increasing and continuous function on the interval $[a,b]$. We would like to find the area under the graph of f. Let $T(x)$ denote the area over the interval $[a,x]$. If $a \leq x < y \leq b$, then

$$f(x)(y-x) \leq T(y) - T(x) \leq f(y)(y-x),$$

since $T(y) - T(x)$ is the area of a domain that contains a rectangle of width $y - x$ and height $f(x)$, and is contained in a rectangle of width $y - x$ and height $f(y)$ (see Figure 13.1). Thus

$$f(x) \leq \frac{T(y) - T(x)}{y - x} \leq f(y),$$

and so

$$\lim_{y \to x} \frac{T(y) - T(x)}{y - x} = f(x),$$

that is, $T'(x) = f(x)$ if $x \in [a,b]$.

Consider, for example, the function $f(x) = x^2$ over the interval $[0,1]$. By the argument above, $T'(x) = x^2$, so $T(x) = \frac{1}{3}x^3 + c$ by the fundamental theorem of integration (Corollary 12.53). However, $T(0) = 0$, so $c = 0$, and $T(x) = \frac{1}{3}x^3$ for all $x \in [0,1]$. Specifically, $T(1) = 1/3$, giving us the theorem of Archimedes regarding the area under a parabola.

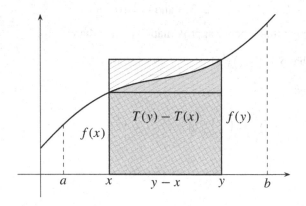

Fig. 13.1

2. With this method, the volume of a sphere can also be determined. Consider a sphere of radius R centered at the origin, and for $0 \le u \le R$, let $V(u)$ denote the volume of the part of the sphere that lies between the horizontal planes $z = 0$ and $z = u$ (Figure 13.2). If $0 \le u < v \le R$, then

$$(R^2 - v^2) \cdot \pi \cdot (v - u) \le V(v) - V(u) \le (R^2 - u^2) \cdot \pi \cdot (v - u). \tag{13.31}$$

Indeed, $V(v) - V(u)$ is the volume of the slice of the sphere determined by the horizontal planes $z = u$ and $z = v$. This slice contains a cylinder of height $v - u$, whose radius is $\sqrt{R^2 - v^2}$, and it is contained in a cylinder of the same height with radius $\sqrt{R^2 - u^2}$. Assuming that the formula for the volume of a cylinder is known, we get inequality (13.31). Now by inequality (13.31), it is clear that $V'(x) = \pi(R^2 - x^2)$ for all $x \in [0, R]$, and thus

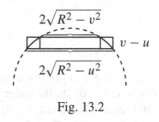

Fig. 13.2

$$V(x) = \pi R^2 x - \frac{1}{3}\pi x^3 + c$$

by the fundamental theorem of integration again. Since $V(0) = 0$, we have $c = 0$. The volume of the hemisphere is thus

$$V(R) = \pi R^3 - \frac{1}{3}\pi R^3 = \frac{2}{3}\pi R^3.$$

The solutions of the examples above required finding a function that has given derivatives.

Definition 13.24. If F is differentiable on the interval I and $F'(x) = f(x)$ for all $x \in I$, then we say that F is a *primitive function* of f on I.

Theorem 13.25. *If F is a primitive function of f on the interval I, then every primitive function of f on I is of the form $F + c$, where c is a constant.*

Proof. This is a restatement of Corollary 12.53.   □

Definition 13.26. The collection of the primitive functions of f is denoted by $\int f\,dx$, and is called the *indefinite integral* or *antiderivative* of f. Thus $\int f\,dx$ is a set of functions. Moreover, $F \in \int f\,dx$ if and only if $F' = f$.

Remarks 13.27. **1.** Theorem 13.25 can be stated as $F \in \int f\,dx$ implies that

$$\int f\,dx = \{F + c;\ c \in \mathbb{R}\}.$$

This statement is often denoted more concisely (and less precisely) by

$$\int f\,dx = F + c.$$

2. A remark on the notation: $F' = f$ can also be written as

$$\frac{dF}{dx} = f,$$

which after "multiplying through" gives us that $dF = f\,dx$. Thus the equation $F = \int f\,dx$ denotes that we are talking about the inverse of differentiation. In any case, dx can be omitted from behind the integral symbol; $\int f$ is an equally usable notation.
3. Does every function have a primitive function? Let, for example,

$$f(x) = \begin{cases} 0, & \text{if } x \le 0, \\ 1, & \text{if } x > 0. \end{cases}$$

Suppose that F is a primitive function of f. Then for $x < 0$, we have $F'(x) = 0$, so $F(x) = c$ if $x \le 0$. On the other hand, if $x > 0$, then $F'(x) = 1$, so $F(x) = x + a$ if $x \ge 0$. It then follows that $F'_-(0) = 0$ and $F'_+(0) = 1$, so F is not differentiable at 0, which is impossible. This shows that f does not have a primitive function.
4. One of the most important *sufficient* conditions for the existence of a primitive function is continuity. That is, *if f is continuous on the interval I, then f has a primitive function on I*. We do not yet have all the tools to prove this important theorem, so we will return to its proof once we have introduced integration (see Theorem 15.5).

We saw in Example 13.23 that if f is nonnegative, monotone, and continuous, then the area function $T(x)$ is a primitive function of f. The proof of the general case also builds on this thought. However, this argument uses the definition and properties of area, and until we make this clear, we cannot come up with a precise proof in this way.
5. Continuity is a sufficient condition for a primitive function to exist. Continuity, however, is *not a necessary* condition. In Example 13.43, we will give a function that is not continuous at 0 but has a primitive function. Later, in the section titled *Properties of Derivative Functions*, we will inspect the necessary conditions for a function to have a primitive function more closely (see Remark 13.45).

When we are looking for the primitive functions of our elementary functions, we use the same method that we used to determine limits and derivatives as well. First of all, we need a list of the primitive functions of the simplest elementary functions (elementary integrals). Besides that, we must familiarize ourselves with the rules that tell us how to find the primitives of functions defined with the help of functions with already known primitives. These theorems are called **integration rules**.

The integrals listed here follow from the differentiation formulas of the corresponding functions.

The Elementary Integrals

$$\int x^\alpha \, dx = \frac{1}{\alpha+1} x^{\alpha+1} + c \quad (\alpha \neq -1) \qquad \int \frac{1}{x} \, dx = \log|x| + c$$

$$\int a^x \, dx = \frac{1}{\log a} \cdot a^x + c \quad (a \neq 1) \qquad \int e^x \, dx = e^x + c$$

$$\int \cos x \, dx = \sin x + c \qquad \int \sin x \, dx = -\cos x + c$$

$$\int \frac{1}{\cos^2 x} \, dx = \operatorname{tg} x + c \qquad \int \frac{1}{\sin^2 x} \, dx = -\operatorname{ctg} x + c$$

$$\int \frac{1}{\sqrt{1-x^2}} \, dx = \arcsin x + c \qquad \int \frac{1}{1+x^2} \, dx = \operatorname{arc tg} x + c$$

$$\int \operatorname{ch} x \, dx = \operatorname{sh} x + c \qquad \int \operatorname{sh} x \, dx = \operatorname{ch} x + c$$

$$\int \frac{1}{\operatorname{ch}^2 x} \, dx = \operatorname{th} x + c \qquad \int \frac{1}{\operatorname{sh}^2 x} \, dx = -\operatorname{cth} x + c$$

$$\int \frac{1}{\sqrt{x^2+1}} \, dx = \operatorname{arsh} x + c = \log(x + \sqrt{x^2+1}) + c$$

$$\int \frac{1}{\sqrt{x^2-1}} \, dx = \operatorname{arch} x + c = \log(x + \sqrt{x^2-1}) + c$$

$$\int \frac{1}{1-x^2} \, dx = \frac{1}{2} \cdot \log\left|\frac{1+x}{1-x}\right| + c$$

These equalities are to be understood that the indefinite integrals are the sets of the functions on the right-hand side on all the intervals where the functions on the right are defined and are differentiable.

For the time being, we will need only the simplest integration rules. Later, in dealing with integration in greater depth, we will get to know further methods and rules.

Theorem 13.28. *If both* f *and* g *have primitive functions on the interval I, then* $f + g$ *and* $c \cdot f$ *do as well, namely*

$$\int (f+g) \, dx = \int f \, dx + \int g \, dx \qquad \text{and} \qquad \int cf \, dx = c \int f \, dx$$

for all $c \in \mathbb{R}$.

The theorem is to be understood as $H \in \int (f+g) \, dx$ if and only if $H = F + G$, where $F \in \int f \, dx$ and $G \in \int g \, dx$. Similarly, $H \in \int cf \, dx$ if and only if $H = cF$, where $F \in \int f \, dx$.

Proof. The theorem is clear by statements (i) and (ii) of Theorem 12.17.   □

Example 13.29.

$$\int (x^3 + 2x - 3)\, dx = \tfrac{x^4}{4} + x^2 - 3x + c.$$

$$\int \frac{(x+1)^2}{x^3}\, dx = \int \frac{x^2 + 2x + 1}{x^3}\, dx = \int \left(\frac{1}{x} + \frac{2}{x^2} + \frac{1}{x^3} \right) dx =$$

$$= \log |x| - \frac{2}{x} - \frac{1}{2x^2} + c.$$

Theorem 13.30. *If* $F \in \int f\, dx$, *then*

$$\int f(ax + b)\, dx = \tfrac{1}{a} F(ax + b) + c$$

for all $a, b \in \mathbb{R}$, *where* $a \neq 0$.

Proof. The statement is clear by the differentiation rules for compositions of functions. □

Example 13.31.

$$\int \sqrt{2x - 3}\, dx = \int (2x - 3)^{1/2}\, dx = \tfrac{1}{2} \cdot \tfrac{2}{3} \cdot (2x - 3)^{3/2} + c = \tfrac{1}{3}(2x - 3)^{3/2} + c.$$

$$\int e^{-x}\, dx = \frac{e^{-x}}{-1} + c = -e^{-x} + c.$$

$$\int \sin \frac{x+1}{3}\, dx = -3 \cos \frac{x+1}{3} + c.$$

$$\int \sin^2 x\, dx = \int \frac{1 - \cos 2x}{2}\, dx = \int \left(\tfrac{1}{2} - \tfrac{1}{2} \cos 2x \right) dx =$$

$$= \frac{1}{2} x - \frac{1}{2} \cdot \frac{\sin 2x}{2} + c = \frac{x}{2} - \frac{\sin 2x}{4} + c.$$

Remark 13.32. With this method we can determine the integrals of every trigonometric function, making use of the identities (11.35)–(11.36). So, for example,

$$\int \sin^3 x\, dx = \int \sin^2 x \cdot \sin x\, dx = \int \frac{1 - \cos 2x}{2} \sin x\, dx =$$

$$= \int \left(\tfrac{1}{2} \sin x - \tfrac{1}{2} \cos 2x \sin x \right) dx =$$

$$= \int \left(\tfrac{1}{2} \sin x - \tfrac{1}{4} \left(\sin 3x + \sin(-x) \right) \right) dx =$$

$$= \int \left(\tfrac{3}{4} \sin x - \tfrac{1}{4} \sin 3x \right) dx = -\tfrac{3}{4} \cos x + \tfrac{1}{12} \cos 3x + c.$$

Theorem 13.33. *If* f *is differentiable and positive everywhere on* I, *then on* I,

$$\int [f(x)]^\alpha f'(x)\, dx = \frac{[f(x)]^{\alpha+1}}{\alpha + 1} + c, \qquad \text{if } \alpha \neq -1.$$

If $f \neq 0$, then

$$\int \frac{f'(x)}{f(x)} = \log|f(x)| + c.$$

Proof. The statement is clear by the differentiation rules for composition of functions. □

Example 13.34.

$$\int \cos x \, \sin^3 x \, dx = \frac{\sin^4 x}{4} + c.$$

$$\int \text{tg} x \, dx = -\int \frac{-\sin x}{\cos x} \, dx = -\log|\cos x| + c.$$

$$\int x\sqrt{1+x^2} \, dx = \int \frac{1}{2}\sqrt{1+x^2} \cdot 2x \, dx = \frac{1}{2} \int (x^2+1)^{1/2} (x^2+1)' \, dx =$$

$$= \frac{1}{2} \cdot \frac{2}{3} \cdot (x^2+1)^{3/2} + c = \frac{1}{3}(x^2+1)^{3/2} + c.$$

$$\int \frac{x}{1+x^2} \, dx = \frac{1}{2} \int \frac{(x^2+1)'}{x^2+1} \, dx = \frac{1}{2}\log(x^2+1) + c.$$

Exercises

13.13. Determine the indefinite integrals of the following functions:

(a) $\cos^2 x$, $\quad \sin 2x$, $\quad 1/\sqrt{x+2}$, $\quad x^2 + e^x - 2$, $\quad 3^x$, $\quad 4^{5x+6}$, $\quad 1/\sqrt[3]{x}$,

(b) $(2^x + 3^x)^2$, $\quad ((1-x)/x)^2$, $\quad (x+1)/\sqrt{x}$, $\quad (1-x)^3/(x^3\sqrt{x})$, $\quad x^3/(x+1)$, $\quad 1/\sqrt{1-2x}$,

(c) $x^2(5-x)^2$, $\quad (e^{3x}+1)/(e^x+1)$, $\quad \text{tg}^2 x$, $\quad \text{ctg}^2 x$, $\quad 1/(x+5)$, $\quad x/(1-x^2)$, $\quad (x^2+3)/(1-x^2)$, $\quad 1/(5x-2)^{5/2}$, $\quad \sqrt[5]{1-2x+x^2}/(1-x)$,

(d) $1/(2+3x^2)$, $\quad e^{-x}+e^{-2x}$, $\quad |x^2-5x+6|$, $\quad xe^{x^2}$, $\quad x/(1+x^2)^{3/2}$, $\quad (1-x^2)^9$,

(e) $x/\sqrt{x+1}$, $\quad (\sqrt{x}+\sqrt[3]{x})^2$, $\quad (\sin x + \cos x)^2$, $\quad \sin x/\sqrt{\cos^3 x}$, $\quad x^2\sqrt{1+x^3}$,

(f) $\sin x/\sqrt{\cos 2x}$, $\quad x/(x^4+1)$, $\quad x/\sqrt{x^2+1}$, $\quad (5x+6)/(2x^2+3)$, $\quad e^x/(e^x+2)$,

(g) $x^2(4x^3+3)^7$, $\quad x^3(4x^2-1)^{10}$, $\quad 1/(x \cdot \log x)$, $\quad (\log x)/x$,

(h) $(e^x - e^{-x})/(e^x + e^{-x})$, $\quad (\sin \sqrt{x})/\sqrt{x}$, $\quad (1+x)^2/(1+x^2)$, $\quad x/(1+x^2)^2$, $\quad x\sqrt{x+1}$.

13.4 Differential Equations

One of the most important applications of differentiation is in expressing a process in mathematical form. Through differentiation, we can obtain a general overview or prediction for how the process will continue. One of the driving reasons for differentiation is make problems of this sort solvable.

An easy example of this application is growth and decay problems. If a quantity is changing at a rate that is proportional to the quantity itself, then we are dealing with a growth or decay problem, depending on whether the quantity is increasing or decreasing. For example, the birth rate in a country in a period when its population is not affected by wars, epidemics, new medical advances, and the quality of life is more or less constant, in which case population changes as outlined above: the number of births is proportional to the population. The same sort of thing is true for the population of rabbits on an island, or a bacteria colony. The amount of material that decays in radioactive decay also follows this law. This is because every molecule of a decaying material decays with probability p after a (short) time h. Thus after time t, the amount that has decayed is about $p \cdot (t/h)$ times the total amount of material. Thus the instantaneous rate of change of material is p/h times the amount of material present at that instant.

How can we express such growth and decay processes mathematically? If the size of the quantity at time t is $f(t)$, then we can write it as

$$f'(t) = k \cdot f(t), \tag{13.32}$$

and we speak of a growth or decay process depending on whether the constant k is positive or negative. Now that we have precisely defined the conditions, it is easy to find the functions satisfying them.

Theorem 13.35. *Let f be differentiable on the interval I, and let $f' = k \cdot f$ on the interval I, where k is constant. Then there exists a constant c such that*

$$f(x) = c \cdot e^{kx} \qquad (x \in I).$$

Proof. Let $g(x) = f(x)e^{-kx}$. Then

$$g'(x) = f'(x)e^{-kx} - kf(x)e^{-kx} = 0,$$

so $g = c$, which gives $f(x) = c \cdot e^{kx}$. □

Growth and decay processes can thus always be written down as a function of the form $c \cdot e^{kx}$.

Example. An interesting (and important) application is radiocarbon dating. Living organisms have a fixed ratio of the radioactive isotope C_{14} to the nonradioactive isotope C_{12}. When the organism perishes, the C_{14} isotope is no longer replenished, and it decays into C_{12} with a half-life of 5730 years (which means that the amount of C_{14} decreases to half its original amount in 5730 years). With the help of this, we can inspect the remains of an organism to estimate about how many years ago it perished. Suppose, for example, that a certain (living) kind of tree possesses C_{14} with a ratio of α to the total amount of carbon it has. If we find a piece of this kind of tree that has a ratio of 0.9α of C_{14} to total carbon, then we can argue in the following way. When the tree fell, in 1 gram of carbon in the tree, α grams were C_{14}. The decay of C_{14} is given by a function of the form $c \cdot e^{kt}$, where $c \cdot e^{k0} = \alpha$. Thus we have $c = \alpha$. We also know that $\alpha \cdot e^{5730k} = \alpha/2$, so $k = -0.000121$. Thus if the tree fell t years ago, then $\alpha \cdot e^{kt} = 0.9\alpha$, which gives $t = (1/k) \cdot \log 0.9 \approx 870$ years.

The equation (13.32) is a **differential equation** in the sense that we defined in Remark 12.39; that is, it defines a relation between the function f and its derivatives. We can write a differential equation symbolically using the equation

$$\Phi\left(x, y, y', \ldots, y^{(n)}\right) = 0. \tag{13.33}$$

The differential equation (13.32) with this notation can be written as $y' - k \cdot y = 0$. The relation (13.33) can generally contain known functions, just as $y' - k \cdot y = 0$ contains the constant k. The function $y = f$ is a **solution** to the differential equation (13.33) if f is n times differentiable on an interval I, and

$$\Phi(x, f(x), f'(x), \ldots, f^{(n)}(x)) = 0$$

for all $x \in I$. Theorem 13.35 can now also be stated by saying that the solutions to the differential equation $y' = ky$ are the functions ce^{kx}.

An important generalization of the differential equation (13.32) is the equation $y' = fy + g$, where f and g are given functions. These are called **first-order linear differential equations.**

Theorem 13.36. *Let f and g be functions defined on the interval I. Suppose that f has a primitive function on I, and let F be one such fixed primitive function. Then every solution of the differential equation $y' = fy + g$ is of the form*

$$y = e^F \int g e^{-F} \, dx.$$

That is, there exists a solution if and only if ge^{-F} has a primitive function, and every solution is of the form $e^F \cdot G$, where $G \in \int g e^{-F} \, dx$.

Proof. If y is a solution, then

$$\left(ye^{-F}\right)' = (fy + g)e^{-F} + ye^{-F}(-f) = ge^{-F},$$

so $y = e^F \int g e^{-F} \, dx$. On the other hand,

$$\left(e^F \int g e^{-F} \, dx\right)' = fe^F \int g e^{-F} \, dx + e^F g e^{-F} = fe^F \int g e^{-F} \, dx + g.$$

$\square$

Example 13.37. (The story of the little ant and the evil gnome.) A little ant starts walking on a 10-cm-long elastic rope, starting at the right-hand side at 1 cm/sec, toward the fixed left-hand endpoint. At the same time, the evil gnome grabs the right-hand endpoint of the elastic rope and starts running away to the right at 100 cm/sec, stretching the rope. Can the little ant ever complete its journey?

If we let $y(t)$ denote the distance of the ant from the left endpoint, then we can easily find the speed of the ant to be

$$y'(t) = 100\frac{y(t)}{10+100t} - 1 = \frac{10}{10t+1} \cdot y(t) - 1.$$

This is a first-order linear *differential equation*, to which we can apply the solution given by Theorem 13.36. Since $F = \log(10t+1)$ is a primitive function of the function $10/(10t+1)$, we have

$$y = e^{\log(10t+1)} \int (-1)e^{-\log(10t+1)} dt = (10t+1) \int -\frac{1}{10t+1} dt =$$
$$= (10t+1)\left(c - \tfrac{1}{10}\log(10t+1)\right).$$

Since $y(0) = 10$, we have $c = 10$, and so

$$y(t) = \left(t + \frac{1}{10}\right)(100 - \log(10t+1)).$$

Thus the little ant *will* make it to the left endpoint in merely t seconds, where $\log(10t+1) = 100$, that is, $10t+1 = e^{100} \approx 2.7 \cdot 10^{43}$. This gives us that $t \approx 2.7 \cdot 10^{42}$ sec $\approx 8.6 \cdot 10^{34}$ years. (While walking, the little ant ends up 10^{26} light years away from the fixed endpoint.)

Theorem 13.38. *Let I and J be intervals, and let $f : I \to \mathbb{R}$, $g : J \to \mathbb{R} \setminus \{0\}$ be functions. Suppose that f and $1/g$ both have primitive functions on I and J respectively, and let these be $F \in \int f$ and $G \in \int(1/g)$. Then a function $y : I_1 \to \mathbb{R}$ is a solution to the differential equation*

$$y' = f(x)g(y)$$

on the interval $I_1 \subset I$ if and only if $G(y(x)) = F(x) + c$ for all $x \in I_1$ with some constant c.

Proof. Let $y : I_1 \to J$ be differentiable. Then

$$y'(x) = f(x)g(y(x)) \iff \left(y'(x)/g(y(x))\right) - f(x) = 0 \iff$$
$$\iff \left(G(y(x)) - F(x)\right)' = 0 \iff$$
$$\iff G(y(x)) - F(x) = c \iff$$
$$\iff G(y(x)) = F(x) + c.$$

$\square$

Remarks 13.39. **1.** Equations of the form $y' = f(x)g(y)$ are called **separable differential equations**. The statement of the theorem above can be expressed by saying that if $g \neq 0$, then we get every solution of the equation if we express y from $G(y) = F(x) + c$. This also means that the following formal process leads us to a correct solution:

$$\frac{dy}{dx} = f(x)g(y), \quad \frac{dy}{g(y)} = f(x)\,dx, \quad \int \frac{dy}{g(y)} = \int f(x)\,dx, \quad G(y) = F(x) + c.$$

2. The solutions of the equation $y' = f(x)g(y)$ are generally not defined on the whole interval I. Consider, for example, the differential equation $y' = y^2$, where $f \equiv 1$, $I = \mathbb{R}$, and $g(x) = x^2$, $J = (0, \infty)$. Then by the above theorem, the solutions satisfy $-1/y = x + c$, $y = -1/x + c$. The solutions are defined on only one open half-line each.

Example 13.40. Consider the graphs of the functions $y = c \cdot x^2$. These are parabolas that cover the plane in one layer except the origin, meaning that every point in the plane, other than the origin, has exactly one parabola of this form containing it. Which are the curves that intersect every $y = c \cdot x^2$ at right angles? If such a curve is the graph of a function $y = f(x)$, then at every point $a \neq 0$, we have

$$f'(a) = -\frac{1}{2c \cdot a} = -\frac{1}{2(f(a)/a^2) \cdot a} = -\frac{a}{2f(a)}.$$

This is so because perpendicular intersections mean (by definition) that at every point $(a, f(a))$ on the curve, the tangent to the curve is perpendicular to the tangent to the parabola that contains this point, and by writing out the slopes, we get the above condition.

Thus f is a solution of the separable differential equation $y' = -x/2y$. Using the method above, we get that

$$2y\,dy = -x\,dx,$$

$$\int 2y\,dy = -\int x\,dx,$$

$$y^2 = -\frac{x^2}{2} + c,$$

so

$$f(x) = \pm\sqrt{c - \frac{x^2}{2}},$$

where c is an arbitrary positive constant. (In the end, the curves we get are the ellipses of the form $\frac{x^2}{2} + y^2 = c$ (Figure 13.3).)

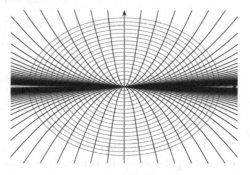

Fig. 13.3

Second-Order Linear Homogeneous Differential Equations. Let g and h be functions defined on the interval I. Below, we will learn all there is to know about solutions of differential equations of the form

$$y'' + gy' + hy = 0. \tag{13.34}$$

We suppose that g has a primitive function on I; let G be one such primitive function.

1. It is easy to check that if y_1 and y_2 are solutions, then $c_1 y_1 + c_2 y_2$ is also a solution for all $c_1, c_2 \in \mathbb{R}$. (That is, the solutions form a vector space.)
2. If y_1 and y_2 are solutions, then $(y_1' y_2 - y_1 y_2')e^G$ is constant on I. (*Proof:* one has to check that its derivative is 0.)
3. If y_1 and y_2 are solutions, then $y_1' y_2 - y_1 y_2'$ is either identically zero on I or nowhere zero on I. (*Proof:* by the previous statement.)
4. If y_1 and y_2 are solutions and there exists an interval $J \subset I$ on which $y_1 \neq 0$ and y_2 / y_1 is not constant, then $y_1' y_2 - y_1 y_2' \neq 0$ on I. (*Proof:* if it were 0, then the derivative of y_2 / y_1 would vanish, and so y_2 / y_1 would be constant on J.)
5. If y_1 and y_2 are solutions for which $y_1' y_2 - y_1 y_2' \neq 0$ on I, then every solution is of the form $c_1 y_1 + c_2 y_2$. (That is, the vector space of the solutions has dimension 2.) (*Proof:* We know that $(y_2 y' - y_2' y)e^G = c_1$, $(y_1 y' - y_1' y)e^G = c_2$, and $(y_2 y_1' - y_2' y_1)e^G = c_3$, where c_1, c_2, and c_3 are constants and $c_3 \neq 0$. Subtract y_2 times the second equality from y_1 times the first equality. We get that $y(y_2 y_1' - y_2' y_1)e^G = c_1 y_1 - c_2 y_2$. Taking into account the third equality, we then have that $y = (c_1/c_3)y_1 - (c_2/c_3)y_2$.)

If condition **5** holds, then we say that y_1 and y_2 form a **basis for the solution space.**

Second-Order Linear Homogeneous Differential Equations with Constant Coefficients. A special case of the differential equations discussed above is that g and h are constant:

$$y'' + ay' + by = 0.$$

Let the roots of the quadratic polynomial $x^2 + ax + b = 0$ be λ_1 and λ_2. It is easy to check that the following solutions y_1 and y_2 satisfy condition 5 above, that is, they form a basis for the solution space.

(i) If λ_1 and λ_2 are distinct and real, then $y_1 = e^{\lambda_1 x}$, $y_2 = e^{\lambda_2 x}$.
(ii) If $\lambda_1 = \lambda_2 = \lambda$, then $y_1 = e^{\lambda x}$, $y_2 = xe^{\lambda x}$.
(iii) If λ_1 and λ_2 are complex, that is, $\lambda_1 = \alpha + i\beta$, $\lambda_2 = \alpha - i\beta$ ($\alpha, \beta \in \mathbb{R}$), then $y_1 = e^{\alpha x} \cos \beta x$, $y_2 = e^{\alpha x} \sin \beta x$.

The Equations of Harmonic Oscillation. Consider a point P that is moving on a straight line, with a force acting on it toward the origin with magnitude proportional to its distance from the origin. Then the equation describing the movement of P is $my'' = cy$, where m is the mass of the point P, and $c < 0$. Let $-c/m = a^2$. Then $y'' + a^2 y = 0$, and so by the above, $y(t) = c_1 \cos at + c_2 \sin at$.

Let $C = \sqrt{c_1^2 + c_2^2}$. Then there exists a b such that $\sin b = c_1/C$, and $\cos b = c_2/C$. Thus $y(t) = C(\sin b \cos at + \cos b \sin at)$, that is,

$$y(t) = C \cdot \sin(at + b).$$

Now suppose that another force acts on the point P, stemming from resistance, which is proportional to the speed of P in the opposite direction. Then the equation describing the motion of the point is $my'' = cy + ky'$, where $c, k < 0$. Let $-c/m = a_0$ and $-k/m = a_1$. Then $y'' + a_1 y' + a_0 y = 0$, where $a_0, a_1 > 0$. Let the roots of the quadratic polynomial $x^2 + a_1 x + a_0 = 0$ be λ_1 and λ_2. Then with the notation $a_1^2 - 4a_0 = d$, we have that

$$\lambda_{1,2} = \frac{-a_1 \pm \sqrt{d}}{2}.$$

Clearly, $d < a_1^2$. If $d > 0$, then $\lambda_1, \lambda_2 < 0$, and every solution is of the form

$$y(t) = c_1 e^{\lambda_1 t} + c_2 e^{\lambda_2 t}.$$

If $d = 0$, then $\lambda_1 = \lambda_2 = \lambda < 0$, and every solution is of the form

$$y(t) = c_1 e^{\lambda t} + c_2 t e^{\lambda t}.$$

Finally, if $d < 0$, then $\lambda_{1,2} = \alpha \pm i\beta$, where $\alpha < 0$. Then every solution is of the form

$$y(t) = e^{\alpha t}(c_1 \cos \beta t + c_2 \sin \beta t) = C \cdot e^{\alpha t} \sin(\beta t + \delta).$$

We observe that in both cases, $y(t) \to 0$ as $t \to \infty$, that is, the vibration is "damped."

Forced oscillation of the point P occurs when aside from a force proportional to its distance from the origin in the opposite direction, a force of magnitude $M \sin \omega t$ also occurs (this force is dependent only on time; for the sake of simplicity, we ignore friction). The equation of this forced oscillation is $my'' = cy + M \sin \omega t$, that is,

$$y'' + a^2 y = \frac{M}{m} \sin \omega t. \tag{13.35}$$

To write down every solution, it is enough to find one solution y_0.

Generally, if g, h, and u are fixed functions, then the equation

$$y'' + gy' + hy = u \tag{13.36}$$

is called a **second-order inhomogeneous linear differential equation.** If we know one solution y_0 of equation (13.36), and in addition, a basis for the solution space y_1, y_2, then every solution of (13.36) is of the form $y_0 + c_1 y_1 + c_2 y_2$.

Returning to forced oscillation, let us find a solution y_0 of the form $c_1 \cos \omega t + c_2 \sin \omega t$. A simple calculation gives that for $\omega \neq a$, the choices $c_1 = 0$ and $c_2 = M/m(a^2 - \omega^2)$ work. Then every solution of (13.35) is of the form

$$y(t) = \frac{M}{m(a^2 - \omega^2)} \sin \omega t + C \sin(at + b).$$

If, however, $\omega = a$, then let us try to find a solution of the form $t(c_1 \cos \omega t + c_2 \sin \omega t)$. Then $c_1 = -M/2m$ and $c_2 = 0$, and thus all of the solutions are given by

$$y(t) = -\frac{M}{2m} t \cos at + C \sin(at + b).$$

We see that the movement is not bounded: at the times $t_k = 2k\pi/a$, the displacement is

$$|y(t_k)| \geq \frac{M\pi}{ma} k - C \to \infty \quad (k \to \infty).$$

This means that if the frequency of the forced oscillation $(\omega/2\pi)$ agrees with its natural frequency $(a/2\pi)$, then *resonance* occurs.

Differential Equations That Can Be Reduced to Simpler Equations.

1. The equation $y'' = f(y, y')$ is called an **autonomous** (not containing x) second-order differential equation. It can be reduced to a first-order differential equation if we first look for the function p for which $y'(x) = p(y(x))$. (Such a p exists if y is strictly monotone.) Then $y'' = p'(y)y' = p'(y) \cdot p$, so $p' \cdot p = f(y, p)$. If we can determine p from this, then we can obtain y from the separable equation $y' = p(y)$.
   A simple example: $y''y^2 = y'$; $p'py^2 = p$; $p' = y^{-2}$; $p = (-1/y) + c$; $y' = (-1/y) + c$, and y can already be determined from this.
2. The equation $y' = f(x + y)$ can be reduced to a separable equation if we introduce the function $z = x + y$. This satisfies $z' - 1 = f(z)$, which is separable.
   Example: $y' = (x + y)^2$; $z' - 1 = z^2$; $z' = z^2 + 1$; $\int dz/(z^2 + 1) = \int dx$; $\arctan z = x + c$; $z = \operatorname{tg}(x + c)$; $y = \operatorname{tg}(x + c) - x$.
3. The equation $y' = f(y/x)$ can be reduced to a separable equation if we introduce the function $z = y/x$. Then $y = zx$, $y' = z + xz' = f(z)$, and $z' = (f(z) - z)/x$, which is separable.

Exercises

13.14. Solve the following equations:

(a) $y' + 2xy = 0$, $\quad y' - 2y \operatorname{ctg} x = 0$, $\quad y' - xy = x^3$, $\quad y' + y = e^{-x}$,
   $y' - (y/x) = x^2 + 3x - 2$, $\quad y' \cos x + y \sin x = 1$;
(b) $y' = y^2$, $\quad xy' = y \log y$, $\quad y' = e^{y-x}$, $\quad xy' + y^2 = -1$,
   $y' - y^2 - 3y + 4 = 0$, $\quad y' = y^2 + 1$, $\quad y' - e^{x-y} - e^x = 0$, $\quad y' = e^x/(y(e^x + 1))$,
   $y'y\sqrt{1 - x^2} = -x\sqrt{1 - y^2}$;
(c) $y' = (y - x)^2$, $\quad y' = \sqrt{y - 2x}$, $\quad xy' = y - x \cdot \cos^2(y/x)$,
   $x \sin(y/x) - y \cos(y/x) + xy' \cos(y/x) = 0$;

(d) $y'' + y = 0$, $y'' - 5y' + 6y = 0$, $y'' - y' - 6y = 0$, $4y'' + 4y' + 37y = 0$;
(e) $y'' = (1 - x^2)^{-1/2}$, $y^3 y'' = 1$, $2xy'' + y' = 0$, $y'' = y^2$.

13.15. Show that the equation $y' = \sqrt[3]{y^2}$ with the initial condition $y(0) = 0$ has both the solutions $y(x) \equiv 0$ and $y(x) = (x/3)^3$.

13.16. We put water boiling at 100 degrees Celsius out to cool in 20-degree weather. The temperature of the water is 30 degrees at 10 a.m., and 25 degrees at 11 a.m. When did we place the water outside? (The speed of cooling is proportional to the difference between the temperature of the body and that of its environment.)

13.17. Draw the integral curves of the following equations (that is, graphs of their solutions): $y' = y/x$, $y' = x/y$, $y' = -y/x$, $y' = -x/y$, $y' = kx^a y^b$.

13.18. We call the solutions of the equation $y' = (a - by)y$ logistic curves. In the special case $a = b = 1$, we get the special logistic equation $y' = (1 - y)y$. Solve this equation. Sketch some integral curves of the equation.

13.19. Find the functions f for which the following statement holds: for an arbitrary point P on graph f, its distance from the origin is equal to the distance of the point Q from the origin, where Q is the intersection of the y-axis with the tangent line of the graph of the function at P.

13.20. Let $f: [0, \infty) \to (0, \infty)$ be differentiable, and suppose that for every $a > 0$, the area of the trapezoid bound by the lines with equations $x = 0$, $x = a$, $y = 0$, and the tangent line to the graph of f at the point $(a, f(a))$ is constant. What can f be?

13.21. A motorboat weighing 300 kg and moving at 16 m/s has its engine turned off. How far does it travel (and in how much time) if the water resistance (which is a force acting in the opposite distance of travel) is v^2 Newtons at a speed of v? What about if the resistance is v Newtons?

13.22. Determine how much time a rocket launched at 100 m/sec initial velocity needs to reach its maximum height if the air resistance creates a negative acceleration of $-v^2/10$. (Assume acceleration due to gravity to be 10 m/sec^2.)

13.23. A body is slowly submerged into a liquid. The resistance is proportional to its speed. Determine the graph of distance against time for the submerging body. (We assume that the initial speed of the body is zero.)

13.24. In a bowl of M liters we have a liquid that is a solution of $a\%$ concentration. We add a solution of $b\%$ concentration at a rate of m liter/sec such that an immediately mixed solution flows out at the same rate. What is the concentration of the solution in our bowl as a function of time?

13.25. What are the curves that intersect the following families of curves at right angles? (a) $y = x^2 + c$; (b) $y = e^x + c$; (c) $y = c \cdot e^x$; (d) $y = \cos x + c$; (e) $y = c \cdot \cos x$.

13.5 The Catenary

Let us suspend an infinitely thin rope that has weight and is homogeneous in the sense that the weight of every arc of the rope of length s is $c \cdot s$, where c is a constant. Our goal is the proof—borrowing some basic ideas from physics—of the fact that *the shape of this suspended homogeneous rope is similar to an arc of the graph of the function* $\operatorname{ch} x$. To prove this, we will need Theorems 10.79 and 10.81 about arc lengths of graphs of functions, as well as one more similar theorem.

Theorem 13.41. *Let f be differentiable on $[a,b]$, and suppose that f' is continuous. Then the graph of f is rectifiable. Let $s(x)$ denote the arc length of graph f over $[a,x]$, that is, let $s(x) = s(f;[a,x])$. Then s is differentiable, and $s'(x) = \sqrt{1 + (f'(x))^2}$ for all $x \in [a,b]$.*

Lemma 13.42. *Let f be differentiable on $[a,b]$, and suppose that*

$$A \le \sqrt{1 + (f'(x))^2} \le B$$

for all $x \in (a,b)$. Then the graph of f is rectifiable, and the bounds

$$A \cdot (b-a) \le s(f;[a,b]) \le B \cdot (b-a) \tag{13.37}$$

hold for the arc length $s(f;[a,b])$.

Proof. Let $a = x_0 < x_1 < \cdots < x_n = b$ be a partition F of the interval $[a,b]$, and let $p_i = (x_i, f(x_i))$ for every $i = 0, \ldots, n$. The distance between the points p_{i-1} and p_i is

$$
|p_i - p_{i-1}| = \sqrt{(x_i - x_{i-1})^2 + (f(x_i) - f(x_{i-1}))^2} =
$$
$$
= \sqrt{1 + \left(\frac{f(x_i) - f(x_{i-1})}{x_i - x_{i-1}} \right)^2} \cdot (x_i - x_{i-1}). \tag{13.38}
$$

By the mean value theorem, there exists a point $c_i \in (x_i, x_{i-1})$ such that

$$f'(c_i) = \frac{f(x_i) - f(x_{i-1})}{x_i - x_{i-1}}.$$

Since by assumption, $A \le \sqrt{1 + (f'(c_i))^2} \le B$, using (13.38), we get the bound

$$A \cdot (x_i - x_{i-1}) \le |p_i - p_{i-1}| \le B \cdot (x_i - x_{i-1}). \tag{13.39}$$

The length of the broken line h_F corresponding to this partition F is the sum of the numbers $|p_i - p_{i-1}|$. Thus if we add inequalities (13.39) for all $i = 1, \ldots, n$, then we get that

$$A \cdot (b-a) \le h_F \le B \cdot (b-a), \tag{13.40}$$

from which (13.37) is clear. $\square$

Proof (Theorem 13.41). By Weierstrass's theorem, f' is bounded on $[a,b]$, so by the previous lemma, we find that graph f is rectifiable. Let $c \in [a,b)$ and $\varepsilon > 0$ be given. Since the functions $\sqrt{1+x^2}$ and f' are continuous, the composition $\sqrt{1+(f'(x))^2}$ must also be continuous. Let D denote the number $\sqrt{1+(f'(c))^2}$. By the definition of continuity, we can find a positive δ such that

$$D - \varepsilon < \sqrt{1 + (f'(x))^2} < D + \varepsilon \qquad (13.41)$$

for all $x \in (c, c+\delta)$. Thus by Lemma 13.42,

$$(D-\varepsilon)\cdot(x-c) \le s(f;[c,x]) = s(x) - s(c) \le (D+\varepsilon)\cdot(x-c),$$

that is,

$$D - \varepsilon \le \frac{s(x) - s(c)}{x - c} \le D + \varepsilon,$$

if $x \in (c, c+\delta)$. Thus we see that $s'_+(c) = D = \sqrt{1+(f'(c))^2}$. We can show similarly that $s'_-(c) = \sqrt{1+(f'(c))^2}$ for all $c \in (a,b]$. $\square$

Consider now a suspended string, and let f be the function whose graph gives us the shape of the string. Graphically, it is clear that f is convex and differentiable.

Due to the tension inside the rope, every point has two equal but opposing forces acting on it whose directions agree with the tangent line. For arbitrary $u < v$, three outer forces act on the arc over $[u,v]$: the two forces pulling the endpoints outward, as well as gravity acting on the arc with magnitude $c \cdot (s(v) - s(u))$ (Figure 13.4). Since the rope is at rest, the sum of these three forces is zero. The gravitational force acts vertically downward, so the horizontal components of the tension forces must cancel each other out. This means that the horizontal components of the tension force agree at any two points. Thus there exists a constant $a > 0$ such that for all u, the horizontal component of tension at u is a, and so the vertical component is $a \cdot f'(u)$, since the direction of tension is parallel to the tangent line. It then follows that the vertical components of the three forces acting on $[u,v]$ are $-a \cdot f'(u)$, $a \cdot f'(v)$, and $-c \cdot (s(v) - s(u))$. Thus we get that

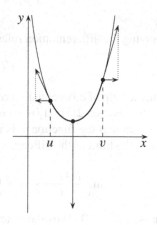

Fig. 13.4

$$a \cdot f'(v) - a \cdot f'(u) - c \cdot (s(v) - s(u)) = 0$$

for all $u < v$. If we divide by $v - u$ and let v tend to u, then we get that $a \cdot f''(u) - c \cdot s'(u) = 0$. Now by the previous theorem, $f''(u) = b \cdot \sqrt{1 + [f'(u)]^2}$, where $b = c/a$.

Since $(\operatorname{arsh} x)' = 1/\sqrt{1 + x^2}$, we have that $(\operatorname{arsh}(f'))' = f''/\sqrt{1 + (f')^2} = b$. Thus $\operatorname{arsh}(f'(x)) = bx + d$ with some constant d. From this, $f'(x) = \operatorname{sh}(bx + d)$, so

$$f(x) = \frac{1}{b} \cdot \operatorname{ch}(bx + d) + e,$$

where b, d, e are constants. This means that the graph of f is similar to an arc of the graph of the function $\operatorname{ch} x$, and this is what we wanted to prove.

13.6 Properties of Derivative Functions

While talking about indefinite integrals, we mentioned that continuity is a sufficient but not necessary condition for the existence of a primitive function. Now we introduce a function g that is not continuous at the point 0 but has a primitive function.

Example 13.43. Consider first the function

$$f(x) = \begin{cases} x^2 \sin(1/x), & \text{if } x \neq 0, \\ 0, & \text{if } x = 0 \end{cases}.$$

Applying the differentiation rules, we get that for $x \neq 0$,

$$f'(x) = 2x \sin \tfrac{1}{x} - \cos \tfrac{1}{x}.$$

With the help of differentiation rules the differentiability of f at 0 cannot be decided, and the function $2x \sin(1/x) - \cos(1/x)$ isn't even defined at 0. This, however, does not mean that the function f is not differentiable at 0. To decide whether it is, we look at the limit of the difference quotient:

$$\lim_{x \to 0} \frac{f(x) - f(0)}{x - 0} = \lim_{x \to 0} \frac{x^2 \sin(1/x) - 0}{x - 0} = \lim_{x \to 0} x \sin \frac{1}{x} \to 0,$$

that is, $f'(0) = 0$. Thus the function f is differentiable everywhere, and its derivative is

$$f'(x) = g(x) = \begin{cases} 2x \sin(1/x) - \cos(1/x), & \text{if } x \neq 0, \\ 0, & \text{if } x = 0. \end{cases} \tag{13.42}$$

Since $\lim_{x \to 0} 2x \sin(1/x) = 0$ and the limit of $\cos(1/x)$ does not exist at 0, the limit of $f' = g$ does not exist at 0 either.

Thus the function $f(x)$ is differentiable everywhere, but its derivative function is not continuous at the point 0. We can also state this by saying that *the function g is not continuous, but it has a primitive function.*

The following theorem states that derivative functions—even though they are not always continuous—possess the properties outlined in the Bolzano–Darboux theorem.

Theorem 13.44 (Darboux's Theorem). *If f is differentiable on $[a,b]$, then f' takes on every value between $f'_+(a)$ and $f'_-(b)$ there.*

Proof. Suppose first that $f'_+(a) < 0 < f'_-(b)$. We show that there exists a $c \in (a,b)$ such that $f'(c) = 0$. By the one-sided variant of Theorem 12.44, f is strictly locally decreasing from the right at a, and strictly locally increasing from the left at b (see Remark 12.45). It then follows that f takes on values in the open interval (a,b) that are smaller than $f(a)$ and $f(b)$.

Now by Weierstrass's theorem, f takes on its absolute minimum at some point $c \in [a,b]$. Since neither $f(a)$ nor $f(b)$ is the smallest value of f in $[a,b]$, we must have $c \in (a,b)$. This means that c is a local extremum, so $f'(c) = 0$ (Figure 13.5).

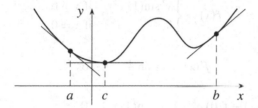

Fig. 13.5

Now let $f'_+(a) < d < f'_-(b)$. Then for the function $g(x) = f'(x) - d \cdot x$, we have $g'_+(a) = f'_+(a) - d < 0 < f'_-(b) - d = g'_-(b)$. Since the function g satisfies the conditions of the previous special case, there exists a $c \in (a,b)$ such that $g'(c) = f'(c) - d = 0$, that is, $f'(c) = d$.

The proof for the case $f'_+(a) > f'_-(b)$ is similar. □

Remarks 13.45. **1.** We say that the function f has the **Darboux property** on the interval I if whenever $a,b \in I$ and $a < b$, then f takes on every value between $f(a)$ and $f(b)$ on $[a,b]$. By the Bolzano–Darboux theorem, a continuous function on a closed and bounded interval always satisfies the Darboux property. Darboux's theorem states that this is also true for derivative functions.

2. Darboux's theorem can be stated thus: *a function has a primitive function only if it has the Darboux property.* This property, however, is not sufficient for a primitive function to exist. One can show that *every derivative function has a point of continuity* (but the proof of this is beyond the scope of this book). This means that for the existence of a primitive function, the function needs to be continuous at a point. It is easy to check that if a function $f : [0,1] \to \mathbb{R}$ takes on every value between 0 and 1 on every subinterval of $[0,1]$ (such as the function we constructed in Exercise 9.13), then f has the Darboux property, but it does not have any points of continuity. Such a function cannot have a primitive function by what we said above. The functions appearing in Exercise 13.33 also have the Darboux property but do not have primitive functions.

3. Continuity in an interval I is a sufficient but not necessary condition for the existence of a primitive function on I; the Darboux property on I is a necessary but not sufficient condition. Currently, we know of no simple condition based on the properties of the function itself that is both necessary and sufficient for a given function to have a primitive function. Many signs indicate that no such condition exists (see the following paper: C. Freiling, On the problem of characterizing derivatives, *Real Analysis Exchange* **23** (2) (1997–98), 805–812).

The question arises whether derivative functions have the same properties as continuous functions outlined in Theorems 10.52 and 10.55. In the following example, we show that the answer is negative. For boundedness, note that the function g constructed in Example 13.43 is bounded on the interval $[-1, 1]$; it is easy to see that for $x \in [-1, 1]$ we have $g(x) \le 2|x| + 1 \le 3$. First we construct a derivative function that is not bounded in $[-1, 1]$.

Example 13.46. Let

$$f(x) = \begin{cases} x^2 \sin(1/x^2), & \text{if } x \neq 0, \\ 0, & \text{if } x = 0. \end{cases}$$

Then for $x \neq 0$,

$$f'(x) = 2x \sin \tfrac{1}{x^2} - \tfrac{2}{x} \cdot \cos \tfrac{1}{x^2}.$$

Since

$$\lim_{x \to 0} \frac{f(x) - f(0)}{x - 0} = \lim_{x \to 0} \frac{x^2 \sin(1/x^2) - 0}{x - 0} = \lim_{x \to 0} x \sin \tfrac{1}{x^2} = 0,$$

f is also differentiable at 0. Now $f'(x)$ is not bounded on $[-1, 1]$, since

$$\lim_{k \to \infty} \left| f'\left(\tfrac{1}{\sqrt{2k\pi}} \right) \right| = \infty.$$

Example 13.47. We show that there exists a function f that is differentiable everywhere, whose derivative is bounded on the interval $I = [0, 1/10]$, but f' does not have a largest value on I. Let

$$f(x) = \begin{cases} (1 - 3x)x^2 \sin(1/x), & \text{if } x \neq 0, \\ 0, & \text{if } x = 0. \end{cases}$$

Then for $x \neq 0$,

$$f'(x) = (2x - 9x^2) \sin \tfrac{1}{x} - (1 - 3x) \cos \tfrac{1}{x}.$$

Since

$$\lim_{x \to 0} \frac{f(x) - f(0)}{x - 0} = \lim_{x \to 0} (1 - 3x)x \sin \tfrac{1}{x} = 0,$$

f is also differentiable at 0. We show that on the interval I, f' does not have a maximum. Indeed, for $x \in I$, $f'(x) \le (2x - 9x^2) + (1 - 3x) < 1 - x$, and

$$\lim_{k \to \infty} f'\left(\tfrac{1}{(2k+1)\pi} \right) = \lim_{k \to \infty} \left(1 - \tfrac{3}{(2k+1)\pi} \right) = 1.$$

It then follows that $\sup_{x \in I} f'(x) = 1$, but f' is less than 1 everywhere on the interval I. Thus f' does not have a largest value in I, even though it is clearly bounded.

Exercises

13.26. Does the function

$$\operatorname{sgn} x = \begin{cases} 1, & \text{if } x > 0 \\ 0, & \text{if } x = 0 \\ -1, & \text{if } x < 0 \end{cases}$$

have a primitive function?

13.27. Does the function $[x]$ (floor function) have a primitive function?

13.28. Does the function

$$g(x) = \begin{cases} 2x + 1, & \text{if } x \leq 1, \\ 3x, & \text{if } x > 1 \end{cases}$$

have a primitive function?

13.29. Prove that if f is differentiable on the interval I, then f' cannot have a removable discontinuity on I.

13.30. Prove that if f is differentiable on the interval I, then f' cannot have a discontinuity of the first type on I.

13.31. Let f be differentiable on an interval I. Prove that $f'(I)$ is also an interval.

13.32. Prove that the functions $h_1(x) = \sin(1/x)$, $h_1(0) = 0$ and $h_2(x) = \cos(1/x)$, $h_2(0) = 0$ have primitive functions. (In the proof, you can use the fact that every continuous function has a primitive function.) (S)

13.33. Show that the squares of the functions h_1 and h_2 appearing in the previous question do not have primitive functions. (S)

13.7 First Appendix: Proof of Theorem 13.20

Proof. Let $f \in C[0,1]$ and $\varepsilon > 0$ be fixed. We begin with the equality

$$f(x) - B_n(x; f) = \sum_{k=0}^{n} \left(f(x) - f\left(\frac{k}{n}\right) \right) \binom{n}{k} x^k (1-x)^{n-k}, \qquad (13.43)$$

which is clear from the definition of $B_n(x; f)$ and from

$$\sum_{k=0}^{n} \binom{n}{k} x^k (1-x)^{n-k} = (x + (1-x))^n = 1$$

from an application of the binomial theorem. By Theorem 10.52, f is bounded, and so there exists an M such that $|f(x)| \leq M$ for all $x \in [0,1]$. On the other hand, by Heine's theorem (Theorem 10.61), there exists a $\delta > 0$ such that if $x, y \in [0,1]$ and $|x - y| < \delta$, then $|f(x) - f(y)| < \varepsilon/2$.

Let $n \in \mathbb{N}^+$ and $x \in [0,1]$ be fixed. Let I and J denote the set of the indices $0 \leq k \leq n$ for which $|x - (k/n)| < \delta$ and $|x - (k/n)| \geq \delta$ respectively. The basic idea of the proof is that if $k \in I$, then by the definition of δ we have $|f(x) - f(k/n)| < \varepsilon/2$, and so on the right-hand side of (13.43), the sum of the terms with indices $k \in I$ is small. We show that the sum of the terms corresponding to indices $k \in J$ is also small for the function $f \equiv 1$, and that the given function f can only increase this sum to at most M times that. Let us see the details. If $|x - (k/n)| < \delta$, then $|f(x) - f(k/n)| < \varepsilon/2$, and so

$$\sum_{k \in I} \left| f(x) - f\left(\frac{k}{n}\right) \right| \cdot \binom{n}{k} x^k (1-x)^{n-k} <$$

$$< \sum_{k \in I} \frac{\varepsilon}{2} \cdot \binom{n}{k} x^k (1-x)^{n-k} \leq \tag{13.44}$$

$$\leq \frac{\varepsilon}{2} \cdot \sum_{k=0}^{n} \binom{n}{k} x^k (1-x)^{n-k} = \frac{\varepsilon}{2}.$$

To approximate the sum $\sum_{k \in J} \binom{n}{k} x^k (1-x)^{n-k}$ we will need the following identities:

$$\sum_{k=0}^{n} \frac{k}{n} \binom{n}{k} x^k (1-x)^{n-k} = x \tag{13.45}$$

and

$$\sum_{k=0}^{n} \frac{k^2}{n^2} \binom{n}{k} x^k (1-x)^{n-k} = x^2 + (x - x^2)/n \tag{13.46}$$

for all $x \in \mathbb{R}$ and $n = 1, 2, \ldots$. Both identities follow easily from the binomial theorem (see Exercises 13.11 and 13.12). For the bound, we actually want the identity

$$\sum_{k=0}^{n} \left(\frac{k}{n} - x \right)^2 \binom{n}{k} x^k (1-x)^{n-k} = \frac{x - x^2}{n}, \tag{13.47}$$

which we get by taking the difference of (13.46) and $2x$ times (13.45), and then adding it to the equality $\sum_{k=0}^{n} x^2 \cdot \binom{n}{k} x^k (1-x)^{n-k} = x^2$. If $k \in J$, then $|x - (k/n)| \geq \delta$, and so

$$\sum_{k \in J} \binom{n}{k} x^k (1-x)^{n-k} \le \frac{1}{\delta^2} \cdot \sum_{k \in J} \left(\frac{k}{n} - x\right)^2 \binom{n}{k} x^k (1-x)^{n-k} <$$

$$< \frac{1}{\delta^2} \cdot \sum_{k=0}^{n} \left(\frac{k}{n} - x\right)^2 \binom{n}{k} x^k (1-x)^{n-k} = \qquad (13.48)$$

$$= \frac{x - x^2}{n\delta^2} \le \frac{1}{4n\delta^2}.$$

Since $|f(x)| \le M$ for all x,

$$\sum_{k \in J} \left|f(x) - f\left(\frac{k}{n}\right)\right| \cdot \binom{n}{k} x^k (1-x)^{n-k} < \frac{M}{2n\delta^2}. \qquad (13.49)$$

Thus by (13.43), using inequalities (13.44) and (13.49), we get that

$$|f(x) - B_n(x; f)| < \frac{\varepsilon}{2} + \frac{M}{2n\delta^2}.$$

Since $x \in [0,1]$ was arbitrary, for $n > M/(\varepsilon\delta^2)$ we have $|f(x) - B_n(x; f)| < (\varepsilon/2) + (\varepsilon/2) = \varepsilon$ for all $x \in [0,1]$, which concludes the proof of the theorem.   □

13.8 Second Appendix: On the Definition of Trigonometric Functions Again

The definitions of trigonometric functions (Definition 11.20) are mostly based on geometry. These definitions directly use the concept of an angle and indirectly of arc length, and to inspect their important properties, we needed to introduce rotations and its properties. At the same time, the trigonometric functions are among the most basic functions in analysis, so the need might arise to separate our definitions from the geometric ideas. Since we defined arc length precisely and proved its important properties, for our purpose only rotations and their role cause trouble. The addition formulas followed from properties of rotations, and the differentiability of the functions $\sin x$ and $\cos x$ used the addition formulas. Our theory of trigonometric functions is not yet complete without a previous background in geometry. We will outline a construction below that avoids this shortcoming.

We will use the notation of Remark 11.21. Let $c(u) = \sqrt{1 - u^2}$, and let $S(u) = s(c; [u, 1])$ for all $u \in [-1, 1]$. (Then $S(u)$ is the length of the arc of the unit circle K that connects the points $(u, c(u))$ and $(1, 0)$.) We know that the function S is strictly monotone decreasing on $[-1, 1]$, and by Remark 11.21, the function $\cos x$ on the interval $[0, \pi]$ is none other than the inverse of S. We also saw that if we measure an arc of length $x + \pi$ onto the unit circle, then we arrive at a point antipodal to $(\cos x, \sin x)$, so its coordinates are $(-\cos x, -\sin x)$. Thus it is clear that the following definition is equivalent to Definition 11.20.

Definition 13.48. (i) Let $c(u) = \sqrt{1-u^2}$ and $S(u) = s\big(c;[u,1]\big)$ for all $u \in [-1,1]$. The inverse of the function S, which is defined on the interval $[0,\pi]$, is denoted by $\cos x$. We extend the function $\cos x$ to the whole real line in such a way that $\cos(x+\pi) = -\cos x$ holds for all real x.

(ii) We define the function $\sin x$ on the interval $[0,\pi]$ by the equation $\sin x = \sqrt{1-\cos^2 x}$. We extend the function $\sin x$ to the whole real line in such a way that $\sin(x+\pi) = -\sin x$ holds for all real x.

The definition above uses the fact that if for a function $f : [0,a] \to \mathbb{R}$ we have $f(a) = -f(0)$, then f can be extended to $\mathbb{R}$ uniquely so that $f(x+a) = -f(x)$ holds for all x. It is easy to check that

$$f(x+ka) = (-1)^k \cdot f(x) \quad (x \in [0,a],\ k \in \mathbb{Z})$$

gives such an extension, and that this is the only possible extension.

Now with the help of the above definition, we can again deduce identities (11.25)–(11.32), (11.38), and (11.39) as we did in Chapter 10. Although the argument we used there used the concept of reflection, it is easy to exclude them from the proofs.

The proof of inequality (11.42) is based on geometric facts in several points, including Lemma 10.82 (in which we used properties of convex n-gons) and the concept of similar triangles. We can prove Theorem 11.26 without these with the following replacements.

Theorem 13.49. *If $x \neq 0$, then*

$$1 - |x| \le \frac{\sin x}{x} \le 1. \tag{13.50}$$

Proof. Since the function $(\sin x)/x$ is even, it suffices to consider the case $x > 0$. The inequality $(\sin x)/x \le 1$ is clear by (11.38).

The inequality $1 - x \le (\sin x)/x$ is evident for $x \ge \pi/2$, since if $(\pi/2) \le x \le \pi$, then $(\sin x)/x \ge 0 \ge 1-x$, and if $x \ge \pi$, then $(\sin x)/x \ge (-1)/\pi \ge -1 > 1-x$.

Finally, suppose that $0 < x \le \pi/2$. Let $\cos x = u$ and $\sin x = v$. Then—again by the definition of $\cos x$ and $\sin x$—the arc length of the graph of the function $c(t) = \sqrt{1-t^2}$ over $[u,1]$ is x. Since the function c is monotone decreasing on the interval $[u,1]$, by Theorem 10.79,

$$x \le (1-u) + \big(c(u)-c(1)\big) = (1-u) + (v-0) = (1-\cos x) + \sin x.$$

Moreover, by (11.39), we have $1 - \cos x \le x^2$, so $x \le x^2 + \sin x$, that is, the first inequality of (13.50) holds in this case too. $\square$

The function $c(x)$ is concave on $[-1,1]$ and differentiable on $(-1,1)$, where its derivative is $-x/\sqrt{1-x^2}$. Since

$$S(u) = s\big(c;[u,1]\big) = \pi - s\big(c;[-1,u]\big)$$

for all $u \in [-1, 1]$, by Theorem 13.41 it follows that S is differentiable on $(-1, 1)$, and its derivative there is

$$-\sqrt{1 + \left(\frac{-x}{\sqrt{1-x^2}}\right)^2} = -\frac{1}{\sqrt{1-x^2}}.$$

By the differentiation rule for inverse functions, it follows that the function $\cos x$ is differentiable on $(0, \pi)$, and its derivative there is

$$\frac{1}{-1/\sqrt{1-\cos^2 x}} = -\sqrt{1-\cos^2 x} = -\sin x.$$

By identities (11.25), this holds for all points $x \neq k\pi$. Since $\cos x$ and $\sin x$ are both continuous, the equality $(\cos x)' = -\sin x$ holds at these points as well (see Exercise 12.75). Using this, it follows from identities (11.32) that $(\sin x)' = \cos x$ for all x.

Finally, we prove the addition formulas. Let $a \in \mathbb{R}$ be arbitrary. The function

$$A(x) = [\sin(a+x) - \sin a \cos x - \cos a \sin x]^2 +$$

$$+ [\cos(a+x) - \cos a \cos x + \sin a \sin x]^2$$

is everywhere differentiable, and its derivative is

$$2 \cdot [\sin(a+x) - \sin a \cos x - \cos a \sin x] \cdot [\cos(a+x) + \sin a \sin x - \cos a \cos x] +$$
$$2 \cdot [\cos(a+x) - \cos a \cos x + \sin a \sin x] \cdot [-\sin(a+x) + \cos a \sin x + \sin a \cos x] = 0.$$

Thus the function A is constant. Since $A(0) = 0$, we have $A(x) = 0$ for all x. This is possible only if for all x and a,

$$\sin(a+x) = \sin a \cos x + \cos a \sin x$$

and

$$\cos(a+x) = \cos a \cos x - \sin a \sin x,$$

which is exactly what we wanted to show.

Chapter 14
The Definite Integral

In the previous chapter of our book, we became acquainted with the concept of the indefinite integral: the collection of primitive functions of f was called the indefinite integral of f. Now we introduce a very different kind of concept that we also call integrals—definite integrals, to be precise. This concept, in contrast to that of the indefinite integral, assigns numbers to functions (and not a family of functions). In the next chapter, we will see that as the name *integral* that they share indicates, there is a strong connection between the two concepts of integrals.

14.1 Problems Leading to the Definition of the Definite Integral

The concept of the definite integral—much like that of differential quotients—arose as a generalization of ideas in mathematics, physics, and other branches of science. We give three examples of this.

Calculating the Area Under the Graph of a Function. Let f be a nonnegative bounded function on the interval $[a,b]$. We would like to find the area A of the region $S_f = \{(x,y) : x \in [a, b], \ 0 \le y \le f(x)\}$ under the graph of f. As we saw in Example 13.23, the area can be easily computed assuming that f is nonnegative, monotone increasing, continuous, and that we know a primitive function of f. If, however, those conditions do not hold, then we need to resort to a different method. Let us return to the argument Archimedes used (page 3) when he computed the area underneath the graph of x^2 over the interval $[0,1]$ by partitioning the interval with base points $x_i = i/n$ and then bounding the area from above and below with approximate areas (see Figure 14.1). Similar processes are often successful. The computation is sometimes easier if we do not use a uniform partitioning.

Example 14.1. Let $0 < a < b$ and $f(x) = 1/x$. In this case, a uniform partitioning will not help us as much as using the base points $x_i = a \cdot q^i$ $(i = 0, \ldots, n)$, where $q = \sqrt[n]{b/a}$.

© Springer New York 2015
M. Laczkovich, V.T. Sós, *Real Analysis*, Undergraduate Texts in Mathematics, DOI 10.1007/978-1-4939-2766-1_14

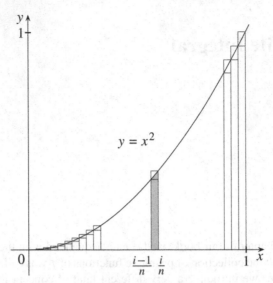

Fig. 14.1

The function is monotone decreasing, so the area over the interval $[x_{i-1}, x_i]$ is at least

$$\frac{1}{x_i} \cdot (x_i - x_{i-1}) = 1 - \frac{x_{i-1}}{x_i} = 1 - q^{-1} \tag{14.1}$$

and at most

$$\frac{1}{x_{i-1}} \cdot (x_i - x_{i-1}) = \frac{x_i}{x_{i-1}} - 1 = q - 1. \tag{14.2}$$

Thus we have the bounds

$$n \left(1 - \sqrt[n]{\tfrac{a}{b}} \right) = n(1 - q^{-1}) \le A \le n(q - 1) = n \left(\sqrt[n]{\tfrac{b}{a}} - 1 \right)$$

for the area A. Since

$$\lim_{n \to \infty} n \left(\sqrt[n]{\tfrac{b}{a}} - 1 \right) = \lim_{n \to \infty} \frac{(b/a)^{1/n} - 1}{1/n} = \left(\left(\tfrac{b}{a} \right)^x \right)'_{x=0} = \log(b/a)$$

and similarly

$$\lim_{n \to \infty} n \left(1 - \sqrt[n]{\tfrac{a}{b}} \right) = -\lim_{n \to \infty} \frac{(a/b)^{1/n} - 1}{1/n} =$$

$$= -\left(\left(\tfrac{a}{b} \right)^x \right)'_{x=0} = -\log(a/b) = \log(b/a),$$

we have $A = \log(b/a)$.

We could have recovered the same result using the method in Example 13.23. Even though the function $1/x$ is monotone decreasing and not increasing, from the point of view of this method, that is of no significance. Moreover, since $\log x$ is a primitive function of $1/x$, the area that we seek is $A = \log b - \log a = \log(b/a)$.

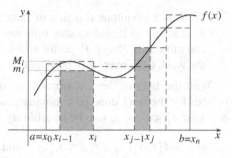

Fig. 14.2

Let us note, however, that the method we used above—bounding the area from above and below using a partition of the interval—did not use the fact that $1/x$ has a primitive function, so we can expect it to work in many more cases than the method seen in Example 13.23.

We argue as follows for the general case. Let $a = x_0 < x_1 < \cdots < x_n = b$ be an arbitrary partition of the interval $[a,b]$. Let

$$m_i = \inf\{f(x) : x_{i-1} \le x \le x_i\} \quad \text{and} \quad M_i = \sup\{f(x) : x_{i-1} \le x \le x_i\}.$$

If S_i denotes the region under the graph over the interval $[x_{i-1}, x_i]$, and A_i the area of S_i, then $m_i(x_i - x_{i-1}) \le A_i \le M_i(x_i - x_{i-1})$, since $m_i(x_i - x_{i-1})$ is the area of the tallest rectangle that can be inscribed into S_i, while $M_i(x_i - x_{i-1})$ is the area of the shortest rectangle that covers S_i. Thus

$$\sum_{i=1}^{n} m_i(x_i - x_{i-1}) \le A \le \sum_{i=1}^{n} M_i(x_i - x_{i-1}).$$

(See Figure 14.2.) Here the left-hand side is the sum of the areas of the inscribed rectangles, while the right-hand side is the sum of the areas of the circumscribed rectangles. For an arbitrary partition of the interval $[a,b]$, we get a lower bound and an upper bound for the area A that we are seeking. If we are lucky (as in the example above), then only one number will satisfy these inequalities, and that will be the value of the area.

The Definition and Computation of Work. Suppose that a point moves along a line with constant speed. If a force acts on the point in the same direction as its motion with constant absolute value P, then the work done by that force is $P \cdot s$, where s is the length of the path traversed.

The question is, how can we generalize the definition and computation of work for the more general case in which the magnitude of the force varies? For the sake of simplicity, we will consider only the case in which the point is moving along a straight line and the force is in the direction of motion. We suppose that work has the following properties:

1. Work is additive: if the point moves from $a = x_0$ to x_1, then from x_1 to x_2, and so on until it moves from x_{n-1} to $x_n = b$, then the work done while moving from a to b is equal to the sum of the work done on each segment $[x_{i-1}, x_i]$.

2. Work is a monotone function of force: if the point moving in $[a,b]$ has a force $P(x)$ acting on it, and another time has $P^*(x)$ acting on it over the same interval, and moreover, $P(x) \le P^*(x)$ for all $x \in [a,b]$, then $L \le L^*$, where L and L^* denote the work done over $[a,b]$ by P and P^* respectively.

With the help of these reasonable assumptions, we can give lower and upper bounds for the work done on a moving point by a force of magnitude $P(x)$ at x. Let $a = x_0 < x_1 < \cdots < x_n = b$ be an arbitrary partition of the interval $[a,b]$. Let

$$m_i = \inf\{P(x): x_{i-1} \le x \le x_i\} \quad \text{and} \quad M_i = \sup\{P(x): x_{i-1} \le x \le x_i\}.$$

If L_i denotes the work done on $[x_{i-1}, x_i]$ by P, then by the monotonicity of work, we have $m_i(x_i - x_{i-1}) \le L_i \le M_i(x_i - x_{i-1})$. Thus the work done on the whole interval $[a,b]$, denoted by L, must satisfy the inequality

$$\sum_{i=1}^{n} m_i(x_i - x_{i-1}) \le L \le \sum_{i=1}^{n} M_i(x_i - x_{i-1})$$

by the additivity of work. For an arbitrary partition of the interval $[a,b]$ we get lower and upper bounds for the work L we are looking for. If we are lucky, only one number satisfies these inequalities, and that will be the value of the work done.

If, for example, $0 < a < b$ and at x a force of magnitude $1/x$ acts on the point, then by the computations done in Example 14.1, we know that the work done by the force is equal to $\log(b/a)$.

Since the method for finding the amount of work uses the same computation as finding the area underneath the graph of a function, we can conclude that the magnitude of the work done by a force P agrees with the area underneath the graph of P. We might, in fact, already know the area beneath the graph, and so the magnitude of work follows. Consider a spring that when stretched to length x exerts a force $c \cdot x$. If we stretch the spring from length a to b, then by what was said above, the work needed for this is the same as the area underneath the graph of $c \cdot x$ over the interval $[a,b]$. This region is a trapezoid with bases ca and cb, and height $b - a$. Thus the amount of work done is $(ca + cb)/2 \cdot (b - a) = c(b^2 - a^2)/2$.

Determining Force Due to Pressure. Consider a rectangular container filled with liquid. How much force is the liquid exerting on the sides of the container? To answer this question, let us use the fact from physics that pressure at every point (that is, the force acting on a unit surface containing that point) of the liquid is independent of the direction. If the pressure were equal throughout the fluid, then it would immediately follow that the force exerted on a side with area A would be equal to A times the constant value of the pressure. Pressure, however, is not constant, but increases with depth. We can overcome this difficulty just as we did in computing work. We suppose that the force due to pressure has the following properties:

1. Pressure depends only on depth.
2. The force due to pressure is additive: if we partition the surface, then the force acting on the whole surface is equal to the sum of the forces acting on the pieces.

3. Force due to pressure is a monotone function of pressure: if we increase pressure on a surface at every point, then the force due to pressure will also increase.

Using these three properties, we can get upper and lower bounds for the force with which the liquid is pushing at the sides of the container. Let the height of the container be b, and let $p(x)$ denote the pressure inside the liquid at depth x. Consider a side whose horizontal length is c.

Let $0 = x_0 < x_1 < \cdots < x_n = b$ be a partition of the interval $[0, b]$, and let

$$m_i = \inf\{p(x) : x_{i-1} \leq x \leq x_i\} \qquad \text{and} \qquad M_i = \sup\{p(x) : x_{i-1} \leq x \leq x_i\}.$$

If F_i denotes the magnitude of the force the liquid exerts on the side of the container between depths x_{i-1} and x_i, then by the monotonicity of such force, $m_i \cdot c \cdot (x_i - x_{i-1}) \leq F_i \leq M_i \cdot c \cdot (x_i - x_{i-1})$. Thus the total value of the force due to pressure must satisfy the inequalities

$$c \cdot \sum_{i=1}^{n} m_i(x_i - x_{i-1}) \leq F \leq c \cdot \sum_{i=1}^{n} M_i(x_i - x_{i-1})$$

due to its additivity. For an arbitrary partition of the interval $[0, b]$, we get lower and upper bounds for the force F due to pressure. If we are lucky, only one number satisfies these inequalities, and that will be the value of the work done.

It is clear that the magnitude of the force due to pressure is the same as c times the area underneath the graph of p. In the simplest case, assuming that the liquid is homogeneous, pressure is proportional to depth, that is, $p(x) = \rho x$ with some constant $\rho > 0$. The magnitude of this force is thus the area underneath the graph of ρx times c. This region is a right triangle whose sides have lengths b and ρb. Thus the force we seek has magnitude $c \cdot (b \cdot \rho b / 2) = c \rho b^2 / 2$.

14.2 The Definition of the Definite Integral

The above examples motivate the following definition of the definite integral (now not only for nonnegative functions). We introduce some notation. We call a **partition** of the interval $[a, b]$ a sequence $F = (x_0, \ldots, x_n)$ such that $a = x_0 < \cdots < x_n = b$. We call the points x_i the **base points** of the partition.

Let $f : [a, b] \to \mathbb{R}$ be a bounded function, and let

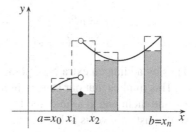

Fig. 14.3

$$m_i = \inf\{f(x) : x_{i-1} \leq x \leq x_i\} \quad \text{and} \quad M_i = \sup\{f(x) : x_{i-1} \leq x \leq x_i\}$$

for all $i = 1, \ldots, n$ (Figure 14.3). The sums

$$s_F(f) = \sum_{i=1}^{n} m_i(x_i - x_{i-1}) \quad \text{and} \quad S_F(f) = \sum_{i=1}^{n} M_i(x_i - x_{i-1})$$

are called the **lower** and **upper sums** of the function f with partition F. If the function f is fixed and it is clear to which function the sums correspond to, sometimes the shorter notation s_F and S_F is used instead of $s_F(f)$ and $S_F(f)$.

As the examples have hinted, the important cases (or functions f) for us are those in which only one number lies between every lower and every upper sum. We will call a function integrable if this condition holds. However, before we turn to the formal definition, let us inspect whether there always exists a number (one or more) that falls between every lower and upper sum. We show that for every bounded function f, there exists such a number.

Definition 14.2. We say that a partition F' is a *refinement* of the partition F if every base point of F is a base point of F'.

Lemma 14.3. *Let f be a bounded function on the interval $[a,b]$, and let F' be a refinement of the partition F. Then $s_F \leq s_{F'}$ and $S_F \geq S_{F'}$. That is, in a refinement of a partition, the lower sum cannot decrease, and the upper sum cannot increase.*

Proof. Consider first the simplest case, that F' can be obtained by adding one new base point to F (Figure 14.4). Let the base points of F be $a = x_0 < \cdots < x_n = b$, and let $x_{k-1} < x' < x_k$. If

$$m'_k = \inf\{f(x) : x_{k-1} \leq x \leq x'\}$$
$$\text{and} \quad m''_k = \inf\{f(x) : x' \leq x \leq x_k\},$$

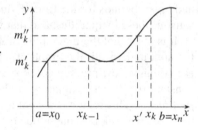

Fig. 14.4

then clearly $m'_k \geq m_k$ and $m''_k \geq m_k$, since the lower bound of a set cannot be larger than a lower bound of a subset of that set.

Since the intervals $[x_{i-1}, x_i]$ not containing x' add the same amount to s_F and $s_{F'}$, we have

$$s_{F'} - s_F = m'_k(x' - x_{k-1}) + m''_k(x_k - x') - m_k(x_k - x_{k-1}) \geq$$
$$\geq m_k(x' - x_{k-1}) + m_k(x_k - x') - m_k(x_k - x_{k-1}) = 0. \qquad (14.3)$$

Thus by adding an extra base point, the lower sum cannot decrease.

This implies that our statement holds for adding several new base points, since if we add them one at a time, the lower sum increases or stays the same at every step, so the last sum $s_{F'}$ is at least as big as s_F.

The statement regarding the upper sum can be proved in the same way.   □

Lemma 14.4. *If F_1 and F_2 are two arbitrary partitions of the set $[a,b]$, then $s_{F_1} \leq S_{F_2}$. That is, for a given function, the lower sum for a partition is less than or equal to the upper sum of any (other) partition.*

Proof. Let F be the union of the partitions F_1 and F_2, that is, let the base points of F be all those points that are base points of F_1 or F_2. Then F is a refinement of both F_1 and F_2. Thus—considering that $s_F \leq S_F$ (since $m_i \leq M_i$ for all i)—Lemma 14.3 gives

$$s_{F_1} \leq s_F \leq S_F \leq S_{F_2}.$$

$\square$

Let $\mathscr{F}$ denote the set of all partitions of the interval $[a,b]$. By the lemma above, for every partition $F_2 \in \mathscr{F}$, the upper sum S_{F_2} is an upper bound for the set $\{s_F : F \in \mathscr{F}\}$. Thus the least upper bound of this set, that is, the value $\sup_{F \in \mathscr{F}} s_F$, is less than or equal to S_{F_2} for every $F_2 \in \mathscr{F}$. In other words, $\sup_{F \in \mathscr{F}} s_F$ is a lower bound of the set $\{S_F : F \in \mathscr{F}\}$, and thus

$$\sup_{F \in \mathscr{F}} s_F \leq \inf_{F \in \mathscr{F}} S_F. \qquad (14.4)$$

Moreover, it is clear that for a real number I, the inequality $s_F \leq I \leq S_F$ holds for every partition F if and only if

$$\sup_{F \in \mathscr{F}} s_F \leq I \leq \inf_{F \in \mathscr{F}} S_F. \qquad (14.5)$$

Thus we have shown that for every bounded function f, there exists a number that falls between every lower and every upper sum. It is also clear that there exists only one such number if and only if $\sup_{F \in \mathscr{F}} s_F = \inf_{F \in \mathscr{F}} S_F$. We accept this condition as a definition.

Definition 14.5. Let $f \colon [a,b] \to \mathbb{R}$ be a bounded function. We call f *Riemann integrable* on the interval $[a,b]$ (or just integrable, for short) if $\sup_{F \in \mathscr{F}} s_F = \inf_{F \in \mathscr{F}} S_F$. We then call the number $\sup_{F \in \mathscr{F}} s_F = \inf_{F \in \mathscr{F}} S_F$ the *definite integral* of f over the interval $[a,b]$, and denote it by $\int_a^b f(x)\, dx$.

It will be useful to have notation for the values $\sup_{F \in \mathscr{F}} s_F$ and $\inf_{F \in \mathscr{F}} S_F$.

Definition 14.6. Let $f \colon [a,b] \to \mathbb{R}$ be a bounded function. We call the value $\sup_{F \in \mathscr{F}} s_F$ the *lower integral* of f, and denote it by $\underline{\int_a^b} f(x)\, dx$. We call the value $\inf_{F \in \mathscr{F}} S_F$ the *upper integral* of f, and denote it by $\overline{\int_a^b} f(x)\, dx$.

With these new definitions, (14.4) and (14.5) can be summarized as follows.

Theorem 14.7.

(i) *For an arbitrary bounded function* $f \colon [a,b] \to \mathbb{R}$, *we have* $\underline{\int_a^b} f(x)\, dx \leq \overline{\int_a^b} f(x)\, dx$.

(ii) *A real number I satisfies the inequalities $s_F \leq I \leq S_F$ for every partition F if and only if* $\underline{\int_a^b} f(x)\, dx \leq I \leq \overline{\int_a^b} f(x)\, dx$.

(iii) f is integrable if and only if $\underline{\int}_a^b f(x)\,dx = \overline{\int}_a^b f(x)\,dx$, and then $\int_a^b f(x)\,dx = \overline{\int}_a^b f(x)\,dx = \underline{\int}_a^b f(x)\,dx$.

Examples 14.8. **1.** If $f \equiv c$ is constant on the interval $[a,b]$, then f is integrable on $[a,b]$, and

$$\int_a^b f\,dx = c(b-a). \tag{14.6}$$

Clearly, for every partition, we have $m_i = M_i = c$ for all i, and so

$$s_F = S_F = c \cdot \sum_{i=1}^n (x_i - x_{i-1}) = c(b-a)$$

makes (14.6) clear.

2. We show that the function $f(x) = x^2$ is integrable on $[0,1]$, and its integral is $1/3$. Let F_n denote the uniform partition of $[0,1]$ into n equal intervals. Then

$$s_{F_n} = \frac{1}{n} \cdot \left(0 + \left(\frac{1}{n}\right)^2 + \cdots + \left(\frac{n-1}{n}\right)^2 \right) = \frac{(n-1) \cdot n \cdot (2n-1)}{6n^3} =$$

$$= \frac{1}{3} \cdot \left(1 - \frac{1}{n}\right)\left(1 - \frac{2}{n}\right),$$

and similarly,

$$S_{F_n} = \frac{1}{n} \cdot \left(\left(\frac{1}{n}\right)^2 + \cdots + \left(\frac{n}{n}\right)^2 \right) = \frac{1}{3} \cdot \left(1 + \frac{1}{n}\right)\left(1 + \frac{2}{n}\right).$$

(See Figure 14.1.) Since $\lim_{n\to\infty} s_{F_n} = 1/3$, we have $\sup_{F \in \mathscr{F}} s_F \geq 1/3$. On the other hand, $\lim_{n\to\infty} S_{F_n} = 1/3$, so $\inf_{F \in \mathscr{F}} S_F \leq 1/3$. Considering that $\sup_{F \in \mathscr{F}} s_F \leq \inf_{F \in \mathscr{F}} S_F$, necessarily $\sup_{F \in \mathscr{F}} s_F = \inf_{F \in \mathscr{F}} S_F = 1/3$, which means that x^2 is integrable on $[0,1]$, and $\int_0^1 x^2\,dx = 1/3$.

We will soon see that most functions in our applications (and so every continuous or monotone function) are integrable. It is important to remind ourselves, however, that there are very simple bounded functions that are not integrable.

Example 14.9. Let f be the Dirichlet function:

$$f(x) = \begin{cases} 0, & \text{if } x \text{ is irrational,} \\ 1, & \text{if } x \text{ is rational.} \end{cases}$$

Let $a < b$ be arbitrary. Since both the set of rational and the set of irrational numbers are everywhere dense in $\mathbb{R}$ (Theorems 3.2 and 3.12), for every partition $F : a = x_0 < x_1 < \cdots < x_n = b$ of the interval $[a,b]$ and all $i = 1,\ldots,n$, we have $m_i = 0$ and $M_i = 1$. It follows from this that

$$s_F = \sum_{i=1}^{n} 0 \cdot (x_i - x_{i-1}) = 0 \qquad \text{and} \qquad S_F = \sum_{i=1}^{n} 1 \cdot (x_i - x_{i-1}) = b - a$$

for every partition. Thus $\underline{\int_a^b} f(x)\,dx = 0$, while $\overline{\int_a^b} f(x)\,dx = b - a$, so the Dirichlet function is not integrable in any interval.

Remarks 14.10. **1.** It is important to note that the m_i and M_i appearing in the lower and upper sums are the infimum and supremum of the set $\{f(x) : x_{i-1} \le x \le x_i\}$, and not its minimum and maximum. We know that a function does not always have a largest or smallest value (even if it is bounded); one of the simplest examples is given by the fractional part function (Example 9.7.8), which does not have a largest value on the interval $[0,1]$. Every bounded and nonempty set does, however, have an infimum and supremum, and this makes it possible to define the lower and upper sums, and through these, the upper and lower integrals for every bounded function. Clearly, if we know that f has a smallest and greatest value on the interval $[x_{i-1}, x_i]$ (if, for example, f is continuous there), then m_i and M_i agree with the minimum and maximum of the set $\{f(x) : x_{i-1} \le x \le x_i\}$.

2. We also emphasize the point that we have defined the concepts of upper and lower sums, upper and lower integrals, and the integral itself *only for bounded functions.* For an arbitrary function $f : [a,b] \to \mathbb{R}$ and partition F, the value $m_i = \inf\{f(x) : x_{i-1} < x \le x_i\}$ is finite or $-\infty$ depending on whether the function f is bounded from below on the interval $[x_{i-1}, x_i]$. If f is not bounded from below on $[a,b]$, then $m_i = -\infty$ for at least one i, and so the lower sum s_F could only be defined to be $s_F = -\infty$. Thus—if we considered unbounded functions in the first place—integrals of functions not bounded from below could only be $-\infty$. A similar statement holds for functions not bounded from above. Thus it is reasonable to study only integrals of bounded functions (for now).

We mention that there are more general forms of integration (for example the improper integral and the Lebesgue integral) that allow us to integrate some unbounded functions as well (see Chapter 19.)

3. It is clear that the integrability of a function f over the interval $[a,b]$ depends on the relationship of the sets $\{s_F : F \in \mathscr{F}\}$ and $\{S_F : F \in \mathscr{F}\}$ to each other. By Theorem 14.7, these sets can be related in two possible ways:

(a) $\sup\{s_F : F \in \mathscr{F}\} = \inf\{S_F : F \in \mathscr{F}\}$.
It is in this case that we call f integrable over $[a,b]$.
(b) $\sup\{s_F : F \in \mathscr{F}\} < \inf\{S_F : F \in \mathscr{F}\}$.
This is the case in which f is not integrable over $[a,b]$. (By Theorem 14.7, the case $\sup\{s_F : F \in \mathscr{F}\} > \inf\{S_F : F \in \mathscr{F}\}$ is impossible.)

4. If the function f is nonnegative, then—as we have already seen—the upper and lower sums correspond to the sum of the areas of the outer and inner rectangles corresponding to the set $B_f = \{(x,y) : a \le x \le b,\ 0 \le y \le f(x)\}$ (see Figure 14.2). We also saw that if only one number falls between every upper and lower sum, then the area of the set B_f must be this number. In other words, if $f \ge 0$ is integrable on $[a,b]$, then the area of the set B_f is $\int_a^b f(x)\,dx$.

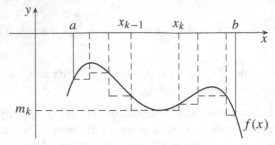

Fig. 14.5

If in $[a,b]$ we have $f(x) \leq 0$ and $B_f = \{(x,y) : a \leq x \leq b, \ f(x) \leq y \leq 0\}$, then $m_i(x_i - x_{i-1})$ is -1 times the area of the rectangle of base $[x_{i-1}, x_i]$ and height $|m_i|$. Thus in this case, $|s_F| = -s_F$ gives the total area of a collection of rectangles that contain B_f; and similarly, $|S_F| = -S_F$ will be the total area of rectangles contained in B_f. Thus we can say that if in $[a,b]$ we have $f(x) \leq 0$, then $\int_a^b f(x)\,dx$ is -1 times the area of the region B_f (Figure 14.5).

5. The question might arise that if $f \geq 0$ is not integrable over $[a,b]$, then how can we compute the area of the set $B_f = \{(x,y) : a \leq x \leq b, \ 0 \leq y \leq f(x)\}$? The question stems from something deeper: can we compute the area of any set? In fact, does every set have an area? In talking about sets previously (concretely, the set $B_f = \{(x,y) : a \leq x \leq b, \ 0 \leq y \leq f(x)\}$), we simply assumed that these sets have a well-defined computable area. This might be true for polygons, but does it still hold for every set? And if it is not true, then how can we say that the area of the set B_f is $\int_a^b f(x)\,dx$?

No matter what we mean by the area of a set, we can agree that

(a) the area of a rectangle with sides parallel to the axes is the product of the lengths of its sides, that is, the area of $[a,b] \times [c,d]$ is $(b-a) \cdot (d-c)$;

(b) if we break a set up into the union finitely many nonoverlapping rectangles, then the area of that set is the sum of the areas of the rectangles; and

(c) if $A \subset B$, then the area of A cannot be larger than the area of B.

When we concluded that the area of the set B_f is $\int_a^b f(x)\,dx$, we used only these properties of area. Thus, if we want to be precise, then we can say only that *if the set B_f has area, then it must be equal to* $\int_a^b f(x)\,dx$. We will deal with the concept of area and its computation in more detail in Chapter 16. We will show that among planar sets, we can easily and naturally identify which sets we can give an area to, and that these areas are well defined (assuming properties (a), (b), and (c)). As for the sets B_f above, we will show that B_f has area if and only if f is integrable on $[a,b]$, and—as we already know—the area of B_f is $\int_a^b f(x)\,dx$.

Exercises

14.1. Determine the upper and lower integrals of the following functions defined on the interval $[0,1]$. Decide whether they are integrable there, and if they are, find their integrals:

(a) $f(x) = x$;
(b) $f(x) = 0$ $(0 \le x < 1/2)$, $f(x) = 1$ $(1/2 \le x \le 1)$;
(c) $f(0) = 1$, $f(x) = 0$ $(0 < x \le 1)$;
(d) $f(1/n) = 1/n$ for all $n \in \mathbb{N}^+$ and $f(x) = 0$ otherwise;
(e) $f(1/n) = 1$ for all $n \in \mathbb{N}^+$ and $f(x) = 0$ otherwise;
(f) $f(x) = x$ if $x \in [0,1] \cap \mathbb{Q}$ and $f(x) = 0$ if $x \in [0,1] \setminus \mathbb{Q}$.

14.2. Let f be integrable on $[-a,a]$. Prove that

(a) if f is an even function, then $\int_{-a}^{a} f(x)\,dx = 2\int_0^a f\,dx$, and
(b) if f is an odd function, then $\int_{-a}^{a} f\,dx = 0$.

14.3. Prove that if f and g are integrable on $[a,b]$ and they agree on $[a,b] \cap \mathbb{Q}$, then $\int_a^b f\,dx = \int_a^b g\,dx$. (H)

14.4. Let f be integrable on $[a,b]$. Is it true that if $g\colon [a,b] \to \mathbb{R}$ is bounded and $f(x) = g(x)$ for all $x \in [a,b] \cap \mathbb{Q}$, then g is also integrable on $[a,b]$?

14.5. Let $f\colon [a,b] \to \mathbb{R}$ be a function such that for every $\varepsilon > 0$, there exist integrable functions $g,h\colon [a,b] \to \mathbb{R}$ such that $g \le f \le h$ and $\int_a^b h(x)\,dx - \int_a^b g(x)\,dx < \varepsilon$ are satisfied. Prove that f is integrable.

14.6. Let $f\colon [a,b] \to \mathbb{R}$ be bounded. Prove that the set of upper sums of f is an interval. (∗ H S)

14.3 Necessary and Sufficient Conditions for Integrability

Unlike differentiability, integrability can be stated in many different ways, and choosing which equivalent definition we will use always depends on the current problem we are facing. We will now focus on necessary and sufficient conditions that more or less follow from the definition. Out of these necessary and sufficient conditions, we could have accepted any of them as the definition of the definite integral.

To shorten the formulas, we will sometimes write $\int f\,dx$ instead of $\int f(x)\,dx$.

Theorem 14.11. *A bounded function* $f\colon [a,b] \to \mathbb{R}$ *is integrable with integral* I *if and only if for arbitrary* $\varepsilon > 0$, *there exists a partition* F *such that*

$$I - \varepsilon < s_F \le S_F < I + \varepsilon. \tag{14.7}$$

Proof. Suppose that f is integrable on $[a,b]$, and its integral is I. Let $\varepsilon > 0$ be given. Since $I = \underline{\int_a^b} f\,dx = \sup_{F \in \mathscr{F}} s_F$, there exists a partition F_1 such that $s_{F_1} > I - \varepsilon$. Similarly, by $I = \overline{\int_a^b} f\,dx = \inf_{F \in \mathscr{F}} S_F$, we can find a partition F_2 such that $S_{F_2} < I + \varepsilon$.

Let F be the union of partitions F_1 and F_2. By Lemma 14.3,

$$I - \varepsilon < s_{F_1} \le s_F \le S_F \le S_{F_2} < I + \varepsilon,$$

so (14.7) holds.

Now suppose that for all $\varepsilon > 0$, there exists a partition F for which (14.7) holds. Since

$$s_F \le \underline{\int_a^b} f\,dx \le \overline{\int_a^b} f\,dx \le S_F \qquad (14.8)$$

for all partitions F, if we choose F such that it satisfies (14.7), then we get that

$$\left| \underline{\int_a^b} f\,dx - I \right| < \varepsilon \qquad \text{and} \qquad \left| \overline{\int_a^b} f\,dx - I \right| < \varepsilon.$$

Since this holds for all $\varepsilon > 0$,

$$\underline{\int_a^b} f\,dx = \overline{\int_a^b} f\,dx = I,$$

so f is integrable on $[a,b]$ with integral I. □

Theorem 14.12. *A bounded function* $f \colon [a,b] \to \mathbb{R}$ *is integrable if and only if for every* $\varepsilon > 0$, *there exists a partition* F *such that* $S_F - s_F < \varepsilon$.

Proof. The "only if" part of the statement is clear from the previous theorem. If the function is not integrable, then $\underline{\int_a^b} f\,dx < \overline{\int_a^b} f\,dx$, and so by (14.8),

$$S_F - s_F \ge \overline{\int_a^b} f\,dx - \underline{\int_a^b} f\,dx > 0$$

for every partition F. This proves the "if" part of the statement. □

By the previous theorem, the integrability of a function depends on whether the value

$$S_F - s_F = \sum_{i=1}^{n} (M_i - m_i)(x_i - x_{i-1})$$

can be made arbitrarily small. It is worth giving a name to the difference $M_i - m_i$ appearing here.

Definition 14.13. Given a bounded function $f \colon [a,b] \to \mathbb{R}$, we call the value

$$\omega(f; [a,b]) = \sup R(f) - \inf R(f),$$

the *oscillation* of the function f, where $R(f)$ denotes the image of f, that is, $R(f) = \{f(x) : a \le x \le b\}$.

It is clear that $\omega(f;[a,b])$ is the length of the shortest interval which covers the image $R(f)$. It is also easy to see that

$$\omega(f;[a,b]) = \sup\{|f(x) - f(y)| : x,y \in [a,b]\} \tag{14.9}$$

(see Exercise 14.7).

Definition 14.14. For a bounded function $f : [a,b] \to \mathbb{R}$, we call the sum

$$\Omega_F(f) = \sum_{i=1}^{n} \omega(f;[x_{i-1},x_i])(x_i - x_{i-1}) = \sum_{i=1}^{n}(M_i - m_i)(x_i - x_{i-1}) = S_F - s_F$$

corresponding to the partition $F : a = x_0 < x_1 < \cdots < x_n = b$ the *oscillatory sum* corresponding to F. (See Figure 14.6.)

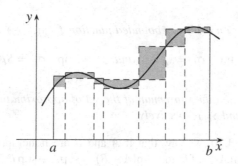

Fig. 14.6

Using this notation, we can rephrase Theorem 14.12:

Theorem 14.15. *A bounded function $f : [a,b] \to \mathbb{R}$ is integrable if and only if for arbitrary $\varepsilon > 0$, there exists a partition F such that $\Omega_F < \varepsilon$.*

Theorems 14.12 and 14.15 remind us of Cauchy's convergence criterion; with their help, we can determine the integrability of a function without having to determine the value of the integral.

Example 14.16. With the help of Theorem 14.15, we can easily see that if f is integrable over $[a,b]$, then so is $|f|$. For arbitrary $x,y \in [a,b]$,

$$\big||f(x)| - |f(y)|\big| \le |f(x) - f(y)|,$$

so by (14.9), $\omega(|f|;[c,d]) \le \omega(f;[c,d])$ for every interval $[c,d] \subset [a,b]$. It follows that $\Omega_F(|f|) \le \Omega_F(f)$ holds for every partition F. Thus if $\Omega_F(f) < \varepsilon$, then $\Omega_F(|f|) < \varepsilon$, and so by Theorem 14.15, $|f|$ is also integrable.

Instead of the values m_i and M_i occurring in the lower and upper sums, in many cases we would like to use the value $f(c_i)$ for some "inner" points $x_{i-1} \le c_i \le x_i$. The sum we get in this way is called an *approximating sum or Riemann sum*.

Definition 14.17. The *approximating, or Riemann, sums* of a function $f: [a,b] \to \mathbb{R}$ corresponding to the partition $F: a = x_0 < x_1 < \cdots < x_n = b$ are the sums

$$\sigma_F(f;(c_i)) = \sum_{i=1}^{n} f(c_i)(x_i - x_{i-1})$$

with any choice of the points $c_i \in [x_{i-1}, x_i]$ $(i = 1, \ldots, n)$.

The value of the approximating sum $\sigma_F(f;(c_i))$ thus depends not only on the partition, but also on the points c_i. However, if it will not cause confusion, we omit the reference to f and c_i in the notation.

The integrability of a function and the value of its integral can be expressed with the help of approximating sums as well. We need to prove only the following statement for this:

Theorem 14.18. *For an arbitrary bounded function* $f: [a,b] \to \mathbb{R}$ *and partition* F,

$$\inf_{(c_1,\ldots,c_n)} \sigma_F = s_F \qquad and \qquad \sup_{(c_1,\ldots,c_n)} \sigma_F = S_F; \qquad (14.10)$$

that is, the infimum and the supremum of the set of approximating sums with every choice of c_i *are* s_F *and* S_F *respectively.*

Proof. In Theorem 3.20, we saw that if A and B are nonempty sets of numbers, then $\inf(A + B) = \inf A + \inf B$ and $\sup(A + B) = \sup A + \sup B$. Then by induction, if $A_1, \ldots, A_n$ are nonempty sets of numbers, then

$$\begin{aligned}\inf(A_1 + \cdots + A_n) = \inf A_1 + \cdots + \inf A_n \quad \text{and} \\ \sup(A_1 + \cdots + A_n) = \sup A_1 + \cdots + \sup A_n,\end{aligned} \qquad (14.11)$$

where $A_1 + \cdots + A_n = \{a_1 + \cdots + a_n : a_i \in A_i \ (i = 1, \ldots, n)\}$.
Let $A_i = \{f(c)(x_i - x_{i-1}) : c \in [x_{i-1}, x_i]\}$ $(i = 1, \ldots, n)$. It is clear that $\inf A_i = m_i(x_i - x_{i-1})$ and $\sup A_i = M_i(x_i - x_{i-1})$ for all i. Then (14.10) is clear by (14.11). $\quad\square$

Now, considering Theorems 14.11 and 14.23, we easily get the following result.

Theorem 14.19. *A bounded function* $f: [a,b] \to \mathbb{R}$ *is integrable with integral* I *if and only if for arbitrary* $\varepsilon > 0$, *there exists a partition* F *such that every Riemann sum* σ_F *satisfies* $|\sigma_F - I| < \varepsilon$.

Exercises

14.7. Prove that if f is bounded on $[a,b]$, then

$$\omega(f;[a,b]) = \sup\{|f(x)-f(y)| : x,y \in [a,b]\}.$$

14.8. Prove that if f is integrable over $[a,b]$, then for all $\varepsilon > 0$, there exists a subinterval $[c,d] \subset [a,b]$ such that $\omega(f;[c,d]) < \varepsilon$.

14.9. Prove that if a function f is integrable over $[a,b]$, then it has a point of continuity. (H)

14.10. Let $f : [a,b] \to \mathbb{R}$ be bounded, and $F \in \mathscr{F}$ fixed. Is it true that the set of values of approximating sums σ_F forms an interval? (S)

14.11. Prove that if f is integrable over $[a,b]$, then so is the function e^f.

14.12. Is it true that if f is positive and integrable over $[a,b]$, then $\int_a^b f\,dx > 0$? (S)

To determine the integrability of a function and the value of the integral, we do not need to know every lower and upper sum. It generally suffices to know the values of the lower and upper sums for a suitable sequence of partitions. Indeed, if $f : [a,b] \to \mathbb{R}$ is integrable with integral I, then by Theorem 14.11, for every $n = 1,2,\ldots$, there exists a partition F_n such that $I - (1/n) < s_{F_n} \le S_{F_n} < I + (1/n)$. Thus the partitions F_n satisfy

$$\lim_{n\to\infty} s_{F_n} = \lim_{n\to\infty} S_{F_n} = I. \tag{14.12}$$

Conversely, if (14.12) holds for the partitions F_n, then for every $\varepsilon > 0$, there exists a partition F that satisfies condition (14.7), namely $F = F_n$, if n is sufficiently large. Thus by Theorem 14.11, f is integrable and has integral I.

The condition for integrability in terms of the oscillatory sum (Theorem 14.15) is clearly equivalent to the existence of partitions F_n such that $\Omega_{F_n} \to 0$. With this, we have the following:

Theorem 14.20.

(i) *A bounded function $f : [a,b] \to \mathbb{R}$ is integrable with integral I if and only if there exists a sequence of partitions $F_1, F_2, \ldots$ for which (14.12) holds.*

(ii) *A bounded function $f : [a,b] \to \mathbb{R}$ is integrable if and only if there exists a sequence of partitions $F_1, F_2, \ldots$ such that $\lim_{n\to\infty} \Omega_{F_n} = 0$.*

Examples 14.21. **1.** We considered the function $1/x$ over the interval $[a,b]$ in Example 14.1, where $0 < a < b$. Let F_n denote the partition with base points $x_i = a \cdot q^i$ ($i = 0,\ldots,n$), where $q = \sqrt[n]{b/a}$. Since the function is monotone decreasing, $m_i = 1/x_i$ and $m_i(x_i - x_{i-1}) = 1 - q^{-1}$, while $M_i = 1/x_{i-1}$ and $M_i(x_i - x_{i-1}) = q - 1$ for all i (see (14.1) and (14.2)). Then $s_{F_n} = n(1 - q^{-1})$, $S_{F_n} = n(q-1)$, and as we computed in Example 14.1, $\lim_{n\to\infty} s_{F_n} = \lim_{n\to\infty} S_{F_n} = \log(b/a)$. Thus according to Theorem 14.20, the function $1/x$ is integrable on the interval $[a, b]$, and its integral there is $\log(b/a)$.

2. Now we show that for $0 < a < b$ and $\alpha > 0$,

$$\int_a^b x^\alpha \, dx = \frac{b^{\alpha+1} - a^{\alpha+1}}{\alpha+1}. \tag{14.13}$$

Consider the partition used in the previous example. The function is monotone increasing, so

$$s_n = \sum_{i=0}^{n-1} x_i^\alpha \cdot (x_{i+1} - x_i) = \sum_{i=0}^{n-1} (aq^i)^\alpha (aq^{i+1} - aq^i) =$$

$$= a^{\alpha+1} \sum_{i=0}^{n-1} (q^{\alpha+1})^i (q-1) = a^{\alpha+1}(q-1) \frac{q^{n(\alpha+1)} - 1}{q^{\alpha+1} - 1} =$$

$$= a^{\alpha+1} \left(\frac{b^{\alpha+1}}{a^{\alpha+1}} - 1 \right) \frac{q-1}{q^{\alpha+1} - 1}.$$

Since $\lim_{n\to\infty} q = \lim_{n\to\infty} \sqrt[n]{b/a} = 1$, we have $\lim_{n\to\infty} \frac{q-1}{q^{\alpha+1} - 1} = \frac{1}{(x^{\alpha+1})'_{x=1}} = \frac{1}{\alpha+1}$, and so

$$\lim_{n\to\infty} s_n = \frac{b^{\alpha+1} - a^{\alpha+1}}{\alpha+1}.$$

We can similarly obtain $\lim_{n\to\infty} S_n = \dfrac{b^{\alpha+1} - a^{\alpha+1}}{\alpha+1}$, which proves (14.13).

3. We show that $\int_a^b e^x \, dx = e^b - e^a$. Consider a uniform partition of the interval $[a, b]$ into n equal parts with the base points $x_i = a + i \cdot (b-a)/n$ $(i = 0, \ldots, n)$. Let s_n and S_n denote lower and upper sums corresponding to this partition. Since e^x is monotone increasing in $[a, b]$, $m_i = e^{x_{i-1}}$ and $M_i = e^{x_i}$ for all i. Then

$$s_n = \sum_{i=0}^{n-1} e^{a+i(b-a)/n} \cdot \frac{b-a}{n} \qquad \text{and} \qquad S_n = \sum_{i=1}^{n} e^{a+i(b-a)/n} \cdot \frac{b-a}{n}.$$

Summing the geometric series, we get

$$s_n = e^a \cdot \frac{e^{((b-a)/n) \cdot n} - 1}{e^{(b-a)/n} - 1} \cdot \frac{b-a}{n} = e^a (e^{b-a} - 1) \cdot \frac{(b-a)/n}{e^{(b-a)/n} - 1}.$$

Then by $\lim_{x\to 0} x/(e^x - 1) = 1$, we have $\lim_{n\to\infty} s_n = e^b - e^a$. We can similarly get that $\lim_{n\to\infty} S_n = e^b - e^a$, so by Theorem 14.20, $\int_a^b e^x \, dx = e^b - e^a$.

4. We show that for $0 < b < \pi/2$,

$$\int_0^b \cos x \, dx = \sin b. \tag{14.14}$$

Consider a uniform partition of the interval $[0, b]$ into n equal parts. Let s_n and S_n denote lower and upper sums corresponding to this partition. Since $\cos x$ is monotone decreasing in the interval $[0, b]$,

$$s_n = \sum_{i=1}^{n} \frac{b}{n} \cdot \cos \frac{ib}{n} \quad \text{and} \quad S_n = \sum_{i=0}^{n-1} \frac{b}{n} \cdot \cos \frac{ib}{n}. \tag{14.15}$$

Now we use the identity

$$\cos \alpha + \cos 2\alpha + \cdots + \cos n\alpha = \frac{\sin(n\alpha/2)}{\sin(\alpha/2)} \cdot \cos \frac{(n+1)\alpha}{2}, \tag{14.16}$$

which can be proved easily by induction. Applying this with $\alpha = b/n$ gives us that

$$s_n = \frac{b}{n} \cdot \frac{\sin(b/2)}{\sin(b/(2n))} \cdot \cos \frac{(n+1)b}{2n}.$$

Since

$$\lim_{n \to \infty} \frac{b/n}{\sin(b/2n)} = 2 \quad \text{and} \quad \lim_{n \to \infty} \cos \frac{(n+1)b}{2n} = \cos \frac{b}{2},$$

we have

$$\lim_{n \to \infty} s_n = 2 \sin \frac{b}{2} \cos \frac{b}{2} = \sin b.$$

Now by (14.15), we know that $S_n - s_n = \frac{b}{n} \cdot (\cos 0 - \cos b) \to 0$ if $n \to \infty$, so $\lim_{n \to \infty} S_n = \lim_{n \to \infty} s_n = \sin b$. Thus Theorem 14.20 gives us (14.14).

If f is integrable, then according to Theorem 14.11, for every ε there exists a partition F such that s_F and S_F approximate the integral of f with an error of less than ε. We now show that every partition with short enough subintervals has this property.

Definition 14.22. Let $F : a = x_0 < x_1 < \cdots < x_n = b$ be a partition of the interval $[a, b]$. We call the value

$$\delta(F) = \max_{1 \leq i \leq n} (x_i - x_{i-1})$$

the *mesh* of the partition F. Sometimes, we say that a partition is *finer* than η if $\delta(F) < \eta$.

Theorem 14.23.

(i) *A bounded function $f : [a, b] \to \mathbb{R}$ is integrable with integral I if and only if for every $\varepsilon > 0$, there exists a $\delta > 0$ such that every partition F with mesh smaller than δ satisfies*

$$I - \varepsilon < s_F \leq S_F < I + \varepsilon. \tag{14.17}$$

(ii) *A bounded function* $f: [a,b] \to \mathbb{R}$ *is integrable if and only if for every* $\varepsilon > 0$, *there exists a* $\delta > 0$ *such that every partition* F *with mesh smaller than* δ *satisfies* $\Omega_F < \varepsilon$.

(iii) *A bounded function* $f: [a,b] \to \mathbb{R}$ *is integrable with integral* I *if and only if for every* $\varepsilon > 0$, *there exists a* $\delta > 0$ *such that every partition* F *with mesh smaller than* δ *and every Riemann sum* σ_F *belonging to the partition* F *satisfies* $|\sigma_F - I| < \varepsilon$.

Proof. If the conditions for the partitions in (i) or (iii) are satisfied, then by Theorems 14.11 and 14.19, f is integrable with integral I. If the conditions hold in (ii), then by Theorem 14.15, f is integrable.

Now suppose that f is integrable. Let $\int_a^b f\,dx = I$, and let $\varepsilon > 0$ be fixed. To prove the theorem, it suffices to show that (14.17) holds for every partition with sufficiently small mesh. This is because if (14.17) is true, then $\Omega_F = S_F - s_F < 2\varepsilon$, and $|\sigma_F - I| < \varepsilon$ for every Riemann sum corresponding to F, so (ii) and (iii) automatically follow from (i).

We first show that $s_F > I - \varepsilon$ for every partition with sufficiently small mesh. To prove this, let us consider how much a lower sum s_F corresponding to a partition with a mesh smaller than δ can increase when a new base point is added. If the new partition is F', then with the notation of Lemma 14.3 and (14.3), we have

$$s_{F'} - s_F = m_k'(x' - x_{k-1}) + m_k''(x_k - x') - m_k(x_k - x_{k-1}) \le 3K\delta,$$

where K is an upper bound of $|f|$ on $[a,b]$. That is, adding a new base point to a partition with mesh smaller than δ can increase the lower sum by at most $3K\delta$.

By definition (of the lower integral), $[a,b]$ has a partition $F_0: a = t_0 < \ldots < t_k = b$ such that $s_{F_0} > I - \varepsilon/2$. If $F: a = x_0 < x_1 < \ldots < x_n = b$ is an arbitrary partition and F_1 denotes the union of F_0 and F, then $s_{F_1} \ge s_{F_0}$ by Lemma 14.3, and so $s_{F_1} > I - \varepsilon/2$. We get the partition F_1 by adding the base points $t_1, \ldots, t_{k-1}$ one after another to F. Let the mesh of F be δ. Then every added base point increases the value of s_F by at most $3K\delta$, so

$$s_{F_1} \le s_F + (k-1) \cdot 3K\delta.$$

Thus if $\delta < \varepsilon/(6K \cdot k)$, then $s_F > s_{F_1} - (\varepsilon/2) > I - \varepsilon$.

This shows that $s_F > I - \varepsilon$ for every partition with mesh smaller than $\varepsilon/(6K \cdot k)$. Let us note that k depends only on ε. We can similarly show that $S_F < I + \varepsilon$ holds for every partition with mesh smaller than $\varepsilon/(6K \cdot k)$, so (14.17) itself holds for every partition with sufficiently small mesh. $\square$

Now we show that if f is integrable, then (14.12) (and so $\lim_{n\to\infty} \Omega_{F_n} = 0$) holds in every case in which the mesh of the partitions tends to zero.

Theorem 14.24.

(i) *A bounded function* $f: [a,b] \to \mathbb{R}$ *is integrable with integral* I *if and only if*

$$\lim_{n\to\infty} s_{F_n} = \lim_{n\to\infty} S_{F_n} = I$$

holds for every sequence of partitions $F_1, F_2, \ldots$ *satisfying* $\lim_{n\to\infty} \delta(F_n) = 0$.

(ii) *A bounded function* $f : [a,b] \to \mathbb{R}$ *is integrable if and only if* $\Omega_{F_n} \to 0$ *holds for every sequence of partitions* $F_1, F_2, \ldots$ *satisfying* $\lim_{n\to\infty} \delta(F_n) = 0$.

Proof. (i) If the condition holds, then f is integrable with integral I by statement (i) of Theorem 14.20. Now suppose that f is integrable, and let $\varepsilon > 0$ be arbitrary. By Theorem 14.23, there exists a δ_0 such that every partition with mesh smaller than δ_0 satisfies (14.17). If $F_1, F_2, \ldots$ is a sequence of partitions such that $\lim_{n\to\infty} \delta(F_n) = 0$, then for every sufficiently large n, we have $\delta(F_n) < \delta_0$, and so $|s_{F_n} - I| < \varepsilon$ and $|S_{F_n} - I| < \varepsilon$.

Statement (ii) can be proved in the same way using statement (ii) of Theorem 14.20. □

Example 14.25. With the help of Theorem 14.24, we give a new proof of the equality

$$\log 2 = 1 - \frac{1}{2} + \frac{1}{3} - \cdots. \tag{14.18}$$

In Example 14.21, we saw that for $0 < a < b$, we have $\int_a^b (1/x)\,dx = \log(b/a)$. For the special case $a = 1$ and $b = 2$, we get $\int_1^2 (1/x)\,dx = \log 2$. Let F_n denote a uniform partition of the interval $[1,2]$ into n equal pieces. Since the sequence of partitions $F_1, F_2, \ldots$ satisfies $\lim_{n\to\infty} \delta(F_n) = 0$, Theorem 14.24 implies $\lim_{n\to\infty} s_{F_n} = \log 2$. Now it is easy to see that the ith term of the sum defining s_{F_n} is

$$m_i(x_i - x_{i-1}) = \frac{1}{1 + i/n} \cdot \frac{1}{n} = \frac{1}{i+n},$$

so $s_{F_n} = \sum_{i=1}^{n} 1/(i+n)$. Observe that

$$1 - \frac{1}{2} + \frac{1}{3} - \cdots + \frac{1}{2n-1} - \frac{1}{2n} = \left(1 + \frac{1}{2} + \cdots + \frac{1}{2n}\right) - 2\left(\frac{1}{2} + \cdots + \frac{1}{2n}\right) =$$

$$= \left(1 + \frac{1}{2} + \cdots + \frac{1}{2n}\right) - \left(1 + \frac{1}{2} + \cdots + \frac{1}{n}\right) =$$

$$= \frac{1}{n+1} + \frac{1}{n+2} + \cdots + \frac{1}{2n} = s_{F_n}.$$

This proves (14.18). This is now the third proof of this equality (see Exercise 12.92 and Remark 13.16).

Exercises

14.13. Prove that if f is differentiable and f' is bounded on $[0,1]$, then there exists a number K such that

$$\left| \frac{1}{n} \sum_{i=1}^{n} f\left(\frac{i}{n}\right) - \int_0^1 f\,dx \right| < \frac{K}{n}.$$

for all n.

14.14. Let F_n be the uniform partition of the interval $[a,b]$ into n equal pieces. Is it true that for every bounded function $f \colon [a,b] \to \mathbb{R}$, the sequence $s_{F_n}(f)$ is monotone increasing and the sequence $S_{F_n}(f)$ is monotone decreasing?

14.15. Let $0 < a < b$, and let $F \colon a = x_0 < x_1 < \cdots < x_n = b$ be an arbitrary partition of the interval $[a,b]$. Compute the approximating sum of F with $c_i = \sqrt{x_{i-1}x_i}$ of the function $1/x^2$, and determine the value of the integral $\int_a^b x^{-2}\,dx$ based on that.

14.16. Prove that if $0 < b < \pi/2$, then $\int_0^b \sin x\,dx = 1 - \cos b$. (H)

14.17. Prove that if f is integrable on $[0,1]$, then

$$\lim_{n \to \infty} \frac{1}{n} \sum_{k=1}^n f\left(\frac{k}{n}\right) = \int_0^1 f\,dx.$$

14.18. Compute the value of

$$\lim_{n \to \infty} \frac{\sqrt{1} + \sqrt{2} + \cdots + \sqrt{n}}{n\sqrt{n}}.$$

14.19. Prove that

$$\sum_{k=1}^n k^2 \sim \frac{n^3}{3}.$$

(Here $a_n \sim b_n$ denotes that $\lim_{n \to \infty} a_n/b_n = 1$; see Definition 5.28.)

14.20. Prove that for every positive α,

$$\sum_{k=1}^n k^\alpha \sim \frac{n^{\alpha+1}}{\alpha+1}.$$

14.21. Compute the value of

$$\lim_{n \to \infty} \frac{1}{n^2} \cdot \sum_{i=1}^n \sqrt{n^2 - i^2}.$$

14.22. To what value does the fraction

$$\frac{\cos(1/n) + \cos(2/n) + \cdots + \cos(n/n)}{n}$$

tend as $n \to \infty$?

14.23. Prove that if f is integrable over $[0,1]$, then

$$\frac{f(0) - f\left(\frac{1}{n}\right) + f\left(\frac{2}{n}\right) - \cdots + (-1)^n f\left(\frac{n}{n}\right)}{n} \to 0, \qquad \text{as } n \to \infty.$$

14.24. Prove that if f is bounded on $[a,b]$, then for every $\varepsilon > 0$, there exists a $\delta > 0$ such that for every partition F with mesh smaller than δ,

$$\underline{\int_a^b} f\,dx - \varepsilon < s_F \qquad \text{and} \qquad S_F < \overline{\int_a^b} f\,dx + \varepsilon.$$

14.25. Prove that if f is integrable on $[0,1]$ and $f(x) \geq c > 0$ there, then

$$\lim_{n \to \infty} \sqrt[n]{f\left(\frac{1}{n}\right) \cdot f\left(\frac{2}{n}\right) \cdots f\left(\frac{n}{n}\right)} = e^{\int_0^1 \log f(x)\,dx}.$$

14.26. Prove that if f is integrable on $[0,1]$ and $f(x) \geq c > 0$ there, then

$$\left(\int_0^1 \frac{dx}{f(x)}\right)^{-1} \leq e^{\int_0^1 \log f(x)\,dx} \leq \int_0^1 f(x)\,dx.$$

14.27. Suppose that f is bounded on $[0,1]$ and that

$$\lim_{n \to \infty} \frac{1}{n} \sum_{k=1}^n f\left(\frac{k}{n}\right) - A.$$

Does it follow from this that f is integrable on $[0,1]$ and that $\int_0^1 f(x)\,dx = A$?

14.28. For each of the functions below and for arbitrary $\varepsilon > 0$, give a number $\delta > 0$ such that if a partition of the given interval has mesh smaller than δ, then the upper and lower sums corresponding to the partition are closer to the integral than ε. (It is not necessary to find the largest such δ.)

(a) e^x, $[0,10]$;

(b) $\cos x$, $[0,2\pi]$;

(c) $\operatorname{sgn} x$, $[-1,1]$;

(d) $f(1/n) = 1/n$ for all $n \in \mathbb{N}^+$ and $f(x) = 0$ otherwise, $[0,1]$;

(e) $f(x) = \sin(1/x)$ if $x \neq 0$, $f(0) = 0$, $[0,1]$.

14.4 Integrability of Continuous Functions and Monotone Functions

One can guess from Theorem 14.15 that the integrability of a function is closely related to its continuity. One can prove that a bounded function is integrable if and only if in a certain well-defined way, it is continuous "almost everywhere." So for example, every bounded function that is continuous everywhere except at countably many points is integrable. The proof of this theorem, however, is far from easy, so we will consider only some very important and often used special cases.

Theorem 14.26. *If f is continuous on $[a,b]$, then f is integrable on $[a,b]$.*

Proof. We will apply Theorem 14.15; but first we refer to Heine's theorem (Theorem 10.61). According to this, if f is continuous on $[a,b]$, then it is uniformly continuous there, that is, for every $\varepsilon > 0$, there exists a $\delta > 0$ such that for arbitrary $x, y \in [a,b]$, if $|x - y| < \delta$, then $|f(x) - f(y)| < \varepsilon$.

Let $\varepsilon > 0$ be fixed, and choose a δ corresponding to ε by the definition of uniform continuity. Let F be a partition of the interval $[a,b]$ in which the distance between any two neighboring base points is less than δ (for example, the uniform partition into n equal pieces where $n > (b-a)/\delta$). By Weierstrass's theorem (Theorem 10.55), the function f has both a largest and a smallest value in each interval $[x_{i-1}, x_i]$, so there are points $u_i, v_i \in [x_{i-1}, x_i]$ such that $f(u_i) = m_i$ and $f(v_i) = M_i$. Then $|u_i - v_i| \le x_i - x_{i-1} < \delta$ by our choice of the partition F, and so

$$M_i - m_i = f(v_i) - f(u_i) < \varepsilon$$

by our choice of δ. Then

$$\Omega_F = \sum_{i=1}^{n}(M_i - m_i)(x_i - x_{i-1}) < \sum_{i=1}^{n} \varepsilon(x_i - x_{i-1}) = \varepsilon(b-a).$$

Since $\varepsilon > 0$ was arbitrary, f is integrable by Theorem 14.15. $\qquad\square$

Corollary 14.27. *The elementary functions (as seen in Chapter 11) are integrable in every interval $[a,b]$ on which they are defined.*

Theorem 14.28. *If f is monotone on $[a,b]$, then f is integrable on $[a,b]$.*

Proof. We use Theorem 14.15 again. Let f be monotone increasing on $[a,b]$. Then for every partition F, we have $m_i = f(x_{i-1})$ and $M_i = f(x_i)$, and so

$$\Omega_F = \sum_{i=1}^{n}(f(x_i) - f(x_{i-1}))(x_i - x_{i-1}).$$

Let $\varepsilon > 0$ be fixed. If the partition is such that $x_i - x_{i-1} \le \varepsilon$ for all $i = 1, \ldots, n$, then

$$\Omega_F \le \sum_{i=1}^{n}(f(x_i) - f(x_{i-1})) \cdot \varepsilon = (f(b) - f(a)) \cdot \varepsilon. \qquad (14.19)$$

Since ε was arbitrary, f is integrable by Theorem 14.15. $\qquad\square$

Remarks 14.29. **1.** For a monotone increasing function with a uniform partition, the oscillatory sum

$$\Omega_F = \sum_{i=1}^{n}(M_i - m_i) \cdot \frac{b-a}{n}$$

can be illustrated easily: this is exactly the area of the rectangle obtained by sliding the "little rectangles"
over the last interval $[x_{n-1}, x_n]$, with
base $(b-a)/n$ and height $f(b) - f(a)$.
That is,

$$\Omega_F = \frac{b-a}{n}(f(b) - f(a)),$$

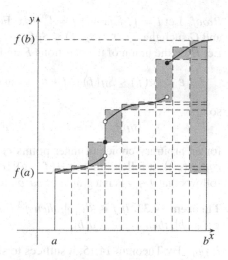

Fig. 14.7

and this is truly arbitrarily small if n is
sufficiently large (Figure 14.7).

2. By Theorem 14.23, if a function is integrable, then for every $\varepsilon > 0$, there exists
a δ such that $\Omega_F = S_F - s_F < \varepsilon$ holds for every partition F with mesh smaller than
δ. Note that for continuous and monotone functions, the proofs of Theorems 14.26
and 14.28 explicitly provide us with such a δ.

14.5 Integrability and Operations

The following theorems tells us that the family of integrable functions is closed under the most frequently used operations. When multiplying a function by a constant
or summing functions, we even get the value of the new integrals. Let us denote the
set of integrable functions on the interval $[a,b]$ by $R[a,b]$.

Theorem 14.30. *If $f \in R[a,b]$ and $c \in \mathbb{R}$, then $cf \in R[a,b]$ and*

$$\int_a^b cf\,dx = c \cdot \int_a^b f\,dx.$$

Proof. The statement follows immediately from the fact that if $c \geq 0$, then for an
arbitrary partition F, we have $s_F(cf) = c \cdot s_F(f)$ and $S_F(cf) = c \cdot S_F(f)$, and if $c < 0$,
then $s_F(cf) = c \cdot S_F(f)$ and $S_F(cf) = c \cdot s_F(f)$. □

Theorem 14.31. *If $f, g \in R[a,b]$, then $f + g \in R[a,b]$, and*

$$\int_a^b (f+g)\,dx = \int_a^b f\,dx + \int_a^b g\,dx.$$

Proof. Let $I = \int_a^b f\,dx$ and $J = \int_a^b g\,dx$. For arbitrary $\varepsilon > 0$, there exist partitions F and G such that $I - \varepsilon < s_F(f) \leq S_F(f) < I + \varepsilon$ and $J - \varepsilon < s_G(g) \leq S_G(g) < J + \varepsilon$. Let H be the union of the partitions F and G. Then

$$I - \varepsilon < s_H(f) \leq S_H(f) < I + \varepsilon \qquad \text{and} \qquad J - \varepsilon < s_H(g) \leq S_H(g) < J + \varepsilon,$$

so

$$|\sigma_H(f;(c_i)) - I| < \varepsilon \qquad \text{and} \qquad |\sigma_H(g;(c_i)) - J| < \varepsilon$$

for an arbitrary choice of inner points c_i. Since $\sigma_H((f+g);(c_i)) = \sigma_H(f;(c_i)) + \sigma_H(g;(c_i))$, it follows that $|\sigma_H((f+g);(c_i)) - (I+J)| < 2\varepsilon$. By Theorem 14.19, it follows that $f + g$ is integrable on $[a,b]$, and its integral is $I + J$. $\qquad \square$

Theorem 14.32. *If $f \in R[a,b]$, then $f^2 \in R[a,b]$. Moreover, if $|f(x)| \geq \delta > 0$ for all $x \in [a,b]$, then $1/f \in R[a,b]$.*

Proof. By Theorem 14.15, it suffices to show that the oscillatory sums of the functions f^2 and $1/f$ (if $|f| \geq \delta > 0$) can be made arbitrarily small.

Since f is integrable, it is also bounded. Let $|f(x)| \leq K$ for $x \in [a,b]$. For arbitrary $x, y \in [a,b]$,

$$|f^2(x) - f^2(y)| = |f(x) - f(y)| \cdot |f(x) + f(y)| \leq 2K \cdot |f(x) - f(y)|,$$

and if $|f| \geq \delta > 0$, we have

$$\left| \frac{1}{f(x)} - \frac{1}{f(y)} \right| = \frac{|f(x) - f(y)|}{|f(x)f(y)|} \leq \frac{1}{\delta^2} \cdot |f(x) - f(y)|.$$

Then by (14.9), we have $\omega(f^2;[u,v]) \leq 2K \cdot \omega(f;[u,v])$, and if $|f| \geq \delta > 0$, then $\omega(1/f;[u,v]) \leq (1/\delta^2) \cdot \omega(f;[u,v])$ for every interval $[u,v] \subset [a,b]$.

It then follows immediately that for an arbitrary partition F, $\Omega_F(f^2) \leq 2K \cdot \Omega_F(f)$, and if $|f| \geq \delta > 0$, then $\Omega_F(1/f) \leq (2K/\delta^2) \cdot \Omega_F(f)$. Since $\Omega_F(f)$ can be made arbitrarily small, the same holds for the oscillatory sums $\Omega_F(f^2)$, and if $|f| \geq \delta > 0$, then it holds for $\Omega_F(1/f)$. $\qquad \square$

Theorem 14.33. *If $f, g \in R[a,b]$, then $fg \in R[a,b]$. Moreover, if $|g(x)| \geq \delta > 0$ for all $x \in [a,b]$, then $f/g \in R[a,b]$.*

Proof. Since

$$fg = \frac{1}{4} \cdot [(f+g)^2 - (f-g)^2],$$

by Theorems 14.30, 14.31, and 14.32, we have that $fg \in R[a,b]$. The second statement of the theorem is clear by $f/g = f \cdot (1/g)$. $\qquad \square$

After the previous theorems, it might be surprising that the composition of two integrable functions is not necessarily integrable.

Example 14.34. Let $f(0) = 0$ and $f(x) = 1$ if $x \neq 0$. It is easy to see that f is integrable on every interval (see Exercise 14.1(c)). Let g denote the Riemann function (see Remark 10.8). We will soon see that the function g is also integrable on every interval (see Example 14.45). On the other hand, $f \circ g$ is exactly the Dirichlet function, which we already saw not to be integrable in any interval (see Example 14.9).

We see that for integrability of compositions of functions, we require something more than just the integrability of the two functions.

Theorem 14.35. *Let g be integrable on $[a,b]$, and let f be continuous on a closed interval $[\alpha, \beta]$ that contains the image of g (that is, the set $g([a,b])$). Then $f \circ g$ is integrable on $[a,b]$.*

Proof. We have to show that the function $f \circ g$ has arbitrarily small oscillatory sums.

By Theorem 10.52, f is bounded, so there exists a $K \geq 0$ such that $|f(t)| \leq K$ for all $t \in [\alpha, \beta]$. By Heine's theorem (Theorem 10.61), f is uniformly continuous on $[\alpha, \beta]$, that is, for arbitrary $\varepsilon > 0$, there exists a $\delta > 0$ such that $|f(t_1) - f(t_2)| < \varepsilon$ if $t_1, t_2 \in [\alpha, \beta]$ and $|t_1 - t_2| < \delta$.

Consider now an arbitrary partition $F : a = x_0 < x_1 < \cdots < x_n = b$ of the interval $[a,b]$. We will find a bound on the oscillatory sum $\Omega_F(f \circ g)$ by separately finding a bound on the terms whose indices satisfy

$$\omega(g; [x_{i-1}, x_i]) < \delta, \tag{14.20}$$

and for the others. Let I denote the set of indices $1 \leq i \leq n$ for which (14.20) holds, and let J be the set of indices $1 \leq i \leq n$ that do not satisfy (14.20). If $i \in I$, then for arbitrary $u, v \in [x_{i-1}, x_i]$, we have $|g(u) - g(v)| < \delta$, so by the choice of δ, $|f(g(u)) - f(g(v))| < \varepsilon$. Let the oscillation $\omega(f \circ g; [x_{i-1}, x_i])$ be denoted by $\omega_i(f \circ g)$. Then

$$\omega_i(f \circ g) = \sup\{|f(g(u)) - f(g(v))| : u, v \in [x_{i-1}, x_i]\} \leq \varepsilon \tag{14.21}$$

for all $i \in I$. On the other hand, if $i \in J$, then $\omega(g; [x_{i-1}, x_i]) \geq \delta$, so

$$\Omega_F(g) = \sum_{i=1}^{n} \omega(g; [x_{i-1}, x_i]) \cdot (x_i - x_{i-1}) \geq \sum_{i \in J} \omega(g; [x_{i-1}, x_i]) \cdot (x_i - x_{i-1}) \geq$$
$$\geq \sum_{i \in J} \delta \cdot (x_i - x_{i-1}),$$

and thus

$$\sum_{i \in J} (x_i - x_{i-1}) \leq \frac{1}{\delta} \cdot \Omega_F(g). \tag{14.22}$$

Now we use the inequalities (14.21) and (14.22) in order to estimate the sum $\Omega_F(f \circ g)$:

$$\Omega_F(f \circ g) = \sum_{i=1}^{n} \omega_i(f \circ g) \cdot (x_i - x_{i-1}) = \sum_{i \in I} + \sum_{i \in J} \leq$$

$$\leq \sum_{i \in I} \varepsilon \cdot (x_i - x_{i-1}) + \sum_{i \in J} 2K \cdot (x_i - x_{i-1}) \leq$$

$$\leq \varepsilon \cdot (b-a) + \frac{2K}{\delta} \cdot \Omega_F(g). \tag{14.23}$$

Since g is integrable on $[a,b]$, we can choose a partition F for which $\Omega_F(g) < \varepsilon\delta/(2K)$. Then by (14.23), $\Omega_F(f \circ g) < \varepsilon(b-a+1)$. Since ε was chosen arbitrarily, it follows from Theorem 14.15 that $f \circ g$ is integrable. This completes the proof. □

Remark 14.36. If in the composition $f \circ g$ we assume g to be continuous and f to be integrable, then we cannot generally expect $f \circ g$ to be integrable. The examples showing this, however, are much more complicated than the one we saw in Example 14.34.

Exercises

14.29. Give an example of a function f for which $|f|$ is integrable on $[a,b]$, but f is not integrable on $[a,b]$.

14.30. Let f and g be bounded on the interval $[a,b]$. Is it true that

$$\overline{\int_a^b} (f+g)\,dx = \overline{\int_a^b} f\,dx + \overline{\int_a^b} g\,dx? \tag{H}$$

14.31. Is it true that if $f : [0,1] \to [0,1]$ is integrable over $[0,1]$, then $f \circ f$ is also integrable on $[0,1]$?

14.6 Further Theorems Regarding the Integrability of Functions and the Value of the Integral

Theorem 14.37. *If a function is integrable on the interval $[a,b]$, then it is also integrable on every subinterval $[c,d] \subset [a,b]$.*

Proof. Let f be integrable on $[a,b]$. Then for every $\varepsilon > 0$, there exists a partition F of the interval $[a,b]$ for which $\Omega_F < \varepsilon$. Let F' be a refinement of the partition F that we get by including the points c and d. By Lemma 14.3, we know that $s_F \leq s_{F'} \leq S_{F'} \leq S_F$ so $\Omega_{F'} \leq \Omega_F < \varepsilon$. On the other hand, if we consider only the base points of F' that belong to $[c,d]$, then we get a partition $\overline{F}$ of $[c,d]$ for which $\Omega_{\overline{F}} \leq \Omega_{F'}$. This is clear, because the sum defining $\Omega_{F'}$ consists of nonnegative terms, and the terms present in $\Omega_{\overline{F}}$ all appear in $\Omega_{F'}$ as well.

This shows that for all $\varepsilon > 0$, there exists a partition $\overline{F}$ of $[c,d]$ for which $\Omega_{\overline{F}} < \varepsilon$. Thus f is integrable over $[c,d]$. $\qquad\square$

Theorem 14.38. *Let $a < b < c$, and let f be defined on $[a,c]$. If f is integrable on both intervals $[a,b]$ and $[b,c]$, then it is also integrable on $[a,c]$, and*

$$\int_a^c f(x)\,dx = \int_a^b f(x)\,dx + \int_b^c f(x)\,dx.$$

Proof. Let $I_1 = \int_a^b f(x)\,dx$ and $I_2 = \int_b^c f(x)\,dx$. Let $\varepsilon > 0$ be given, and consider partitions F_1 and F_2 of $[a,b]$ and $[b,c]$ respectively that satisfy $I_1 - \varepsilon < s_{F_1} \leq S_{F_1} < I_1 + \varepsilon$ and $I_2 - \varepsilon < s_{F_2} \leq S_{F_2} < I + \varepsilon$.

Taking the union of the base points of the partitions F_1 and F_2 yields a partition F of the interval $[a,c]$ for which

$$s_F = s_{F_1} + s_{F_2} \qquad \text{and} \qquad S_F = S_{F_1} + S_{F_2}.$$

Thus

$$I_1 + I_2 - 2\varepsilon < s_F \leq S_F < I_1 + I_2 + 2\varepsilon.$$

Since ε was arbitrary, f is integrable on $[a,c]$ by Theorem 14.11, and its integral is $I_1 + I_2$ there. $\qquad\square$

Remark 14.39. Let f be integrable on $[a,b]$. As we saw, in this case, f is integrable over every subinterval $[c,d] \subset [a,b]$ as well. The function

$$[c,d] \mapsto \int_c^d f(x)\,dx, \qquad\qquad (14.24)$$

which is defined on the set of closed subintervals of $[a,b]$, has the property that the sum of its values at two adjacent intervals is equal to its value at the union of those two intervals. This is usually expressed by saying that the formula (14.24) defines an **additive interval function**.

Thus far, we have given meaning to the expression $\int_a^b f(x)\,dx$ only when $a < b$ holds. It is worthwhile, however, to extend the definition to the case $a \geq b$.

Definition 14.40. If $a > b$ and f is integrable on $[b,a]$, then let

$$\int_a^b f(x)\,dx = -\int_b^a f(x)\,dx.$$

If f is defined at the point a, then let

$$\int_a^a f(x)\,dx = 0.$$

Theorem 14.41. *For arbitrary a,b,c,*

$$\int_a^c f(x)\,dx = \int_a^b f(x)\,dx + \int_b^c f(x)\,dx, \qquad (14.25)$$

if the integrals in question exist.

Proof. Inspecting the possible cases, we see that the theorem immediately follows by Definition 14.40 and Theorem 14.38. If, for example, $a < c < b$, then

$$\int_a^c f(x)\,dx + \int_c^b f(x)\,dx = \int_a^b f(x)\,dx,$$

which we can rearrange to give us (14.25). The other cases are similar.   □

We now turn our attention to theorems that ensure integrability of functions having properties weaker than continuity.

Theorem 14.42. *Let $a < b$, and let f be bounded in $[a,b]$.*

(i) *If f is integrable on $[a+\delta,b]$ for all $\delta > 0$, then f is integrable on the interval $[a,b]$ as well.*

(ii) *If f is integrable on $[a,b-\delta]$ for every $\delta > 0$, then f is integrable on the interval $[a,b]$ as well.*

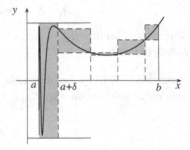

Fig. 14.8

Proof. We prove only (i). Let $|f| \le K$ in $[a,b]$. Consider a partition of the interval $[a+\delta,b]$, and let the oscillatory sum for this partition be Ω_δ (Figure 14.8). The same partition extended by the interval $[a,a+\delta]$ will be a partition of $[a,b]$. The oscillation Ω corresponding to this will satisfy $\Omega \le \Omega_\delta + 2K\delta$, since the contribution of the new interval $[a,a+\delta]$ is $(M-m)\delta \le 2K\delta$. Thus if we choose δ to be sufficiently small (for example, $\delta < \varepsilon/(4K)$), and consider a partition of $[a+\delta,b]$ for which Ω_δ is small enough (for example, $\Omega_\delta < \varepsilon/2$), then Ω will also be small (smaller than ε).   □

Theorem 14.43. *If f is bounded in $[a,b]$ and is continuous there except at finitely many points, then f is integrable on $[a,b]$.*

Proof. Let $a = c_0 \le c_1 < \cdots < c_k = b$, and suppose that f is continuous everywhere except at the points $c_0,\ldots,c_k$ in $[a,b]$. Let us take the points $d_i = \dfrac{c_{i-1}+c_i}{2}$ ($i = 1,\ldots,k$) as well; this gives us intervals $[c_{i-1},d_i]$ and $[d_i,c_i]$ that satisfy the conditions of Theorem 14.42. Thus f is integrable on these intervals, and then by applying Theorem 14.38 repeatedly, we see that f is integrable on the whole interval $[a,b]$.
          □

Remark 14.44. Theorem 14.43 is clearly a consequence of the theorem we mentioned earlier that if a function is bounded and is continuous everywhere except at countably many points, then it is integrable. Moreover, the integrability of monotone functions also follows from that theorem, since a monotone function can have only countably many points of discontinuity, as seen in Theorem 10.70.

Example 14.45. We define the Riemann function as follows (see Example 10.7 and Remark 10.8). Let

$$f(x) = \begin{cases} 0 & \text{if } x \text{ is irrational;} \\ \frac{1}{q} & \text{if } x = \frac{p}{q}, \text{ where } p \text{ and } q \text{ are integers, } q > 0, \text{ and } (p,q) = 1. \end{cases}$$

In Example 10.7, we saw that the function f is continuous at every irrational point but discontinuous at every rational point. Thus the set of discontinuities of f is $\mathbb{Q}$, which is countable by Theorem 8.2. Since f is bounded, f is integrable on every interval $[a,b]$ by the general theorem we mentioned. We will prove this directly too. Moreover, we will show that the integral of f is zero over every interval.

Since every interval contains an irrational number, for an arbitrary partition F : $a = x_0 < x_1 < \cdots < x_n = b$ and $i = 1,\ldots,n$, we have $m_i = 0$. Thus $s_F = 0$ for all $F \in \mathscr{F}$. All we need to show is that for arbitrary $\varepsilon > 0$, there exists a partition for which $S_F < \varepsilon$ holds. By the definition of the integral, this will imply $\int_a^b f(x)\,dx = 0$.

Let $\eta > 0$ be fixed. Notice that the function takes on values greater than η at only finitely many points $x \in [a,b]$. This is because if $f(x) > \eta$, then $x = p/q$, where $0 < q < 1/\eta$. However, for each $0 < q < 1/\eta$, there are only finitely many integers p such that $p/q \in [a,b]$. Let $c_1,\ldots,c_N$ be all the points in $[a,b]$ where the value of f is greater than η. Let $F : a = x_0 < x_1 < \cdots < x_n = b$ be a partition with mesh smaller than η/N. We show that $S_F < (2+b-a) \cdot \eta$.

The number of indices i for which $[x_{i-1},x_i]$ contains one of the points c_j is at most $2N$, since every point c_j belongs to at most two intervals. In the sum S_F, the terms from these intervals are $M_i(x_i - x_{i-1}) \leq 1 \cdot (\eta/N)$, so the sum of these terms is at most $2N \cdot (\eta/N) = 2\eta$. The rest of the terms satisfy $M_i(x_i - x_{i-1}) \leq \eta \cdot (x_i - x_{i-1})$, so their sum is at most

$$\sum_{i=1}^{n} \eta \cdot (x_i - x_{i-1}) = \eta \cdot (b-a).$$

Adding these two bounds, we obtain $S_F \leq (2+b-a) \cdot \eta$. Thus if we choose an η such that $\eta < \varepsilon/(2+b-a)$, then the construction above gives an upper sum that is smaller than ε.

Theorem 14.46. *Let the functions f and g be defined on the interval $[a,b]$. If f is integrable on $[a,b]$, and $g = f$ everywhere except at finitely many points, then g is integrable on $[a,b]$, and*

$$\int_a^b g(x)\,dx = \int_a^b f(x)\,dx.$$

Proof. Since f is bounded on $[a,b]$, so is g. Let $|f(x)| \leq K$ and $|g(x)| \leq K$ for all $x \in [a,b]$. Suppose first that f and g differ only at the point a. Let $I = \int_a^b f(x)\,dx$. By Theorem 14.11, for every $\varepsilon > 0$, there exists a partition F such that $I - \varepsilon < s_F(f) \leq S_F(f) \leq I + \varepsilon$. Since adding a new base point does not decrease $s_F(f)$ and does not increase $S_F(f)$, we can suppose that $a + (\varepsilon/K)$ is a base point, and so $x_1 - x_0 \leq \varepsilon/K$.

If f and g are equal in $(a,b]$, then the sums $s_F(f)$ and $s_F(g)$ differ from each other in only the first term. The absolute value of these terms can be at most $K \cdot (x_1 - x_0) \leq \varepsilon$, and so $|s_F(g) - s_F(g)| \leq 2\varepsilon$ and $s_F(g) > I - 3\varepsilon$. We similarly get that $S_F(g) < I + 3\varepsilon$. Since ε was chosen arbitrarily, the function g is integrable with integral I by Theorem 14.11.

We can argue similarly if f and g differ only at the point b. Then to get the general statement, we can apply Theorem 14.38 repeatedly, just as in the proof of Theorem 14.43.   □

Remark 14.47. The above theorem gives us an opportunity to extend the concept of integrability and the values of integrals to functions that are undefined at finitely many points of the interval $[a,b]$.

Let f be a function defined on $[a,b]$ except for at most finitely many points. If there exists a function g integrable on $[a,b]$ for which $g(x) = f(x)$ with the exception of finitely many points, then we say that f is integrable, and

$$\int_a^b f(x)\,dx = \int_a^b g(x)\,dx.$$

If such a g does not exist, then f is not integrable. By Theorem 14.46, it is clear that integrability and the value of the integral do not depend on our choice of g.

Summarizing our observations above, we conclude that the integrability and the value of the integral of a function f over the interval $[a,b]$ *do not change* if

(i) we change the values of f at finitely many points,
(ii) we extend the definition of f at finitely many points,
(iii) if we make f undefined at finitely many points.

Example 14.48. The function $f \colon [a,b] \to \mathbb{R}$ is called a **step function** if there exists a partition $c_0 = a < c_1 < \cdots < c_n = b$ such that f is constant on every open interval (c_{i-1}, c_i) (while the value of f can be arbitrary at the base points c_i). We show that every step function is integrable, and if $f(x) = d_i$ for all $x \in (c_{i-1}, c_i)$ $(i = 1, \ldots, n)$, then

$$\int_a^b f(x)\,dx = \sum_{i=1}^n d_i(c_i - c_{i-1}).$$

This is quite clear by Example 14.8 and the remark above, which say that $\int_{c_{i-1}}^{c_i} f(x)\,dx = d_i(c_i - c_{i-1})$ for all i, and the rest follows by Theorem 14.38.

Exercises

14.32. Let f and g be integrable on $[a,b]$, and suppose that $s_F(f) = s_F(g)$ for all partitions F. Prove that $\int_c^d f\,dx = \int_c^d g\,dx$ for all $[c,d] \subseteq [a,b]$.

14.33. Let f be a continuous nonnegative function defined on $[a,b]$. Prove that $\int_a^b f\,dx = 0$ if and only if $f \equiv 0$. (S)

14.34. Is the function $\sin(1/x)$ integrable on the interval $[-1,1]$?

14.35. Let $\{x\}$ denote the fractional part function. Is $\{1/x\}$ integrable over the interval $[0,1]$?

14.36. Let $f(x) = n^2\,(x - 1/n)$ if $1/n + 1 < x \leq 1/n$ $(n \in \mathbb{N}^+)$. Is f integrable on the interval $[0,1]$?

14.37. Let $f(x) = (-1)^n$ if $1/n + 1 < x \leq 1/n$ $(n \in \mathbb{N}^+)$. Prove that f is integrable in $[0,1]$. What is the value of its integral?

14.38. Let $f\colon [a,b] \to \mathbb{R}$ be a bounded function such that $\lim_{x \to c} f(x) = 0$ for all $c \in (a,b)$. Prove that f is integrable on $[a,b]$. What is the value of its integral? (H)

14.39. Let $f(x) = x$ if x is irrational, and $f(p/q) = (p+1)/q$ if $p,q \in \mathbb{Z}$, $q > 0$, and $(p,q) = 1$. What are the lower and upper integrals of f in $[0,1]$?

14.40. Let $a < b < c$, and let f be bounded on the interval $[a,c]$. Prove that

$$\overline{\int_a^c} f(x)\,dx = \overline{\int_a^b} f(x)\,dx + \overline{\int_b^c} f(x)\,dx.$$

14.7 Inequalities for Values of Integrals

We begin this section with some simple but often used inequalities.

Theorem 14.49. *Let $a < b$.*

(i) *If f is integrable on $[a,b]$ and $f(x) \geq 0$ for all $x \in [a,b]$, then $\int_a^b f(x)\,dx \geq 0$.*

(ii) *If f and g are integrable on $[a,b]$ and $f(x) \leq g(x)$ for all $x \in [a,b]$, then*

$$\int_a^b f(x)\,dx \leq \int_a^b g(x)\,dx.$$

(iii) *If f is integrable on $[a,b]$ and $m \leq f(x) \leq M$ for all $x \in [a,b]$, then*

$$m(b-a) \leq \int_a^b f(x)\,dx \leq M(b-a). \tag{14.26}$$

(iv) *If f is integrable on $[a,b]$ and $|f(x)| \le K$ for all $x \in [a,b]$, then*

$$\left| \int_a^b f(x)\,dx \right| \le K(b-a).$$

(v) *If h is integrable on $[a,b]$, then*

$$\left| \int_a^b h(x)\,dx \right| \le \int_a^b |h(x)|\,dx.$$

Proof. (i) The statement is clear, because if f is nonnegative, then each of its lower sums is also nonnegative.
(ii) If $f \le g$, then $g - f \ge 0$, so the statement follows from (i).
(iii) Both inequalities of (14.26) are clear from (ii).
(iv) Apply (iii) with $m = -K$ and $M = K$.
(v) Apply (ii) with the choices $f = h$ and $g = |h|$ first, then with the choices $f = -|h|$ and $g = f$. □

Remark 14.50. The definite integral is a mapping that assigns numbers to specific functions. We have determined this map to be a linear (Theorems 14.30 and 14.31), additive interval function (Theorem 14.38), which is monotone (Theorem 14.49) and assigns the value $b - a$ to the constant function 1 on the interval $[a,b]$. It is an important fact that these properties characterize the integral. To see the precise statement and proof of this statement, see Exercises 14.50 and 14.51.

The following theorem is a simple corollary of inequality (14.26).

Theorem 14.51 (First Mean Value Theorem for Integration). *If f is continuous on $[a,b]$, then there exists a $\xi \in [a,b]$ such that*

$$\int_a^b f(x)\,dx = f(\xi) \cdot (b-a). \qquad (14.27)$$

Proof. If $m = \min f([a,b])$ and $M = \max f([a,b])$, then

$$m \le \frac{1}{b-a} \cdot \int_a^b f(x)\,dx \le M. \quad (14.28)$$

Since by the Bolzano–Darboux theorem (Theorem 10.57), the function f takes on every value between m and M in $[a,b]$, there must be a $\xi \in [a,b]$ for which (14.27) holds (Figure 14.9).

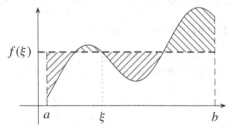

Fig. 14.9

□

Remark 14.52. The graphical meaning of the theorem above is that if f is nonnegative and continuous on $[a,b]$, then there exists a value ξ such that the area of the rectangle with height $f(\xi)$ and base $[a,b]$ is equal to the area underneath the graph of f.

We consider the value in (14.28) to be a generalization of the arithmetic mean. If, for example, $a = 0$, $b = n \in \mathbb{N}^+$ and $f(x) = a_i$ for all $x \in (i-1,i)$, $(i = 1,\dots,n)$, then the value is exactly the arithmetic mean of $a_1,\dots,a_n$.

Theorem 14.53 (Abel's[1] Inequality). *Let f be monotone decreasing and nonnegative, and let g be integrable on $[a,b]$. If*

$$
m \le \int_a^c g(x)\,dx \le M
$$

for all $c \in [a,b]$, then

$$
f(a) \cdot m \le \int_a^b f(x)g(x)\,dx \le f(a) \cdot M. \tag{14.29}
$$

To prove the theorem, we require an analogous inequality for sums.

Theorem 14.54 (Abel's Inequality). *If $a_1 \ge a_2 \ge \cdots \ge a_n \ge 0$ and $m \le b_1 + \cdots + b_k \le M$ for all $k = 1,\dots,n$, then*

$$
a_1 \cdot m \le a_1 b_1 + \cdots + a_n b_n \le a_1 \cdot M. \tag{14.30}
$$

Proof. Let $s_k = b_1 + \cdots + b_k$ $(k = 1,\dots,n)$. Then

$$
\begin{aligned}
a_1 b_1 + \cdots + a_n b_n &= a_1 s_1 + a_2(s_2 - s_1) + \cdots + a_n(s_n - s_{n-1}) = \\
&= (a_1 - a_2)s_1 + (a_2 - a_3)s_2 + \cdots + (a_{n-1} - a_n)s_{n-1} + a_n s_n. \tag{14.31}
\end{aligned}
$$

(This rearrangement is called an **Abel rearrangement**.) If we replace each s_k by M here, then we are increasing the sum, since $s_k \le M$ for all k, and the coefficients $a_i - a_{i+1}$ and a_n are nonnegative. The number we get in this way is

$$
(a_1 - a_2)M + (a_2 - a_3)M + \cdots + (a_{n-1} - a_n)M + a_n M = a_1 \cdot M,
$$

which proves the second inequality of (14.30). The first inequality can be proved in a similar way. $\qquad\square$

Proof (Theorem 14.53). Let $\varepsilon > 0$ be given, and let us choose a partition $F : a = x_0 < \cdots < x_n = b$ such that $\Omega_F(g) < \varepsilon$ and $\Omega_F(f \cdot g) < \varepsilon$. Then for arbitrary $1 \le k \le n$,

$$
m - \varepsilon < \sum_{i=1}^{k} g(x_{i-1})(x_i - x_{i-1}) < M + \varepsilon. \tag{14.32}
$$

---

[1] Niels Henrik Abel (1802–1829), Norwegian mathematician.

Indeed, if F_k denotes the partition $a = x_0 < \cdots < x_k$ of the interval $[a, x_k]$, then the oscillatory sum of g corresponding to F_k is at most $\Omega_F(g)$, which is smaller than ε. Thus $s_{F_k}(g)$ and $S_{F_k}(g)$ are both closer to $\int_a^{x_k} g\, dx$ than ε. By the condition, this last integral falls between m and M, so

$$m - \varepsilon \leq \int_a^{x_k} g\, dx - \varepsilon < s_{F_k}(g) \leq \sum_{i=1}^k g(x_{i-1})(x_i - x_{i-1}) \leq S_{F_k}(g) <$$

$$< \int_a^{x_k} g\, dx + \varepsilon \leq M + \varepsilon,$$

which proves (14.32).

Let $S = \sum_{i=1}^n f(x_{i-1})g(x_{i-1})(x_i - x_{i-1})$. Then $\left| S - \int_a^b fg\, dx \right| < \varepsilon$, since $\Omega_F(f \cdot g) < \varepsilon$. Since $f(x_0) \geq f(x_1) \geq \ldots \geq f(x_{n-1}) \geq 0$, by Abel's inequality (Theorem 14.54) applied to the sum S, we have $f(x_0) \cdot (m - \varepsilon) \leq S \leq f(x_0) \cdot (M + \varepsilon)$. Then by $a = x_0$, we see that $f(a) \cdot (m - \varepsilon) \leq S \leq f(a) \cdot (M + \varepsilon)$, so

$$f(a) \cdot (m - \varepsilon) - \varepsilon < \int_a^b fg\, dx < f(a) \cdot (M + \varepsilon) + \varepsilon.$$

This is true for every $\varepsilon > 0$, so we have proved (14.29).   □

Inequalities about sums can often be generalized to integrals. One of these is **Hölder's inequality,** which corresponds to Theorem 11.18.

Theorem 14.55 (Hölder's Inequality). *Let p and q be positive numbers such that $1/p + 1/q = 1$. If f and g are integrable on $[a, b]$, then*

$$\left| \int_a^b f(x)g(x)\, dx \right| \leq \sqrt[p]{\int_a^b |f(x)|^p\, dx} \cdot \sqrt[q]{\int_a^b |g(x)|^q\, dx}. \qquad (14.33)$$

Proof. By Theorems 14.33 and 14.35, the functions $|fg|$, $|f|^p$, and $|g|^q$ are all integrable on $[a, b]$. Let $F : a = x_0 < \cdots < x_n = b$ be a partition such that $\Omega_F(|fg|) < \varepsilon$, $\Omega_F(|f|^p) < \varepsilon$ and $\Omega_F(|g|^q) < \varepsilon$.
Then we can say that the Riemann sum $\sum_{i=1}^n f(x_i)g(x_i) \cdot (x_i - x_{i-1})$ is less than ε away from the integral $A = \int_a^b f(x)g(x)\, dx$, the Riemann sum $\sum_{i=1}^n |f(x_i)|^p \cdot (x_i - x_{i-1})$ is less than ε away from the integral $B = \int_a^b |f(x)|^p\, dx$, and the Riemann sum $\sum_{i=1}^n |g(x_i)|^q \cdot (x_i - x_{i-1})$ is less than ε away from the integral $C = \int_a^b |g(x)|^q\, dx$.
Now by Hölder's inequality, for these sums (Theorem 11.18), we have

$$|A| - \varepsilon < \left| \sum_{i=1}^{n} f(x_i)g(x_i) \cdot (x_i - x_{i-1}) \right| =$$

$$= \left| \sum_{i=1}^{n} (f(x_i)(x_i - x_{i-1})^{1/p}) \cdot (g(x_i)(x_i - x_{i-1})^{1/q}) \right| \le$$

$$\le \sqrt[p]{\sum_{i=1}^{n} |f(x_i)|^p (x_i - x_{i-1})} \cdot \sqrt[q]{\sum_{i=1}^{n} |g(x_i)|^q (x_i - x_{i-1})} \le$$

$$\le \sqrt[p]{B + \varepsilon} \cdot \sqrt[q]{C + \varepsilon}.$$

Since ε was arbitrary, (14.33) holds. □

For the case $p = q = 2$, we get the following famous inequality, which is the analogue of Theorem 11.19 for integrals.

Theorem 14.56 (Schwarz Inequality). *If f and g are integrable in $[a,b]$, then*

$$\left(\int_a^b f(x)g(x)\,dx \right)^2 \le \int_a^b f^2(x)\,dx \cdot \int_a^b g^2(x)\,dx.$$

Remark 14.57. The Schwarz inequality forms the basis of an important analogy between functions integrable on $[a,b]$ and vectors.

If $x = (x_1, x_2)$ and $y = (y_1, y_2)$ are vectors in $\mathbb{R}^2$, then the number $x_1 y_1 + x_2 y_2$ is called the **dot** or **scalar product** of x and y, and is denoted by $\langle x, y \rangle$. The Cauchy–Schwarz–Bunyakovsky inequality (Theorem 11.19) states that

$$|\langle x, y \rangle| \le |x| \cdot |y| \quad \text{for all} \quad x, y \in \mathbb{R}^2.$$

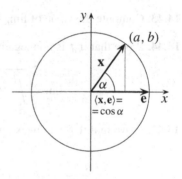

Fig. 14.10

In fact, $\langle x, y \rangle = |x| \cdot |y| \cdot \cos \alpha$, where α denotes the angle between the rays pointing toward x and y. We can prove this in the following way. Since

$$\langle \lambda x, y \rangle = \langle x, \lambda y \rangle = \lambda \cdot \langle x, y \rangle$$

for all $\lambda \in \mathbb{R}$, we can suppose that x and y are unit vectors. We have to show that in this case, $\langle x, y \rangle$ is equal to the cosine of the subtended angle. By the equality

$$\langle x, y \rangle = \frac{1}{4} \cdot \left(|x+y|^2 - |x-y|^2 \right),$$

we know that congruences (that is, isometries—mappings preserving distances) do not change the value of the dot product. A suitable isometry maps the vector x into the vector $(1,0)$, and the vector y into the vector (a,b), where $\sqrt{a^2 + b^2} = 1$. The dot product of these vectors is a. On the other hand, (a,b) is a point on the unit circle, and so by the definition of the cosine function, $a = \cos\alpha$, where α is the angle enclosed by the rays containing (a,b) and $(1,0)$. That is, $\langle x,y \rangle = a = \cos\alpha$, which is what we wanted to show (Figure 14.10). The above also shows that two vectors are perpendicular to each other if and only if their scalar product is zero.

Theorem 14.56 provides the analogy to consider the number $\int_a^b f(x)g(x)\,dx$ the **scalar product** of the functions f and g, and the number $\sqrt{\int_a^b f^2(x)\,dx}$ the **absolute value** of the function f. Hence we can consider two functions f and g **perpendicular (or orthogonal)** if their scalar product $\int_a^b f(x)g(x)\,dx$ is zero.

This analogy works quite well, and leads to the theory of Hilbert[2] spaces.

Exercises

14.41. Prove that if $a < b$ and f is continuous on $[a,b]$, then $\lim_{h\to 0+0}\int_a^{b-h}[f(x+h)-f(x)]\,dx = 0$.

14.42. Let $f\colon [0,\infty) \to \mathbb{R}$ be a continuous function, and suppose that $\lim_{x\to\infty} f(x) = c$. Prove that $\lim_{t\to\infty}\int_0^1 f(tx)\,dx = c$. (H)

14.43. Compute the value of $\lim_{n\to\infty}\int_0^1 (1-x)^n\,dx$.

14.44. Prove that if f is nonnegative and continuous in $[a,b]$, then

$$\lim_{n\to\infty} \sqrt[n]{\int_a^b f^n(x)\,dx} = \max f([a,b]). \text{ (H)}$$

14.45. Prove that if f is convex on $[a,b]$, then

$$f\left(\frac{a+b}{2}\right)\cdot(b-a) \le \int_a^b f(x)\,dx \le \frac{f(a)+f(b)}{2}\cdot(b-a).$$

14.46. Prove that if f is differentiable on $[a,b]$ and $f(a) = f(b) = 0$, then there exists a $c \in [a,b]$ such that

$$f'(c) \ge \frac{2}{(b-a)^2}\cdot\int_a^b f(x)\,dx.$$

(We can interpret this exercise as follows: if a point moves along a straight line through the time interval $[a,b]$ with zero initial and final velocity, then for it to travel a distance d, it must have reached an acceleration of $2d/(b-a)^2$ along its route.)

---

[2] David Hilbert (1862–1943), German mathematician.

14.47. Prove that if f is nonnegative, continuous, and concave on $[0,1]$, and if furthermore, $f(0) = 1$, then

$$\int\limits_0^1 x \cdot f(x)\,dx \le \frac{2}{3} \cdot \left(\int\limits_0^1 f(x)\,dx\right)^2.$$

When does equality hold?

14.48. When does equality hold in (14.33)?

14.49. Let $f\colon \mathbb{R} \to \mathbb{R}$ be a continuous function, and suppose that

$$A = \lim_{x \to -\infty} f(x) \qquad \text{and} \qquad B = \lim_{x \to \infty} f(x).$$

Determine the limit

$$\lim_{a \to \infty} \int\limits_{-a}^a [f(x+1) - f(x)]\,dx.$$

14.50. Suppose that there is a number $\Phi(f;[a,b])$ assigned to every interval $[a,b]$ and integrable function f with the following properties:

  (i) If f is integrable on $[a,b]$, then $\Phi(cf;[a,b]) = c \cdot \Phi(f;[a,b])$ for all $c \in \mathbb{R}$.
  (ii) If f and g are integrable on $[a,b]$, then $\Phi(f+g;[a,b]) = \Phi(f;[a,b]) + \Phi(g; [a,b])$.
  (iii) If $a < b < c$ and f is integrable on $[a,c]$, then $\Phi(f;[a,c]) = \Phi(f;[a,b]) + \Phi(f;[b,c])$.
  (iv) If f and g are integrable on $[a,b]$ and $f(x) \le g(x)$ for all $x \in [a,b]$, then $\Phi(f;[a,b]) \le \Phi(g;[a,b])$.
  (v) If $e(x) = 1$ for all $x \in [a,b]$, then $\Phi(e;[a,b]) = b - a$.

Prove that $\Phi(f;[a,b]) = \int_a^b f(x)\,dx$ for every function f integrable over $[a,b]$. (H)

14.51. Assume Φ to be as in the previous question, except replace (v) with the following condition:

(vi) If f is integrable on $[a,b]$, then $\Phi(f_c;[a-c,b-c]) = \Phi(f;[a,b])$ for all $c \in \mathbb{R}$, where $f_c(x) = f(x+c)$ $(x \in [a-c,b-c])$.

Prove that there exists a constant $\alpha \ge 0$ such that $\Phi(f;[a,b]) = \alpha \cdot \int_a^b f(x)\,dx$ for every function f integrable over $[a,b]$. (H)

Chapter 15
Integration

In this chapter, we will familiarize ourselves with the most important methods for computing integrals, which will also make the link between definite and indefinite integrals clear.

15.1 The Link Between Integration and Differentiation

Examples 14.8 and 14.21 both provide equalities of the form $\int_a^b f(x)\,dx = F(b) - F(a)$ with the following cast:

$$f(x) \equiv c, \qquad F(x) = c \cdot x;$$
$$f(x) = 1/x, \qquad F(x) = \log x;$$
$$f(x) = x^\alpha, \qquad F(x) = \frac{1}{\alpha+1} \cdot x^{\alpha+1} \qquad (\alpha > 0);$$
$$f(x) = e^x, \qquad F(x) = e^x;$$
$$f(x) = \cos x, \qquad F(x) = \sin x.$$

As we can see, in each example, $F' = f$, that is, F is a primitive function of f. These examples illustrate an important link between integration and differentiation, and are special cases of a famous general theorem.

Theorem 15.1 (Fundamental Theorem of Calculus). *Let f be integrable on $[a,b]$. If the function F is continuous on $[a,b]$, differentiable on (a,b), and $F'(x) = f(x)$ for all $x \in (a,b)$ (that is, F is a primitive function of f on (a,b)), then*

$$\int_a^b f(x)\,dx = F(b) - F(a).$$

Proof. Let $a = x_0 < x_1 < \cdots < x_n = b$ be an arbitrary partition of $[a,b]$. By the mean value theorem (Theorem 12.50), for all i, there exists a point $c_i \in (x_{i-1}, x_i)$ such that

$$F(x_i) - F(x_{i-1}) = F'(c_i)(x_i - x_{i-1}) = f(c_i)(x_i - x_{i-1})$$

© Springer New York 2015
M. Laczkovich, V.T. Sós, *Real Analysis*, Undergraduate Texts
in Mathematics, DOI 10.1007/978-1-4939-2766-1_15

holds. If we sum these equalities for all $i = 1, \ldots, n$, then every term cancels out on the left-hand side except for the terms $F(x_n) = F(b)$ and $F(x_0) = F(a)$, and so we get that

$$F(b) - F(a) = \sum_{i=1}^{n} f(c_i)(x_i - x_{i-1}).$$

This means that for every partition, there exist inner points such that the Riemann sum with those points is equal to $F(b) - F(a)$. Thus the number $F(b) - F(a)$ lies between the lower and upper sums for every partition. Since f is integrable, there is only one such number: the integral of f. Thus $F(b) - F(a) = \int_a^b f(x)\,dx$.   □

Remark 15.2. While making clear the definition of differentiability back in Chapter 12, we concluded that if the function $s(t)$ defines the position of a moving point, then its instantaneous velocity is $v(t) = s'(t)$. Since $s(b) - s(a)$ is the distance the point travels during the time interval $[a,b]$, the physical interpretation of the fundamental theorem of calculus says that the distance traveled is equal to the integral of the velocity.

As we saw in Chapter 13, deciding whether a function has a primitive function is generally a hard task (see Remarks 13.27 and 13.45). However, if the function f is integrable, then deciding this question—with the help of the fundamental theorem of calculus—is quite easy. Suppose, for example, that f is integrable on $[a,b]$, and that F is a primitive function of f. We can assume that $F(a) = 0$, since if this does not hold, we can just consider the function $F(x) - F(a)$ instead of $F(x)$. Let $x \in [a,b]$, and apply the fundamental theorem of calculus to the interval $[a,x]$. We get that $\int_a^x f(t)\,dt = F(x) - F(a) = F(x)$, that is, $F(x) = \int_a^x f(t)\,dt$ for all $x \in [a,b]$. This means that if f has a primitive function, then the function $x \mapsto \int_a^x f(t)\,dt$ must also be a primitive function. We will introduce a name for this function.

Definition 15.3. Let f be integrable on $[a,b]$. The function

$$I(x) = \int_a^x f(t)\,dt \qquad (x \in [a,b])$$

is called the *integral function* of f.

With the use of this new concept, we can summarize the results of our previous argument as follows.

Theorem 15.4. *An integrable function has a primitive function if and only if its integral function is its primitive function.*

The most important properties of the integral function are expressed by the following theorem.

Theorem 15.5. *Let f be integrable on $[a,b]$, and let $I(x)$ be its integral function.*

(i) *The function I is continuous and even has the Lipschitz property on the interval $[a,b]$.*

(ii) *If f is continuous at the point $x_0 \in [a,b]$, then I is differentiable there, and $I'(x_0) = f(x_0)$.*

(iii) *If f is continuous on $[a,b]$, then I is differentiable on $[a,b]$, and $I' = f$. It follows that if f is continuous on $[a,b]$, then it has a primitive function there.*

Proof. (i) Let $|f(x)| \le K$ for all $x \in [a,b]$. If $a \le x < y \le b$, then by Theorem 14.38, we have

$$I(y) - I(x) = \int_a^y f(t)\,dt - \int_a^x f(t)\,dt = \int_x^y f(t)\,dt,$$

so $|I(y) - I(x)| \le K \cdot |y-x|$ by statement (iv) of Theorem 14.49.

(ii) Again by Theorem 14.38, we have

$$I(x) - I(x_0) = \int_a^x f(t)dt - \int_a^{x_0} f(t)dt = \int_{x_0}^x f(t)dt$$

so the difference quotient of the function I corresponding to the points x and x_0 is

$$\frac{I(x) - I(x_0)}{x - x_0} = \frac{1}{x - x_0} \int_{x_0}^x f(t)\,dt.$$

Since f is continuous at x_0, for arbitrary $\varepsilon > 0$ there exists a $\delta > 0$ such that

$$f(x_0) - \varepsilon < f(t) < f(x_0) + \varepsilon \quad \text{if } |t - x_0| < \delta.$$

First let $x_0 < x < x_0 + \delta$. For all such x, it follows from statement (iii) of Theorem 14.49 that

$$(f(x_0) - \varepsilon)(x - x_0) \le \int_{x_0}^x f(t)\,dt \le (f(x_0) + \varepsilon)(x - x_0)$$

holds, that is,

$$f(x_0) - \varepsilon \le \frac{I(x) - I(x_0)}{x - x_0} \le f(x_0) + \varepsilon.$$

The same can be said when $x < x_0$ by rearranging $(I(x_0) - I(x))/(x_0 - x)$, so we have

$$I'(x_0) = \lim_{x \to x_0} \frac{I(x) - I(x_0)}{x - x_0} = f(x_0).$$

Statement (iii) is clear from (ii). □

Remarks 15.6. **1.** We can see from the proof that if f is continuous from the right or the left at x_0, then $I'_+(x_0) = f(x_0)$ or $I'_-(x_0) = f(x_0)$ respectively.

2. The proof of statement (ii) above uses an argument we have already seen before. In Example 10.7, when we determined the area under the graph of a nonnegative monotone increasing and continuous function $f : [a,b] \to \mathbb{R}$, we showed that if $T(x)$ denotes the area over the interval $[a,x]$, then $T'(x) = f(x)$. Statement (ii) of Theorem 15.5 is actually a rephrasing of this, in which we replace area—which we still have not clearly defined—with the integral, and the function $T(x)$ with the integral function.

3. The fundamental theorem of calculus implies that if a function F is continuously differentiable,[1] then differentiating F and integrating the derivative gives us F back (more precisely, its increment on the interval $[a,b]$). By statement (iii) of 15.5, if we integrate a continuous function f from a to x, and then we differentiate the integral function we get, then we obtain f. These two statements express that *integration and differentiation are inverse operations in some sense.*

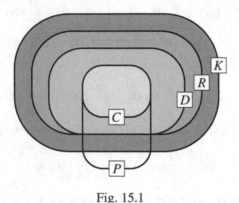

Fig. 15.1

In talking about the theory of integration, several different properties of functions came into play: boundedness, integrability, continuity, and the property of having a primitive function. We will use the following notation for functions that have the corresponding properties.

$$K[a,b] = \{f : [a,b] \to \mathbb{R} \text{ and } f \text{ is bounded in } [a,b]\},$$
$$R[a,b] = \{f : [a,b] \to \mathbb{R} \text{ and } f \text{ is Riemann integrable on } [a,b]\},$$
$$C[a,b] = \{f : [a,b] \to \mathbb{R} \text{ and } f \text{ is continuous on } [a,b]\},$$
$$P[a,b] = \{f : [a,b] \to \mathbb{R} \text{ and } f \text{ has a primitive function on } [a,b]\}.$$

We introduce separate notation for the set of integrable functions whose integral function is differentiable:

$$D[a,b] = \{f \in R[a,b] \text{ and the integral function of } f \text{ is differentiable in } [a,b]\}.$$

By Theorem 15.5 (and the proper definitions), the containment relations

$$C[a,b] \subset D[a,b] \subset R[a,b] \subset K[a,b] \quad \text{and} \quad C[a,b] \subset P[a,b] \tag{15.1}$$

---

[1] By this we mean that the function is differentiable and its derivative is continuous.

hold for these classes of functions. Moreover,

$$R[a,b] \cap P[a,b] \subset D[a,b]. \tag{15.2}$$

This is a straightforward corollary of Theorem 15.4 (Figure 15.1).

We now show that aside from what is listed in (15.1), no other containment relations exist between these classes of functions.

Examples 15.7. **1.** $f \in K[0,1] \not\Rightarrow f \in R[0,1]$: The Dirichlet function is an example.
2. $f \in R[0,1] \not\Rightarrow f \in D[0,1]$: Let $f(x) = 0$ if $0 \leq x < 1/2$, and $f(x) = 1$ if $1/2 \leq x \leq 1$.
3. $f \in D[0,1] \not\Rightarrow f \in C[0,1]$: By (15.2), it suffices to give a function that is integrable, has a primitive function, but is not continuous. Let

$$f(x) = \begin{cases} 2x\sin(1/x) - \cos(1/x), & \text{if } x \neq 0, \\ 0, & \text{if x=0.} \end{cases}$$

Since f is bounded and continuous everywhere except at a point, it is integrable. The function f has a primitive function, too, namely the function

$$F(x) = \begin{cases} x^2\sin(1/x), & \text{if } x \neq 0, \\ 0, & \text{if } x = 0. \end{cases}$$

On the other hand, f is not continuous at 0.
4. $f \in D[0,1] \not\Rightarrow f \in P[0,1]$: Let $f(0) = 1$ and $f(x) = 0$ $(0 < x \leq 1)$.
5. $f \in P[0,1] \not\Rightarrow f \in K[0,1]$: See Example 13.46.

We mention that there exists a bounded function that has a primitive function but is not integrable (that is, $f \in K[0,1] \cap P[0,1] \not\Rightarrow f \in R[0,1]$). Constructing such a function is significantly more difficult than constructing the previous ones, so we will skip that for now.

Combining the continuity of integral functions with Abel's inequality yields an important result.

Theorem 15.8 (Second Mean Value Theorem for Integration).

(i) *Let f be monotone decreasing and nonnegative, and let g be integrable in $[a,b]$. Then there exists a $\xi \in [a,b]$ such that*

$$\int_a^b f(x)g(x)\,dx = f(a) \cdot \int_a^\xi g(x)\,dx. \tag{15.3}$$

(ii) *Let f be monotone and let g be integrable in $[a,b]$. Then there exists a $\xi \in [a,b]$ such that*

$$\int_a^b f(x)g(x)\,dx = f(a) \cdot \int_a^\xi g(x)\,dx + f(b) \cdot \int_\xi^b g(x)\,dx. \tag{15.4}$$

Proof. (i) The integral function $G(x) = \int_a^x g(t)\,dt$ is continuous in $[a,b]$ by The-orem 15.5, so its range in $[a,b]$ has a smallest and a greatest element. Let $m = \min G[a,b]$, $M = \max G[a,b]$, and $I = \int_a^b fg\,dx$. Then $f(a)\cdot m \leq I \leq f(a)\cdot M$ by Theorem 14.53. Since $f(a)\cdot G$ takes on every value between $f(a)\cdot m$ and $f(a)\cdot M$ by the Bolzano–Darboux theorem, there exists a $\xi \in [a,b]$ such that $f(a)\cdot G(\xi) = I$, which is exactly (15.3).

(ii) We can suppose that f is monotone decreasing, since otherwise, we can switch to the function $-f$. Then $f - f(b)$ is monotone decreasing and nonnegative in $[a,b]$, so by (i), there exists a $\xi \in [a,b]$ such that

$$\int_a^b (f(x) - f(b))g(x)\,dx = (f(a) - f(b))\cdot \int_a^\xi g(x)\,dx,$$

from which we get

$$\int_a^b f(x)g(x)\,dx = (f(a) - f(b))\cdot \int_a^\xi g(x)\,dx + f(b)\cdot \int_a^b g(x)\,dx =$$

$$= f(a)\cdot \int_a^\xi g(x)\,dx + f(b)\cdot \int_\xi^b g(x)\,dx.$$

$\square$

Exercises

15.1. Give every primitive function, integral function, indefinite integral (see Defi-nition 13.24), and definite integral of the functions below over the interval $[-2,3]$:

(a) $|x|$;

(b) $\mathrm{sgn}(x)$;

(c) $f(x) = \begin{cases} 1+x^2, & \text{if } x \geq 0, \\ 1-x^2, & \text{if } x < 0. \end{cases}$

15.2. Let $f(x) = |x| - 2$ $(x \in [-2,1])$. Does there exist a function whose integral function is f? Decide the same for the function $g(x) = [x]$ $(x \in [-2,1])$.

15.3. Does there exist a function on $[0,1]$ whose integral function is $\sqrt{x}$? (H)

15.4. Let $f : [a,b] \to \mathbb{R}$ be bounded, and let F be a primitive function of f. Prove that

$$\underline{\int_a^b} f(x)\,dx \leq F(b) - F(a) \leq \overline{\int_a^b} f(x)\,dx. \qquad \text{(H)}$$

15.5. Let

$$G(x) = \int_0^{x^4} e^{t^3}\cdot \sin t\,dt \qquad (x \in \mathbb{R}).$$

Determine the derivative of G.

15.6. Prove that there are only two continuous functions defined on $[a,b]$ that satisfy

$$\int_a^x f(t)\,dt = \int_a^x f^2(t)\,dt$$

for all $x \in [a,b]$.

15.7. Prove that if f is continuous in $[0,1]$ and $f(x) < 1$ for all $x \in [0,1]$, then the equation

$$2x - \int_0^x f(t)\,dt = 1$$

has exactly one root in $[0,1]$.

15.8. For which values of x is the value of

$$\int_0^x \frac{\sin t}{\sqrt{t}}\,dt$$

maximized?

15.9. Let f be integrable on $[a,b]$, and let the integral function of f be I. Is it possible for I to be differentiable everywhere and $I'(x) \neq f(x)$ for all $x \in [a,b]$? (H)

15.10. Prove that

$$\lim_{n\to\infty} n\left(\frac{1}{1+n^2} + \frac{1}{2^2+n^2} + \cdots + \frac{1}{n^2+n^2}\right) = \frac{\pi}{4}.$$

15.11. Determine the limits of the following sequences:

(a) $a_n = \sum_{k=1}^{2n} \frac{n}{k^2+n^2}$,

(b) $a_n = \sum_{k=n}^{2n} \frac{n}{k(n+k)}$,

(c) $a_n = \sum_{k=1}^{n} \frac{k}{k^2+n^2}$,

(d) $a_n = \sum_{k=2n}^{3n} \frac{k}{n^2} e^{\frac{k}{n}}$,

(e) $a_n = \left(\left(1+\frac{1}{n}\right)\left(1+\frac{2}{n}\right)\cdots\left(1+\frac{n}{n}\right)\right)^{\frac{1}{n}}$.

15.12. Let

$$G(x) = \int_x^{2x} \frac{dt}{t} \qquad (x > 0).$$

Determine $G'(x)$ without using the fundamental theorem of calculus. How can we interpret the result?

15.2 Integration by Parts

The fundamental theorem of calculus is significant not only from a theoretical standpoint (in which it outlines a link between differentiation and integration), but in terms of applications as well, since it tells us the value of the definite integral whenever we know a primitive function of the function we are integrating. Thus we can use the methods for computing indefinite integrals to compute definite integrals. In order to have the formulas in a more concise form, we introduce the following notation: if the function F is defined on the interval $[a,b]$, then we denote the difference $F(b) - F(a)$ by $[F]_a^b$.

We now extend our toolkit from Chapter 13 (Theorems 13.28, 13.30, and 13.33) with two new methods that greatly increase the number of integrals we can compute.

Theorem 15.9 (Integration by Parts). *Suppose the functions f and g are differentiable on the interval I, and fg' has a primitive function there. Then $f'g$ has a primitive function on I as well, and*

$$\int f'g \, dx = fg - \int fg' \, dx. \tag{15.5}$$

Proof. Let $F \in \int fg' \, dx$. Since $(fg)' = f'g + fg'$, we have

$$(fg - F)' = f'g + fg' - fg' = f'g,$$

which is exactly (15.5). $\qquad\square$

Examples 15.10. A few examples of integration by parts follow.

1. $\int x \cdot \cos x \, dx = \int x \cdot (\sin x)' \, dx = x \sin x - \int x' \cdot \sin x \, dx =$
$= x \sin x - \int 1 \cdot \sin x \, dx = x \sin x + \cos x + C.$

2. $\int x \cdot e^x \, dx = \int x \cdot (e^x)' \, dx = x \cdot e^x - \int x' \cdot e^x \, dx = x \cdot e^x - \int 1 \cdot e^x \, dx =$
$= (x - 1)e^x + C.$

3. $\int x \cdot \log x \, dx = \int \left(\frac{x^2}{2}\right)' \cdot \log x \, dx = \frac{x^2}{2} \cdot \log x - \int \frac{x^2}{2} \cdot (\log x)' \, dx =$
$= \frac{x^2}{2} \cdot \log x - \int \frac{x^2}{2} \cdot \frac{1}{x} \, dx = \frac{x^2}{2} \log x - \frac{x^2}{4} + C \;\; (x > 0).$

4. $\int e^x \cdot \cos x \, dx = \int e^x \cdot (\sin x)' \, dx = e^x \sin x - \int (e^x)' \cdot \sin x \, dx =$
$= e^x \sin x - \int e^x \sin x \, dx = e^x \sin x - \int e^x (-\cos x)' \, dx =$
$= e^x \sin x + e^x \cos x - \int e^x \cos x \, dx,$
so $\int e^x \cos x \, dx = \frac{1}{2} \cdot (e^x \sin x + e^x \cos x) + C.$

5. $\int \log x \, dx = \int x' \cdot \log x \, dx = x \log x - \int x \cdot (\log x)' \, dx =$
$= x \log x - \int x \cdot \frac{1}{x} \, dx = (x \cdot \log x) - x + C \;\; (x > 0).$

6. $\int \operatorname{arc tg} x \, dx = \int x' \cdot \operatorname{arc tg} x \, dx = x \cdot \operatorname{arc tg} x - \int x \cdot (\operatorname{arc tg} x)' \, dx =$
$= x \cdot \operatorname{arc tg} x - \int \frac{x}{1+x^2} \, dx = x \cdot \operatorname{arc tg} x - \frac{1}{2} \log(1 + x^2) + C.$

Applying integration by parts repeatedly allows us to compute various integrals such as the following:

$$\int x^k \cos x \, dx, \quad \int x^k \sin x \, dx, \quad \int x^k e^x \, dx,$$

$$\int x^k \log^n x \, dx, \quad \int x^k e^x \cos x \, dx, \quad \int x^k e^x \sin x \, dx.$$

The following theorem gives us integration by parts for definite integrals.

Theorem 15.11. *Suppose f and g are differentiable functions, while f' and g' are integrable over $[a, b]$. Then*

$$\int_a^b f' g \, dx = [fg]_a^b - \int_a^b fg' \, dx. \tag{15.6}$$

Proof. Since f and g are differentiable, they are continuous, so by Theorem 14.26, the are also integrable on $[a, b]$. Thus $f'g$ and fg' are both integrable on $[a, b]$ by Theorem 14.33. Since $(fg)' = f'g + fg'$, we have

$$\int_a^b (f'g + fg') \, dx = [fg]_a^b$$

by the fundamental theorem of calculus. Applying Theorem 14.31 and rearranging what we get yields (15.6). $\qquad\square$

As an interesting application of the previous theorem, we get the following formulas.

Theorem 15.12.

$$\int_0^\pi \sin^{2n} x \, dx = \frac{1 \cdot 3 \cdots (2n-1)}{2 \cdot 4 \cdots 2n} \cdot \pi \qquad (n \in \mathbb{N}^+), \tag{15.7}$$

and

$$\int_0^\pi \sin^{2n+1} x \, dx = \frac{2 \cdot 4 \cdots 2n}{1 \cdot 3 \cdots (2n+1)} \cdot 2 \qquad (n \in \mathbb{N}). \tag{15.8}$$

Proof. Let $I_k = \int_0^\pi \sin^k x \, dx$ for all $k \in \mathbb{N}$. Then $I_0 = \pi$ and $I_1 = \cos 0 - \cos \pi = 2$. If $k \geq 1$, then

$$I_{k+1} = \int_0^\pi \sin^2 x \cdot \sin^{k-1} x \, dx = \int_0^\pi (1 - \cos^2 x) \cdot \sin^{k-1} x \, dx =$$

$$= \int_0^\pi \left[\sin^{k-1} x - \cos^2 x \cdot \sin^{k-1} x \right] dx =$$

$$= I_{k-1} - \int_0^\pi \cos x \cdot \left[\sin^{k-1} x \cdot \cos x \right] dx. \tag{15.9}$$

Now, using integration by parts, we get that

$$\int_0^{\pi} \cos x \cdot \left[\sin^{k-1} x \cdot \cos x \right] dx = \int_0^{\pi} \cos x \cdot \left(\frac{1}{k} \cdot \sin^k x \right)' dx =$$

$$= \left[\cos x \cdot \frac{1}{k} \cdot \sin^k x \right]_0^{\pi} - \int_0^{\pi} \frac{1}{k} \cdot \sin^k x \cdot (-\sin x) \, dx =$$

$$= 0 + \frac{1}{k} \cdot I_{k+1}.$$

Combining this with (15.9), we obtain $I_{k+1} = I_{k-1} - \frac{1}{k} \cdot I_{k+1}$, so $I_{k+1} = \frac{k}{k+1} \cdot I_{k-1}$. Thus

$$I_{2n} = \frac{2n-1}{2n} \cdot I_{2n-2} = \cdots = \frac{2n-1}{2n} \cdot \frac{2n-3}{2n-2} \cdots \frac{1}{2} \cdot I_0,$$

which is exactly (15.7). Similarly,

$$I_{2n+1} = \frac{2n}{2n+1} \cdot I_{2n-1} = \cdots = \frac{2n}{2n+1} \cdot \frac{2n-2}{2n-1} \cdots \frac{2}{3} \cdot I_1,$$

which is (15.8).   □

   The equations above make possible the proof of a fundamentally important identity that expresses the number π as the limit of a simple product.

Theorem 15.13 (Wallis' Formula[2]).

$$\pi = \lim_{n \to \infty} \left[\frac{2 \cdot 4 \cdots 2n}{1 \cdot 3 \cdots (2n-1)} \right]^2 \cdot \frac{1}{n}.$$

Proof. Since $\sin^{2n-1} x \geq \sin^{2n} x \geq \sin^{2n+1} x$ for all $x \in [0, \pi]$, we have $I_{2n-1} \geq I_{2n} \geq I_{2n+1}$. Thus

$$\frac{2 \cdot 4 \cdots (2n-2)}{1 \cdot 3 \cdots (2n-1)} \cdot 2 \geq \frac{1 \cdot 3 \cdots (2n-1)}{2 \cdot 4 \cdots 2n} \cdot \pi \geq \frac{2 \cdot 4 \cdots 2n}{1 \cdot 3 \cdots (2n+1)} \cdot 2,$$

which gives

$$\left[\frac{2 \cdot 4 \cdots 2n}{1 \cdot 3 \cdots (2n-1)} \right]^2 \cdot \frac{1}{n} \geq \pi \geq \left[\frac{2 \cdot 4 \cdots 2n}{1 \cdot 3 \cdots (2n-1)} \right]^2 \cdot \frac{2}{2n+1}.$$

Let W_n denote the product $[(2 \cdot 4 \cdots 2n)/(1 \cdot 3 \cdots (2n-1))]^2 \cdot 1/n$. Then $W_n \geq \pi \geq W_n \cdot 2n/2n+1$, that is, $\pi \leq W_n \leq \pi \cdot 2n+1/2n$. Thus, by the squeeze theorem, $\lim_{n \to \infty} W_n = \pi$.   □

---

[2] John Wallis (1616–1703), English mathematician.

Remark 15.14. Since

$$\frac{2 \cdot 4 \cdots 2n}{1 \cdot 3 \cdots (2n-1)} = \frac{(2 \cdot 4 \cdots 2n)^2}{1 \cdot 2 \cdots (2n)} = \frac{[2^n \cdot n!]^2}{(2n)!} = \frac{4^n}{\binom{2n}{n}},$$

Wallis's formula gives

$$\lim_{n \to \infty} \frac{4^n}{\binom{2n}{n}\sqrt{n}} = \sqrt{\pi}.$$

With our asymptotic notation, we can express this by saying that

$$\binom{2n}{n} \sim \frac{4^n}{\sqrt{n\pi}}. \tag{15.10}$$

By the binomial theorem, the sum of the binomial coefficients

$$\binom{2n}{0}, \binom{2n}{1}, \ldots, \binom{2n}{2n} \tag{15.11}$$

is 4^n, so their mean is $4^n/(2n+1)$. Now (15.10) says that the middle term in (15.11) (which is the largest term as well) is about $c \cdot \sqrt{n}$ times the mean, where $c = 2/\sqrt{\pi}$.

We now prove an important theorem as an application of Wallis's formula that itself has many applications in many fields of mathematics, especially in probability.

Theorem 15.15 (Stirling's Formula). $n! \sim \left(\frac{n}{e}\right)^n \cdot \sqrt{2\pi n}.$

Proof. First we show that the sequence $a_n = (n/e)^n \sqrt{2\pi n}/n!$ is strictly monotone increasing and bounded. A simple computation yields

$$\frac{a_{n+1}}{a_n} = \frac{\left(1+\frac{1}{n}\right)^{n+(1/2)}}{e},$$

so

$$\log a_{n+1} - \log a_n = \left(n+\frac{1}{2}\right) \cdot \log\left(1+\frac{1}{n}\right) - 1 \tag{15.12}$$

for all n. We know that for $x > 0$, we have $\log(1+x) > 2x/(x+2)$ (as seen in Example 12.56). Apply this for $x = 1/n$, then multiply through by $n+(1/2)$ to get that

$$\left(n+\frac{1}{2}\right) \cdot \log\left(1+\frac{1}{n}\right) > 1.$$

This proves that (a_n) is strictly monotone increasing by (15.12).

Next, we will use the inequality

$$\log(1+x) \leq x - \frac{x^2}{2} + \frac{x^3}{3} \qquad (x > 0) \tag{15.13}$$

to find an upper bound for the right-hand side of (15.12). (We refer to Exercise 12.91 or (13.26) to justify (15.13).) If we substitute $x = 1/n$ into (15.13) and multiply through by $n + (1/2)$, then with the help of (15.12), we get

$$\log a_{n+1} - \log a_n \le \frac{1}{12n^2} + \frac{1}{6n^3} \le \frac{1}{12n^2} + \frac{1}{6n^2} = \frac{1}{4n^2}. \tag{15.14}$$

Thus

$$\log a_n - \log a_1 = \sum_{i=1}^{n-1} (\log a_{i+1} - \log a_i) \le \sum_{i=1}^{n-1} \frac{1}{4i^2} < \frac{1}{2}$$

for all n, so it is clear that the sequence (a_n) is bounded. Since we have shown that (a_n) is monotone increasing and bounded, it must be convergent. Let $\lim_{n \to \infty} a_n = a$. Since every term of the sequence is positive, $a > 0$. It is clear that $a_n^2/a_{2n} \to a$. On the other hand, a simple computation gives

$$\frac{a_n^2}{a_{2n}} = \frac{\binom{2n}{n} \cdot \sqrt{\pi n}}{4^n}$$

for all n. Thus by Wallis's formula (or more precisely by (15.10)), $a_n^2/a_{2n} \to 1$, so $a = 1$.   □

Remark 15.16. One can show that

$$\left(\frac{n}{e}\right)^n \cdot \sqrt{2\pi n} < n! < \left(\frac{n}{e}\right)^n \cdot \sqrt{2\pi n} \cdot e^{1/(12n)}$$

for every positive integer n. A somewhat weaker statement is presented in Exercise 15.24.

Exercises

15.13. Compute the following integrals:

(a) $\int_0^1 \sqrt{x} \cdot e^{\sqrt{x}} \, dx$;   (b) $\int_2^3 \frac{\sqrt{\log x}}{x} \, dx$;

(c) $\int_0^{\pi^2} \sin \sqrt{x} \, dx$;   (d) $\int_0^1 \arctan x \, dx$;

(e) $\int_0^1 \arctan \sqrt{x} \, dx$;   (f) $\int_0^1 \log(1 + x^2) \, dx$;

(g) $\int_0^1 \sqrt{x^3 + x^2} \, dx$;   (h) $\int e^{ax} \cos(bx) \, dx$.

15.14. Apply integration by parts to get the equation

$$\int \frac{1}{x} \cdot \frac{1}{\log x} \, dx = \int (\log x)' \cdot \frac{1}{\log x} \, dx =$$

$$= \log x \cdot \frac{1}{\log x} - \int \log x \cdot \frac{1}{x} \cdot \frac{-1}{\log^2 x} \, dx = 1 + \int \frac{1}{x} \cdot \frac{1}{\log x} \, dx.$$

Thus $0 = 1$. Where did we make a mistake?

15.15. Prove that if f is strictly monotone and differentiable in the interval I, $\varphi = f^{-1}$, and $\int f(x)\,dx = F(x) + c$, then

$$\int \varphi(y)\,dy = y\varphi(y) - F(\varphi(y)) + c.$$

15.16. Check the correctness of the following computation:

$$2n \cdot \int \frac{x^2}{(x^2+1)^{n+1}}\,dx = -\int x \cdot \left(\frac{1}{(x^2+1)^n}\right)'\,dx =$$

$$= -\frac{x}{(x^2+1)^n} + \int \frac{dx}{(x^2+1)^n} + c.$$

15.17. Prove that if f and g are n times continuously differentiable in an interval I, then

$$\int f\,g^{(n)}\,dx =$$

$$= f\,g^{(n-1)} - f'\,g^{(n-2)} + \cdots + (-1)^{n-1}\,f^{(n-1)}g + (-1)^n \int f^{(n)}g\,dx. \quad \text{(H)}$$

15.18. Prove that if p is a polynomial of degree n, then

$$\int e^{-x}\,p(x)\,dx = -e^{-x} \cdot \left[p(x) + p'(x) + \cdots + p^{(n)}(x)\right] + c.$$

15.19. Prove that if f is twice differentiable and f'' is integrable in $[a,b]$, then

$$\int_a^b xf''(x)\,dx = (bf'(b) - f(b)) - (af'(a) - f(a)).$$

15.20. Prove that

$$\int_0^1 x^m(1-x)^n\,dx = \frac{m!\,n!}{(m+n+1)!} \quad (m,n \in \mathbb{N}).$$

15.21. Compute the value of $\int_0^1 (1-x^2)^n\,dx$ for every $n \in \mathbb{N}$.

15.22. Prove that if

$$f_1(x) = \int_0^x f(t)\,dt,\ f_2(x) = \int_0^x f_1(t)\,dt,\ \ldots,\ f_k(x) = \int_0^x f_{k-1}(t)\,dt$$

then

$$f_k(x) = \frac{1}{(k-1)!} \cdot \int_0^x f(t)(x-t)^{k-1}\,dt.$$

15.23. (a) Prove that for every $n \in \mathbb{N}$, there exist integers a_n and b_n such that $\int_0^1 x^n e^x\,dx = a_n \cdot e + b_n$ holds.

(b) Prove that

$$\lim_{n\to\infty} \int_0^1 x^n e^x \, dx = 0.$$

(c) Prove that e is irrational.

15.24. Prove the following stronger version of Stirling's formula:

$$\left(\frac{n}{e}\right)^n \cdot \sqrt{2\pi n} < n! \leq \left(\frac{n}{e}\right)^n \cdot \sqrt{2\pi n} \cdot e^{1/(4(n-1))} \tag{15.15}$$

for every integer $n > 1$. (S)

15.3 Integration by Substitution

We obtained the formulas in Theorems 13.30 and 13.33 by differentiating $f(ax+b)$, $f(x)^{\alpha+1}$, and $\log f(x)$, and using the differentiation rules for compositions of functions. These formulas are special cases of the following theorem, which is called **integration by substitution**.

Theorem 15.17. *Suppose the function g is differentiable on the interval I, f is defined on the interval $J = g(I)$, and f has a primitive function on J.[3] Then the function $(f \circ g) \cdot g'$ also has a primitive function on I, and*

$$\int f(g(t)) \cdot g'(t) \, dt = F(g(t)) + c, \tag{15.16}$$

where $\int f \, dx = F(x) + c$.

Proof. The theorem is rather clear from the differentiation rule for compositions of functions. □

We can use equation (15.16) in both directions. We use it "left to right" when we need to compute an integral of the form $\int f(g(t)) \cdot g'(t) \, dt$. Then the following formal procedure automatically changes the integral we want to compute to the right-hand side of (15.16):

$$g(t) = x; \quad g'(t) = \frac{dx}{dt}; \quad g'(t) dt = dx; \quad \int f(g(t)) \cdot g'(t) \, dt = \left(\int f(x) \, dx \right)_{x=g(t)}.$$

Examples 15.18. **1.** The integral $\int t \cdot e^{t^2} \, dt$ is changed into the form $\int f(g(t)) \cdot g'(t) \, dt$ if we divide it and multiply it by 2 at the same time:

$$\int t \cdot e^{t^2} \, dt = \frac{1}{2} \cdot \int e^{t^2} \cdot (2t) \, dt = \frac{1}{2} \cdot \int e^{t^2} \cdot (t^2)' \, dt = F(t^2) + c,$$

---

[3] Here we use the fact that the image of an interval under a continuous function is also an interval; see Corollary 10.58 and the remark following it.

where $F(x) = \int e^x dx = e^x + c$. Thus the integral is equal to $(1/2) \cdot e^{t^2} + c$. With the help of the formalism above, we can get the same result faster: $x = t^2$, $dx/dt = 2t$, $2t\,dt = dx$,

$$\int t \cdot e^{t^2} dt = \int \frac{1}{2} \cdot e^{t^2} 2t\,dt = \int \frac{1}{2} \cdot e^x dx = \frac{1}{2} \cdot e^x + c = \frac{1}{2} \cdot e^{t^2} + c.$$

2. $\int \operatorname{tg} t\,dt = \int (\sin t / \cos t)\,dt = -\int (1/\cos t) \cdot (\cos t)'\,dt = -F(\cos t) + c$, where $F(x) = \int (1/x)\,dx = \log|x| + c$. Thus the integral is equal to $-\log|\cos t| + c$. The same result can be obtained with the formal procedure we introduced above: $\cos t = x$, $dx/dt = -\sin t$, $-\sin t\,dt = dx$,

$$\int \frac{\sin t}{\cos t}\,dt = \int -\frac{dx}{x} = -\log|x| + c = -\log|\cos t| + c.$$

Let us see some examples when we apply (15.16) "right to left" that is, when we want to determine an integral of the form $\int f\,dx$, and we are looking for a g with which we can compute the left-hand side of (15.16), that is, the integral $\int f(g(t)) \cdot g'(t)\,dt$. To achieve this goal, we usually look for a function g for which $f \circ g$ is simpler than f (and then we hope that the $g'(t)$ factor does not make our integral too complicated). If $\int f(g(t)) \cdot g'(t)\,dt = G(t) + c$, then by (15.16), the primitive function F of the function f we seek satisfies $G(t) = F(g(t)) + c$. Therefore, we have $\int f\,dx = F(x) + c = G(g^{-1}(x)) + c$, assuming that g has an inverse.

Examples 15.19. **1.** We can attempt to solve the integral $\int dx/(1 + \sqrt{x})$ with the help of the function $g(t) = t^2$, since then for $t > 0$, we have $f(g(t)) \cdot g'(t) = 2t/(1 + t)$, whose integral can be easily computed. If this is $G(t) + c$, then the integral we seek is $g(\sqrt{x}) + c$, since the inverse of g is the function $\sqrt{x}$. With the formalism above, the computation looks like this: $x = t^2$, $dx/dt = 2t$, $dx = 2t\,dt$,

$$\int \frac{dx}{1 + \sqrt{x}} = \int \frac{1}{1 + t} \cdot 2t\,dt = 2 \cdot \int \left(1 - \frac{1}{1 + t}\right) dt =$$
$$= 2t - 2\log(1 + t) + c = 2\sqrt{x} - 2\log(1 + \sqrt{x}) + c.$$

Here we get the last inequality by substituting $t = \sqrt{x}$, that is, $g(t) = t^2$.

2. We can use the substitution $e^x = t$, i.e., $x = \log t$, for the integral $\int e^{2x}/(e^x + 1)\,dx$. We get that $dx/dt = 1/t$, $dx = dt/t$,

$$\int \frac{e^{2x}}{e^x + 1}\,dx = \int \frac{t^2}{t + 1} \cdot \frac{1}{t}\,dt = \int \frac{t}{t + 1}\,dt = \int \left(1 - \frac{1}{t + 1}\right) dt =$$
$$t - \log(t + 1) + c = e^x - \log(e^x + 1) + c.$$

3. Let us compute the integral $\int \sqrt{1 - x^2}\,dx$. Let $x = \sin t$, where $t \in [-\pi/2, \pi/2]$. Then $dx/dt = \cos t$, $dx = \cos t\,dt$,

$$\int \sqrt{1-x^2}\,dx = \int \sqrt{1-\sin^2 t}\cdot \cos t\,dt =$$

$$= \int \cos^2 t\,dt = \int \frac{1+\cos 2t}{2}\,dt = \frac{t}{2} + \frac{\sin 2t}{4} + c =$$

$$= \frac{\arcsin x}{2} + \frac{\sin(2\arcsin x)}{4} + c.$$

Here the second term can be simplified if we notice that $\sin 2t = 2\sin t \cdot \cos t$, and so $\sin(2\arcsin x) = 2x \cdot \cos(\arcsin x) = 2x\sqrt{1-x^2}$. In the end, we get that

$$\int \sqrt{1-x^2}\,dx = \frac{1}{2}\cdot \arcsin x + \frac{1}{2}\cdot x\sqrt{1-x^2} + c. \tag{15.17}$$

Examples 15.20. **1.** Let $r > 0$. By (15.17) and applying Theorem 13.30, we get

$$\int \sqrt{r^2-x^2}\,dx = \frac{r^2}{2}\cdot \arcsin\frac{x}{r} + \frac{rx}{2}\cdot \sqrt{1-\left(\frac{x}{r}\right)^2} + c.$$

Thus by the fundamental theorem of calculus,

$$\int_{-r}^{r} \sqrt{r^2-x^2}\,dx = \left[\frac{r^2}{2}\cdot \arcsin\frac{x}{r} + \frac{rx}{2}\cdot \sqrt{1-\left(\frac{x}{r}\right)^2}\right]_{-r}^{r} = r^2 \cdot \arcsin 1 = \frac{r^2 \pi}{2},$$

that is, the area of the semicircle with radius r is $r^2\pi/2$. (Recall that we defined π to be the circumference of the unit semicircle on p. 163.) We have recovered the well-known formula for the area of a circle with radius r (namely $r^2\pi$), which Archimedes stated as *the area of a circle agrees with the area of the right triangle whose legs* (sides adjacent to the right angle) *are equal to the radius and the circumference of the circle.*
2. With the help of the integral (15.17), we can determine the area of an ellipse as well. The equation of an ellipse with axes a and b is

$$\frac{x^2}{a^2} + \frac{y^2}{b^2} = 1,$$

so the graph of the function $f(x) = b\cdot \sqrt{1-\left(\frac{x}{a}\right)^2}$ ($x \in [-a,a]$) bounds half of the area of the ellipse. Now by (15.17) and Theorem 13.30,

$$\int b\cdot \sqrt{1-\left(\frac{x}{a}\right)^2}\,dx = \frac{ba}{2}\cdot \arcsin\frac{x}{a} + \frac{bx}{2}\cdot \sqrt{1-\left(\frac{x}{a}\right)^2} + c,$$

so by the fundamental theorem of calculus,

$$\int_{-a}^{a} b\cdot \sqrt{1-\left(\frac{x}{a}\right)^2}\,dx = ba\cdot \arcsin 1 = \frac{ab\pi}{2},$$

that is, *the area of the ellipse is $ab\pi$.*

The following theorem gives us a version of integration by substitution for definite integrals.

Theorem 15.21. *Suppose that g is differentiable and g' is integrable on the interval $[a,b]$. If f is continuous on the image of g, that is, on the interval[4] $g([a,b])$, then*

$$\int_a^b f(g(t)) \cdot g'(t)\, dt = \int_{g(a)}^{g(b)} f(x)\, dx. \qquad (15.18)$$

Proof. Since g is differentiable, it is continuous. Thus $f \circ g$ is also continuous, so it is integrable on $[a,b]$, which implies that $(f \circ g) \cdot g'$ is also integrable on $[a,b]$. On the other hand, statement (iii) of Theorem 15.5 ensures that f has a primitive function. If $F' = f$, then by the fundamental theorem of calculus, the right-hand side of (15.18) is $F(g(b)) - F(g(a))$. Now by the differentiation rules for compositions of functions, $(F \circ g)' = (f \circ g) \cdot g'$, so by applying the fundamental theorem of calculus again, we get that the left-hand side of (15.18) is also $F(g(b)) - F(g(a))$, meaning that (15.18) is true. □

We note that in the theorem above, we can relax the condition of continuity of f and assume only that f is integrable on the image of g. In other words, the following theorem also holds.

Theorem 15.22. *Suppose that g is differentiable and g' is integrable on the interval $[a,b]$. If f is integrable on the image of g, that is, on the interval $g([a,b])$, then $(f \circ g) \cdot g'$ is integrable on $[a,b]$, and (15.18) holds.*

This more general theorem is harder to prove, since the integrability of $(f \circ g) \cdot g'$ does not follow as easily as in the case of Theorem 15.21, and the fundamental theorem of calculus cannot be applied either. The proof can be found in the appendix of this chapter.

Exercises

15.25. Prove that

$$\int_0^\pi \frac{\sin 2kx}{\sin x}\, dx = 0$$

holds for every integer k. (H)

15.26. Prove that if f is integrable on $[0,1]$, then

$$\int_0^\pi f(\sin x) \cos x\, dx = 0.$$

---

[4] Corollary 10.58 ensures that $g([a,b])$ is an interval.

15.4 Integrals of Elementary Functions

In Chapter 11, we became acquainted with the elementary functions. These are the polynomials, rational, exponential, power, and logarithmic functions, trigonometric and hyperbolic functions, their inverses, and every function that can be expressed from these using a finite sequence of basic operations and compositions.[5] We will familiarize ourselves with methods that allow us to determine the indefinite integrals of numerous elementary functions.

15.4.1 Rational Functions

Definition 15.23. We define *elementary rational functions* to be

(i) quotients of the form $A/(x-a)^n$, where $n \in \mathbb{N}^+$ and $A, a \in \mathbb{R}$; as well as

(ii) quotients of the form $(Ax+B)/(x^2+ax+b)^n$, where $n \in \mathbb{N}^+$ and $A, B, a, b \in \mathbb{R}$ are constants such that $a^2 - 4b < 0$ holds.

(This last condition is equivalent to saying that $x^2 + ax + b$ does not have any real roots.)

We will first determine the integrals of the elementary rational functions. The first type does not give us any trouble, since $\int \frac{A}{(x-a)} dx = A \cdot \log|x-a| + c$, and $n > 1$ implies $\int A/(x-a)^n \, dx = \left(A/(1-n)\right)/(x-a)^{n-1} + c$.

To find the integrals of the second type of elementary rational functions we will first show that computing the integral of $(Ax+B)/(x^2+ax+b)^n$ can be reduced to computing the integral $\int dx/(x^2+1)^n$. This can be seen as

$$\int \frac{Ax+B}{(x^2+ax+b)^n} dx = \int \frac{(2x+a)\cdot(A/2)+(B-(aA/2))}{(x^2+ax+b)^n} =$$

$$= \frac{A}{2} \cdot \int \frac{(x^2+ax+b)'}{(x^2+ax+b)^n} dx + \left(B - \frac{aA}{2}\right) \cdot \int \frac{dx}{(x^2+ax+b)^n}.$$

Here

$$\int \frac{(x^2+ax+b)'}{(x^2+ax+b)} dx = \log(x^2+ax+b) + c,$$

and if $n > 1$, then

$$\int \frac{(x^2+ax+b)'}{(x^2+ax+b)^n} dx = \frac{1}{1-n} \cdot \frac{1}{(x^2+ax+b)^{n-1}} + c.$$

---

[5] What we call elementary functions is partially based on history and tradition, partially based on usefulness, and partially based on a deeper reason that comes to light through complex analysis. In some investigations, it proves to be reasonable to list algebraic functions among the elementary functions. (Algebraic functions were defined in Exercise 11.45.)

Moreover,

$$\int \frac{dx}{(x^2+ax+b)^n} = \int \frac{dx}{\left((x+(a/2))^2+b-(a^2/4)\right)^n} =$$

$$= \int \frac{dx}{\left((x+(a/2))^2+d^2\right)^n},$$

where $d = \sqrt{b-(a^2/4)}$. (By the conditions, $b-(a^2/4) > 0$.) Now if

$$\int \frac{dx}{(x^2+1)^n} = F_n(x) + c,$$

then by Theorem 13.30,

$$\int \frac{dx}{\left((x+(a/2))^2+d^2\right)^n} = \frac{1}{d^{2n}} \cdot \int \frac{dx}{\left((\frac{x}{d}+\frac{a}{2d})^2+1\right)^n} = \frac{1}{d^{2n-1}} \cdot F_n\left(\frac{x}{d}+\frac{a}{2d}\right) + c.$$

As for the functions F_n, we know that $F_1(x) = \operatorname{arctg} x + c$. On the other hand, for every $n \geq 1$, the equality

$$F_{n+1} = \frac{1}{2n} \cdot \frac{x}{(x^2+1)^n} + \frac{2n-1}{2n} \cdot F_n + c \qquad (15.19)$$

holds. This is easy to check by differentiating both sides (see also Exercise 15.16). Applying the recurrence formula (15.19) repeatedly gives us the functions F_n. So for example,

$$\int \frac{dx}{(x^2+1)^2} = \frac{1}{2} \cdot \frac{x}{x^2+1} + \frac{1}{2} \cdot \operatorname{arctg} x + c,$$

$$\int \frac{dx}{(x^2+1)^3} = \frac{1}{4} \cdot \frac{x}{(x^2+1)^2} + \frac{3}{8} \cdot \frac{x}{x^2+1} + \frac{3}{8} \cdot \operatorname{arctg} x + c. \qquad (15.20)$$

By the following theorem every rational function can be expressed as the sum of a polynomial and finitely many elementary rational functions, and if we know how to integrate these, we know how to determine the integral of any rational function (at least theoretically).

We will need the concept of divisibility for polynomials. We say that the polynomial p is **divisible** by the polynomial q, and we denote this by $q \mid p$, if there exists a polynomial r such that $p = q \cdot r$. It is known that if the polynomials q_1 and q_2 do not have a nonconstant common divisor, then there exist polynomials p_1 and p_2 such that $p_1 q_1 + p_2 q_2 \equiv 1$. (We can find such a p_1 and p_2 by repeatedly applying the Euclidean algorithm to q_1 and q_2.)

Moreover, we will use the fundamental theorem of algebra (see page 201) and the corollary that every polynomial with real coefficients can be written as the product of polynomials with real coefficients of degree one and two.

Theorem 15.24 (Partial Fraction Decomposition). *Every rational function R can be written as the sum of a polynomial and finitely many elementary rational functions such that the denominators of these elementary rational functions all divide the denominator of R.*

Proof. Let $R = p/q$, where p and $q \neq 0$ are polynomials. Let the degree of q be n; we will prove the theorem by induction on n. If $n = 0$, that is, q is constant, then R is a polynomial, and the statement holds (without any elementary rational functions in the decomposition).

Let $n > 0$, and assume the statement is true for every rational function whose denominator has degree smaller than n. Factor q into a product of polynomials of degree one and two. We can assume that the degree-two polynomials here do not have any real roots, since otherwise, we could factor them further. We distinguish three cases.

1. Two different terms appear in the factorization of the polynomial q. Then q can be written as $q_1 q_2$, where $q_1 \neq c$, $q_2 \neq c$, and the polynomials q_1, q_2 do not have any nonconstant common divisors. Then there exist polynomials p_1 and p_2 such that $p_1 q_1 + p_2 q_2 \equiv 1$;

$$\frac{p}{q} = \frac{pp_1 q_1 + pp_2 q_2}{q_1 q_2} = \frac{pp_1}{q_2} + \frac{pp_2}{q_1}.$$

Here we can apply the inductive hypothesis to both pp_1/q_2 and pp_2/q_1, which immediately gives us the statement of the theorem.

2. $q = c(x-a)^n$. In this case, let us divide p by $(x-a)$ with remainder: $p = p_1(x-a) + A$, from which

$$\frac{p}{q} = \frac{p_1}{c(x-a)^{n-1}} + \frac{A/c}{(x-a)^n}.$$

follows. Here $(A/c)/(x-a)^n$ is an elementary rational function, and we can apply the induction hypothesis to the term $p_1/(c(x-a)^{n-1})$.

3. $q = c \cdot (x^2 + ax + b)^k$, where $a^2 - 4b < 0$ and $n = 2k$. Then dividing p by $(x^2 + ax + b)$ with remainder gives us $p = p_1 \cdot (x^2 + ax + b) + (Ax + B)$, from which

$$\frac{p}{q} = \frac{p_1}{c(x^2 + ax + b)^{k-1}} + \frac{\frac{A}{c}x + \frac{B}{c}}{(x^2 + ax + b)^k}.$$

follows. Here we can apply the inductive hypothesis to the first term, while the second term is an elementary rational function, proving the theorem. □

Remarks 15.25. **1.** One can show that the decomposition of rational functions in Theorem 15.24 is unique (see Exercise 15.32). This is the **partial fraction decomposition** of a rational function.
2. If the degree of p is smaller than q, then only elementary rational functions appear in the partial fraction decomposition of p/q (and no polynomial). This is because then $\lim_{x\to\infty} p/q = 0$. Since every elementary rational function tends to 0 at ∞, this

must hold for the polynomial appearing in the decomposition as well, which implies it must be identically zero.

How can we find a partial fraction decomposition? We introduce three methods.

1. Follow the proof of the theorem. If, for example,

$$\frac{p}{q} = \frac{x+2}{x(x^2+1)^2},$$

then $1 \cdot (x^2+1)^2 - (x^3+2x) \cdot x = 1$ and

$$\frac{p}{q} = \frac{x+2}{x} - \frac{x^4+2x^3+2x^2+4x}{(x^2+1)^2} =$$

$$= 1 + \frac{2}{x} - \frac{(x^2+1)(x^2+2x+1)+2x-1}{(x^2+1)^2} =$$

$$= \frac{2}{x} - \frac{2x}{x^2+1} - \frac{2x-1}{(x^2+1)^2}.$$

2. The method of indeterminate coefficients. From the theorem and by remark 15.25, we know that

$$\frac{p}{q} = \frac{A}{x} + \frac{Bx+C}{x^2+1} + \frac{Dx+E}{(x^2+1)^2}. \qquad (15.21)$$

Bringing this to a common denominator yields

$$x+2 = A(x^4+2x^2+1) + (Bx+C)(x^2+1)x + (Dx+E)x.$$

This gives us a system of equalities for the unknown coefficients: $A+B=0$, $C=0$, $2A+B+D=0$, $C+E=1$, and $A=2$. Then we can compute that $A=2$, $B=-2$, $C=0$, $D=-2$, and $E=1$.

3. If a term of the form $A/(x-a)^n$ appears in the decomposition, then (assuming that n is the largest such power for a) we can immediately determine A if we multiply everything by $(x-a)^n$ and we substitute $x=a$ into the equation. So for example, for (15.21) we have

$$A = \left.\frac{x+2}{(x^2+1)^2}\right|_{x=0} = 2.$$

If we subtract the known terms from both sides, we reduce the question to finding the partial fraction decomposition of a simpler rational function:

$$\frac{x+2}{x(x^2+1)^2} - \frac{2}{x} = \frac{x+2-2x^4-4x^2-2}{x(x^2+1)^2} = \frac{-2x^3-4x+1}{(x^2+1)^2}.$$

Here $-2x^3 - 4x + 1 = (-2x)(x^2+1) + (-2x+1)$, so we get the same decomposition.

15.4.2 Integrals Containing Roots

From now on, $R(u,v)$ will denote a two-variable rational function. This means that $R(u,v)$ is constructed from the variables u and v and from constants by the four basic operations. One can easily show that this holds exactly when

$$R(u,v) = \frac{\sum_{i=0}^{n}\sum_{j=0}^{n}a_{ij}\,u^i\,v^j}{\sum_{i=0}^{n}\sum_{j=0}^{n}b_{ij}\,u^i\,v^j}, \tag{15.22}$$

where $n \geq 0$ is an integer and a_{ij} and b_{ij} are constants.

1. We show that the integral

$$\int R\left(x,\ \sqrt[n]{\frac{ax+b}{cx+d}}\right)dx$$

can be reduced to the integral of a (one-variable) rational function with the substitution $\sqrt[n]{(ax+b)/(cx+d)} = t$. Clearly, with this substitution, $(ax+b)/(cx+d) = t^n$, $ax+b = ct^nx+dt^n$ and $x = (dt^n - b)/(a - ct^n)$, so dx/dt is a rational function.

Example 15.26. Compute the integral $\int x^{-2} \cdot \sqrt[3]{x+1/x}\,dx$. With the substitution $\sqrt[3]{x+1/x} = t$, $x+1 = t^3x$, $x = 1/(t^3 - 1)$, and $dx/dt = -3t^2/(t^3 - 1)^2$, so

$$\int \frac{1}{x^2} \cdot \sqrt[3]{\frac{x+1}{x}}\,dx = \int (t^3 - 1)^2 \cdot t \cdot \frac{-3t^2}{(t^3 - 1)^2}\,dt =$$

$$= \int -3t^3\,dt = -\frac{3}{4}t^4 + c = -\frac{3}{4} \cdot \left(\frac{x+1}{x}\right)^{4/3} + c.$$

2. The integral $R(x,\sqrt{ax^2+bx+c})\,dx$ (where $a \neq 0$) can also be reduced to an integral of a rational function with a suitable substitution.

   a. If $ax^2 + bx + c$ has a root, then $ax^2 + bx + c = a(x - \alpha)(x - \beta)$, so

   $$\sqrt{ax^2+bx+c} = \sqrt{a(x-\alpha)(x-\beta)} = |x - \alpha|\sqrt{\frac{a(x-\beta)}{x-\alpha}},$$

   and this leads us back to an integral in the previous part.

   b. If $ax^2 + bx + c$ does not have real roots, then it must be positive everywhere, since otherwise, the integrable function is not defined anywhere. Thus $a > 0$ and $c > 0$. In this case, we can use the substitution $\sqrt{ax^2+bx+c} - \sqrt{a} \cdot x = t$. We get that

   $$ax^2 + bx + c = t^2 + 2t\sqrt{a} \cdot x + ax^2,$$

   $x = (c - t^2)/(2\sqrt{a} \cdot t - b)$, and so dx/dt is also a rational function.

We can apply the substitution $\sqrt{ax^2 + bx + c} - \sqrt{c} = tx$ as well. This gives us $ax^2 + bx + c = x^2 t^2 + 2\sqrt{c}tx + c$, $ax + b = xt^2 + 2\sqrt{c}t$, $x = (2\sqrt{c}t - b)/(a - t^2)$, so dx/dt is a rational function.

Example 15.27. Compute the integral $\int \sqrt{x^2 + 1}\,dx$. Substituting $\sqrt{x^2 + 1} - x = t$, we obtain $x = (1 - t^2)/(2t)$, $\sqrt{x^2 + 1} = x + t = (1 + t^2)/(2t)$, $dx/dt = -(1 + t^2)/(2t^2)$, so

$$
\int \sqrt{x^2 + 1}\,dx = \int \frac{1+t^2}{2t} \cdot \frac{-(1+t^2)}{2t^2}\,dt = -\frac{1}{4}\int \frac{1 + 2t^2 + t^4}{t^3}\,dt =
$$

$$
= \frac{1}{8} \cdot \frac{1}{t^2} - \frac{1}{2}\log|t| - \frac{t^2}{8} + c =
$$

$$
= \frac{1}{8} \cdot \frac{1}{(\sqrt{x^2+1}-x)^2} - \frac{1}{2}\log(\sqrt{x^2+1}-x) - \frac{(\sqrt{x^2+1}-x)^2}{8} + c =
$$

$$
= \frac{1}{2} \cdot x \cdot \sqrt{x^2+1} - \frac{1}{2}\log(\sqrt{x^2+1}-x) + c. \tag{15.23}
$$

15.4.3 Rational Functions of e^x

Let $R(x)$ be a one-variable rational function. To compute the integral $\int R(e^x)\,dx$, let us use the substitution $e^x = t$, $x = \log t$, $dx/dt = 1/t$. The integral becomes the integral of a rational function. See, e.g, Example 15.19.

15.4.4 Trigonometric Functions

a. Integration of an expression of the form $R(\sin x, \cos x)$ can always be done with the substitution $\operatorname{tg}(x/2) = t$. Indeed, $\sin^2 x + \cos^2 x = 1$ gives $\operatorname{tg}^2 x + 1 = 1/\cos^2 x$, and thus

$$
\cos x = \frac{1}{\pm\sqrt{1 + \operatorname{tg}^2 x}}, \qquad \sin x = \frac{\operatorname{tg} x}{\pm\sqrt{1 + \operatorname{tg}^2 x}}, \tag{15.24}
$$

so

$$
\sin x = 2\sin\frac{x}{2}\cos\frac{x}{2} = \frac{2\operatorname{tg}(x/2)}{1 + \operatorname{tg}^2(x/2)} \quad \text{and} \quad \cos x = \cos^2\frac{x}{2} - \sin^2\frac{x}{2} = \frac{1 - \operatorname{tg}^2(x/2)}{1 + \operatorname{tg}^2(x/2)}.
$$

Thus with the substitution $\operatorname{tg}(x/2) = t$, we have

$$
\sin x = 2t/(1 + t^2), \quad \cos x = (1 - t^2)/(1 + t^2), \quad x = 2\arctan\operatorname{tg} t,
$$

and $dx/dt = 2/(1 + t^2)$.

Example 15.28.

$$\int \frac{dx}{\sin x} = \int \frac{1+t^2}{2t} \frac{2}{1+t^2} dt = \int \frac{dt}{t} = \log|t| + c = \log\left|\operatorname{tg}\frac{x}{2}\right| + c.$$

Then

$$\int \frac{dx}{\cos x} = \int \frac{dx}{\sin\left(\frac{\pi}{2}-x\right)} = -\log\left|\operatorname{tg}\left(\frac{\pi}{4}-\frac{x}{2}\right)\right| + c.$$

b. In some cases, the substitution $\operatorname{tg} x = t$ will also lead us to our goal. By (15.24),

$$\sin x = \frac{t}{\pm\sqrt{1+t^2}}, \qquad \cos x = \frac{1}{\pm\sqrt{1+t^2}} \qquad \text{and} \qquad \frac{dx}{dt} = \frac{1}{1+t^2},$$

so this substitution also leads to a rational function if the exponents of $\sin x$ and $\cos x$ are of the same parity in every term of the denominator and every term of the numerator.

Example 15.29.

$$\int \frac{dx}{1+\cos^2 x} = \int \frac{1}{1+1/(1+t^2)} \cdot \frac{dt}{1+t^2} = \int \frac{dt}{2+t^2} = \frac{1}{2} \cdot \int \frac{dt}{1+\left(t/\sqrt{2}\right)^2} =$$

$$= \frac{\sqrt{2}}{2} \cdot \operatorname{arctg}\frac{t}{\sqrt{2}} + c = \frac{1}{\sqrt{2}} \cdot \operatorname{arctg}\left(\frac{1}{\sqrt{2}}\operatorname{tg} x\right) + c.$$

c. Applying the substitution $\sin x = t$ on the interval $[-\pi/2, \pi/2]$ gives us

$$\cos x = \sqrt{1-t^2}, \qquad x = \operatorname{arc\,sin} t \qquad \text{and} \qquad \frac{dx}{dt} = \frac{1}{\sqrt{1-t^2}},$$

so we get a rational function if the power of $\cos x$ is even in the numerator and odd in the denominator, or vice versa.

Example 15.30.

$$\int \frac{dx}{\cos x} = \int \frac{1}{\sqrt{1-t^2}} \cdot \frac{dt}{\sqrt{1-t^2}} = \int \frac{dt}{1-t^2} =$$

$$= \frac{1}{2} \cdot \log\left|\frac{1+t}{1-t}\right| + c = \frac{1}{2} \cdot \log\left|\frac{1+\sin x}{1-\sin x}\right| + c.$$

(Check that this agrees with the result from Example 15.28. Also check that the right-hand side is the primitive function of $1/\cos x$ in every interval where $\cos x \neq 0$, not just in $(-\pi/2, \pi/2)$.)

d. The substitution $\cos x = t$ also leads us to the integral of a rational function if the power of $\sin x$ is even in the numerator and odd in the denominator, or vice versa.

Let us note that the integrals of the form $\int R(x, \sqrt{ax^2+bx+c})\,dx$ can also be computed using a method different from the one seen on page 354. With a linear substitution, we can reduce the integral to one of the integrals

$$\int R(x, \sqrt{1-x^2})\,dx, \quad \int R(x, \sqrt{x^2-1})\,dx, \quad \text{or} \int R(x, \sqrt{x^2+1})\,dx.$$

In the first case, the substitution $x = \sin t$ gives us an integral that we encountered in the last section. In the second case, the substitution $x = \mathrm{ch}\,t$ gives us the integral $\int R(\mathrm{ch}\,t, \mathrm{sh}\,t)\,\mathrm{sh}\,t$, which we can tackle as in the third section if we recall the definitions of $\mathrm{ch}\,t$ and $\mathrm{sh}\,t$. The third integral can be computed with the substitution $x = \mathrm{sh}\,t$.

Exercises

15.27. Compute the following definite integrals.

(a) $\displaystyle\int_0^1 \frac{x}{x^4+1}\,dx$;

(b) $\displaystyle\int_1^2 \frac{e^x+2}{e^x+e^{2x}}\,dx$;

(c) $\displaystyle\int_1^2 \frac{dx}{4^x-2^x}\,dx$;

(d) $\displaystyle\int_{\pi/4}^{\pi/2} \frac{dx}{\sin x(2+\cos x)}$;

(e) $\displaystyle\int_0^1 \sqrt{2^x-1}\,dx$;

(f) $\displaystyle\int_0^{\pi/4} (\mathrm{tg}\,x)^2\,dx$;

(g) $\displaystyle\int_1^2 \arcsin(1/x)\,dx$;

(h) $\displaystyle\int_2^3 x\cdot\log(x^2-x)\,dx$;

(i) $\displaystyle\int_0^{\pi/4} \frac{dx}{\cos^{10}x}\,dx$;

(j) $\displaystyle\int_2^4 \frac{x^2}{\sqrt{x^2-1}}\,dx$;

(k) $\displaystyle\int_0^{\pi/4} \frac{dx}{\sin^4 x+\cos^4 x}\,dx$.

15.28. Compute the following indefinite integrals.

(a) $\displaystyle\int \frac{2x+3}{x^2-5x+6}\,dx$;

(b) $\displaystyle\int \frac{x^3-2x^2+5x+1}{x^2+1}\,dx$;

(c) $\displaystyle\int \frac{x^{100}}{x-1}\,dx$;

(d) $\displaystyle\int \frac{dx}{x^3+8}\,dx$;

(e) $\displaystyle\int \frac{dx}{\sqrt{x+1}+\sqrt{x-1}}\,dx$;

(f) $\displaystyle\int \frac{1+\sqrt[3]{x}}{1-\sqrt[3]{x}}\,dx$;

(g) $\displaystyle\int \frac{dx}{\log\log x}$;

(h) $\displaystyle\int \log(x+\sqrt{1+x^2})\,dx$;

(i) $\int \dfrac{e^x}{\sqrt{1+e^x}}\,dx;$

(j) $\int \dfrac{dx}{1+\sin x}\,dx;$

(k) $\int \dfrac{dx}{1+\cos x}\,dx;$

(l) $\int \dfrac{\sin x \cos x}{1+\sin^2 x}\,dx;$

(m) $\int \dfrac{dx}{1+\operatorname{tg} x};$

(n) $\int \sin x \cdot \log(\operatorname{tg} x)\,dx.$

15.29. Compute the integral $\int \sqrt{x^2+1}\,dx$ with the substitution $x = \operatorname{sh} t$, and compare the result with (15.23).

15.30. Compute the integral $\int \sqrt{x^2-1}\,dx$ with the substitution $x = \operatorname{ch} t$. (S)

15.31. The radius of a regular cylindrical container filled with water is r. The container is lying horizontally, that is, the curved part is on the ground. What force does the water exert on the vertical flat circular sides of the container due to pressure if the pressure at depth x is ρx?

15.32. Show that the partial fraction decomposition of rational functions is unique. (H)

15.33. Suppose that p and q are polynomials, $a \in \mathbb{R}$, and $q(a) \neq 0$. We know that terms of the form $A_k/(x-a)^k$ appear in the partial fraction decomposition of the rational function $p(x)/(q(x) \cdot (x-a)^n)$ for all $k = 0,\ldots,n$.

(a) Prove that $A_n = p(a)/q(a)$.
(b) Express the rest of the A_k with p and q as well.

15.34. Prove (15.19) with the help of exercise 15.16.

15.5 Nonelementary Integrals of Elementary Functions

Not all integrals of elementary functions can be computed with the help of the methods above. It might sound surprising at first, but there are some elementary functions whose primitive functions cannot be expressed by elementary functions.[6] This is a significant difference between differentiation and integration, since—as we saw in Chapter 12—the derivative of an elementary function is always elementary. But if an operation (in this case differentiation) does not lead us out of a class of objects, why should we expect its inverse operation (integration) to do the same? For example, addition keeps us within the class of positive numbers, but subtraction does not; the set of integers is closed under multiplication but not under division; squaring numbers keeps us in the realm of rational numbers but taking roots does not. It appears that inverse operations are more complicated.

This phenomenon can be observed with differentiation and integration as well if we consider subclasses of the elementary functions. For example, the derivatives of

---

[6] Some examples are e^x/x, $1/\log x$, and e^{-x^2}; see Example 15.32.

rational functions are rational functions. However, the integral of a rational function
is not necessarily a rational function:

$$\int \frac{dx}{x} = \log|x| + c,$$

and $\log|x|$ is not a rational function (which can easily be shown from the fact that
$\lim_{x\to\infty} x^\beta \cdot \log x = \infty$ and $\lim_{x\to\infty} x^{-\beta} \cdot \log x = 0$ for all $\beta > 0$; see Example 13.2
regarding this last statement).

The same is the case with rational functions formed from trigonometric functions: the derivative of a function of the form $R(\cos x, \sin x)$ has the same form, but
the integral can be different. This is clear from the fact that every function of the
form $R(\cos x, \sin x)$ is periodic with period 2π, while, for example,

$$\int (1 + \cos x)\, dx = x + \sin x + c$$

is not periodic. A less trivial example:

$$\int \frac{dx}{\sin x} = \log\left|\text{tg}\,\frac{x}{2}\right| + c,$$

and here the right-hand side cannot be expressed as a rational function of $\cos x$ and
$\sin x$.

Thinking about it more carefully, we see that it is not that surprising that the
integrals of some elementary functions are not elementary functions. It is a different matter that a rigorous proof of this is surprisingly hard. Joseph Liouville[7] was
the first to show—with the help of complex-analytic methods—that such elementary functions exist. In fact, Liouville proved that if the integral of an elementary
function is elementary, then the formula for that integral cannot be much more complicated than the original function. Unfortunately, even the precise expression of this
is achieved with great difficulty. Some special cases are easier to express, as is the
following theorem of Liouville (but the proof still exceeds the scope of this book).

Theorem 15.31. *Let f and g be rational functions, and suppose that f is not constant. If $\int e^f g\, dx$ can be expressed as an elementary function, then $\int e^f g\, dx =
e^f h + c$, where h is a rational function and c is a constant.*

With the help of this theorem, we can find several functions that do not have an
elementary integral.

Examples 15.32. **1.** Let us show that $\int \frac{e^x}{x}\, dx$ cannot be expressed in terms of
elementary functions.

If $\int (e^x/x)\, dx$ could be expressed in terms of elementary functions, then by Liouville's theorem, there would exist a rational function S such that $(S \cdot e^x)' = e^x/x$.
Let $S = p/q$, where the polynomials p and q do not have any nonconstant common
divisors. Then

---

[7] Joseph Liouville (1809–1882), French mathematician.

$$\left(\frac{p}{q} \cdot e^x\right)' = \frac{p'q - pq'}{q^2} \cdot e^x + \frac{p}{q} \cdot e^x = \frac{e^x}{x},$$

so $x(p'q - pq' + pq) = q^2$. We will show that this is impossible. First of all, q must be divisible by x. Let $q = x^k q_1$, where q_1 is not divisible by x. Then p cannot be divisible by x either, since $x \mid q$, and we assumed that p and q do not have any nonconstant common divisors. Thus the polynomial

$$P = p'q - pq' + pq = p'x^k q_1 - p(kx^{k-1}q_1 + x^k q_1') + px^k q_1$$

is not divisible by x^k, since every term on the right-hand side except for one is divisible by x^k. On the other hand, $P = q^2/x = x^{2k-1}q_1^2$ is divisible by x^k, since $2k - 1 \geq k$. This is a contradiction, which shows that $\int (e^x/x)\,dx$ cannot be elementary.

2. We can immediately deduce that the integral $\int (1/\log x)\,dx$ cannot be expressed in terms of elementary functions either. With the substitution $e^x = t$, we get $x = \log t$, $dx = dt/t$, so

$$\int \frac{e^x}{x}\,dx = \int \frac{dt}{\log t}.$$

Thus if $\int (1/\log t)\,dt$ were elementary, then so would $\int (e^x/x)\,dx$, which is impossible.

The integral $\int (1/\log t)\,dt$ appears often in various fields of mathematics, so it has its own notation:

$$\mathrm{Li}\,x = \int_2^x \frac{dt}{\log t} \qquad (x \geq 2).$$

The function $\mathrm{Li}\,x$ (which is called the **logarithmic integral function,** or sometimes the **integral logarithm**) plays an important role in number theory. Let $\pi(x)$ denote the number of primes less than or equal to x. According to the **prime number theorem,**

$$\pi(x) \sim \frac{x}{\log x} \qquad \text{if} \qquad x \to \infty,$$

that is, the function $x/\log x$ approximates $\pi(x)$ asymptotically well. One can show that the function $\mathrm{Li}\,x$ is an even better approximation, in that

$$|\pi(x) - \mathrm{Li}\,x| = o\left(\frac{x}{\log^k x}\right)$$

for all k, while

$$\left|\pi(x) - \frac{x}{\log x}\right| > \frac{cx}{\log^2 x}$$

for a suitable constant $c > 0$. The logarithmic integral function is just one of many important functions that are defined as integrals of elementary functions but are not elementary themselves.

3. Another such function of great importance in probability theory is $\Phi(x) = \int_0^x e^{-t^2}\,dt$, the function that describes the so-called normal distribution.

By Liouville's theorem above, we can easily deduce that this function is not an elementary functions (see Exercise 15.38).

4. Another example is the **elliptic integrals**. In Remark 15.20, we saw that the area of an ellipse with axes a and b is $ab\pi$. Determining the circumference of an ellipse is a harder problem. To simplify computation, assume that $a = 1$ and $b < a$. The graph of the function $f(x) = b \cdot \sqrt{1 - x^2}$ over the interval $[0,1]$ gives us the portion of the ellipse with axes 1 and b lying in the quadrant $\{(x,y) : x, y \geq 0\}$. The arc length of this graph is thus a quarter of the circumference of the ellipse. The function f is monotone decreasing on the interval $[0,1]$, so by Theorem 10.79, the graph is rectifiable there. Let $s(x)$ denote the arc length of the graph over the interval $[0,x]$. Since the derivative of f,

$$f'(x) = b \cdot \frac{-2x}{2 \cdot \sqrt{1 - x^2}} = -\frac{bx}{\sqrt{1 - x^2}},$$

is continuous on $[0,1)$, s is differentiable on this interval by Theorem 13.41, and $s'(x) = \sqrt{1 + (f'(x))^2}$ for all $0 \leq x < 1$. Let us introduce the notation $k = \sqrt{1 - b^2}$. Then $0 < k < 1$, and

$$1 + (f'(x))^2 = 1 + \frac{b^2 x^2}{1 - x^2} = \frac{1 - (1 - b^2)x^2}{1 - x^2} = \frac{1 - k^2 x^2}{1 - x^2}.$$

Thus s is a primitive function of $\sqrt{(1 - k^2 t^2)/(1 - t^2)}$ in the interval $(0,1)$, that is,

$$s \in \int \frac{\sqrt{1 - k^2 \cdot t^2}}{\sqrt{1 - t^2}} \, dt = \int \frac{1 - k^2 \cdot t^2}{\sqrt{(1 - t^2)(1 - k^2 t^2)}} \, dt. \tag{15.25}$$

Liouville showed that the integral appearing in (15.25) cannot be expressed with elementary functions, so the circumference of the ellipse cannot generally be expressed in a "closed" form (not containing an integral symbol).

Let us substitute $t = 1 - (1/u)$ in the integral above. We get that $dt = du/u^2$ and

$$\int \frac{\sqrt{1 - k^2 \cdot t^2}}{\sqrt{1 - t^2}} \, dt = \int \frac{\sqrt{(1 - k^2) + (2k^2/u) - (k^2/u^2)}}{\sqrt{(2/u) - (1/u^2)}} \cdot \frac{1}{u^2} \, du =$$

$$= \int \frac{b^2 u^2 + 2k^2 u - k^2}{u^2 \cdot \sqrt{(2u - 1)(b^2 u^2 + 2k^2 u - k^2)}} \, du. \tag{15.26}$$

Of course, we still cannot express this integral in terms of elementary functions as in (15.25), but it appears simpler than that one, since here, the polynomial inside the square root has degree three (and not four). These integrals motivate the following naming convention.

5. An **elliptic integral** is an integral of the form $\int R(x, \sqrt{f}) \, dx$, where f is a polynomial of degree three or four and $R(x, \sqrt{f})$ is a rational function with arguments x and

$\sqrt{f}$. If the degree of f is greater than four, then we call the integral a **hyperelliptic integral.**

The (hyper)elliptic integrals usually cannot be expressed in terms of elementary functions (but they can in some special cases, for example,

$$\int \frac{f'}{\sqrt{f}} dx = 2\sqrt{f} + c$$

for every positive f).

Since (hyper)elliptic integrals appear often in various applications, it is best to reduce them to simpler integrals. Write $R(u,v)$ in the form seen in (15.22), and then replace u by x and v by $\sqrt{f(x)}$. After taking powers and combining all that we can, we get that

$$R(x, \sqrt{f}) = \frac{A + B\sqrt{f}}{C + D\sqrt{f}},$$

where A, B, C, and D are polynomials. If we multiply both the numerator and denominator by $C - D\sqrt{f}$ here, then we get an expression of the form $(E + F\sqrt{f})/G$. Since we already know the integrals of rational functions, it suffices to find the integral of

$$\frac{F\sqrt{f}}{G} = \frac{Ff}{G\sqrt{f}}.$$

Now apply Theorem 15.24 and decompose the rational function Ff/G into the sum of a polynomial and finitely many elementary rational functions. If we divide this decomposition by $\sqrt{f}$, then we deduce that it suffices to find the integrals

$$I_k = \int \frac{x^k}{\sqrt{f(x)}} dx \qquad \text{and} \qquad J_r = \int \frac{r(x)}{\sqrt{f(x)}} dx,$$

where $k \in \mathbb{N}$ and r is an arbitrary elementary rational function. One can show that if the degree of f is n, then every I_k can be expressed as a linear combination of elementary functions and the integrals $I_0, I_1, \ldots, I_{n-2}$. A similar recurrence holds for the integrals J_r (see Exercise 15.39).

Exercises

15.35. Prove that $\mathrm{Li}\, x \sim \frac{x}{\log x}$ if $x \to \infty$.

15.36. Prove that for all $n \in \mathbb{N}^+$,

$$\mathrm{Li}\, x = \frac{x}{\log x} + \frac{x}{\log^2 x} + \cdots + (n-1)! \frac{x}{\log^n x} + n! \int_2^x \frac{dt}{\log^{n+1} t} + c_n \qquad (x \geq 2)$$

with a suitable constant c_n.

15.37. Prove that for all $n \in \mathbb{N}^+$,

$$\left| \mathrm{Li}\, x - \sum_{k=1}^{n} (k-1)! \frac{x}{\log^k x} \right| = o\left(\frac{x}{\log^n x} \right) \quad \text{if} \quad x \to \infty. \tag{H}$$

15.38. Prove that the function $\Phi(x) = \int_0^x e^{-t^2}\, dt$ is not elementary.

15.39. Let $I_k = \int (x^k / \sqrt{f(x)})\, dx\ (k \in \mathbb{N})$, where f is a polynomial of degree n. Prove that for all $k > n-2$, I_k can be expressed as a linear combination of an elementary function and the integrals $I_0, I_1, \ldots, I_{n-2}$. (S)

15.6 Appendix: Integration by Substitution for Definite Integrals (Proof of Theorem 15.22)

Proof (Theorem 15.22). **I.** First, we assume that g is monotone increasing. If g is constant, then on the one hand, $g' = 0$, so the left-hand side of (15.18) is zero, and on the other hand, $g(a) = g(b)$, so the right-hand side of (15.18) is also zero. Thus we can assume that g is not constant, that is, $g(a) < g(b)$.

By our assumptions, f is integrable on the image of g, that is, on the interval $[g(a), g(b)]$ (which must be the image of g by the Bolzano–Darboux theorem). Let $\varepsilon > 0$ be fixed, and let F be a partition of the interval $[g(a), g(b)]$ such that $\Omega_F(f) < \varepsilon$. Since g' is integrable on $[a, b]$, we can choose a partition $\Phi : a = t_0 < t_1 < \cdots < t_n = b$ such that $\Omega_\Phi(g') < \varepsilon$. By adding new base points (which does not increase the value of $\Omega_\Phi(g')$), we can ensure that the points $g(t_0), \ldots, g(t_n)$ include every base point of F. Let $h = (f \circ g) \cdot g'$. We will show that if $c_i \in [t_{i-1}, t_i]\ (i = 1, \ldots, n)$ are arbitrary inner points, then the approximating sum

$$\sigma_\Phi(h; (c_i)) = \sum_{i=1}^{n} f(g(c_i)) \cdot g'(c_i) \cdot (t_i - t_{i-1})$$

is close to the value of $I = \int_{g(a)}^{g(b)} f\, dx$.

Let F_1 denote the partition with the base points $g(t_0), \ldots, g(t_n)$. Then F_1 is a refinement of F. Let us introduce the notation $g(t_i) = u_i$, where $(i = 0, \ldots, n)$. Then the points $g(a) = u_0 \le u_1 \le \cdots \le u_n = g(b)$ list the base points of F_1, possibly more than once (if g is not strictly monotone). By the mean value theorem, for every $i = 1, \ldots, n$, there exists a point $d_i \in (t_{i-1}, t_i)$ such that

$$u_i - u_{i-1} = g'(d_i) \cdot (t_i - t_{i-1}).$$

Then

$$\sum_{i=1}^{n} f(g(c_i))(u_i - u_{i-1}) = \sum_{i=1}^{n} f(g(c_i)) \cdot g'(d_i) \cdot (t_i - t_{i-1}). \tag{15.27}$$

If we drop the terms where $u_{i-1} = u_i$ (and are thus zero) from the left-hand side of (15.27), then we get the approximating sum for the function f corresponding to the partition F_1, since $g(c_i) \in [g(t_{i-1}), g(t_i)] = [u_{i-1}, u_i]$ for all i. Since $\Omega_{F_1}(f) \le \Omega_F(f) < \varepsilon$, every such approximating sum must differ from I by less than ε. Thus by (15.27), we get that

$$\left| \sum_{i=1}^{n} f(g(c_i)) \cdot g'(d_i) \cdot (t_i - t_{i-1}) - I \right| < \varepsilon. \tag{15.28}$$

Let $\omega_i(g')$ denote the oscillation of the function g' over the interval $[t_{i-1}, t_i]$. Then $|g'(c_i) - g'(d_i)| \le \omega_i(g')$ for all i, so

$$|\sigma_\Phi(h; (c_i)) - I| \le \left| \sum_{i=1}^{n} f(g(c_i)) \left(g'(c_i) - g'(d_i)\right) (t_i - t_{i-1}) \right| +$$

$$+ \left| \sum_{i=1}^{n} f(g(c_i)) \cdot g'(d_i) \cdot (t_i - t_{i-1}) - I \right| <$$

$$< K \cdot \sum_{i=1}^{n} \omega_i(g')(t_i - t_{i-1}) + \varepsilon = K \cdot \Omega_\Phi(g') + \varepsilon <$$

$$< (K+1)\varepsilon,$$

where K denotes an upper bound of $|f|$ on the interval $[g(a), g(b)]$. Since this holds for an arbitrary choice of inner points c_i, by Theorem 14.19, $(f \circ g) \cdot g'$ is integrable on $[a, b]$, and its integral is I there. This proves (15.18) for the case that g is monotone increasing.

II. If g is monotone decreasing, then the proof goes the same way, using the fact that $g(a) \ge g(b)$ implies

$$\int_{g(a)}^{g(b)} f \, dx = -\int_{g(b)}^{g(a)} f \, dx.$$

III. Now consider the general case. Let $\varepsilon > 0$ be given, and let $\Phi \colon a = t_0 < t_1 < \cdots < t_n = b$ be a partition such that $\Omega_\Phi(g') < \varepsilon$. Let $g(t_i) = u_i$ $(i = 0, \ldots, n)$. If $I = \int_{g(a)}^{g(b)} f \, dx$ and $I_i = \int_{u_{i-1}}^{u_i} f \, dx$ for all $i = 1, \ldots, n$, then $I_1 + \cdots + I_n = I$ by Theorem 14.41.

Let J_1 denote the set of indices i such that the function g is monotone on the interval $[t_{i-1}, t_i]$. If $i \in J_1$, then by the previous cases that we have already proved, $(f \circ g) \cdot g'$ is integrable on $[t_{i-1}, t_i]$, and

$$\int_{t_{i-1}}^{t_i} f(g(t)) \cdot g'(t)\, dt = \int_{u_{i-1}}^{u_i} f\, dx = I_i.$$

Thus the interval $[t_{i-1}, t_i]$ has a partition Φ_i such that

$$I_i - (\varepsilon/n) < s_{\Phi_i} \le S_{\Phi_i} < I_i + (\varepsilon/n), \tag{15.29}$$

where s_{Φ_i} and S_{Φ_i} denote the lower and upper sums of the function $(f \circ g) \cdot g'$ restricted to the interval $[t_{i-1}, t_i]$ over the partition Φ_i. Let Φ' be the union of the partitions Φ and Φ_i $(i \in J_1)$. We will show that the lower and upper sums $s_{\Phi'}$ and $S_{\Phi'}$ of $(f \circ g) \cdot g'$ over the partition Φ' are close to I. Consider first the upper sum. Clearly,

$$S_{\Phi'} = \sum_{i \in J_1} S_{\Phi_i} + \sum_{i \in J_2} M_i \cdot (t_i - t_{i-1}), \tag{15.30}$$

where $J_2 = \{1, \ldots, n\} \setminus J_1$ and $M_i = \sup\{f(g(x))g'(x) : x \in [t_{i-1}, t_i]\}$. If $i \in J_2$, then g is not monotone on the interval $[t_{i-1}, t_i]$, and so there exists a point $d_i \in [t_{i-1}, t_i]$ such that $g'(d_i) = 0$. If this weren't the case, then by Darboux's theorem (Theorem 13.44), g' would have a constant sign on the interval $[t_{i-1}, t_i]$, and so by Theorem 12.54, g would be monotone there, which is a contradiction to what we just assumed.

Let $\omega_i(g')$ denote the oscillation of the function g' over the interval $[t_{i-1}, t_i]$. Then for arbitrary inner points $c_i \in [t_{i-1}, t_i]$, we have

$$|g'(c_i)| = |g'(c_i) - g'(d_i)| \le \omega_i(g'), \tag{15.31}$$

so $|M_i| \le K \cdot \omega_i(g')$, where K is an upper bound of $|f|$ on the image of g, that is, on the interval $g([a,b])$. Then using (15.29) and (15.30), we get that

$$
\begin{aligned}
|S_{\Phi'} - I| &\le \sum_{i \in J_1} |S_{\Phi_i} - I_i| + \sum_{i \in J_2} |M_i| \cdot (t_i - t_{i-1}) + \sum_{i \in J_2} |I_i| < \\
&< n \cdot (\varepsilon/n) + \sum_{i \in J_2} K \cdot \omega_i(g')(t_i - t_{i-1}) + \sum_{i \in J_2} |I_i| \le \\
&\le \varepsilon + K \cdot \Omega_\Phi(g') + \sum_{i \in J_2} |I_i| < \\
&< (K+1)\varepsilon + \sum_{i \in J_2} |I_i|.
\end{aligned}
\tag{15.32}
$$

Now by the mean value theorem, for all $i = 1, \ldots, n$, there exists a point $c_i \in (t_{i-1}, t_i)$ such that

$$u_i - u_{i-1} = g'(c_i) \cdot (t_i - t_{i-1}).$$

Thus if $i \in J_2$ then by (15.31),

$$|u_i - u_{i-1}| \le \omega_i(g') \cdot (t_i - t_{i-1}),$$

so by statement (iv) of Theorem 14.49,

$$|I_i| \le K \cdot |u_i - u_{i-1}| \le K \cdot \omega_i(g') \cdot (t_i - t_{i-1}).$$

Then

$$\sum_{i \in J_2} |I_i| \leq \sum_{i \in J_2} K \cdot \omega_i(g') \cdot (t_i - t_{i-1}) \leq K \cdot \Omega_\Phi(g') < K\varepsilon.$$

Comparing this with (15.32), we find that $|S_{\Phi'} - I| < (2K+1)\varepsilon$. The same argument gives that $|s_{\Phi'} - I| < (2K+1)\varepsilon$. Since this inequality holds for all ε, $(f \circ g) \cdot g'$ is integrable, and its integral is I.   □

Chapter 16
Applications of Integration

One of the main goals of mathematical analysis, besides applications in physics, is to compute the measure of sets (arc length, area, surface area, and volume). We have already spent time computing arc lengths, but only for graphs of functions. We saw examples of computing the area of certain shapes (mostly regions under graphs), and at the same time, we got a taste of computing volumes when we determined the volume of a sphere (see item 2 in Example 13.23). We also noted, however, that in computing area, some theoretical problems need to be addressed (as mentioned in point 5 of Remark 14.10). In this chapter, we turn to a systematic discussion of these questions.

When computing area, we obviously deal with sets in the plane, while when computing volume, we deal with sets in the space. However, when we deal with arc length we need to concern ourselves with both sets in the plane and in space, since some curves lie in the plane while others do not. Therefore, it is best to tackle questions concerning the plane and space simultaneously whenever possible. In mathematical analysis, points of the plane are associated with ordered pairs of real numbers, and the plane itself is associated with the set $\mathbb{R} \times \mathbb{R} = \mathbb{R}^2$ (see the appendix of Chapter 9). We will proceed analogously for representing three-dimensional space as well. We consider three lines in space intersecting at a point that are mutually perpendicular, which we call the x-, y-, and z-**axes.** We call the plane spanned by the x- and y-axes the xy-plane, and we have similar definitions for the xz- and yz-planes. We assign an ordered triple (a,b,c) to every point P in space, in which a, b, and c denote the distance (with positive or negative sign) of the point from the yz-, xz-, and xy-planes respectively. We call the numbers a, b, and c the **co-ordinates** of P. The geometric properties of space imply that the map $P \mapsto (a,b,c)$ that we obtain in this way is a bijection. This justifies our representation of three-dimensional space by ordered triples of real numbers.

Thus if we want to deal with questions both in the plane and in space, we need to deal with sets that consist of ordered d-tuples of real numbers, where $d = 2$ or $d = 3$. We will see that the specific value of d does not usually play a role in the definitions and proofs coming up. Therefore, for every positive integer d, we can define d-**dimensional Euclidean space,** by which we simply mean the set of all

© Springer New York 2015
M. Laczkovich, V.T. Sós, *Real Analysis*, Undergraduate Texts
in Mathematics, DOI 10.1007/978-1-4939-2766-1_16

sequences of real numbers of length d, with the appropriately defined addition, multiplication by a constant, absolute value, and distance. If $d = 1$, then the Euclidean space is exactly the real line; if $d = 2$, then it is the plane; and if $d = 3$, then it is 3-dimensional space. For $d > 3$, a d-dimensional space does not have an observable meaning, but it is very important for both theory and applications.

Definition 16.1. $\mathbb{R}^d$ denotes the set of ordered d-tuples of real numbers, that is, the set

$$\mathbb{R}^d = \{(x_1,\ldots,x_d) \colon x_1,\ldots,x_d \in \mathbb{R}\}.$$

The points of the set $\mathbb{R}^d$ are sometimes called *d-dimensional vectors*. The *sum of the vectors* $x = (x_1,\ldots,x_d)$ and $y = (y_1,\ldots,y_d)$ is the vector

$$x+y = (x_1+y_1,\ldots,x_d+y_d),$$

and the *product of the vector x and a real number c* is the vector

$$c \cdot x = (cx_1,\ldots,cx_d).$$

The *absolute value* of the vector x is the nonnegative real number

$$|x| = \sqrt{x_1^2 + \cdots + x_d^2}.$$

It is clear that for all $x \in \mathbb{R}^d$ and $c \in \mathbb{R}$, $|cx| = |c| \cdot |x|$. It is also easy to see that if $x = (x_1,\ldots,x_d)$, then

$$|x| \leq |x_1| + \cdots + |x_d|. \tag{16.1}$$

The **triangle inequality** also holds:

$$|x+y| \leq |x| + |y| \qquad (x,y \in \mathbb{R}^d). \tag{16.2}$$

To prove this, it suffices to show that $|x+y|^2 \leq (|x|+|y|)^2$, since both sides are nonnegative. By the definition of the absolute value, this is exactly

$$(x_1+y_1)^2 + \cdots + (x_d+y_d)^2 \leq$$
$$\leq (x_1^2 + \cdots + x_n^2) + 2 \cdot \sqrt{x_1^2 + \cdots + x_d^2} \cdot \sqrt{y_1^2 + \cdots + y_d^2} + y_1^2 + \cdots + y_d^2,$$

that is,

$$x_1 y_1 + \cdots + x_d y_d \leq \sqrt{x_1^2 + \cdots + x_d^2} \cdot \sqrt{y_1^2 + \cdots + y_d^2},$$

which is the Cauchy–Schwarz–Bunyakovsky inequality (Theorem 11.19).

The **distance** between the vectors x and y is the number $|x - y|$. By (16.2), it is clear that

$$\big||x| - |y|\big| \leq |x-y| \qquad \text{and} \qquad |x-y| \leq |x-z| + |z-y|$$

for all $x, y, z \in \mathbb{R}^d$. We can consider these to be variants of the triangle inequality.

If we apply (16.1) to the difference of the vectors $x = (x_1, \ldots, x_d)$ and $y = (y_1, \ldots, y_d)$, then we get that

$$\big| |x| - |y| \big| \le |x - y| \le |x_1 - y_1| + \cdots + |x_d - y_d|. \tag{16.3}$$

The **scalar product** of the vectors $x = (x_1, \ldots, x_d)$ and $y = (y_1, \ldots, y_d)$ is the real number $\sum_{i=1}^{d} x_i y_i$, which we denote by $\langle x, y \rangle$. With the help of the arguments in Remark 14.57, it is easy to see that $\langle x, y \rangle = |x| \cdot |y| \cdot \cos \alpha$, where α denotes the angle enclosed by the two vectors.

16.1 The General Concept of Area and Volume

We deal with the concepts of area and volume at once; we will use the word *measure* instead. We will actually define measure in every space $\mathbb{R}^d$, and area and volume will be the special cases $d = 2$ and $d = 3$.

Most of the concepts we define in the plane and in space can be generalized—purely through analogy—for the space $\mathbb{R}^d$, independent of the value of d. That includes, first of all, the concepts of axis-parallel rectangles or rectangular boxes. Since these are sets of the form $[a_1, b_1] \times [a_2, b_2]$ and $[a_1, b_1] \times [a_2, b_2] \times [a_3, b_3]$ in the plane and in space respectively, by an **axis-parallel rectangle** in $\mathbb{R}^d$, or just a **rectangle** for short, we will mean the set

$$[a_1, b_1] \times \cdots \times [a_d, b_d],$$

where $a_i < b_i$ for all $i = 1, \ldots, d$. (Here we use the Cartesian product with a finite number of terms. This means that $A_1 \times \cdots \times A_d$ denotes the set of sequences $(x_1, \ldots, x_d)$ that satisfy $x_1 \in A_1, \ldots, x_d \in A_d$.) For the case $d = 1$, the definition of a rectangle agrees with the definition of a nondegenerate closed and bounded interval.

We get (open) balls in $\mathbb{R}^d$ in the same way through analogy. The **open ball** $B(a, r)$ with center $a \in \mathbb{R}^d$ and radius $r > 0$ is the set of points that are less than distance r away from a, that is,

$$B(a, r) = \{x \in \mathbb{R}^d : |x - a| < r\}.$$

For the case $d = 1$, $B(a, r)$ is exactly the open interval $(a - r, a + r)$, while when $d = 2$, it is an open disk of radius r centered at a (where "open" means that the boundary of the disk does not belong to the set).

We call the set $A \subset \mathbb{R}^d$ **bounded** if there exists a rectangle $[a_1, b_1] \times \ldots \times [a_d, b_d]$ that contains it. It is easy to see that a set is bounded if and only if it is contained in a ball (see exercise 16.1).

We say that x is an **interior point** of the set $H \subset \mathbb{R}^d$ if H contains a ball centered at x; that is, if there exists an $r > 0$ such that $B(x, r) \subset H$. Since every ball contains a rectangle and every rectangle contains a ball, a set A has an interior point if and only if A contains a rectangle.

We call the sets A and B **nonoverlapping** if they do not share any interior points.

If we want to convert the intuitive meaning of measure into a precise notion, then we should first list our expectations for the concept. Measure has numerous properties that we consider natural. We choose three of these (see Remark 14.10.5):

(a) The measure of the rectangle $R = [a_1, b_1] \times \cdots \times [a_d, b_d]$ equals the product of its side lengths, that is, $(b_1 - a_1) \cdots (b_d - a_d)$.
(b) If we decompose a set into the union of finitely many nonoverlapping sets, then the measure of the set is the sum of the measures of the parts.
(c) If $A \subset B$, then the measure of A is not greater than the measure of B.

We will see that these requirements naturally determine to which sets we can assign a measure, and what that measure should be.

Definition 16.2. If $R = [a_1, b_1] \times \cdots \times [a_d, b_d]$, then we let $m(R)$ denote the product $(b_1 - a_1) \cdots (b_d - a_d)$.

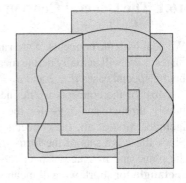

Let A be an arbitrary bounded set in $\mathbb{R}^d$. Cover A in every possible way by finitely many rectangles $R_1, \ldots, R_K$, and form the sum $\sum_{i=1}^{K} m(R_i)$ for each cover. The *outer measure* of the set A is defined as the infimum of the set of all the sums we obtain in this way (Figure 16.1). We denote the outer measure of the set A by $\overline{m}(A)$.

If A does not have an interior point, then we define the *inner measure* to be zero. If A does

Fig. 16.1

have an interior point, then choose every combination of finitely many rectangles $R_1, \ldots, R_K$ each in A such that they are mutually nonoverlapping, and form the sum $\sum_{i=1}^{K} m(R_i)$ each time. The inner measure of A is defined as the supremum of the set of all such sums. The inner measure of the set A will be denoted by $\underline{m}(A)$.

It is intuitively clear that for every bounded set A, the values $\underline{m}(A)$ and $\overline{m}(A)$ are finite. Moreover, $0 \leq \underline{m}(A) \leq \overline{m}(A)$. Now by restrictions (a) and (c) above, it is clear that the measure of the set A should fall between $\underline{m}(A)$ and $\overline{m}(A)$. If $\underline{m}(A) < \overline{m}(A)$, then without further inspection, it is not clear which number (between $\underline{m}(A)$ and $\overline{m}(A)$) we should consider the measure of A to be. We will do what we did when we considered integrals, and restrict ourselves to sets for which $\underline{m}(A) = \overline{m}(A)$, and this shared value will be called the measure of A.

Definition 16.3. We call the bounded set $A \subset \mathbb{R}^d$ *Jordan[1] measurable* if $\underline{m}(A) = \overline{m}(A)$. The *Jordan measure* of the set A (the measure of A, for short) is the shared value $\underline{m}(A) = \overline{m}(A)$, which we denote by $m(A)$.

If $d \geq 3$, then instead of Jordan measure, we can say **volume**; if $d = 2$, **area**; and if $d = 1$, **length** as well.

---

[1] Camille Jordan (1838–1922), French mathematician.

If we want to emphasize that we are talking about the inner, outer, or Jordan measure of a d-dimensional set, then instead of $\underline{m}(A), \overline{m}(A)$, or $m(A)$, we may write $\underline{m}_d(A), \overline{m}_d(A)$, or $m_d(A)$.

Exercises

16.1. Prove that for every set $A \subset \mathbb{R}^d$, the following statements are equivalent.

(a) The set A is bounded.
(b) There exists an $r > 0$ such that $A \subset B(0,r)$.
(c) For all $i = 1,\ldots,d$, the ith coordinates of the points of A form a bounded set in $\mathbb{R}$.

16.2. Prove that the set

$$A = \{(x,y): 0 \le x \le 1,\, 0 \le y \le 1,\, y \ne 1/n\ (n \in \mathbb{N}^+)\}$$

is Jordan measurable.

16.3. Prove that the set

$$A = \{(x,y): 0 \le x \le 1,\, 0 \le y \le 1,\, x,y \in \mathbb{Q}\}$$

is not Jordan measurable.

16.4. Prove that if $A \subset \mathbb{R}^p$ and $B \subset \mathbb{R}^q$ are measurable sets, then $A \times B \subset \mathbb{R}^{p+q}$ is also measurable, and that $m_{p+q}(A \times B) = m_p(A) \cdot m_q(B)$. (S)

16.2 Computing Area

With the help of Definition 16.3, we can now prove that the area under the graph of a function agrees with the integral of the function (see Remark 14.10.5). We will actually determine the areas of slightly more general regions, and then the area under the graph of a function, as well as the reflection of that region in the x-axis, will be special cases.

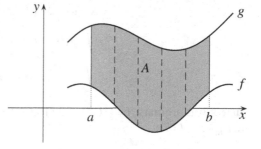

Fig. 16.2

Definition 16.4. We call the set $A \subset \mathbb{R}^2$ a *normal domain* if

$$A = \{(x,y) : x \in [a,b], \ f(x) \leq y \leq g(x)\}, \tag{16.4}$$

where f and g are integrable on $[a,b]$ and $f(x) \leq g(x)$ for all $x \in [a,b]$ (Figure 16.2).

Theorem 16.5. *If f and g are integrable on $[a,b]$ and $f(x) \leq g(x)$ for all $x \in [a,b]$, then the normal domain in (16.4) is measurable, and its area is*

$$m_2(A) = \int_a^b (g - f) \, dx.$$

Proof. For a given $\varepsilon > 0$, choose partitions F_1 and F_2 such that $\Omega_{F_1}(f) < \varepsilon$ and $\Omega_{F_2}(g) < \varepsilon$. If $F = (x_0, \ldots, x_n)$ is the union of the partitions F_1 and F_2, then $\Omega_F(f) < \varepsilon$ and $\Omega_F(g) < \varepsilon$. Let $m_i(f), m_i(g), M_i(f)$, and $M_i(g)$ be the infimum and supremum of the functions f and g respectively on the interval $[x_{i-1}, x_i]$. Then the rectangles $[x_{i-1}, x_i] \times [m_i(f), M_i(g)]$ $(i = 1, \ldots, n)$ cover the set A, so

$$\overline{m}_2(A) \leq \sum_{i=1}^{n} (M_i(g) - m_i(f)) \cdot (x_i - x_{i-1}) =$$

$$= S_F(g) - s_F(f) <$$

$$< \int_a^b g \, dx + \varepsilon - \left(\int_a^b f \, dx - \varepsilon \right) =$$

$$= \int_a^b (g - f) \, dx + 2\varepsilon. \tag{16.5}$$

Let I denote the set of indices i that satisfy $M_i(f) \leq m_i(g)$. Then the rectangles $[x_{i-1}, x_i] \times [M_i(f), m_i(g)]$ $(i \in I)$ are contained in A and are nonoverlapping, so

$$\underline{m}_2(A) \geq \sum_{i \in I} (m_i(g) - M_i(f)) \cdot (x_i - x_{i-1}) \geq$$

$$\geq \sum_{i=1}^{n} (m_i(g) - M_i(f)) \cdot (x_i - x_{i-1}) =$$

$$= s_F(g) - S_F(f) >$$

$$> \int_a^b g \, dx - \varepsilon - \int_a^b f \, dx - \varepsilon =$$

$$= \int_a^b (g - f) \, dx - 2\varepsilon. \tag{16.6}$$

Since ε was arbitrary, by (16.5) and (16.6) it follows that A is measurable and has area $\int_a^b (g - f) \, dx$. $\qquad\square$

Example 16.6. With the help of the theorem above, we can conclude that the domain bounded by the ellipse with equation $x^2/a^2 + y^2/b^2 = 1$ is a measurable set, whose area is $ab\pi$. Clearly, the set A in question is a normal domain given by the continuous functions $f(x) = -b \cdot \sqrt{1 - \left(\frac{x}{a}\right)^2}$ and $g(x) = b \cdot \sqrt{1 - \left(\frac{x}{a}\right)^2}$ over the interval $[-a,a]$. Then by Theorem 16.5, A is measurable and $m_2(A) = \int_a^b (g-f)\,dx = 2 \cdot \int_a^b g\,dx$, so by the computation done in Remark 15.20.2, we obtain $m_2(A) = ab\pi$.

We now turn to a generalization of Theorem 16.5 that can (theoretically) be used to compute the measure of any measurable plane set.

Definition 16.7. The *sections* of the set $A \subset \mathbb{R}^2$ are the sets

$$A_x = \{y \in \mathbb{R} : (x,y) \in A\}$$

and

$$A^y = \{x \in \mathbb{R} : (x,y) \in A\}$$

for every $x, y \in \mathbb{R}$ (Figure 16.3).

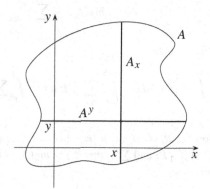

Fig. 16.3

Theorem 16.8. *Let* $A \subset \mathbb{R}^2$ *be a measurable set such that* $A \subset [a,b] \times [c,d]$. *Then the functions* $x \mapsto \overline{m}_1(A_x)$ *and* $x \mapsto \underline{m}_1(A_x)$ *are integrable in* $[a,b]$, *and*

$$m_2(A) = \int_a^b \overline{m}_1(A_x)\,dx = \int_a^b \underline{m}_1(A_x)\,dx. \tag{16.7}$$

Similarly, the functions $y \mapsto \overline{m}_1(A^y)$ *and* $y \mapsto \underline{m}_1(A^y)$ *are integrable in* $[c,d]$, *and*

$$m_2(A) = \int_c^d \overline{m}_1(A^y)\,dy = \int_c^d \underline{m}_1(A^y)\,dy.$$

Proof. It suffices to prove (16.7). Since $A \subset [a,b] \times [c,d]$, we have $A_x \subset [c,d]$ for all $x \in [a,b]$. It follows that if $x \in [a,b]$, then $\underline{m}_1(A_x) \le \overline{m}_1(A_x) \le d-c$, so the functions $\underline{m}_1(A_x)$ and $\overline{m}_1(A_x)$ are bounded in $[a,b]$.

Let $\varepsilon > 0$ be given, and choose rectangles $T_i = [a_i,b_i] \times [c_i,d_i]$ ($i = 1,\ldots,n$) such that $A \subset \bigcup_{i=1}^n T_i$ and $\sum_{i=1}^n m_2(T_i) < m_2(A) + \varepsilon$. We can assume that $[a_i,b_i] \subset [a,b]$ for all $i = 1,\ldots,n$. Let

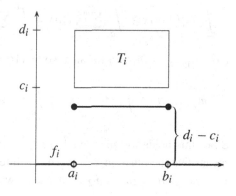

Fig. 16.4

$$f_i(x) = \begin{cases} 0, & \text{if } x \notin [a_i, b_i], \\ d_i - c_i, & \text{if } x \in [a_i, b_i] \end{cases} \qquad (i = 1, \ldots, n).$$

(See Figure 16.4.) Then f_i is integrable in $[a, b]$, and $\int_a^b f_i \, dx = m_2(T_i)$. For arbitrary $x \in [a, b]$, the sections A_x are covered by the intervals $[c_i, d_i]$ that correspond to indices i for which $x \in [a_i, b_i]$. Thus by the definition of the outer measure,

$$\overline{m}_1(A_x) \le \sum_{x \in [a_i, b_i]} (d_i - c_i) = \sum_{i=1}^n f_i(x).$$

It follows that

$$\overline{\int}_a^b \overline{m}_1(A_x) \, dx \le \overline{\int}_a^b \sum_{i=1}^n f_i \, dx = \int_a^b \sum_{i=1}^n f_i \, dx = \sum_{i=1}^n m_2(T_i) < m_2(A) + \varepsilon. \tag{16.8}$$

Now let $R_i = [p_i, q_i] \times [r_i, s_i]$ $(i = 1, \ldots, m)$ be nonoverlapping rectangles such that $A \supset \bigcup_{i=1}^m R_i$ and $\sum_{i=1}^m m_2(R_i) > m_2(A) - \varepsilon$. Then $[p_i, q_i] \subset [a, b]$ for all $i = 1, \ldots, m$. Let

$$g_i(x) = \begin{cases} 0, & \text{if } x \notin [p_i, q_i], \\ s_i - r_i, & \text{if } x \in [p_i, q_i] \end{cases} \qquad (i = 1, \ldots, m).$$

Then g_i is integrable in $[a, b]$, and $\int_a^b g_i \, dx = m_2(R_i)$. If $x \in [a, b]$, then the section A_x contains all the intervals $[r_i, s_i]$ whose indices i satisfy $x \in [a_i, b_i]$. We can also easily see that if x is distinct from all points p_i, q_i, then these intervals are nonoverlapping. Then by the definition of the inner measure,

$$\underline{m}_1(A_x) \ge \sum_{x \in [p_i, q_i]} (s_i - r_i) = \sum_{i=1}^m g_i(x).$$

It follows that

$$\underline{\int}_a^b \underline{m}_1(A_x) \, dx \ge \underline{\int}_a^b \sum_{i=1}^m g_i \, dx = \int_a^b \sum_{i=1}^m g_i \, dx = \sum_{i=1}^m m_2(R_i) > m_2(A) - \varepsilon. \tag{16.9}$$

Now $\underline{m}_1(A_x) \le \overline{m}_1(A_x)$ for all x, so by (16.8) and (16.9), we get that

$$m_2(A) - \varepsilon < \underline{\int}_a^b \underline{m}_1(A_x) \, dx \le \overline{\int}_a^b \underline{m}_1(A_x) \, dx \le \overline{\int}_a^b \overline{m}_1(A_x) \, dx < m_2(A) + \varepsilon.$$

Since this holds for all ε, we have $\underline{\int}_a^b \underline{m}_1(A_x) \, dx = \overline{\int}_a^b \underline{m}_1(A_x) \, dx = m_2(A)$, which means that the function $x \mapsto \underline{m}_1(A_x)$ is integrable on $[a, b]$ with integral $m_2(A)$. We get that $\int_a^b \overline{m}_1(A_x) \, dx = m_2(A)$ the same way. $\qquad \square$

Remark 16.9. Observe that we did not assume the measurability of the sections A_x and A^y in Theorem 16.8 (that is, that $\underline{m}_1(A_x) = \overline{m}_1(A_x)$ and $\underline{m}_1(A^y) = \overline{m}_1(A^y)$).

Corollary 16.10. *Let f be a nonnegative and bounded function on the interval $[a,b]$. The set $B_f = \{(x,y): x \in [a,b], 0 \le y \le f(x)\}$ is measurable if and only if f is integrable, and then*

$$m_2(B_f) = \int_a^b f \, dx.$$

Proof. With the help of Theorem 16.5. we need to prove only that if B_f is measurable, then f is integrable. This, however, is clear by Theorem 16.8, since $(B_f)_x = [0, f(x)]$, and so $\underline{m}_1((B_f)_x) = \overline{m}_1((B_f)_x) = f(x)$ for all $x \in [a,b]$. □

Exercises

16.5. Determine the area of the set $\{(x,y): 2 - x \le y \le 2x - x^2\}$.

16.6. Determine the area of the set $\{(x,y): 2^x \le y \le x + 1\}$.

16.7. For a given $a > 0$, determine the area of the set $\{(x,y): y^2 \le x^2(a^2 - x^2)\}$.

16.8. Let $0 < \delta < \pi/2$, $r > 0$, and $x_0 = r \cos \delta$. Prove, by computing the area under the graph of the function

$$f(x) = \begin{cases} (\operatorname{tg} \delta) \cdot x, & \text{if } 0 \le x \le x_0, \\ \sqrt{r^2 - x^2}, & \text{if } x_0 \le x \le r, \end{cases}$$

that a circular sector with central angle δ and radius r is measurable and has area $r^2 \delta/2$. (S)

16.9. Let $u > 1$ and $v = \sqrt{u^2 - 1}$. The two segments connecting the origin to the points (u, v) and $(u, -v)$ and the hyperbola $x^2 - y^2 = 1$ between the points (u, v) and $(u, -v)$ define a region A_u. Determine the area of A_u. (S)

To determine the **center of mass** of a region, we will borrow the fact from physics that if we break up a region into smaller parts, then the "weighed average" of the centers of mass of the parts, where the weights are equal to the area of each part, gives us the center of mass of the whole region. More precisely, this means that if we break the region A into regions $A_1, \ldots, A_n$, and p_i is the center of mass of A_i, then the center of mass of A is

$$\frac{\sum_{i=1}^n m(A_i) p_i}{m(A)}.$$

Let $f: [a,b] \to \mathbb{R}$ be a nonnegative continuous function, and consider the region under the graph of f, $B_f = \{(x,y): x \in [a,b], 0 \le y \le f(x)\}$. Let $F: a = x_0 < x_1 < \cdots < x_n = b$ be a fine partition, and break up B_f into the regions $A_i = \{(x,y): x \in [x_{i-1}, x_i], 0 \le y \le f(x)\}$ $(i = 1, \ldots, n)$. If $x_i - x_{i-1}$ is small, then A_i can be well approximated by the rectangle $T_i = [x_{i-1}, x_i] \times [0, f(c_i)]$, where $c_i = (x_{i-1} + x_i)/2$. Its center of mass is the point $p_i = (c_i, f(c_i)/2)$, and its area is $f(c_i) \cdot (x_i - x_{i-1})$. Thus the center of mass of the region $\bigcup_{i=1}^{n} T_i$ approximating B_f is the weighed average of the points p_i with weights $f(c_i) \cdot (x_i - x_{i-1})$, that is, the point

$$\frac{1}{\sigma_F} \cdot \sum_{i=1}^{n} f(c_i) \cdot (x_i - x_{i-1}) \cdot (c_i, f(c_i)/2), \tag{16.10}$$

where $\sigma_F = \sum_{i=1}^{n} f(c_i) \cdot (x_i - x_{i-1})$. The first coordinate of the point (16.10) is

$$x_F = \sigma_F{}^{-1} \cdot \sum_{i=1}^{n} f(c_i) \cdot c_i \cdot (x_i - x_{i-1}),$$

and its second coordinate is $y_F = (1/2) \cdot \sigma_F^{-1} \cdot \sum_{i=1}^{n} f^2(c_i) \cdot (x_i - x_{i-1})$. Let $I = \int_a^b f \, dx$. If the partition F is fine enough (has small enough mesh), then σ_F is close to I, x_F is close to the value of $x_s = I^{-1} \int_a^b f(x) \cdot x \, dx$, and y_F is close to the value of $y_s = (1/2) \cdot I^{-1} \int_a^b f^2(x) \, dx$.

This motivates the following definition. Let $f: [a,b] \to \mathbb{R}$ be nonnegative and integrable, and suppose that $I = \int_a^b f \, dx > 0$. Then the **center of mass** of the region $B_f = \{(x,y): x \in [a,b], \ 0 \le y \le f(x)\}$ is the point (x_s, y_s), where

$$x_s = \frac{1}{I} \int_a^b f(x) \cdot x \, dx \quad \text{and} \quad y_s = \frac{1}{2I} \int_a^b f(x)^2 \, dx.$$

Exercise

16.10. Determine the centers of mass of the following regions.

(a) $\{(x,y): x \in [0,m], \ 0 \le y \le c \cdot x\} \ (c, m > 0)$;
(b) $\{(x,y): x \in [-r,r], \ 0 \le y \le \sqrt{r^2 - x^2}\} \ (r > 0)$;
(c) $\{(x,y): x \in [0,a], \ 0 \le y \le x^n\} \ (a, n > 0)$.

16.3 Computing Volume

Theorem 16.8 can be generalized to higher dimensions without trouble, and the proof of the generalization is the same. Consider, for example, the three-dimensional version. If $A \subset \mathbb{R}^3$, then let A_x denote the set $\{(y,z): (x,y,z) \in A\}$.

Theorem 16.11. *Let $A \subset \mathbb{R}^3$ be a measurable set such that $A \subset [a,b] \times [c,d] \times [e,f]$. Then the functions $x \mapsto \overline{m}_2(A_x)$ and $x \mapsto \underline{m}_2(A_x)$ are integrable in $[a,b]$, and*

$$m_3(A) = \int_a^b \overline{m}_2(A_x) \, dx = \int_a^b \underline{m}_2(A_x) \, dx. \tag{16.11}$$

In the theorem above—just as in Theorem 16.8—the variable x can be replaced by the variable y or z.

With the help of equation (16.11), we can easily compute the volume of measurable sets whose sections are simple geometric shapes, for example rectangles or disks. One such family of sets are called solids of revolution, which we obtain by rotating the region under the graph of a function around the x-axis. More precisely, if the function f is nonnegative on the interval $[a,b]$, then the set

$$B_f = \{(x,y,z): a \le x \le b,\ y^2 + z^2 \le f^2(x)\}$$

is called the **solid of revolution** determined by the function f.

Theorem 16.12. *If f is nonnegative and integrable on the interval $[a,b]$, then the solid of revolution determined by f is measurable, and its volume is*

$$m_3(B_f) = \pi \cdot \int_a^b f^2(x)\,dx. \tag{16.12}$$

Proof. By (16.11), it is clear that (16.12) gives us the volume (assuming that B_f is measurable). The measurability of B_f is, however, not guaranteed by Theorem 16.11, so we give a direct proof of this fact that uses the idea of squeezing the solid of revolution between solids whose volumes we know. We can obtain such solids if we rotate the inner and outer rectangles corresponding to the curve $\{(x,y): a \le x \le b,\ 0 \le y \le f(x)\}$, which give us so-called inner and outer cylinders (Figure 16.5).

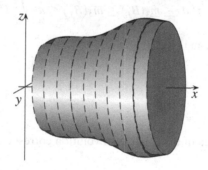

Fig. 16.5

Consider a partition $F: a = x_0 < \cdots < x_n = b$, and let

$$m_i = \inf\{f(x): x \in [x_{i-1},x_i]\} \quad \text{and} \quad M_i = \sup\{f(x): x \in [x_{i-1},x_i]\} \ (i = 1,\ldots,n).$$

The cylinders

$$\underline{C}_i = \{(x,y,z): x_{i-1} \le x \le x_i,\ y^2 + z^2 \le m_i\},$$

and

$$\overline{C}_i = \{(x,y,z) : x_{i-1} \leq x \leq x_i, \; y^2 + z^2 \leq M_i\}$$

clearly satisfy

$$\bigcup_{i=1}^{n} \underline{C}_i \subset B_f \subset \bigcup_{i=1}^{n} \overline{C}_i. \tag{16.13}$$

Now we use the fact that the cylinder $\{(x,y,z) : c \leq x \leq d, \; y^2 + z^2 \leq r\}$ is measurable and has area $r^2 \pi \cdot (d - c)$. This is a simple corollary of the fact that if $A \subset \mathbb{R}^p$ and $B \subset \mathbb{R}^q$ are measurable sets, then $A \times B$ is also measurable, and $m_{p+q}(A \times B) = m_p(A) m_q(B)$ (see Exercise 16.4). Then by (16.13),

$$\overline{m}_3(B_f) \leq m_3 \left(\bigcup_{i=1}^{n} \overline{C}_i \right) = \pi \sum_{i=1}^{n} M_i^2 (x_i - x_{i-1}) = \pi \cdot S_F(f^2),$$

and

$$\underline{m}_3(B_f) \geq m_3 \left(\bigcup_{i=1}^{n} \underline{C}_i \right) = \pi \sum_{i=1}^{n} m_i^2 (x_i - x_{i-1}) = \pi \cdot s_F(f^2).$$

Since

$$\inf_F S(f^2) = \sup_F s(f^2) = \int_a^b f^2 \, dx,$$

we have that

$$\pi \cdot \int_a^b f^2 \, dx \leq \underline{m}_3(B_f) \leq \overline{m}_3(B_f) \leq \pi \cdot \int_a^b f^2 \, dx.$$

Thus B_f is measurable, and (16.12) holds.   □

Exercises

16.11. Compute the area of the solids of revolution corresponding to the following functions:

(a) $\arcsin x \; (x \in [0,1])$;
(b) $f(x) = e^{-x} \cdot \sqrt{\sin x} \; (x \in [0,\pi])$;
(c) $f(x) = \mathrm{ch}\, x \; (x \in [-a,a])$.

16.12. Prove that the ellipsoid that is a result of rotating the ellipse $\frac{x^2}{a^2} + \frac{y^2}{b^2} = 1$ around the x-axis has volume $(4/3)ab^2\pi$.

16.13. Consider two right circular cylinders of radius R whose axes intersect and are perpendicular. Compute the volume of their intersection. (H)

16.14. Consider a right circular cylinder of radius R. Compute the volume of the part of the cylinder bounded by the side, the base circle, and a plane passing through a diameter of the base circle and forming an angle of $\frac{\pi}{4}$ with it.

16.15. Compute the volumes of the following solids (taking for granted that they are measurable):

(a) $\{(x,y,z): 0 \leq x \leq y \leq 1, \ 0 \leq z \leq 2x+3y+4\}$;
(b) $\{(x,y,z): x^2 \leq y \leq 1, \ 0 \leq z \leq x^2 + y^2\}$;
(c) $\{(x,y,z): x^4 \leq y \leq 1, \ 0 \leq z \leq 2\}$;
(d) $\{(x,y,z): (x^2/a^2) + (y^2/b^2) + (z^2/c^2) \leq 1\}$ (ellipsoid);
(e) $\{(x,y,z): |x| + |y| + |z| \leq 1\}$;
(f) $\{(x,y,z): x,y,z \geq 0, (x+y)^2 + z^2 \leq 1\}$.

16.16. Check, with the help of Theorem 16.11, that both of the integrals

$$\int_0^1 \left(\int_{y^2}^1 y \cdot e^{-x^2} \, dx \right) dy \quad \text{and} \quad \int_0^1 \left(\int_0^{\sqrt{x}} y \cdot e^{-x^2} \, dy \right) dx$$

give us the volume of the set

$$\{(x,y,z) : y \geq 0, \ y^2 \leq x \leq 1, \ 0 \leq z \leq y \cdot e^{-x^2}\}$$

(taking for granted that it is measurable); therefore, their values agree. Compute this common value.

16.17. Using the idea behind the previous question, compute the following integrals:

(a) $\int_0^1 \left(\int_x^1 \frac{x \sin y}{y} \, dy \right) dx$;
(b) $\int_0^1 \left(\int_{\sqrt{y}}^1 \sqrt{1+x^3} \, dx \right) dy$;
(c) $\int_0^1 \left(\int_{y^{2/3}}^1 y \cos x^2 \, dx \right) dy$.

16.18. Let f be nonnegative and integrable on $[a,b]$. Prove that the volume of the solid of revolution

$$\{(x,y,z): a \leq x \leq b, \ y^2 + z^2 \leq f^2(x)\}$$

is equal to the area of the set

$$A = \{(x,y): a \leq x \leq b, \ 0 \leq y \leq f(x)\}$$

times the circumference of the circle described by rotating the center of mass of A. (This is sometimes called **Guldin's[2] second Theorem.**)

---

[2] Paul Guldin (1577–1643), Swiss mathematician.

16.4 Computing Arc Length

We defined the arc length and rectifiability of graphs of functions in Definition 10.78. Let the function $f : [a,b] \to \mathbb{R}$ be continuously differentiable. Then by Theorem 13.41, the graph of f is rectifiable. Moreover, if $s(x)$ denotes the arc length of the graph of the function over the interval $[a,x]$, then s is differentiable, and $s'(x) = \sqrt{1 + (f'(x))^2}$ for all $x \in [a,b]$. Since the arc length of the graph of f is $s(b) = s(b) - s(a)$, the fundamental theorem of calculus gives us the following theorem.

Theorem 16.13. *If the function $f : [a,b] \to \mathbb{R}$ is continuously differentiable, then the arc length of the graph is $\int_a^b \sqrt{1 + (f'(x))^2}\, dx$.*

Example 16.14 (The Arc Length of a Parabola). Let s denote the arc length of the function x^2 over the interval $[0,a]$. By the previous theorem, $s = \int_0^a \sqrt{1 + 4x^2}\, dx$. Since by (15.23), $\int \sqrt{x^2 + 1}\, dx = \frac{1}{2} \cdot x \cdot \sqrt{x^2 + 1} - \frac{1}{2} \log(\sqrt{x^2 + 1} - x) + c$, with the help of a linear substitution and the fundamental theorem of calculus, we obtain

$$s = \left[\frac{1}{2} \cdot x \cdot \sqrt{4x^2 + 1} - \frac{1}{4} \log(\sqrt{4x^2 + 1} - 2x) \right]_0^a =$$
$$= \frac{a}{2} \cdot \sqrt{4a^2 + 1} - \frac{1}{4} \log(\sqrt{4a^2 + 1} - 2a).$$

If we want to compute the arc length of curves more general than graphs of functions, we first need to clarify the notion of a curve. We can think of a curve as the path of a moving particle, and we can determine that path by defining the particle's position vector at every time t. Thus the movement of a particle is defined by a function that assigns a vector in whatever space the particle is moving to each point of the time interval $[a,b]$. If the particle is moving in d-dimensional space, then this means that we assign a d-dimensional vector to each $t \in [a,b]$.

We will accept this idea as the definition of a curve, that is, a **curve** is a map of the form $g : [a,b] \to \mathbb{R}^d$. If $d = 2$, then we are talking about **planar curves**, and if $d = 3$, then we are talking about **space curves**.

Consider a curve $g : [a,b] \to \mathbb{R}^d$, and let the coordinates of the vector $g(t)$ be denoted by $g_1(t), \ldots, g_d(t)$ for all $t \in [a,b]$. This defines a function $g_i : [a,b] \to \mathbb{R}$ for each $i = 1, \ldots, d$, which is called the ith **coordinate function** of the curve g. Thus the curve is the map

$$t \mapsto (g_1(t), \ldots, g_d(t)) \qquad (t \in [a,b]).$$

We say that a curve g is **continuous, differentiable, continuously differentiable, Lipschitz,** etc. if each coordinate function of g has the corresponding property.

We emphasize that when we talk about curves, we are talking about the mapping itself, and not the path the curve traces (that is, its image). More simply: a curve is a map, not a set in $\mathbb{R}^d$. If the set H agrees with the image of the curve $g : [a,b] \to \mathbb{R}^d$, that is, $H = g([a,b])$, then we say that g is a **parameterization** of H. A set can have several parameterizations. Let us see some examples.

The **segment** determined by the points $u, v \in \mathbb{R}^d$ is the set

$$[u, v] = \{u + t \cdot (v - u) : t \in [0, 1]\}.$$

The segment $[u, v]$ is *not* a curve. On the other hand, the map $g \colon [0, 1] \to \mathbb{R}^d$, which is defined by $g(t) = u + t \cdot (v - u)$ for all $t \in [0, 1]$, is a curve that traces out the points of the segment $[u, v]$, that is, g is a parameterization of the segment $[u, v]$. Another curve is the map $h \colon [0, 1] \to \mathbb{R}^d$, which is defined as $h(t) = u + t^2 \cdot (v - u)$ $(t \in [0, 1])$. The curve h is also a parameterization of the segment $[u, v]$. Nevertheless, the curves g and h are different, since $g(1/2) = (u + v)/2$, while $h(1/2) = (3u + v)/4$.

Consider now the map $g \colon [0, 2\pi] \to \mathbb{R}^2$, for which

$$g(t) = (\cos t, \sin t) \qquad (t \in [0, 2\pi]).$$

The planar curve g defines the path of a particle that traces out the unit circle C centered at the origin, that is, g is a parameterization of the circle C. The same holds for the curve $g_1 \colon [0, 2\pi] \to \mathbb{R}^2$, where

$$g_1(t) = (\cos 2t, \sin 2t) \qquad (t \in [0, 2\pi]).$$

The curve g_1 also traces out C, but "twice over." Clearly, $g \neq g_1$. Since the length of g is 2π, while the length of g_1 is 4π, this example shows that arc length should be assigned to the curve (that is, the map), and not to the image of the curve.

Note that the graph of any function $f \colon [a, b] \to \mathbb{R}$ can be parameterized with the planar curve $t \mapsto (t, f(t)) \in \mathbb{R}^2$ $(t \in [a, b])$.

We define the arc length of curves similarly to how we defined the arc length of graphs of functions. A broken or polygonal line is a set that is the union of connected segments. If $a_0, \ldots, a_n$ are arbitrary points of the space $\mathbb{R}^d$, then the polygonal line connecting the points a_i (in this order) consists of the segments $[a_0, a_1], [a_1, a_2], \ldots, [a_{n-1}, a_n]$. The length of a polygonal line is the sum of the lengths of the segments that constitute it, that is, $|a_1 - a_0| + |a_2 - a_1| + \cdots + |a_n - a_{n-1}|$ (Figure 16.6).

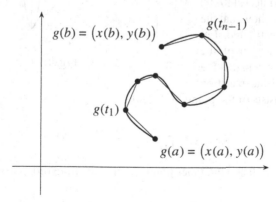

Fig. 16.6

Definition 16.15. An *inscribed polygonal path* of the curve $g \colon [a,b] \to \mathbb{R}^d$ is a polygonal line connecting the points $g(t_0), g(t_1), \ldots, g(t_n)$, where $a = t_0 < t_1 < \ldots < t_n = b$ is a partition of the interval $[a,b]$. The *arc length* of the curve g is the least upper bound of the set of lengths of inscribed polygonal paths of g (which can be infinite). We denote the arc length of the curve g by $s(g)$. Thus

$$s(g) = \sup\left\{ \sum_{i=1}^{n} |g(t_i) - g(t_{i-1})| : a = t_0 < t_1 < \cdots < t_n = b,\ n = 1, 2, \ldots \right\}.$$

We say that a curve g is *rectifiable* if $s(g) < \infty$.

Not every curve is rectifiable. It is clear that if the image of the curve $g \colon [a,b] \to \mathbb{R}^d$ is unbounded, then there exist arbitrarily long inscribed polygonal paths of g, and so $s(g) = \infty$. The following example shows that it is not enough for a curve to be continuous or even differentiable in order for it to be rectifiable.

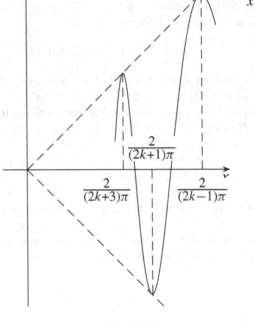

Example 16.16. Consider the planar curve $g(t) = (t, f(t))$ ($t \in [0,1]$), where

$$f(t) = \begin{cases} t \cdot \sin(1/t) & \text{if } t \neq 0, \\ 0 & \text{if } t = 0. \end{cases}$$

(The curve g parameterizes the graph of f on the interval $[0,1]$) (Figure 16.7). We show that the curve g is not rectifiable. Let us compute the length of the inscribed polygonal path of g

Fig. 16.7

corresponding to the partition F_n, where F_n consists of the points 0, 1, and

$$x_i = \frac{2}{(2i-1)\pi} \qquad (i = 1, \ldots, n).$$

(We happen to have listed the inner points x_i in decreasing order.) Since

$$f(x_i) = (-1)^{i+1} \frac{2}{(2i-1)\pi} \qquad (i = 1, \ldots, n)$$

if $1 \leq i \leq n-1$, the length of the segment $[g(x_i), g(x_{i+1})]$ is

$$|g(x_{i+1}) - g(x_i)| \geq |f(x_{i+1}) - f(x_i)| = \frac{2}{\pi} \left(\frac{1}{2i-1} + \frac{1}{2i+1} \right) > \frac{2}{\pi} \cdot \frac{1}{i+1}.$$

Thus the length of the inscribed polygonal path is at least

$$\sum_{i=1}^{n} |g(x_i) - g(x_{i-1})| > \frac{2}{\pi} \sum_{i=1}^{n} \frac{1}{i}.$$

Since $\sum_{i=1}^{n} (1/i) \to \infty$ if $n \to \infty$ (see Theorem 7.8), the set of lengths of inscribed polygons is indeed unbounded. Thus the curve g is not rectifiable, even though the functions t and $f(t)$ defining g are continuous.

Now consider the planar curve $h(t) = (t^2, f(t^2))$ ($t \in [0,1]$), where f is the function above. The curve h also parameterizes the graph of f and has the same inscribed polygonal paths as g. Thus h is not rectifiable either, even though the functions t^2 and $f(t^2)$ are both differentiable on $[0,1]$ (see Example 13.46).

Remark 16.17. As we have seen, the arc length of the curve $g: [a,b] \to \mathbb{R}^d$ depends on the map g, since the inscribed polygonal paths already depend on g. This means that we cannot generally speak of the arc length of a set H—even when H is the image of a curve, or in other words, even if it is parameterizable. This is because H can have multiple parameterizations, and the arc lengths of these different parameterizations could be different. So for example, the curves $f(t) = (t^2, 0)$ ($t \in [0,1]$) and $g(t) = (t^2, 0)$ ($t \in [-1,1]$) parameterize the same set (namely the interval $[0,1]$ of the x-axis), but their arc lengths are different.

In some important cases, however, we do give sets H a unique arc length. We call sets that have a bijective and continuous parameterization **simple curves**. Some examples of simple curves are the segments, arcs of a circle, and the graph of any continuous function.

Theorem 16.18. *If $H \subset \mathbb{R}^d$ is a simple curve, then every bijective and continuous parameterization of H defines the same arc length.*

Proof. We outline a sketch of the proof. Let $\beta: [a,b] \to H$ and $\gamma: [c,d] \to H$ be bijective parameterizations. Then the function $h = \gamma^{-1} \circ \beta$ maps the interval $[a,b]$ onto the interval $[c,d]$ bijectively, and one can show that h is also continuous. (This step—which belongs to multivariable calculus—is not detailed here.) It follows that in this case, h is strictly monotone (see Exercise 10.54). Thus $\beta = \gamma \circ h$, where h is a strictly monotone bijection of $[a,b]$ onto $[c,d]$.

This property ensures that the curves $\beta: [a,b] \to H$ and $\gamma: [c,d] \to H$ have the same inscribed polygonal paths. If $F: a = t_0 < t_1 < \ldots < t_n = b$ is a partition of the interval $[a,b]$, then either $c = h(a) < h(t_1) < \cdots < h(t_n) = d$ or $c = h(t_n) < h(t_{n-1}) < \cdots < h(t_0) = d$, depending on whether h is increasing or decreasing. One of the two will give a partition of the interval $[c,d]$ that will give the same inscribed polygonal path as given by F under the map $\beta = \gamma \circ h$. That is, every inscribed polygonal path of $\beta\colon [a,b] \to H$ is an inscribed polygonal path of $\gamma\colon [c,d] \to H$. In the same way, every inscribed polygonal path of $\gamma\colon [c,d] \to H$ is also an inscribed polygonal path of $\beta\colon [a,b] \to H$. It then follows that the arc lengths of the curves $\beta\colon [a,b] \to H$ and $\gamma\colon [c,d] \to H$ are the suprema of the same set, so the arc lengths agree.   □

According to what we said above, we can talk about arc lengths of simple curves: by this, we mean the arc length of a parameterization that is bijective and continuous.

The following theorem gives us simple sufficient conditions for a curve to be rectifiable.

Theorem 16.19. *Consider a curve* $g\colon [a,b] \to \mathbb{R}^d$.

  (i) *If the curve is Lipschitz, then it is rectifiable.*
 (ii) *If the curve g is differentiable, and the derivatives of the coordinate functions of g are bounded on* $[a,b]$, *then g is rectifiable.*
(iii) *If the curve is continuously differentiable, then it is rectifiable.*

Proof. (i) Let the curve g be Lipschitz, and suppose that $|g_i(x) - g_i(y)| \le K \cdot |x-y|$ for all $x \in [a,b]$ and $i = 1,\ldots,d$. Then using (16.3), we get that $|g(x) - g(y)| \le Kd \cdot |x-y|$ for all $x,y \in [a,b]$. Then it immediately follows that every inscribed polygonal path of g has length at most $Kd \cdot (b-a)$, so g is rectifiable.
(ii) By the mean value theorem, if the function $g_i\colon [a,b] \to \mathbb{R}$ is differentiable on the interval $[a,b]$ and its derivative is bounded, then g_i is Lipschitz. Thus the rectifiability of g follows from (i).

Statement (iii) is clear from (ii), since continuous functions on the interval $[a,b]$ are necessarily bounded (Theorem 10.52).   □

Theorem 16.20. *Suppose that the curve* $g\colon [a,b] \to \mathbb{R}^d$ *is differentiable, and the derivatives of the coordinate functions of g are integrable on* $[a,b]$. *Then g is rectifiable, and*

$$s(g) = \int_a^b \sqrt{\left(g_1'(t)\right)^2 + \cdots + \left(g_d'(t)\right)^2}\, dt. \tag{16.14}$$

We give a proof of this theorem in the appendix of the chapter.

Remark 16.21. Let $f\colon [a,b] \to \mathbb{R}$, and apply the above theorem to the curve given by $g(t) = (t, f(t))$ $(t \in [a,b])$. We get that in Theorem 16.13, instead of having to assume the continuous differentiability of the function f, it is enough to assume that f is differentiable and that f' is integrable on $[a,b]$.

Remark 16.22. Suppose that the curve $g\colon [a,b] \to \mathbb{R}^d$ is differentiable. Let the coordinate functions of g be $g_1,\ldots,g_d$. If t_0 and t are distinct points of the interval $[a,b]$, then

$$\frac{g(t) - g(t_0)}{t - t_0} = \left(\frac{g_1(t) - g_1(t_0)}{t - t_0}, \ldots, \frac{g_d(t) - g_d(t_0)}{t - t_0} \right).$$

Here if we let t approach t_0, the jth coordinate of the right-hand side tends to $g_j'(t_0)$. Thus it is reasonable to call the vector $(g_1'(t_0),\ldots,g_d'(t_0))$ the **derivative** of the curve g at the point t_0. We denote it by $g'(t_0)$. With this notation, (16.14) takes on the form

$$s(g) = \int_a^b |g'(t)| \, dt. \tag{16.15}$$

The physical meaning of the derivative g' is the velocity vector of a particle moving along the curve g. Clearly, the displacement of the particle between the times t_0 and t is $g(t) - g(t_0)$. The vector $(g(t) - g(t_0))/(t - t_0)$ describes the average displacement of the particle during the time interval $[t_0, t]$. As $t \to t_0$, this average tends to the velocity vector of the particle. Since $(g(t) - g(t_0))/(t - t_0)$ tends to $g'(t_0)$ in each coordinate, $g'(t_0)$ is exactly the velocity vector.

On the other hand, the value $|(g(t) - g(t_0))/(t - t_0)|$ denotes the average magnitude of the displacement of the moving particle during the time interval $[t_0, t]$. The limit of this as $t \to t_0$ is the instantaneous velocity of the particle. Thus the absolute value of the velocity vector, $|g'(t_0)|$, is the instantaneous velocity. Thus the physical interpretation of (16.15) is that during movement (along a curve) of a particle, the distance traversed by the point is equal to the integral of its instantaneous velocity. We already saw this for motions along a straight path: this was the physical statement of the fundamental theorem of calculus (Remark 15.2). Thus we can consider (16.14), that is, (16.15), to be an analogue of the fundamental theorem of calculus for curves.

Example 16.23. Consider a circle of radius a that is rolling along the x-axis. The path traced out by a point P on the rolling circle is called a **cycloid**. Suppose that the point P was at the origin at the start of the movement. The circle rolls along the x-axis (without slipping), meaning that at each moment, the length of the circular arc between the point A of the circle touching the axis and P is equal to the length of the segment OA.

Let t denote the angle between the rays CA and CP, where C denotes the center of the circle. Then the length of the line segment AP is at, that is, $\overline{OA} = at$. In the triangle CPR seen in the figure, $\overline{PR} = a \sin t$ and $\overline{CR} = -a \cos t$, so the coordinates of the point P are $(at - a \sin t, a - a \cos t)$. After a full revolution of the circle, the point P is touching the axis again. Thus the parameterization of the cycloid is

$$g(t) = (at - a \sin t, a - a \cos t) \qquad (t \in [0, 2\pi]).$$

Since $g'(t) = (a - a \cos t, a \sin t)$ and

$$|g'(t)| = \sqrt{(a - a \cos t)^2 + (a \sin t)^2} = a \cdot \sqrt{2 - 2 \cos t} =$$

$$= a \cdot \sqrt{4 \sin^2 \frac{t}{2}} = 2a \sin \frac{t}{2},$$

by (16.15), the arc length of a cycloid is

$$\int_0^{2\pi} 2a \sin \frac{t}{2} \, dt = \left[-4a \cos \frac{t}{2} \right]_0^{2\pi} = 8a.$$

Thus the arc length of a cycloid is eight times the radius of the rolling circle (Figure 16.8).

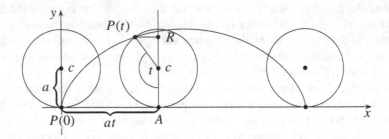

Fig. 16.8

Exercises

16.19. Construct (a) a segment; (b) the boundary of a square as the image of both a differentiable and a nondifferentiable curve.

16.20. Compute the arc lengths of the graphs of the following functions:

(a) $f(x) = x^{3/2}$ $(0 \leq x \leq 4)$;
(b) $f(x) = \log(1 - x^2)$ $(0 \leq x \leq a < 1)$;
(c) $f(x) = \log \cos x$ $(0 \leq x \leq a)$;
(d) $f(x) = \log \frac{e^x + 1}{e^x - 1}$ $(a \leq x \leq b)$.

16.21. Let the arc length of the graph $f : [a,b] \to \mathbb{R}$ be denoted by L, and the arc length of $g(t) = (t, f(t))$ $(t \in [a,b])$ by S. Prove that $L \leq S \leq L + (b - a)$. Show that the graph of f is rectifiable if and only if the curve $g(t)$ $(t \in [a,b])$ is rectifiable.

16.22. In the following exercises, by the planar curve with parameterization $x = x(t)$, $y = y(t)$ $(t \in [a,b])$ we mean the curve $g(t) = (x(t), y(t))$ $(t \in [a,b])$. Compute the arc lengths of the following planar curves:

(a) $x = a \cdot \cos^3 t$, $y = a \cdot \sin^3 t$ $(0 \leq t \leq 2\pi)$ (**astroid**);
(b) $x = a \cdot \cos^4 t$, $y = a \cdot \sin^4 t$ $(0 \leq t \leq \pi/2)$;
(c) $x = e^t (\cos t + \sin t)$, $y = e^t (\cos t - \sin t)$ $(0 \leq t \leq a)$;
(d) $x = t - \text{th} t$, $y = 1/\text{ch} t$ $(t \in [0,1])$;
(e) $x = \text{ctg} t, y = 1/(2 \sin^2 t)$ $(\pi/4 \leq t \leq \pi/2)$.

16.23. Let $n > 0$, and consider the curve $g = (\cos(t^n), \sin(t^n))$ $(t \in [0, \sqrt[n]{2\pi}])$ (which is a parameterization of the unit circle). Check that the arc length of this curve is 2π (independent of the value of n).

16.24. For a given $b, d > 0$, compute the arc length of the catenary, that is, the graph of the function $f(x) = b^{-1} \cdot \text{ch}(bx)$ $(0 \leq x \leq d)$.

16.25. Let a and b be fixed positive numbers. For which c will the arc length of the ellipse with semiaxes a and b be equal to the arc length of the function $c \cdot \sin x$ over the interval $[0, \pi]$?

16.26. How large can the arc length of the graph of a (a) monotone; (b) monotone and continuous; (c) strictly monotone; (d) strictly monotone and continuous function $f: [0,1] \to [0,1]$ be?

16.27. Let f be the Riemann function in $[0,1]$. For which $c > 0$ will the graph of f^c be rectifiable? (∗ H S)

16.28. Prove that if $f: [a,b] \to \mathbb{R}$ is differentiable and f' is bounded on $[a,b]$, then

(a) the graph of f is rectifiable, and
(b) the arc length of the graph of f lies between

$$\underline{\int_a^b} \sqrt{1 + (f'(x))^2}\, dx \quad \text{and} \quad \overline{\int_a^b} \sqrt{1 + (f'(x))^2}\, dx.$$

16.29. Prove that if the curve $g: [a,b] \to \mathbb{R}^2$ is continuous and rectifiable, then for every $\varepsilon > 0$, the image $g([a,b])$ can be covered by finitely many disks whose total area is less than ε. (∗ H S)

The center of mass of a curve. Imagine a curve $g: [a,b] \to \mathbb{R}^d$ made up of some homogeneous material. Then the weight of every arc of g is ρ times the length of that arc, where ρ is some constant (density). Consider a partition $F: a = t_0 < t_1 < \cdots < t_n = b$, and let $c_i \in [t_{i-1}, t_i]$ be arbitrary inner points. If the curve is continuously differentiable and the partition is fine enough, then the length of the arc $g([t_{i-1}, t_i])$ is close to the length of the segment $[g(t_{i-1}), g(t_i)]$, so the weight of the arc is close to $\rho \cdot |g(t_i) - g(t_{i-1})|$. Thus if for every i, we concentrate a weight $\rho \cdot |g(t_i) - g(t_{i-1})|$ at the point $g(c_i)$, then the weight distribution of the points of weights we get in this way will be close to the weight distribution of the curve itself. We can expect the center of mass of the collection of these points to be close to the center of mass of the curve.

The center of mass of the system of points above is the point $\frac{1}{L_F} \cdot \sum_{i=1}^n |g(t_i) - g(t_{i-1})| \cdot g(c_i)$, where $L_F = \sum_{i=1}^n |g(t_i) - g(t_{i-1})|$. If the partition is fine enough, then L_F is close to the arc length L.

Let the coordinate functions of g be $g_1, \ldots, g_d$. Then the length $|g(t_i) - g(t_{i-1})|$ is well approximated by the value $\sqrt{g_1'(c_i)^2 + \cdots + g_d'(c_i)^2} \cdot (t_i - t_{i-1})$, so the jth coordinate of the point $\sum_{i=1}^n |g(t_i) - g(t_{i-1})| \cdot g(c_i)$ will be close to the sum

$$\sum_{i=1}^n \sqrt{g_1'(c_i)^2 + \cdots + g_d'(c_i)^2} \cdot g_j(c_i) \cdot (t_i - t_{i-1}).$$

If the partition is fine enough, then this sum is close to the integral

$$s_j = \int_a^b \sqrt{g_1'(t)^2 + \cdots + g_d'(t)^2} \cdot g_j(t)\, dt. \tag{16.16}$$

This motivates the following definition.

Definition 16.24. If the curve $g\colon [a,b] \to \mathbb{R}^p$ is differentiable and the derivatives of the coordinate functions of g are integrable on $[a,b]$, then the **center of mass** of g is the point $(s_1/L,\ldots,s_d/L)$, where L is the arc length of the curve, and the s_j are defined by (16.16) for all $j = 1,\ldots,d$.

Exercise

16.30. Compute the center of mass of the following curves:

(a) $g(t) = (t,t^2)$ $(t \in [0,1])$;
(b) $g(t) = (t,\sin t)$ $(y \in [0,\pi])$;
(c) $g(t) = (a \cdot (1+\cos t)\cos t, a \cdot (1+\cos t)\sin t)$ $(0 \le t \le 2\pi)$, where $a > 0$ is constant (cardioid).

16.5 Polar Coordinates

The **polar coordinates** of a point P distinct from the origin are given by the ordered pair (r,φ), where r denotes the distance of P from the origin, and φ denotes the angle between $\overrightarrow{OP}$ and the positive direction of the x-axis (Figure 16.9). From the figure, it is clear that if the polar coordinates of P are (r,φ), then the usual (Cartesian) coordinates are $(r\cos\varphi, r\sin\varphi)$. The polar

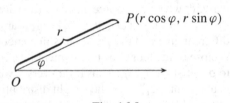

Fig. 16.9

coordinates of the origin are given by $(0,\varphi)$, where φ can be arbitrary.

If $[\alpha,\beta] \subset [0,2\pi)$, then every function $r\colon [\alpha,\beta] \to [0,\infty)$ describes a curve, the collection of points $(r(\varphi),\varphi)$ for $\varphi \in [\alpha,\beta]$. Since the Cartesian coordinates of the point $(r(\varphi),\varphi)$ are $(r(\varphi)\cos\varphi, r(\varphi)\sin\varphi)$, using our old notation we are actually talking about the curve

$$g(t) = (r(t)\cos t, r(t)\sin t) \qquad (t \in [\alpha,\beta]). \tag{16.17}$$

Definition 16.25. The function $r: [\alpha,\beta] \to [0,\infty)$ is called the *polar coordinate form* of the curve (16.17).

In this definition, we do not assume $[\alpha,\beta]$ to be part of the interval $[0,2\pi]$. This is justified in that for arbitrary $t \in \mathbb{R}$, if $r > 0$, then the polar coordinate form of the point $P = (r\cos t, r\sin t)$ is $(r, t - 2k\pi)$, where k is an integer such that $0 \le t - 2k\pi < 2\pi$. In other words, t is equal to *one* of the angles between $\overrightarrow{OP}$ and the positive half of the x-axis, so in this more general sense, we can say that the points $(r(\varphi), \varphi)$ given in polar coordinate form give us the curve (16.17).

Examples 16.26. **1.** If $[\alpha,\beta] \subset [0,2\pi)$, then the function $r \equiv a$ ($\varphi \in [\alpha,\beta]$), where $a > 0$ is constant, is the polar coordinate form of a subarc of the circle of radius a centered at the origin.

2. The function

$$r(\varphi) = a \cdot \varphi \qquad (\varphi \in [0,\beta]) \qquad (16.18)$$

describes what is called the **Archimedean spiral**. The Archimedean spiral is the path of a particle that moves uniformly along a ray starting from the origin while the ray rotates uniformly about the origin (Figure 16.10).

Theorem 16.27. *Suppose that the function* $r: [\alpha,\beta] \to [0,\infty)$ *is differentiable, and its derivative is integrable on* $[\alpha,\beta]$. *Then the curve given by* r *in its polar coordinate form is rectifiable, and its arc length is*

$$\int_\alpha^\beta \sqrt{(r')^2 + r^2}\, d\varphi. \qquad (16.19)$$

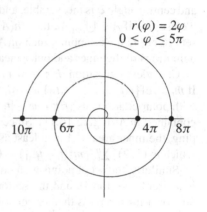

Fig. 16.10

Proof. The curve g given by (16.17) is differentiable, and the derivatives of the coordinate functions, $r'\cos t - r\sin t$ and $r'\sin t + r\cos t$, are integrable on $[\alpha,\beta]$. Thus by Theorem 16.20, the curve is rectifiable. Since

$$|g'(t)| = \sqrt{(r'\cos t - r\sin t)^2 + (r'\sin t + r\cos t)^2} = \sqrt{(r')^2 + r^2},$$

by (16.15), we get (16.19). $\qquad\qquad\square$

Example 16.28. The arc length of the Archimedean spiral given by (16.18) is

$$\int_0^\beta \sqrt{a^2 t^2 + a^2}\, dt = a \cdot \left[\frac{1}{2} \cdot \beta \cdot \sqrt{\beta^2 + 1} - \frac{1}{2}\log(\sqrt{\beta^2 + 1} - \beta) \right].$$

Consider a curve given in its polar coordinate form
$r\colon [\alpha,\beta] \to [0,\infty)$. The union of the segments connecting every point of the curve to the origin is
called a **sectorlike region**. By the definition of polar
coordinates, the region in question is exactly the set

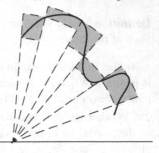

$$A = \{(r\cos\varphi,\ r\sin\varphi) : 0 \le r \le r(\varphi),\ \alpha \le \varphi \le \beta\}.$$
$$(16.20)$$

Theorem 16.29. *Let* $0 \le \alpha < \beta \le 2\pi$. *If the*   Fig. 16.11
function f *is nonnegative and integrable on* $[\alpha,\beta]$,
then the sectorlike region given in (16.20) *is measurable, and its area is*
$\frac{1}{2}\int_\alpha^\beta r^2(\varphi)\,d\varphi$.

Proof. To prove this theorem, we use the fact that the circular sector with radius r
and central angle δ is measurable, and has area $r^2\delta/2$ (see Exercise 16.8). Moreover,
we use that if $A \subset \bigcup_{i=1}^n A_i$, then $\overline{m}(A) \le \sum_{i=1}^n \overline{m}(A_i)$, and if $A \supset \bigcup_{i=1}^n B_i$, where the
sets B_i are nonoverlapping, then $\underline{m}(A) \ge \sum_{i=1}^n \underline{m}(B_i)$. These follow easily from the
definitions of the inner and outer measure.

Consider a partition $F : \alpha = t_0 < t_1 < \cdots < t_n = \beta$ of the interval $[\alpha,\beta]$.
If $m_i = \inf\{r(t) : t \in [t_{i-1},t_i]\}$ and $M_i = \sup\{r(t) : t \in [t_{i-1},t_i]\}$, then the set of points
with polar coordinates (r,φ) $(\varphi \in [t_{i-1},t_i],\ 0 \le r \le m_i)$ is a circular sector B_i that is
contained in A (Figure 16.11). Since the circular sectors $B_1,\ldots,B_n$ are nonoverlap-
ping, the inner area of A is at least as large as the sum of the areas of these sectors,
which is $(1/2)\cdot\sum_{i=1}^n m_i^2(t_i-t_{i-1}) = (1/2)\cdot s_F(r^2)$.

Similarly, the set of points with polar coordinates (r,φ) $(\varphi \in [t_{i-1},t_i],\ 0 \le r \le M_i)$
is a circular sector A_i, and the sectors $A_1,\ldots,A_n$ together cover A. Thus the outer
area of A must be less than or equal to the sum of the areas of these sectors, which
is $\frac{1}{2}\cdot\sum_{i=1}^n M_i^2(t_i-t_{i-1}) = \frac{1}{2}\cdot S_F(r^2)$. Thus

$$\tfrac{1}{2}s_F(r^2) \le \underline{m}(A) \le \overline{m}(A) \le \tfrac{1}{2}S_F(r^2)$$

for every partition F. Since r^2 is integrable,

$$\sup_F s_F(r^2) = \inf_F S_F(r^2) = \int_\alpha^\beta r^2(\varphi)\,d\varphi,$$

and

$$\frac{1}{2}\int_\alpha^\beta r^2(\varphi)\,d\varphi \le \underline{m}(A) \le \overline{m}(A) \le \frac{1}{2}\int_\alpha^\beta r^2(\varphi)\,d\varphi.$$

This shows that A is measurable, and its area is equal to half of the integral. $\square$

Exercises

16.31. Compute the arc lengths of the following curves given in polar coordinate form:

(a) $r = a \cdot (1 + \cos \varphi)$ $(0 \le \varphi \le 2\pi)$, where $a > 0$ is constant (cardioid);
(b) $r = a/\varphi$ $(\pi/2 \le \varphi \le 2\pi)$, where $a > 0$ is constant;
(c) $r = a \cdot e^{c \cdot \varphi}$ $(0 \le \varphi \le \alpha)$, where $a > 0$, $c \in \mathbb{R}$, and $\alpha > 0$ are constants;

(d) $r = \frac{p}{1 + \cos \varphi}$ $(0 \le \varphi \le \pi/2)$, where $p > 0$ is constant; what is this curve? (H)

(e) $r = \frac{p}{1 - \cos \varphi}$ $(\pi/2 \le \varphi \le \pi)$, where $p > 0$ is constant; what is this curve? (H)

16.32. Let $a > 0$ be constant. The set of planar points whose distance from $(-a, 0)$ times its distance from $(a, 0)$ is equal to a^2 is called a *lemniscate* (Figure 16.12). Show that $r^2 = 2a^2 \cdot \cos 2\varphi$ $(-\pi/4 \le \varphi \le \pi/4$ or $3\pi/4 \le \varphi \le 5\pi/4)$ is a parameterization of the lemniscate in polar coordinate form. Compute the area of the region bounded by the lemniscate.

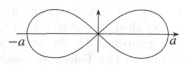

Fig. 16.12

16.33. Compute the area of the set of points satisfying $r^2 + \varphi^2 \le 1$.

16.34. Compute the area of the region bounded by the curve $r = \sin \varphi + e^\varphi$ $(0 \le \varphi \le \pi)$ given in polar coordinate form, and the segment $[-e^\pi, 1]$ of the x-axis.

16.35. The curve $r = a \cdot \varphi$ $(0 \le \varphi \le \pi/4)$ given in polar coordinate form is the graph of a function f.

(a) Compute the area of the region under the graph of f.
(b) Revolve the region under this graph about the x-axis. Compute the volume of the solid of revolution we obtain in this way. (H)

16.36. The cycloid with parameter a over $[0, 2a\pi]$ is the graph of a function g.

(a) Compute the area of the region under the graph of g.
(b) Revolve the region under this graph about the x-axis. Compute the volume of the solid of revolution we obtain in this way. (H)

16.37. Express the curve satisfying the conditions

(a) $x^4 + y^4 = a^2(x^2 + y^2)$; and
(b) $x^4 + y^4 = a \cdot x^2 y$

in polar coordinate form, and compute the area of the enclosed region.

16.6 The Surface Area of a Surface of Revolution

Determining the surface area of surfaces is a much harder task than finding the area of planar regions or the volume of solids; the definition of surface area itself already causes difficulties. To define surface area, the method used to define area—bounding the value from above and below—does not work. We could try to copy the method of defining arc length and use the known surface area of inscribed polygonal surfaces, but this already fails in the simplest cases: one can show that the inscribed polygonal surfaces of a right circular cylinder can have arbitrarily large surface area. To precisely define surface area, we need the help of differential geometry, or at least multivariable differentiation and integration, which we do not yet have access to.

Determining the surface area of a surface of revolution is a simpler task. Let $f : [a,b] \to \mathbb{R}$ be a nonnegative function, and let A^f denote the set we get by rotating graph f about the x-axis. It is an intuitive assumption that the surface area of A^f is well approximated by the surface area of the rotation of an inscribed polygonal path about the x-axis. Before we inspect this assumption in more detail, let us compute the surface area of the rotated inscribed polygonal paths.

Let $F : a = x_0 < x_1 < \cdots < x_n = b$ be an arbitrary partition. Rotating the inscribed polygonal paths corresponding to F about x gives us a set P^F, which consists of n parts: the ith part, which we will denote by P_i^F, is the rotated segment over the interval $[x_{i-1}, x_i]$ (Figure 16.13). We can see in the figure that the set P_i^F is the side of a right conical frustum with height $x_i - x_{i-1}$, and radii of bases $f(x_i)$ and $f(x_{i-1})$.

It is intuitively clear that if we unroll the side of a right conical frustum, then the area of the region we get is equal to the surface area of the side of the frustum. Since the un-rolled side is the difference of two circular sectors, the area of this can be computed easily, using that the area of the sector is half the radius times the arc length.

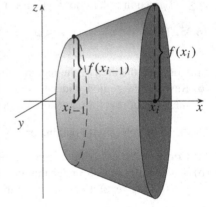

In the end, we get that if the height of the frustum is m, and the bases have radii r and R, then the lateral surface area is $\pi(R + r)\sqrt{(R-r)^2 + m^2}$ (see Exercise 16.38). Thus the surface area of the side P_i^F is

$$\pi \cdot (f(x_i) + f(x_{i-1})) \cdot h_i,$$

Fig. 16.13

and the surface area of the set P^F is

$$\Phi_F = \pi \cdot \sum_{i=1}^{n} (f(x_i) + f(x_{i-1})) \cdot h_i, \qquad (16.21)$$

where $h_i = \sqrt{(f(x_i) - f(x_{i-1}))^2 + (x_i - x_{i-1})^2}$. Here we should note that the sum $\sum_{i=1}^{n} h_i$ is equal to the length of the inscribed polygonal path (corresponding to F). Therefore, $\sum_{i=1}^{n} h_i \leq L$, where L denotes the arc length of graph f.

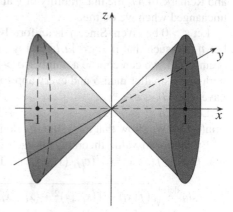

Fig. 16.14

Now let us return to figuring out in what sense the value Φ_F approximates the surface area of the set A^f. Since the length of the graph of f is equal to the supremum of the arc lengths of the inscribed polygonal paths, our first thought might be that the surface area of A^f needs to be equal to the supremum of the values Φ_F. However, this is already not the case with simple functions. Consider, for example, the function $|x|$ on the interval $[-1,1]$. In this case, the set A^f is the union of two sides of cones, and its surface area is $2 \cdot (2\pi \cdot \sqrt{2}/2) = 2\sqrt{2}\pi$ (Figure 16.14). But if the partition F consists only of the points -1 and 1, then P^F is a cylinder whose surface area is $2\pi \cdot 2 = 4\pi$, which is larger than the surface area of A^f.

By the example above, we can rule out being able to define the surface area of A^f as the supremum of the set of values Φ_F. However, the example hints at the correct definition, since an arbitrary partition F of $[-1,1]$ makes P^F equal either to A^f (if 0 is a base point of F) or to the union of the sides of three frustums. If the mesh of the partition is small, then the surface area of the middle frustum will be small, and the two other frustums will be close to the two cones making up A^f. This means that the surface area of P^F will be arbitrarily close to the surface area of A^f if the partition becomes fine enough. This observation motivates the following definition.

Definition 16.30. Let f be nonnegative on $[a,b]$. We say that the *surface area* of

$$A^f = \{(x,y,z) : a \leq x \leq b, \ \sqrt{y^2 + z^2} = f(x)\}$$

exists and equals Φ if for every $\varepsilon > 0$, there exists a $\delta > 0$ such that for every partition F of $[a,b]$ with mesh smaller than δ, we have $|\Phi_F - \Phi| < \varepsilon$, where Φ_F is the surface area of the set we get by rotating the inscribed polygonal path corresponding to F about the x-axis, defined by (16.21).

Theorem 16.31. *Let f be a nonnegative and continuous function on the interval $[a,b]$ whose graph is rectifiable. Suppose that f is differentiable on (a,b), and $f \cdot \sqrt{1 + (f')^2}$ is integrable on $[a,b]$. Then the surface area of A^f exists, and its value is*

$$2\pi \int_a^b f(x)\sqrt{1 + (f'(x))^2}\, dx.$$

Proof. We did not assume the function f to be differentiable at the points a and b, so the function $g = f \cdot \sqrt{1 + (f')^2}$ might not be defined at these points. To prevent any ambiguity, let us define g to be zero at the points a and b; by Theorem 14.46 and Remark 14.47, the integrability of g and the value of the integral $I = \int_a^b g\,dx$ are unchanged when we do this.

Let $\varepsilon > 0$ be given. Since f is uniformly continuous on $[a, b]$, there exists a number $\eta > 0$ such that if $x, y \in [a, b]$ and $|y - x| < \eta$, then $|f(y) - f(x)| < \varepsilon$. By Theorem 14.23, we can choose a number $\delta > 0$ such that for every partition F of $[a, b]$ with mesh smaller than δ and every approximating sum σ_F corresponding to F, we have $|\sigma_F - I| < \varepsilon$.

Let $F : a = x_0 < x_1 < \cdots < x_n = b$ be a partition with mesh smaller than $\min(\eta, \delta)$. We show that the value Φ_F defined by (16.21) is close to $2\pi I$.

By the mean value theorem, for each i, there exists a point $c_i \in (x_{i-1}, x_i)$ such that $f(x_i) - f(x_{i-1}) = f'(c_i) \cdot (x_i - x_{i-1})$. Then

$$h_i \overset{\text{def}}{=} \sqrt{(f(x_i) - f(x_{i-1}))^2 + (x_i - x_{i-1})^2} = \sqrt{(f'(c_i))^2 + 1} \cdot (x_i - x_{i-1})$$

for all i. Thus the approximating sum of g with inner points c_i is

$$\sigma_F(g; (c_i)) = \sum_{i=1}^n f(c_i) \cdot \sqrt{(f'(c_i))^2 + 1} \cdot (x_i - x_{i-1}) = \sum_{i=1}^n f(c_i) \cdot h_i.$$

Since the partition F has smaller mesh than δ, we have $|\sigma_F(g; (c_i)) - I| < \varepsilon$. The partition F has smaller mesh than η, too, so by the choice of η, we have $|f(x_i) - f(c_i)| < \varepsilon$ and $|f(x_{i-1}) - f(c_i)| < \varepsilon$, so $|f(x_i) + f(x_{i-1}) - 2f(c_i)| < 2\varepsilon$ for all i. Thus

$$\left| \frac{1}{2\pi} \cdot \Phi_F - \sigma_F(g; (c_i)) \right| = \left| \sum_{i=1}^n \frac{f(x_i) + f(x_{i-1}) - 2f(c_i)}{2} \cdot h_i \right| \leq$$

$$\leq \varepsilon \cdot \sum_{i=1}^n h_i \leq \varepsilon \cdot L,$$

where L denotes the arc length of graph f. This gives

$$\left| \frac{1}{2\pi} \cdot \Phi_F - I \right| \leq \left| \frac{1}{2\pi} \cdot \Phi_F - \sigma_F(g; (c_i)) \right| + |\sigma_F(g; (c_i)) - I| < (L + 1) \cdot \varepsilon. \quad (16.22)$$

Since $\varepsilon > 0$ was arbitrary and (16.22) holds for every partition with small enough mesh, we have shown that the surface area of A^f is $2\pi I$. $\qquad\square$

Example 16.32. We compute the surface area of a **spherical segment**, that is, part of a sphere centered at the origin with radius r that falls between the planes $x = a$ and $x = b$, where $-r \leq a < b \leq r$.

The sphere centered at the origin with radius r is given by the rotated graph of the function $f(x) = \sqrt{r^2 - x^2}$ ($x \in [-r, r]$) about the x-axis.

Since f is monotone on the intervals $[-r,0]$ and $[0,r]$, its graph is rectifiable on $[-r,r]$ and thus on $[a,b]$, too. On the other hand, f is continuous on $[-r,r]$ and differentiable on $(-r,r)$, where we have

$$f(x)\sqrt{1+(f'(x))^2} = \sqrt{r^2 - x^2} \cdot \sqrt{1 + \frac{x^2}{r^2 - x^2}} = r.$$

Thus we can apply Theorem 16.31. We get that the surface area we are looking for is

$$2\pi \cdot \int_a^b r\,dx = 2\pi r \cdot (b-a),$$

so *the area of a spherical segment agrees with the distance between the planes that define it times the circumference of a great circle of the sphere.* As a special case, *the surface area of a sphere of radius r is $4r^2\pi$.*

Remark 16.33. Suppose that f is a nonnegative continuous function on the interval $[a,b]$ whose graph is rectifiable. Denote the arc length of graph f over the interval $[a,x]$ by $s(x)$. Then s is strictly monotone increasing and continuous on $[a,b]$, and so it has an inverse function s^{-1}.

By a variant of the proof of Theorem 16.31, one can show that *with these conditions, A^f has a surface area, and it is equal to $2\pi \cdot \int_0^L (f \circ s^{-1})\,dx$, where L denotes the arc length of graph f.*

Exercises

16.38. Prove that if the height of a frustum is m, and the radii of the lower and upper circles are r and R, then flattening the side of the frustum gives us a region whose area is $\pi(R+r)\sqrt{(R-r)^2+m^2}$.

16.39. Compute the surface area of the surfaces that we get by revolving the graphs of the following functions about the x-axis:

(a) e^x over $[a,b]$;
(b) $\sqrt{x}$ over $[a,b]$, where $0 < a < b$;
(c) $\sin x$ over $[0,\pi]$;
(d) $\mathrm{ch}\,x$ over $[-a,a]$.

16.40. Call a region falling between two parallel lines a strip. By the width of the strip, we mean the distance between the two lines. Prove that if we cover a circle with finitely many strips, then the sum of the widths of the strips we used is at least as large as the diameter of the circle. (H S)

16.41. Let f be nonnegative and continuously differentiable on $[a,b]$. Prove that the surface area of the surface of revolution $A^f = \{(x,y,z) : a \le x \le b,\ y^2 + z^2 = f^2(x)\}$

equals the length of graph f times the circumference of the circle traced by the center of mass of graph f during its revolution (this is sometimes called **Guldin's first theorem**).

16.7 Appendix: Proof of Theorem 16.20

Proof. Since every integrable function is already bounded, statement (ii) of Theorem 16.19 implies that g is rectifiable. Let $f = \sqrt{(g_1')^2 + \cdots + (g_d')^2}$. We will show that for every $\varepsilon > 0$, there exists a partition F such that the inscribed polygonal path of g corresponding to F has length ℓ_F, which differs from $s(g)$ by less than ε, and also that every Riemann sum $\sigma_F(f)$ of f differs from ℓ_F by less than ε. It will follow from this that $|\sigma_F(f) - s(g)| < 2\varepsilon$ for every Riemann sum corresponding to the partition F, and so by Theorem 14.19, f is integrable with integral $s(g)$.

Let $F : a = t_0 < \cdots < t_n = b$ be a partition of the interval $[a,b]$, and let ℓ_F denote the length of the corresponding inscribed polygonal path, that is, let $\ell_F = \sum_{i=1}^n |g(t_i) - g(t_{i-1})|$. For all $i = 1, \ldots, n$,

$$|g(t_i) - g(t_{i-1})| = \sqrt{(g_1(t_i) - g_1(t_{i-1}))^2 + \cdots + (g_d(t_i) - g_d(t_{i-1}))^2}.$$

By the mean value theorem, there exist points $c_{i,1}, \ldots, c_{i,d} \in (t_{i-1}, t_i)$ such that

$$g_j(t_i) - g_j(t_{i-1}) = g_j'(c_{i,j})(t_i - t_{i-1}) \qquad (j = 1, \ldots, d).$$

Then

$$\ell_F = \sum_{i=1}^n |g(t_i) - g(t_{i-1})| =$$
$$= \sum_{i=1}^n \sqrt{(g_1'(c_{i,1}))^2 + \cdots + (g_d'(c_{i,d}))^2} \cdot (t_i - t_{i-1}). \qquad (16.23)$$

Now let $e_i \in [t_{i-1}, t_i]$ $(i = 1, \ldots, n)$ be arbitrary inner points, and consider the corresponding Riemann sum of f:

$$\sigma_F(f; (e_i)) = \sum_{i=1}^n f(e_i)(t_i - t_{i-1}) =$$
$$= \sum_{i=1}^n \sqrt{(g_1'(e_i))^2 + \cdots + (g_d'(e_i))^2} \cdot (t_i - t_{i-1}). \qquad (16.24)$$

By the similarities of the right-hand sides of the equalities (16.23) and (16.24), we can expect that for a suitable partition F, ℓ_F and $\sigma_F(f; (e_i))$ will be close to each other. By inequality (16.3),

$$\left| \sqrt{\left(g_1'(c_{i,1})\right)^2 + \cdots + \left(g_d'(c_{i,d})\right)^2} - \sqrt{\left(g_1'(e_i)\right)^2 + \cdots + \left(g_d'(e_i)\right)^2} \right| \le$$

$$\le \sum_{j=1}^{d} |g_j'(c_{i,d}) - g_j'(e_i)| \le \sum_{j=1}^{d} \omega(g_j'; [t_{i-1}, t_i]).$$

Thus subtracting (16.23) and (16.24), we get that

$$|\ell_F - \sigma_F(f;(e_i))| \le \sum_{j=1}^{d} \sum_{i=1}^{n} \omega(g_j'; [t_{i-1}, t_i])(t_i - t_{i-1}) \le \sum_{j=1}^{d} \Omega_F(g_j') \qquad (16.25)$$

for every partition F and every e_i. Let $\varepsilon > 0$ be fixed. Since $s(g)$ is the supremum of the numbers ℓ_F, there exists a partition F_0 such that $s(g) - \varepsilon < \ell_{F_0} \le s(g)$. It is easy to check that if we add new base points to a partition, then the value of ℓ_F cannot decrease. Clearly, if we add another base point, then we replace a term $|g(t_{k-1}) - g(t_k)|$ in the sum by $|g(t_{k-1}) - g(t_k')| + |g(t_k') - g(t_k)|$. The triangle inequality ensures that the value of ℓ_F does not decrease with this. Thus if F is a refinement of F_0, then

$$s(g) - \varepsilon < \ell_{F_0} \le \ell_F \le s(g). \qquad (16.26)$$

Since the functions g_j' are integrable, there exist partitions F_j such that $\Omega_{F_j}(g_j') < \varepsilon$ $(j = 1, \ldots, d)$. Let F be the union of partitions $F_0, F_1, \ldots, F_d$. Then

$$\Omega_F(g_j') \le \Omega_{F_j}(g_j') < \varepsilon$$

for all $j = 1, \ldots, d$. If we now combine (16.25) and (16.26), we get that

$$|\sigma_F(f;(e_i)) - s(g)| \le |\sigma_F(f;(e_i)) - \ell_F| + |\ell_F - s(g)| < (d+1)\varepsilon. \qquad (16.27)$$

In the end, for every $\varepsilon > 0$, we have constructed a partition F that satisfies (16.27) with an arbitrary choice of inner points e_i. Then by Theorem 14.19, f is integrable, and its integral is $s(g)$. $\qquad \square$

Chapter 17
Functions of Bounded Variation

We know that if f is integrable, then the lower and upper sums of every partition F approximate its integral from below and above, and so the difference between either sum and the integral is at most $S_F - s_F = \Omega_F$, the oscillatory sum corresponding to F.

Thus the oscillatory sum is an upper bound for the difference between the approximating sums and the integral.

We also know that if f is integrable, then the oscillating sum can become smaller than any fixed positive number for a sufficiently fine partition (see Theorem 14.23).

If the function f is monotone, we can say more: $\Omega_F(f) \le |f(b) - f(a)| \cdot \delta(F)$ for all partitions F, where $\delta(F)$ denotes the mesh of the partition F (see Theorem 14.28 and inequality (14.19) in the proof of the theorem). A similar inequality holds for Lipschitz functions: if $|f(x) - f(y)| \le K \cdot |x - y|$ for all $x, y \in [a, b]$, then $\Omega_F(f) \le K \cdot (b - a) \cdot \delta(F)$ for every partition F (see Exercise 17.1). We can state this condition more concisely by saying that $\Omega_F(f) = O(\delta(F))$ holds for f if there exists a number C such that for an arbitrary partition F, $\Omega_F(f) \le C \cdot \delta(F)$. (Here we used the big-oh notation seen on p. 141.) By the above, this condition holds for both monotone and Lipschitz functions.

Is it true that the condition $\Omega_F(f) = O(\delta(F))$ holds for every integrable function? The answer is no: one can show that the function

$$f(x) = \begin{cases} x \cdot \sin(1/x), & \text{if } 0 < x \le 1; \\ 0, & \text{if } x = 0 \end{cases}$$

does not satisfy the condition (see Exercise 17.3). It is also true that for an arbitrary sequence ω_n that tends to zero, there exists a continuous function $f: [0, 1] \to \mathbb{R}$ such that $\Omega_{F_n}(f) \ge \omega_n$ for all n, where F_n denotes a partition of $[0, 1]$ into n equal subintervals (see Exercise 17.4). That is, monotone functions "are better behaved" than continuous functions in this aspect.

We characterize below the class of functions for which $\Omega_F(f) = O(\delta(F))$ holds. By what we stated above, every monotone and every Lipschitz function is included

© Springer New York 2015
M. Laczkovich, V.T. Sós, *Real Analysis*, Undergraduate Texts
in Mathematics, DOI 10.1007/978-1-4939-2766-1_17

in this class, but not every continuous function is. The elements of this class are the so-called functions of bounded variation, and they play an important role in analysis.

Definition 17.1. Let the function f be defined on the interval $[a,b]$. If we have a partition of the interval $[a,b]$ given by $F : a = x_0 < x_1 < \cdots < x_n = b$, let $V_F(f)$ denote the sum $\sum_{i=1}^{n} |f(x_i) - f(x_{i-1})|$. The *total variation* of f over $[a,b]$ is the supremum of the set of sums $V_F(f)$, where F ranges over all partitions of the interval $[a,b]$. We denote the total variation of f on $[a,b]$ by $V(f;[a,b])$ (which can be infinite). We say that the function $f \colon [a,b] \to \mathbb{R}$ is of *bounded variation* if $V(f;[a,b]) < \infty$.

Remarks 17.2. **1.** Suppose that the graph of the function f consists of finitely many monotone segments. Let f be monotone on each of the intervals $[c_{i-1},c_i]$ $(i = 1,\ldots,k)$, where $F_0 : a = c_0 < c_1 < \cdots < c_k = b$ is a suitable partition of $[a,b]$. It is easy to check that for an arbitrary partition F, we have

$$V_F(f) \leq \sum_{i=1}^{k} |f(c_i) - f(c_{i-1})| = V_{F_0}(f).$$

Thus the total variation of f is equal to $V_{F_0}(f)$, and so the supremum defining the total variation is actually a maximum. This statement can be turned around: if there is a largest value among $V_F(f)$, then the graph of f consists of finitely many monotone segments (see Exercise 17.5).
2. Suppose again that f is monotone on each of the intervals $[c_{i-1},c_i]$ $(i = 1,\ldots,k)$, where $F_0 : a = c_0 < c_1 < \cdots < c_k = b$. Consider the graph of f to be the crest of a mountain along which a tourist is walking. Suppose that for this tourist, the effort required to change altitude is proportional to the change in altitude, independent of whether the tourist is ascending or descending (and thus the tourist floats effortlessly when the mountain crest is horizontal). Then the value $V_{F_0}(f)$ measures the required effort for the tourist to traverse the crest of the mountain.

Generalizing this interpretation, we can say that the total variation of an arbitrary function is the effort required to "climb" the graph, and so a function is of bounded variation if the graph can be climbed with a finite amount of effort.

Theorem 17.3.

(i) *If f is monotone on $[a,b]$, then f is of bounded variation there, and $V(f;[a,b]) = |f(b) - f(a)|$.*
(ii) *Let f be Lipschitz on $[a,b]$, and suppose that $|f(x) - f(y)| \leq K \cdot |x - y|$ for all $x,y \in [a,b]$. Then f is of bounded variation on $[a,b]$, and $V(f;[a,b]) \leq K \cdot (b - a)$.*

Proof. (i) If f is monotone, then for an arbitrary partition $a = x_0 < x_1 < \cdots < x_n = b$,

$$\sum_{i=1}^{n} |f(x_i) - f(x_{i-1})| = \left| \sum_{i=1}^{n} (f(x_i) - f(x_{i-1})) \right| = |f(b) - f(a)|.$$

(ii) If $|f(x) - f(y)| \leq K \cdot |x - y|$ for all $x, y \in [a, b]$, then for an arbitrary partition $a = x_0 < x_1 < \cdots < x_n = b$,

$$\sum_{i=1}^{n} |f(x_i) - f(x_{i-1})| \leq \sum_{i=1}^{n} K \cdot (x_i - x_{i-1}) = K \cdot (b - a).$$

□

As the example mentioned in the introduction above demonstrates, not every continuous function is of bounded variation.

Example 17.4. Let $f(x) = x \cdot \sin(1/x)$ if $0 < x \leq 1$, and $f(0) = 0$. We show that f is not of bounded variation on $[0, 1]$. Let F_n be the partition that consists of the base points 0, 1, and $x_i = 2/((2i - 1)\pi)$ $(i = 1, \ldots, n)$. Then $V_{F_n}(f) \geq \sum_{i=2}^{n} |f(x_i) - f(x_{i-1})|$. In Example 16.16, we saw that this sum can be arbitrarily large if we choose n to be sufficiently large, so f is not of bounded variation.

Theorem 17.5.

(i) *For every $f : [a, b] \to \mathbb{R}$ and $c \in \mathbb{R}$, we have*

$$V(c \cdot f; [a, b]) = |c| \cdot V(f; [a, b]). \tag{17.1}$$

(ii) *For arbitrary functions $f, g : [a, b] \to \mathbb{R}$,*

$$V(f + g; [a, b]) \leq V(f; [a, b]) + V(g; [a, b]). \tag{17.2}$$

(iii) *If both f and g are of bounded variation on $[a, b]$, then $a \cdot f + b \cdot g$ is also of bounded variation there for every $a, b \in \mathbb{R}$.*

Proof. (i) For an arbitrary partition $a = x_0 < x_1 < \cdots < x_n = b$,

$$\sum_{i=1}^{n} |c \cdot f(x_i) - c \cdot f(x_{i-1})| = |c| \cdot \sum_{i=1}^{n} |f(x_i) - f(x_{i-1})|.$$

Taking the supremum of both sides over all partitions, we obtain (17.1).
(ii) Let $h = f + g$. Then for an arbitrary partition $a = x_0 < x_1 < \cdots < x_n = b$,

$$\sum_{i=1}^{n} |h(x_i) - h(x_{i-1})| \leq \sum_{i=1}^{n} |f(x_i) - f(x_{i-1})| + \sum_{i=1}^{n} |g(x_i) - g(x_{i-1})| \leq$$
$$\leq V(f; [a, b]) + V(g; [a, b]).$$

Since this is true for every partition, 17.2 holds. The third statement of the theorem follows from (i) and (ii). □

Theorem 17.6. *If f is of bounded variation on $[a, b]$, then it also is of bounded variation in every subinterval.*

Proof. If $[c, d] \subset [a, b]$, then extending an arbitrary partition $F : c = x_0 < x_1 < \cdots < x_n = d$ of $[c, d]$ to a partition F' of $[a, b]$, we obtain that $\sum_{i=1}^{n} |f(x_i) - f(x_{i-1})| \leq V_{F'}(f) \leq V(f; [a, b])$. Since this holds for every partition of $[c, d]$, $V(f; [c, d]) \leq V(f; [a, b]) < \infty$. □

Theorem 17.7. *Let $a < b < c$. If f is of bounded variation on $[a,b]$ and $[b,c]$, then it is of bounded variation in $[a,c]$ as well, and*

$$V(f;[a,c]) = V(f;[a,b]) + V(f;[b,c]). \tag{17.3}$$

Proof. Let $F : a = x_0 < x_1 < \cdots < x_n = c$ be an arbitrary partition of the interval $[a,c]$, and let $x_{k-1} \leq b \leq x_k$. It is clear that

$$V_F(f) \leq \sum_{i=1}^{k-1} |f(x_i) - f(x_{i-1})| + |f(b) - x_{k-1}| +$$

$$+ |f(x_k) - f(b)| + \sum_{i=k+1}^{n} |f(x_i) - f(x_{i-1})| \leq V(f;[a,b]) + V(f;[b,c]),$$

and so $V(f;[a,c]) \leq V(f;[a,b]) + V(f;[b,c])$.

Now let $\varepsilon > 0$ be given, and let $a = x_0 < x_1 < \cdots < x_n = b$ and $b = y_0 < y_1 < \cdots < y_k = c$ be partitions of $[a,b]$ and $[b,c]$ respectively such that $\sum_{i=1}^{n} |f(x_i) - f(x_{i-1})| > V(f;[a,b]) - \varepsilon$ and $\sum_{i=1}^{k} |f(y_i) - f(y_{i-1})| > V(f;[b,c]) - \varepsilon$. Then

$$V(f;[a,c]) \geq \sum_{i=1}^{n} |f(x_i) - f(x_{i-1})| + \sum_{i=1}^{k} |f(y_i) - f(y_{i-1})| >$$

$$> V(f;[a,b]) + V(f;[b,c]) - 2\varepsilon.$$

Since ε was arbitrary, $V(f;[a,c]) \geq V(f;[a,b]) + V(f;[b,c])$ follows, and so 17.3 holds. $\square$

The following theorem gives a simple characterization of functions of bounded variation.

Theorem 17.8. *The function $f : [a,b] \to \mathbb{R}$ is of bounded variation if and only if it can be expressed as the difference of two monotone increasing functions.*

Proof. Every monotone function is of bounded variation (Theorem 17.3), and the difference of two functions of bounded variation is also of bounded variation (Theorem 17.5), so the "if" part of the theorem is clearly true.

Now suppose that f is of bounded variation. Let $g(x) = V(f;[a,x])$ for all $x \in (a,b]$, and let $g(a) = 0$. If $a \leq x < y \leq b$, then by Theorem 17.7,

$$g(y) = g(x) + V(f;[x,y]) \geq g(x) + |f(y) - f(x)| \geq g(x) - f(y) + f(x).$$

Thus on the one hand, $g(y) \geq g(x)$, while on the other hand, $g(y) + f(y) \geq g(x) + f(x)$, so both g and $g + f$ are monotone increasing in $[a,b]$. Since $f = (g+f) - g$, this ends the proof of the theorem. $\square$

Corollary 17.9. *If f is of bounded variation on $[a,b]$, then it is integrable there.*

In Example 17.4, we saw that not every continuous function is of bounded variation. Thus the corollary above cannot be turned around.

The following theorem can often be applied to compute total variation. We leave its proof to the reader (in Exercise 17.12).

Theorem 17.10. *If f is differentiable and f' is integrable on $[a,b]$, then f is of bounded variation and $V(f; [a,b]) = \int_a^b |f'| dx$.*

The following theorem clarifies the condition for rectifiability.

Theorem 17.11.

(i) *The graph of a function $f: [a,b] \to \mathbb{R}$ is rectifiable if and only if f is of bounded variation.*

(ii) *A curve $g: [a,b] \to \mathbb{R}^d$ is rectifiable if and only if its coordinate functions are of bounded variation.*

Proof. (i) Let the arc length of the graph f be denoted by $s(f; [a,b])$ (see Definition 10.78). Then for an arbitrary partition $F: a = x_0 < x_1 < \cdots < x_n = b$, we have

$$V_F(f) = \sum_{i=1}^n |f(x_i) - f(x_{i-1})| \leq$$

$$\leq \sum_{i=1}^n \sqrt{(x_i - x_{i-1})^2 + (f(x_i) - f(x_{i-1}))^2} \leq s(f; [a,b])$$

and

$$\sum_{i=1}^n \sqrt{(x_i - x_{i-1})^2 + (f(x_i) - f(x_{i-1}))^2} \leq$$

$$\leq \sum_{i=1}^n (x_i - x_{i-1}) + \sum_{i=1}^n |f(x_i) - f(x_{i-1})| =$$

$$= (b - a) + V_F(f) \leq (b - a) + V(f; [a,b]).$$

Since these hold for every partition, we have $V(f; [a,b]) \leq s(f; [a,b])$ and $s(f; [a,b]) \leq (b - a) + V(f; [a,b])$. It is then clear that $s(f; [a,b])$ is finite if and only if $V(f; [a,b])$ is finite, that is, the graph of f is rectifiable if and only if f is of bounded variation.

(ii) Let the coordinate functions of g be $g_1, \ldots, g_d$, and let $F: a = t_0 < t_1 < \cdots < t_n = b$ be a partition of the interval $[a,b]$. If $p_i = g(t_i)$ $(i = 0, \ldots, n)$, then for every $i = 1, \ldots, n$ and $j = 1, \ldots, d$,

$$|g_j(x_i) - g_j(x_{i-1})| \leq$$

$$\leq |p_i - p_{i-1}| = \sqrt{(g_1(x_i) - g_1(x_{i-1}))^2 + \cdots + (g_d(x_i) - g_d(x_{i-1}))^2} \leq$$

$$\leq |g_1(x_i) - g_1(x_{i-1})| + \cdots + |g_d(x_i) - g_d(x_{i-1})|.$$

If we sum these equations for $i = 1, \ldots, n$, then we get that

$$V_F(g_j) \leq \ell_F \leq V_F(g_1) + \cdots + V_F(g_d) \leq V(g_1; [a,b]) + \cdots + V(g_d; [a,b]),$$

where ℓ_F denotes the length of the inscribed polygon corresponding to the partition F. Since this holds for every partition, we have

$$V(g_j; [a,b]) \leq s(g) \leq V(g_1; [a,b]) + \cdots + V(g_d; [a,b])$$

for all $j = 1, \ldots, d$, where $s(g)$ is the arc length of the curve. It is then clear that $s(g)$ is finite if and only if $V(g_j; [a,b])$ is finite for all $j = 1, \ldots, d$, that is, g is rectifiable if and only if its coordinate functions are of bounded variation.   □

Now we prove the statement from the introduction.

Theorem 17.12. *A function $f : [a,b] \to \mathbb{R}$ satisfies $\Omega_F(f) = O(\delta(F))$ for every partition F if and only if f is of bounded variation.*

Proof. We first prove that if f is of bounded variation, then $\Omega_F(f) = O(\delta(F))$ holds. At the beginning of the chapter, we saw that if g is monotone in $[a,b]$, then $\Omega_F(g) \leq |g(b) - g(a)| \cdot \delta(F)$ for every partition F. We also know that if f is of bounded variation in $[a,b]$, then it can be expressed as $f = g - h$, where g and h are monotone increasing functions (Theorem 17.8). Thus for an arbitrary partition F,

$$\Omega_F(f) \leq \Omega_F(g) + \Omega_F(h) \leq |g(b) - g(a)| \cdot \delta(F) + |h(b) - h(a)| \cdot \delta(F) = C \cdot \delta(F),$$

where $C = |g(b) - g(a)| + |h(b) - h(a)|$. Thus the condition $\Omega_F(f) = O(\delta(F))$ indeed holds.

Now we show that if f is not of bounded variation, then the property $\Omega_F(f) = O(\delta(F))$ does not hold, that is, for every real number A, there exists a partition F such that $\Omega_F(f) > A \cdot \delta(F)$. The proof relies on the observation that for an arbitrary bounded function $f : [a,b] \to \mathbb{R}$ and partition $F : a = x_0 < x_1 < \cdots < x_n = b$,

$$\Omega_F(f) = \sum_{i=1}^{n} \omega(f; [x_{i_1}, x_i]) \cdot (x_i - x_{i-1}) \geq$$

$$\geq \left[\sum_{i=1}^{n} |f(x_i) - f(x_{i-1})| \right] \cdot \min_{1 \leq i \leq n} (x_i - x_{i-1}) =$$

$$= V_F(f) \cdot \rho(F),$$

where $\rho(F) = \min_{1 \leq i \leq n}(x_i - x_{i-1})$.

If f is not of bounded variation, then for every real number A, there exists a partition F_0 such that $V_{F_0}(f) > A$. However, we know only that this partition F_0 satisfies $\Omega_{F_0}(f) \geq V_{F_0}(f) \cdot \rho(F_0) > A \cdot \rho(F_0)$, while we want $\Omega_F(f) > A \cdot \delta(F)$ to hold for some F.

So consider a refinement F of F_0 such that $\rho(F) \geq \delta(F)/2$. We can get such a refinement by further subdividing the intervals in F_0 into pieces whose lengths are between $\rho(F_0)/2$ and $\rho(F_0)$. In this case, $\delta(F) \leq \rho(F_0)$ and $\rho(F) \geq \rho(F_0)/2$, so $\rho(F) \geq \delta(F)/2$ holds. Since F is a refinement of F_0, we easily see that $V_F(f) \geq V_{F_0}(f) > A$, so $\Omega_F(f) \geq V_F(f) \cdot \rho(F) > A \cdot \delta(F)/2$. Since A was arbitrary, this concludes the proof.   □

With the theorems above in hand, we might ask whether there exist functions for which we can say more than $\Omega_F(f) = O(\delta(F))$? Could it be possible for $\Omega_F(f) \leq C \cdot \delta(F)^2$ to hold for every partition with some constant C? The answer to this question is no. If f is constant, then of course $\Omega_F(f) = 0$ for every partition. If, however, f is not constant, then there exists a $c > 0$ such that $\Omega_{F_n}(f) \geq c \cdot \delta(F_n)$ for all n, where F_n denotes the uniform partition of $[a,b]$ into n equal subintervals (see Exercise 17.2).

Exercises

17.1. Prove that if $|f(x) - f(y)| \leq K \cdot |x - y|$ for all $x, y \in [a,b]$, then $\Omega_F(f) \leq K \cdot (b-a) \cdot \delta(F)$ for every partition F of $[a,b]$. (S)

17.2. Prove that if f is bounded in $[a,b]$ and F_n denotes the uniform partition of $[a,b]$ into n equal subintervals, then

$$\Omega_{F_n}(f) \geq \omega(f; [a,b]) \cdot (b-a)/n,$$

where $\omega(f; [a,b])$ is the oscillation of the function f on the interval $[a,b]$. (H)

17.3. Let $f(x) = x \cdot \sin(1/x)$ if $0 < x \leq 1$, and $f(0) = 0$. Prove that there exists a constant $c > 0$ such that $\Omega_{F_n}(f) \geq c \cdot (\log n)/n = c \cdot (\log n) \cdot \delta(F_n)$ for all n, where F_n denotes the uniform partition of $[0, 1]$ into n equal subintervals. ($*$H)

17.4. Show that if an arbitrary sequence ω_n tends to zero, then there exists a continuous function $f: [0,1] \to \mathbb{R}$ such that $\Omega_{F_n}(f) \geq \omega_n$ for all n, where F_n denotes the uniform partition of $[0,1]$ into n equal subintervals. ($*$)

17.5. Let $f: [a,b] \to \mathbb{R}$ and suppose that there is a largest value among $V_F(f)$ (where F runs over the partitions of $[a,b]$). Show that in this case, the graph of f is made up of finitely many monotone segments. (H)

17.6. Show that if f is of bounded variation in $[a,b]$, then so is f^2.

17.7. Show that if f and g are of bounded variation in $[a,b]$, then so is $f \cdot g$. Moreover, if $\inf |g| > 0$, then so is f/g.

17.8. Let $f(x) = x^\alpha \cdot \sin(1/x)$ if $0 < x \leq 1$ and $f(0) = 0$. For what α will f be of bounded variation in $[0,1]$?

17.9. Give an example for a function f that is differentiable in $[0,1]$ but is not of bounded variation there.

17.10. For what $c > 0$ will the cth power of the Riemann function be of bounded variation in $[0, 1]$? (H)

17.11. Prove that if f is differentiable on $[a, b]$ and f' is bounded there, then f is of bounded variation on $[a, b]$.

17.12. Prove Theorem 17.10. (S)

17.13. Let $\alpha > 0$ be given. We say that f is **Hölder** α in the interval $[a, b]$ if there exists a number C such that $|f(x) - f(y)| \leq C \cdot |x - y|^\alpha$ for all $x, y \in [a, b]$. Show that if $\alpha > 0$, then the function x^α is Hölder β in the interval $[0, 1]$, where $\beta = \min(\alpha, 1)$. (S)

17.14. Show that if f is Hölder α in the interval $[a, b]$, where $\alpha > 1$, then f is constant. (H)

17.15. Let $f(x) = x^\alpha \cdot \sin x^{-\beta}$ if $0 < x \leq 1$ and $f(0) = 0$, where α and β are positive constants. Show that f is Hölder γ in the interval $[0, 1]$, where

$$\gamma = \min\left(\frac{\alpha}{\beta + 1}, 1\right). \ (*H\,S)$$

17.16. For what α can we say that if f is Hölder α, then f is of bounded variation in $[a, b]$? (H)

17.17. Prove that a function of bounded variation in $[a, b]$ has at most countably many points of discontinuity.

17.18. Let $f : [a, b] \to \mathbb{R}$ be continuous. Prove that for every $\varepsilon > 0$, there exists a $\delta > 0$ such that every partition F with mesh smaller than δ satisfies $V_F(f) > V(f; [a, b]) - \varepsilon$.

17.19. Prove that a function defined on $[a, b]$ is not of bounded variation in $[a, b]$ if and only if there exists a strictly monotone sequence (x_n) in $[a, b]$ such that $\sum_{i=1}^\infty |f(x_{i+1}) - f(x_i)| = \infty$. $(*H)$

Chapter 18
The Stieltjes Integral

In this chapter we discuss a generalization of the Riemann integral that is often used in both theoretical and applied mathematics. Stieltjes[1] originally introduced this concept to deal with infinite continued fractions,[2] but it was soon apparent that the concept is useful in other areas of mathematics—and thus in mathematical physics, probability, and number theory, independently of its role in continued fractions. We illustrate the usefulness of the concept with two simple examples.

Example 18.1. Consider a planar curve parameterized by $\gamma(t) = (x(t), y(t))$ ($t \in [a,b]$), where the x-coordinate function is strictly monotone increasing and continuous, and the y-coordinate function is nonnegative on $[a,b]$. The problem is to find the area under the region bounded by the curve. If $a = t_0 < t_1 < \cdots < t_n = b$ is a partition of the interval $[a,b]$ and $c_i \in [t_{i-1}, t_i]$ for all i, then the area can be approximated by the sum

$$\sum_{i=1}^{n} y(c_i) \big(x(t_i) - x(t_{i-1}) \big).$$

We can expect the area to be the limit—in a suitable sense—of these sums.

Example 18.2. Consider a metal rod of negligible thickness but not negligible mass $M > 0$. Suppose that the rod lies on the interval $[a,b]$, and let the mass of the rod over the subinterval $[a,x]$ be $m(x)$ for all $x \in [a,b]$. Our task is to find the center of mass of the rod.

We know that if we place weights $m_1, \ldots, m_n$ at the points $x_1, \ldots, x_n$, then the center of mass of this system of points $\{x_1, \ldots, x_n\}$ is

$$\frac{m_1 x_1 + \cdots + m_n x_n}{m_1 + \cdots + m_n}.$$

---

[1] Thomas Joannes Stieltjes (1856–1894), Dutch mathematician.
[2] For more on continued fractions, see [5].

© Springer New York 2015
M. Laczkovich, V.T. Sós, *Real Analysis*, Undergraduate Texts
in Mathematics, DOI 10.1007/978-1-4939-2766-1_18

Consider a partition $a = t_0 < t_1 < \cdots < t_n = b$ and choose points $c_i \in [t_{i-1}, t_i]$ for all i. If we suppose that the mass distribution of the rod is continuous (meaning that the mass of the rod at every single point is zero), then the mass of the rod over the interval $[t_{i-1}, t_i]$ is $m(t_i) - m(t_{i-1})$. Concentrating this weight at the point c_i, the center of mass of the system of points $\{c_1, \ldots, c_n\}$ is

$$\frac{c_1 \big(m(t_1) - m(t_0)\big) + \cdots + c_n \big(m(t_n) - m(t_{n-1})\big)}{M}.$$

This approximates the center of mass of the rod itself, and once again, we expect that the limit of these numbers in a suitable sense will be the center of mass.

We can see that in both examples, a sum appears that depends on two functions. In these sums, we multiply the value of the first function (which, in Example 18.2, was the function x) at the inner points by the increments of the second function.

We use the following notation and naming conventions. Let $f, g : [a, b] \to \mathbb{R}$ be given functions, let $F : a = x_0 < x_1 < \cdots < x_n = b$ be a partition of the interval $[a, b]$, and let $c_i \in [x_{i-1}, x_i]$ $(i = 1, \ldots, n)$ be arbitrary inner points. Then the sum

$$\sum_{i=1}^{n} f(c_i) \cdot \big(g(x_i) - g(x_{i-1})\big)$$

is denoted by $\sigma_F(f, g; (c_i))$, and is called the **approximating sum of f with respect to g**.

Definition 18.3. Let $f, g : [a, b] \to \mathbb{R}$ be given functions. We say that the *Stieltjes integral $\int_a^b f\, dg$ of f with respect to g exists and has value I* if for every $\varepsilon > 0$, there exists a $\delta > 0$ such that if $F : a = x_0 < x_1 < \cdots < x_n = b$ is a partition of $[a, b]$ with mesh smaller than δ and $c_i \in [x_{i-1}, x_i]$ $(i = 1, \ldots, n)$ are arbitrary inner points, then

$$\big|\sigma_F(f, g; (c_i)) - I\big| < \varepsilon. \qquad (18.1)$$

Remarks 18.4. **1.** Let $g(x) = x$ for all $x \in [a, b]$. It is clear that the Stieltjes integral $\int_a^b f\, dg$ exists exactly when the Riemann integral $\int_a^b f\, dx$ does, and in this case, their values agree.

2. If the function g is constant, then the Stieltjes integral $\int_a^b f\, dg$ always exists, and its value is zero. If the function f is constant and has value c, then the Stieltjes integral $\int_a^b f\, dg$ always exists and has value $c \cdot \big(g(b) - g(a)\big)$.

3. The existence of the Stieltjes integral $\int_a^b f\, dg$ is not guaranteed by f and g being Riemann integrable in $[a, b]$. One can show that *if f and g share a point of discontinuity, then the Stieltjes integral $\int_a^b f\, dg$ does not exist.* (See Exercise 18.4.) Thus if f and g are bounded functions in $[a, b]$ and are continuous everywhere except at a common point, then they are both Riemann integrable in $[a, b]$, while the Stieltjes integral $\int_a^b f\, dg$ does not exist.

4. For the Stieltjes integral $\int_a^b f\, dg$ to exist, it is not even sufficient for f and g to be continuous in $[a, b]$. See Exercise 18.5.

Now we show that if g is strictly monotone and continuous, then the Stieltjes integral $\int_a^b f\,dg$ can be reduced to a Riemann integral.

Theorem 18.5. *If $g\colon [a,b] \to \mathbb{R}$ is strictly monotone increasing and continuous, then the Stieltjes integral $\int_a^b f\,dg$ exists if and only if the Riemann integral $\int_{g(a)}^{g(b)} f \circ g^{-1}\,dx$ does, and then*

$$\int_a^b f\,dg = \int_{g(a)}^{g(b)} f \circ g^{-1}\,dx.$$

Proof. If $F\colon a = x_0 < x_1 < \cdots < x_n = b$ is a partition and $c_i \in [x_{i-1}, x_i]$ $(i = 1, \ldots, n)$ are arbitrary inner points, then the points $t_i = g(x_i)$ $(i = 1, \ldots, n)$ give us a partition $\overline{F}$ of the interval $[g(a), g(b)]$, and we have $g(c_i) \in [g(x_{i-1}), g(x_i)]$ for all $i = 1, \ldots, n$. Then

$$\sum_{i=1}^n f(c_i) \cdot (g(x_i) - g(x_{i-1})) = \sum_{i=1}^n (f \circ g^{-1})(g(c_i)) \cdot (t_i - t_{i-1}) \qquad (18.2)$$

is an approximating sum of the Riemann sum $\int_{g(a)}^{g(b)} f \circ g^{-1}\,dx$. Conversely, if we have $\overline{F}\colon g(a) = t_0 < t_1 < \cdots < t_n = g(b)$ as a partition of the interval $[g(a), g(b)]$ and $d_i \in [t_{i-1}, t_i]$ for all $i = 1, \ldots, n$, then the points $x_i = g^{-1}(t_i)$ $(i = 1, \ldots, n)$ create a partition F of the interval $[a, b]$. If $c_i = g^{-1}(d_i)$, then $c_i \in [x_{i-1}, x_i]$ for all $i = 1, \ldots, n$, and (18.2) holds.

By the uniform continuity of g, we know that for every $\delta > 0$, if the partition F has small enough mesh, then $\overline{F}$ has mesh smaller than δ. Thus comparing statement (iii) of Theorem 14.23 with equality (18.2), we get that if the Riemann integral $\int_{g(a)}^{g(b)} f \circ g^{-1}\,dx$ exists, then the Stieltjes integral $\int_a^b f\,dg$ also exists, and they are equal.

Since the function g^{-1} is also uniformly continuous, it follows that for every $\delta > 0$, if the partition $\overline{F}$ has small enough mesh, then F has mesh smaller than δ. Thus by the definition of the Stieltjes integral, by statement (iii) of Theorem 14.23, and by equality (18.2), it follows that if the Stieltjes integral $\int_a^b f\,dg$ exists, then the Riemann integral $\int_{g(a)}^{g(b)} f \circ g^{-1}\,dx$ exists as well, and they have the same value. $\square$

The following statements can be deduced easily from the definition of the Stieltjes integral. We leave their proofs to the reader.

Theorem 18.6.

(i) *If the Stieltjes integrals $\int_a^b f_1\,dg$ and $\int_a^b f_2\,dg$ exist, then for all $c_1, c_2 \in \mathbb{R}$, the Stieltjes integral $\int_a^b (c_1 f_1 + c_2 f_2)\,dg$ exists as well, taking on the value $c_1 \cdot \int_a^b f_1\,dg + c_2 \cdot \int_a^b f_2\,dg$.*

(ii) *If the Stieltjes integrals $\int_a^b f\,dg_1$ and $\int_a^b f\,dg_2$ exist, then for all $c_1, c_2 \in \mathbb{R}$, the Stieltjes integral $\int_a^b f\,d(c_1 g_1 + c_2 g_2)$ exists as well, and has value $c_1 \cdot \int_a^b f\,dg_1 + c_2 \cdot \int_a^b f\,dg_2$.*

The following theorem gives us a necessary and sufficient condition for the existence of Stieltjes integrals. For the proof of the theorem, see Exercise 18.6.

Theorem 18.7 (Cauchy's Criterion). *The Stieltjes integral $\int_a^b f\,dg$ exists if and only if for every $\varepsilon > 0$, there exists a $\delta > 0$ such that if F_1 and F_2 are partitions of $[a,b]$ with mesh smaller than δ, then*

$$\left| \sigma_{F_1}\left(f,g;(c_i)\right) - \sigma_{F_2}\left(f,g;(d_j)\right) \right| < \varepsilon$$

with an arbitrary choice of the inner points c_i and d_j.

With the help of Cauchy's criterion, it is easy to prove the following theorem. We leave the proof to the reader once again.

Theorem 18.8. *If the Stieltjes integral $\int_a^b f\,dg$ exists, then for all $a < c < b$, the integrals $\int_a^c f\,dg$ and $\int_c^b f\,dg$ also exist, and $\int_a^b f\,dg = \int_a^c f\,dg + \int_c^b f\,dg$.*

We should note that the existence of the Stieltjes integrals $\int_a^c f\,dg$ and $\int_c^b f\,dg$ alone does not imply the existence of $\int_a^b f\,dg$ (see Exercise 18.2). If, however, at least one of f and g is continuous at c, and the other function is bounded then the existence of $\int_a^b f\,dg$ already follows (see Exercise 18.3).

The following important theorem shows that the roles of f and g in the Stieltjes integral are symmetric in some sense. We remind our readers that $[F]_a^b$ denotes the difference $F(b) - F(a)$.

Theorem 18.9 (Integration by Parts). *If the Stieltjes integral $\int_a^b f\,dg$ exists, then $\int_a^b g\,df$ also exists, and $\int_a^b f\,dg + \int_a^b g\,df = [f \cdot g]_a^b$.*

Proof. The proof relies on Abel's rearrangement (see equation (14.31)).

Let $F: a = x_0 < x_1 < \cdots < x_n = b$ be a partition, and let $c_i \in [x_{i-1}, x_i]$ ($i = 1, \ldots, n$) be inner points. If we apply Abel's rearrangement to the approximating sum $\sigma_F(g,f;(c_i))$, we get that

$$\sigma_F\left(g,f;(c_i)\right) = \sum_{i=1}^{n} g(c_i) \cdot \left(f(x_i) - f(x_{i-1})\right) =$$

$$= f(b)g(b) - f(a)g(a) - \sum_{i=0}^{n} f(x_i)\left(g(c_{i+1}) - g(c_i)\right), \qquad (18.3)$$

where $c_0 = a$ and $c_{n+1} = b$. Since $a = c_0 \leq c_1 \leq \ldots \leq c_{n+1} = b$ and $x_i \in [c_i, c_{i+1}]$ for all $i = 0, \ldots, n$, we have

$$\sum_{i=0}^{n} f(x_i)\left(g(c_{i+1}) - g(c_i)\right) = \sigma_{F'}\left(f,g;(d_i)\right), \qquad (18.4)$$

where F' denotes the partition defined by the points c_i ($0 \leq i \leq n$) with the corresponding inner points. (We list each of the c_i points only once. Note that on the left-hand side of (18.4), we can leave out the terms in which $c_i = c_{i+1}$.) Then by (18.3),

$$\sigma_F\left(g,f;(c_i)\right) = [f \cdot g]_a^b - \sigma_{F'}\left(f,g;(d_i)\right).$$

Let $\varepsilon > 0$ be given, and suppose that $\delta > 0$ satisfies the condition of Definition 18.3. It is easy to see that if the mesh of the partition F is smaller than $\delta/2$, then the mesh of F' is smaller than δ, and so $|\sigma_{F'}(f,g;(d_i)) - I| < \varepsilon$, where $I = \int_a^b f\,dg$. This shows that if the mesh of F is smaller than $\delta/2$, then $|\sigma_F(g,f;(c_i)) - ([fg]_a^b - I)| < \varepsilon$ for an arbitrary choice of the inner points. It follows that the integral $\int_a^b g\,df$ exists, and that its value is $[fg]_a^b - I$. $\qquad\qquad\qquad\qquad\qquad\qquad\qquad\qquad\qquad\qquad\square$

Since Cauchy's criterion (Theorem 18.7) is hard to apply in deciding whether a specific Stieltjes integral exists, we need other conditions guaranteeing the existence of the Stieltjes integral that can be easier to check. The simplest such condition is the following.

Theorem 18.10. *If one of the functions f and g defined on the interval $[a,b]$ is continuous, while the other is of bounded variation, then the Stieltjes integrals $\int_a^b f\,dg$ and $\int_a^b g\,df$ exist.*

Proof. By Theorem 18.9, it suffices to prove the existence of the integral $\int_a^b f\,dg$, and we can also assume that f is continuous and g is of bounded variation. By Theorems 17.8 and 18.6, it suffices to consider the case that g is monotone increasing.

For an arbitrary partition $F: a = x_0 < x_1 < \cdots < x_n = b$, let

$$s_F = \sum_{i=1}^{n} m_i \cdot (g(x_i) - g(x_{i-1})) \quad \text{and} \quad S_F = \sum_{i=1}^{n} M_i \cdot (g(x_i) - g(x_{i-1})),$$

where $m_i = \min\{f(x): x \in [x_{i-1}, x_i]\}$ and $M_i = \max\{f(x): x \in [x_{i-1}, x_i]\}$ for all $i = 1, \ldots, n$. Since $g(x_i) - g(x_{i-1}) \geq 0$ for all i,

$$s_F \leq \sigma_F(f,g;(c_i)) \leq S_F \qquad\qquad (18.5)$$

with any choice of inner points c_i.

It is easy to see that $s_{F_1} \leq S_{F_2}$ for any partitions F_1 and F_2 (by repeating the proofs of Lemmas 14.3 and 14.4, using that g is monotone increasing). Thus the set of "lower sums" s_F is nonempty and bounded from above. If I denotes the supremum of this set, then $s_F \leq I \leq S_F$ for every partition F.

Now we show that $\int_a^b f\,dg$ exists, and its value is I. Let $\varepsilon > 0$ be given. By Heine's theorem, f is uniformly continuous on $[a,b]$, so there exists a $\delta > 0$ such that $|f(x) - f(y)| < \varepsilon$ whenever $x, y \in [a,b]$ and $|x-y| < \delta$. Let $F: a = x_0 < x_1 < \cdots < x_n = b$ be an arbitrary partition with mesh smaller than δ. By Weierstrass's theorem, for each i, there are points $c_i, d_i \in [x_{i-1}, x_i]$ such that $f(c_i) = m_i$ and $f(d_i) = M_i$. Then $|d_i - c_i| \leq x_i - x_{i-1} < \delta$, so by our choice of δ, we have $M_i - m_i = f(d_i) - f(c_i) < \varepsilon$. Thus

$$S_F - s_F = \sum_{i=1}^{n} (M_i - m_i) \cdot (g(x_i) - g(x_{i-1})) \leq \varepsilon \cdot \sum_{i=1}^{n} (g(x_i) - g(x_{i-1})) =$$

$$= (g(b) - g(a)) \cdot \varepsilon.$$

Now using (18.5), we get that

$$I - \big(g(b) - g(a)\big) \cdot \varepsilon < s_F \leq \sigma_F\big(f, g; (c_i)\big) \leq S_F < I + \big(g(b) - g(a)\big) \cdot \varepsilon$$

for arbitrary inner points c_i. This shows that $\int_a^b f \, dg$ exists and that its value is I. □

Remark 18.11. One can show that the class of continuous functions and the class of functions of bounded variation are "dual classes" in the sense that a function f is continuous if and only if $\int_a^b f \, dg$ exists for all functions g that are of bounded variation, and a function g is of bounded variation if and only if $\int_a^b f \, dg$ exists for every continuous function f. (See Exercises 18.8 and 18.9.)

The following theorem can often be applied to computing Stieltjes integrals.

Theorem 18.12. *If f is Riemann integrable, g is differentiable, and g' is Riemann integrable on $[a,b]$, then the Stieltjes integral of f with respect to g exists, and*

$$\int_a^b f \, dg = \int_a^b f \cdot g' \, dx. \tag{18.6}$$

Proof. Since f and g' are integrable on $[a,b]$, the Riemann integral on the right-hand side of (18.6) exists. Let its value be I. We want to show that for an arbitrary $\varepsilon > 0$, there exists a $\delta > 0$ such that for every partition $a = x_0 < x_1 < \cdots < x_n = b$ with mesh smaller than δ, (18.1) holds with an arbitrary choice of inner values $c_i \in [x_{i-1}, x_i]$ ($i = 1, \ldots, n$).

Let $\varepsilon > 0$ be given. Since f is Riemann integrable on $[a,b]$, there must exist a $\delta_1 > 0$ such that $\Omega_F(f) < \varepsilon$ whenever the partition F has mesh smaller than δ_1. By Theorem 14.23, there exists a $\delta_2 > 0$ such that whenever $a = x_0 < x_1 < \cdots < x_n = b$ is a partition with mesh smaller than δ_2,

$$\left| I - \sum_{i=1}^n f(d_i) g'(d_i)(x_i - x_{i-1}) \right| < \varepsilon \tag{18.7}$$

holds for arbitrary inner points $d_i \in [x_{i-1}, x_i]$ ($i = 1, \ldots, n$).

Let $\delta = \min(\delta_1, \delta_2)$, and consider an arbitrary partition $F: a = x_0 < x_1 < \cdots < x_n = b$ with mesh smaller than δ. Let $c_i \in [x_{i-1}, x_i]$ ($i = 1, \ldots, n$) be arbitrary inner points.

Since g is differentiable on $[a,b]$, by the mean value theorem,

$$\sum_{i=1}^n f(c_i)\big(g(x_i) - g(x_{i-1})\big) = \sum_{i=1}^n f(c_i) g'(d_i)(x_i - x_{i-1}) \tag{18.8}$$

for suitable numbers $d_i \in [x_{i-1}, x_i]$. Let K denote an upper bound of $|g'|$ on $[a,b]$. Then

$$\left| I - \sum_{i=1}^{n} f(c_i)\big(g(x_i) - g(x_{i-1})\big) \right| = \left| I - \sum_{i=1}^{n} f(c_i) g'(d_i)(x_i - x_{i-1}) \right| \le$$

$$\le \left| I - \sum_{i=1}^{n} f(d_i) g'(d_i)(x_i - x_{i-1}) \right| + \left| \sum_{i=1}^{n} \big(f(c_i) - f(d_i)\big) g'(d_i)(x_i - x_{i-1}) \right| <$$

$$< \varepsilon + K \cdot \sum_{i=1}^{n} \omega(f; [x_{i-1}, x_i]) \cdot (x_i - x_{i-1}) = \varepsilon + K \cdot \Omega_F(f) < (1+K) \cdot \varepsilon,$$

which concludes the proof of the theorem.   □

Remarks 18.13. **1.** The conditions for the existence of the integral in Theorem 18.12 can be significantly weakened. So for example, the integral $\int_a^b f \, dg$ is guaranteed to exist if f is Riemann integrable and g is Lipschitz (see Exercise 18.11).
2. In the statement above, the Lipschitz property of g can be weakened further. We say that the function $f: [a,b] \to \mathbb{R}$ is **absolutely continuous** if for each $\varepsilon > 0$, there exists a $\delta > 0$ such that whenever $[a_1, b_1], \ldots, [a_n, b_n]$ are nonoverlapping subintervals of $[a,b]$ such that $\sum_{i=1}^{n} (b_i - a_i) < \delta$, then $\sum_{i=1}^{n} |f(b_i) - f(a_i)| < \varepsilon$. One can show that *if f is Riemann integrable and g is absolutely continuous in $[a,b]$, then the Stieltjes integral $\int_a^b f \, dg$ exists.*
3. The class of Riemann integrable functions and the class of absolutely continuous functions are also dual: *a function f is Riemann integrable if and only if $\int_a^b f \, dg$ exists for every absolutely continuous function g, and a function g is absolutely continuous if and only if $\int_a^b f \, dg$ exists for every Riemann integrable function f.* The proof of this theorem, however, uses concepts from measure theory that we do not deal with in this book.

A number-theoretic application. In the introduction of the chapter we mentioned that Stieltjes integrals pop up in many areas of mathematics, such as in number theory. Dealing with an important problem—namely the distribution of the prime numbers—we often need to approximate sums that consist of the values of specific functions at the prime numbers. For example, $L(x) = \sum_{p \le x} (\log p)/p$ and $R(x) = \sum_{p \le x} 1/p$ are such sums. In these sums, we need to add the numbers $(\log p)/p$ or $1/p$ for all prime numbers p less than or equal to x. Transforming these sums (often using Abel's rearrangement) can be efficiently done with the help of the Stieltjes integral, as shown by the following theorem.

Let a sequence $a_1 < a_2 < \cdots$ that tends to infinity and the function φ defined at the numbers a_n be given. Let $A(x) = \sum_{a_n \le x} \varphi(a_n)$. (If (a_n) is the sequence of prime numbers and $\varphi(x) = (\log x)/x$, then $A(x) = L(x)$, and if $\varphi(x) = 1/x$, then $A(x) = R(x)$.)

Theorem 18.14. *Suppose that f is differentiable and f' is integrable on $[a,b]$, where $a < a_1 \le b$. Then*

$$\sum_{a_n \le b} f(a_n) \cdot \varphi(a_n) = f(b) \cdot A(b) - \int_a^b A(x) \cdot f'(x) \, dx. \tag{18.9}$$

Proof. We show that $\sum_{a_n \le b} f(a_n) \cdot \varphi(a_n) = \int_a^b f \, dA$. The function $A(x)$ is constant on the interval $[a_{n-1}, a_n)$, and has a jump discontinuity at the point a_n, and there, $A(a_n) - \lim_{x \to a_n - 0} A(x) = \varphi(a_n)$. Thus if we take a partition of the interval $[a, b]$ with mesh small enough, then any approximating sum of the Stieltjes integral $\int_a^b f \, dA$ will consist of mostly zero terms, except for the terms that correspond to a subinterval containing one of $a_n \le b$, and the nth such term will be close to $f(a_n) \cdot \varphi(a_n)$ by the continuity of f.

Now integration by parts (Theorem 18.9) gives

$$\sum_{a_n \le b} f(a_n) \cdot \varphi(a_n) = \int_a^b f \, dA = f(b) \cdot A(b) - \int_a^b A \, df = f(b) \cdot A(b) - \int_a^b A \cdot f' \, dx,$$

also using Theorem 18.12. $\qquad\qquad\qquad\qquad\qquad\qquad\qquad\qquad\qquad\qquad\qquad\qquad\qquad$ □

If (a_n) is the sequence of prime numbers and $\varphi \equiv 1$, then the value of $A(x)$ is the sum of prime numbers up to x, that is $\pi(x)$. This gives us the following corollary.

Corollary 18.15. *Suppose that f is differentiable and f' is integrable on the interval* $[1, x]$ $(x \ge 2)$. *Then*

$$\sum_{p \le x} f(p) = f(x) \cdot \pi(x) - \int_1^x \pi(t) \cdot f'(t) \, dt. \qquad (18.10)$$

If, for example, $f(x) = 1/x$, then we get that $\sum_{p \le x} 1/p \ge \int_2^x (\pi(t)/t^2) \, dt$. Now there exists a constant $c > 0$ such that $\pi(x) \ge c \cdot x / \log x$. (See [5], Corollary 8.6.) Since $\int_e^x dt / (t \cdot \log t) = \log \log x$, we get that $\sum_{p \le x} 1/p \ge c \cdot \log \log x$. This proves the following theorem.

Corollary 18.16.

$$\sum_p \frac{1}{p} = \infty.$$

We can get a much better approximation for the partial sums if we use the fact that the difference between the function $L(x) = \sum_{p \le x} (\log p)/p$ and $\log x$ is bounded. (A proof of this fact can be found in [5, Theorem 8.8(b)].) Let $\eta(x) = L(x) - \log x$.

Let (a_n) be the sequence of prime numbers, and apply (18.9) with the choices $\varphi(x) = (\log x)/x$ and $f(x) = 1/\log x$. Then $A(x) = L(x)$, and so

$$\sum_{p \le x} \frac{1}{p} = \frac{L(x)}{\log x} + \int_2^x \frac{L(t)}{t \cdot \log^2 t} \, dt =$$

$$= 1 + \frac{\eta(x)}{\log x} + \int_2^x \frac{1}{t \cdot \log t} \, dt + \int_2^x \frac{\eta(t)}{t \cdot \log^2 t} \, dt =$$

$$= \log \log x - \log \log 2 + 1 + \frac{\eta(x)}{\log x} + \int_2^x \frac{\eta(t)}{t \cdot \log^2 t} \, dt.$$

Here if $x \to \infty$, then $\eta(x)/\log x$ tends to zero, and we can also show that the integral $\int_2^x (\eta(t)/(t \cdot \log^2 t))\, dt$ has a finite limit as $x \to \infty$; this follows easily by the theory of improper integrals; see the next chapter. Comparing all of this, we get the following.

Theorem 18.17. *The limit* $\lim_{x \to \infty} \left(\sum_{p \le x} \frac{1}{p} - \log\log x \right)$ *exists and is finite.*

Exercises

18.1. Let

$$\alpha(x) = \begin{cases} 0, & (0 \le x < 1) \\ 1, & (x = 1) \end{cases}; \quad \beta(x) = \begin{cases} 1, & (x = 0) \\ 0, & (0 < x \le 1) \end{cases}; \quad \gamma(x) = \begin{cases} 0, & (x \ne 1/2) \\ 1, & (x = 1/2) \end{cases}.$$

Show that the following Stieltjes integrals exist, and compute their values.

(a) $\int_0^1 \sin x\, d\alpha$;   (b) $\int_0^1 \alpha\, d\sin x$;   (c) $\int_0^1 e^x\, d\beta$;

(d) $\int_0^1 \beta\, de^x$;   (e) $\int_0^1 x^2\, d\gamma$;   (f) $\int_0^1 \gamma\, dx^2$;

(g) $\int_0^2 e^x\, d[x]$;   (h) $\int_0^2 [x]\, de^x$.

18.2. Let

$$f(x) = \begin{cases} 0, & \text{if } -1 \le x \le 0 \\ 1, & \text{if } 0 < x \le 1 \end{cases} \quad \text{and} \quad g(x) = \begin{cases} 0, & \text{if } -1 \le x < 0 \\ 1, & \text{if } 0 \le x \le 1. \end{cases}$$

Prove that the Stieltjes integrals $\int_{-1}^0 f\, dg$ and $\int_0^1 f\, dg$ exist, but $\int_{-1}^1 f\, dg$ does not.

18.3. Prove that if the Stieltjes integrals $\int_a^c f\, dg$ and $\int_c^b f\, dg$ exist, at least one of f or g is continuous at c, and the other function is bounded then the Stieltjes integral $\int_a^b f\, dg$ also exists.

18.4. Prove that if the functions f and g share a point of discontinuity, then the Stieltjes integral $\int_a^b f\, dg$ does not exist. (H S)

18.5. Let $f(x) = \sqrt{x} \cdot \sin(1/x)$ if $x \ne 0$, and $f(0) = 0$. Prove that the Stieltjes integral $\int_0^1 f\, df$ does not exist. (H S)

18.6. Prove Theorem 18.7. (H)

18.7. Prove that if $\int_a^b f\, df$ exists, then its value is $(f(b)^2 - f(a)^2)/2$.

18.8. Prove that if $\int_a^b f\, dg$ exists for every function g that is of bounded variation, then f is continuous. (H)

18.9. Prove that if $\int_a^b f\, dg$ exists for every continuous function f, then g is of bounded variation. (H)

18.10. Let $\mathscr{F}$ be an arbitrary set of functions defined on the interval $[a,b]$. Let $\mathscr{G}$ be the set of those functions $g\colon [a,b] \to \mathbb{R}$ for which the integral $\int_a^b f\,dg$ exists for all $f \in \mathscr{F}$. Furthermore, let $\mathscr{H}$ denote the set of functions $h\colon [a,b] \to \mathbb{R}$ whose integral $\int_a^b h\,dg$ exists for all $g \in \mathscr{G}$. Show that $\mathscr{H}$ and $\mathscr{G}$ are dual classes, that is, a function h satisfies $h \in \mathscr{H}$ if and only if $\int_a^b h\,dg$ exists for all $g \in \mathscr{G}$, and a function g satisfies $g \in \mathscr{G}$ if and only if $\int_a^b h\,dg$ exists for all $h \in \mathscr{H}$.

18.11. Prove that if f is Riemann integrable and g is Lipschitz on $[a,b]$, then the Stieltjes integral $\int_a^b f\,dg$ exists. (H)

18.12. Show that

(a) if f is Lipschitz, then it is absolutely continuous;
(b) if f is absolutely continuous, then it is continuous and of bounded variation.

Chapter 19
The Improper Integral

19.1 The Definition and Computation of Improper Integrals

Until now, we have dealt only with integrals of functions that are defined in some closed and bounded interval (except, perhaps, for finitely many points of the interval) and are bounded on that interval. These restrictions are sometimes too strict; there are problems whose solutions require us to integrate functions on unbounded intervals, or that themselves might not be bounded.

Suppose that we want to compute the integral $\int_0^{\pi/2} \sqrt{1+\sin x}\,dx$ with the substitution $\sin x = t$. The formulas $x = \arcsin t$, $dx/dt = 1/\sqrt{1-t^2}$ give

$$\int_0^{\pi/2} \sqrt{1+\sin x}\,dx = \int_0^1 \frac{\sqrt{1+t}}{\sqrt{1-t^2}}\,dt = \int_0^1 \sqrt{\frac{1}{1-t}}\,dt. \qquad (19.1)$$

The integral on the right-hand side, however, is undefined (for now), since the function $\sqrt{1/(1-t)}$ is unbounded, and so it is not integrable in $[0,1]$. (This problem is caused by the application of the substitution formula, Theorem 15.21, with the choices $f(x) = \sqrt{1+\sin x}$, $g(t) = \arcsin t$, $[a,b] = [0,1]$. The theorem cannot be applied, since its conditions are not satisfied, the function $\arcsin t$ not being differentiable at 1.)

Wanting to compute the integral $\int_0^{\pi} dx/(1+\sin x)$ with the substitution $\operatorname{tg}(x/2)=t$, we run into trouble with the limits of integration. Since $\lim_{x\to\pi-0} \operatorname{tg}(x/2) = \infty$, the result of the substitution is

$$\int_0^{\pi} \frac{1}{1+\sin x} = \int_0^{\infty} \frac{1}{1+2t/(1+t^2)} \cdot \frac{2}{1+t^2}\,dt = \int_0^{\infty} 2\cdot \frac{dt}{(1+t)^2}, \qquad (19.2)$$

and the integral on the right-hand side is undefined (for now). (Here the problem is caused by the function $\operatorname{tg}(x/2)$ not being defined—and so not being differentiable—at π.)

© Springer New York 2015
M. Laczkovich, V.T. Sós, *Real Analysis*, Undergraduate Texts
in Mathematics, DOI 10.1007/978-1-4939-2766-1_19

We know that if the function $f: [a,b] \to \mathbb{R}$ is differentiable and f' is integrable on $[a,b]$, then the graph of f is rectifiable, and its arc length is $\int_a^b \sqrt{1+(f'(x))^2}\,dx$ (see Theorem 16.13 and Remark 16.21). We can expect this equation to give us the arc length of the graph of f whenever the graph is rectifiable and f is differentiable with the exception of finitely many points. This, however, is not always so. Consider, for example, the function $f(x) = \sqrt{x}$ on $[0,1]$. Since f is monotone, its graph is rectifiable. On the other hand, f is differentiable at every point $x > 0$, and

$$\sqrt{1+(f'(x))^2} = \sqrt{1+\tfrac{1}{4x}}.$$

We expect the arc length of $\sqrt{x}$ over $[0,1]$ to be

$$\int_0^1 \sqrt{1+\tfrac{1}{4x}}\,dx. \tag{19.3}$$

However, this integral is not defined (for now), because the function we are integrating is not bounded.

We can avoid all of these problems by extending the concept of the integral.

Definition 19.1. Let f be defined on the interval $[a,\infty)$, and suppose that f is integrable on $[a,b]$ for all $b > a$. If the limit

$$\lim_{b\to\infty} \int_a^b f(x)\,dx = I$$

exists and is finite, then we say that the *improper integral of f in $[a,\infty)$ is convergent and has value I.* We denote this by $\int_a^\infty f(x)\,dx = I$.

If the limit $\lim_{b\to\infty} \int_a^b f(x)\,dx$ does not exist or exists but is not finite, then we say that the *improper integral is divergent.* If

$$\lim_{b\to\infty} \int_a^b f(x)\,dx = \infty \qquad (\text{or } -\infty),$$

then we say that the improper integral $\int_a^b f(x)\,dx$ exists, and its value is ∞ (or $-\infty$).

We define the improper integral $\int_{-\infty}^a f(x)\,dx$ similarly.

Example 19.2. By the definition above, the improper integral $\int_0^\infty dx/(1+x)^2$ is convergent, and its value is 1. Indeed, the function $1/(1+x)^2$ is integrable on the interval $[0,b]$ for all $b > 0$, and

$$\lim_{b\to\infty} \int_0^b \frac{dx}{(1+x)^2} = \lim_{b\to\infty}\left[-\frac{1}{(1+x)}\right]_0^b = \lim_{b\to\infty}\left[1-\frac{1}{(1+b)}\right] = 1.$$

Once we know this, it is easy to see that equality (19.2) truly holds, and the value of the integral on the left is 2. Indeed, for all $0 < b < \pi$, the function $\mathrm{tg}(x/2)$ is differentiable, and its derivative is integrable on $[0,b]$, so the substitution $\mathrm{tg}(x/2) = t$ can be used here. Thus

$$\int_0^b \frac{dx}{1+\sin x} = \int_0^{\text{tg}(b/2)} \frac{1}{1+2t/(1+t^2)} \cdot \frac{2}{1+t^2} \, dt = \int_0^{\text{tg}(b/2)} 2 \cdot \frac{dt}{(1+t)^2}.$$

Now using the theorems regarding the continuity of the integral function (Theorem 15.5) and regarding the limit of compositions of functions, we get that

$$\int_0^\pi \frac{dx}{1+\sin x} = \lim_{b\to\pi-0} \int_0^b \frac{dx}{1+\sin x} =$$

$$= \lim_{b\to\pi-0} \int_0^{\text{tg}(b/2)} 2 \cdot \frac{dt}{(1+t)^2} = \int_0^\infty 2 \cdot \frac{dt}{(1+t)^2} = 2.$$

We define the integrals appearing in the right-hand side of (19.1) and those in (19.3) in a similar way.

Definition 19.3. Let $[a,b]$ be a bounded interval, let f be defined on $[a,b)$, and suppose that f is integrable on the interval $[a,c]$ for all $a \le c < b$. If the limit

$$\lim_{c\to b-0} \int_a^c f(x)\,dx = I$$

exists and is finite, then we say that the *improper integral* of the function f is convergent and its value is I. We denote this by $\int_a^b f(x)\,dx = I$.

If the limit $\lim_{c\to b-0} \int_a^c f(x)\,dx$ does not exist, or exists but is not finite, then we say that the *improper integral is divergent*. If

$$\lim_{c\to b-0} \int_a^c f(x)\,dx = \infty \ (\text{or} \ -\infty),$$

then we say that the improper integral $\int_a^b f(x)\,dx$ exists, and that its value is ∞ (or $-\infty$).

We define the improper integral $\int_a^b f(x)\,dx$ similarly in the case that f is integrable on the intervals $[c,b]$ for all $a < c \le b$.

Remarks 19.4. **1.** We say that an improper integral exists if it is convergent or if its value is ∞ or $-\infty$. (In these last two cases, the improper integral exists but is divergent.)

2. If f is integrable on the bounded interval $[a,b]$, then the expression $\int_a^b f(x)\,dx$ now denotes three numbers at once: the Riemann integral of the function f over the interval $[a,b]$, the limit $\lim_{c\to b-0} \int_a^c f(x)\,dx$, and the limit $\lim_{c\to a+0} \int_c^b f(x)\,dx$. Luckily, these three numbers agree; by Theorem 15.5, the integral function $I(x) = \int_a^x f(t)\,dt$ is continuous on $[a,b]$, so

$$\lim_{c\to b-0} \int_a^c f(x)\,dx = \lim_{c\to b-0} I(c) = I(b) = \int_a^b f(x)\,dx,$$

and

$$\lim_{c\to a+0} \int_c^b f(x)\,dx = \lim_{c\to a+0} [I(b) - I(c)] = I(b) - I(a) = \int_a^b f(x)\,dx.$$

In other words, if the Riemann integral of the function f exists on the interval $[a,b]$, then (both of) its improper integrals exist there, and their values agree with those of the Riemann integral. Thus the concept of the improper integral is an extension of the Riemann integral.

Even though this is true, when we say that f is integrable on the closed interval $[a,b]$, we will still mean that f is Riemann integrable on $[a,b]$.

Example 19.5. The integral on the right-hand side of equality (19.1) is convergent, and its value is 2. Clearly, we have

$$\lim_{c\to 1-0}\int_0^c \sqrt{\frac{1}{1-t}}\,dt = \lim_{c\to 1-0}\left[-2\sqrt{1-t}\right]_0^c = \lim_{c\to 1-0}\left(2-2\sqrt{1-c}\right) = 2.$$

Using the argument of Example 19.2, we can easily check that equality (19.1) is true.

Now for some further examples.

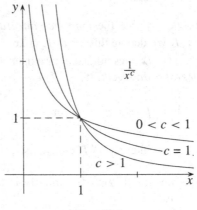

Fig. 19.1

1. The integral $\int_1^\infty dx/x^c$ is convergent if and only if $c > 1$. More precisely,

$$\int_1^\infty \frac{dx}{x^c} = \begin{cases} 1/(c-1), & \text{if } c > 1, \\ \infty, & \text{if } c \le 1. \end{cases} \qquad (19.4)$$

For $c \ne 1$, we have

$$\int_1^\infty \frac{dx}{x^c} = \lim_{\omega\to\infty}\int_1^\omega \frac{dx}{x^c} =$$

$$= \lim_{\omega\to\infty}\left[\frac{1}{1-c}\cdot x^{1-c}\right]_1^\omega = \lim_{\omega\to\infty}\left[\frac{1}{c-1} + \frac{1}{1-c}\cdot \omega^{1-c}\right].$$

Now

$$\lim_{x \to \infty} x^{1-c} = \begin{cases} 0, & \text{if } c > 1, \\ \infty, & \text{if } c < 1, \end{cases}$$

so if $c \neq 1$, then (19.4) is true. On the other hand,

$$\int_1^\infty \frac{dx}{x} = \lim_{\omega \to \infty} \int_1^\omega \frac{dx}{x} = \lim_{\omega \to \infty} [\log x]_1^\omega = \lim_{\omega \to \infty} \log \omega = \infty,$$

so (19.4) also holds when $c = 1$.

2. The integral $\int_0^1 dx/x^c$ is convergent if and only if $c < 1$. More precisely,

$$\int_0^1 \frac{dx}{x^c} = \begin{cases} 1/(1-c), & \text{if } c < 1, \\ \infty, & \text{if } c \geq 1. \end{cases} \tag{19.5}$$

For $c \neq 1$, we have

$$\int_0^1 \frac{dx}{x^c} = \lim_{\delta \to 0+0} \int_\delta^1 \frac{dx}{x^c} =$$

$$= \lim_{\delta \to 0+0} \left[\frac{1}{1-c} \cdot x^{1-c} \right]_\delta^1 = \lim_{\delta \to 0+0} \left[\frac{1}{1-c} - \frac{1}{1-c} \cdot \delta^{1-c} \right]$$

for all $0 < \delta < 1$. Now

$$\lim_{x \to 0+0} x^{1-c} = \begin{cases} 0, & \text{if } c < 1, \\ \infty, & \text{if } c > 1, \end{cases}$$

so if $c \neq 1$, then (19.5) is true. On the other hand,

$$\int_\delta^1 \frac{dx}{x} = \lim_{\delta \to 0+0} [\log x]_\delta^1 = \lim_{\delta \to 0+0} [-\log \delta] = \infty,$$

so (19.5) also holds when $c = 1$ (Figure 19.1).

Incidentally, the substitution $x = 1/t$ gives us that

$$\int_1^\infty \frac{dx}{x^c} = \int_0^1 \frac{1}{t^{-c}} \cdot \frac{1}{t^2} \, dt = \int_0^1 \frac{dt}{t^{2-c}}.$$

This shows that statements (19.4) and (19.5) follow each from the other. (We will justify this statement soon, in Theorem 19.12.)

3. The integral $\int_2^\infty dx/(x \cdot \log^c x)$ is convergent if and only if $c > 1$. This is because if $c \neq 1$, then

$$\int_2^\infty \frac{dx}{x \cdot \log^c x} = \lim_{\omega \to \infty} \int_2^\omega \frac{dx}{x \cdot \log^c x} = \lim_{\omega \to \infty} \left[\frac{1}{1-c} \cdot \log^{1-c} x \right]_2^\omega =$$

$$= \lim_{\omega \to \infty} \left[\frac{1}{c-1} \cdot \log^{1-c} 2 + \frac{1}{1-c} \cdot \log^{1-c} \omega \right].$$

Now

$$\lim_{x\to\infty} \log^{1-c} x = \begin{cases} 0, & \text{if } c > 1, \\ \infty, & \text{if } c < 1, \end{cases}$$

so if $c > 1$, then the integral is convergent, while if $c < 1$, then it is divergent. In the case $c = 1$,

$$\int_2^\infty \frac{dx}{x \cdot \log x} = \lim_{\omega\to\infty} \int_2^\omega \frac{dx}{x \cdot \log x} = \lim_{\omega\to\infty} \left[\log\log x\right]_2^\omega =$$

$$= \lim_{\omega\to\infty} \left[\log\log \omega - \log\log 2\right] = \infty,$$

so the integral is also divergent when $c = 1$.

4. $\int_a^\infty e^{-x} dx = e^{-a}$ for all a. Clearly,

$$\int_a^\infty e^{-x} dx = \lim_{\omega\to\infty} \int_a^\omega e^{-x} dx = \lim_{\omega\to\infty} \left[e^{-a} - e^{-\omega}\right] = e^{-a}.$$

5. $\int_0^\infty dx/(1+x^2) = \pi/2$, since $\lim_{\omega\to\infty} \int_0^\omega dx/(1+x^2) = \lim_{\omega\to\infty} \left[\text{arc tg} x\right]_0^\omega = \pi/2$. We can similarly see that $\int_{-\infty}^0 dx/(1+x^2) = \pi/2$.

6. $\int_0^1 \log x dx = -1$, since

$$\int_0^1 \log x dx = \lim_{\delta\to 0+0} \int_\delta^1 \log x dx = \lim_{\delta\to 0+0} \left[x\log x - x\right]_\delta^1 =$$

$$= \lim_{\delta\to 0+0} \left[-\delta\log\delta + \delta - 1\right] = -1.$$

7. The integral $\int_0^1 dx/(1-x^2)$ is divergent, since

$$\lim_{c\to 1-0} \int_0^c \frac{dx}{1-x^2} = \lim_{c\to 1-0} \left[\frac{1}{2}\cdot\log\left|\frac{1+x}{1-x}\right|\right]_0^c = \lim_{c\to 1-0} \left[\frac{1}{2}\cdot\log\left|\frac{1+c}{1-c}\right|\right] = \infty.$$

8. $\int_0^1 dx/\left(\sqrt{1-x^2}\right) = \pi/2$, since

$$\lim_{c\to 1-0} \int_0^c \frac{dx}{\sqrt{1-x^2}} = \lim_{c\to 1-0} \text{arc sin} c = \frac{\pi}{2}.$$

We sometimes need to compute integrals that are "improper" at both end-points, and possibly at numerous interior points as well. The integrals $\int_{-\infty}^\infty e^{-x^2} dx$, $\int_0^\infty \sqrt{x}\cdot e^{-x} dx$, and $\int_{-1}^1 dx/\sqrt{|x|}$ are examples of this. We define these in the following way.

Definition 19.6. Let the function f be defined on the finite or infinite interval (α, β), except at at most finitely many points. Suppose that there exists a partition $\alpha = c_0 < c_1 < \cdots < c_{n-1} < c_n = \beta$ with the property that for all $i = 1, \ldots, n$, the improper integral $\int_{c_{i-1}}^{c_i} f(x)\,dx$ is convergent by Definition 19.1 or by Definition 19.3. Then we say that the *improper integral $\int_\alpha^\beta f(x)\,dx$ is convergent, and its value is*

$$\int_\alpha^\beta f(x)\,dx = \int_{c_0}^{c_1} f(x)\,dx + \cdots + \int_{c_{n-1}}^{c_n} f(x)\,dx.$$

To justify this definition, we need to check that if $F' \colon \alpha = d_0 < d_1 < \cdots < d_{k-1} < d_k = \beta$ is another partition such that the integrals $\int_{d_{i-1}}^{d_i} f(x)\,dx$ are convergent by Definition 19.1 or Definition 19.3, then

$$\sum_{i=1}^n \int_{c_{i-1}}^{c_i} f(x)\,dx = \sum_{j=1}^k \int_{d_{j-1}}^{d_j} f(x)\,dx. \tag{19.6}$$

Suppose first that we obtain the partition F' by adding one new base point to the partition $F \colon \alpha = c_0 < c_1 < \cdots < c_{n-1} < c_n = \beta$. If this base point is $d \in (c_{i-1}, c_i)$, then to prove (19.6), we must show that

$$\int_{c_{i-1}}^{c_i} f(x)\,dx = \int_{c_{i-1}}^d f(x)\,dx + \int_d^{c_i} f(x)\,dx.$$

This is easy to check by distinguishing the cases appearing in Definitions 19.1 and 19.3, and using Theorem 14.38.

Once we know this, we can use induction to find that (19.6) also holds if we obtain F' by adding any (finite) number of new base points. Finally, for the general case, consider the partition F'', the union of the partitions F and F'. By what we just said, both sides of (19.6) are equal to the sum of integrals corresponding to the partition F'', so they are equal to each other.

Examples 19.7. **1.** The integral $\int_{-\infty}^\infty dx/(1+x^2)$ is convergent, and its value is π. Clearly, $\int_0^\infty dx/(1+x^2) = \pi/2$ and $\int_{-\infty}^0 dx/(1+x^2) = \pi/2$ by Example 19.5.5.

2. The integral $\int_{-1}^1 dx/\left(\sqrt{|x|}\right)$ is convergent, and its value is 4. Clearly, $\int_0^1 dx/(\sqrt{x}) = 2$ by Example 19.5.2. Similarly, we can show (or use the substitution $x = -t$ to reduce it to the previous integral) that $\int_{-1}^0 dx/(\sqrt{-x}) = 2$.

The methods used in Example 19.5 to compute improper integrals can be used in most cases in which the function we want to integrate has a primitive function: in such cases, the convergence of the improper integral depends on the existence of the limit of the primitive function. We introduce the following notation. Let F be defined on the finite or infinite interval (α, β). If the limits $\lim_{x\to\alpha+0} F(x)$ and $\lim_{x\to\beta-0} F(x)$ exist and are finite, then let

$$[F]_\alpha^\beta = \lim_{x\to\beta-0} F(x) - \lim_{x\to\alpha+0} F(x).$$

Here, when $\alpha = -\infty$, the notation $\lim_{x\to\alpha+0}$ means the limit at negative infinity, while if $\beta = \infty$, then $\lim_{x\to\beta-0}$ is the limit at positive infinity.

Theorem 19.8. *Let* $f: (\alpha, \beta) \to \mathbb{R}$ *be integrable on every closed and bounded subinterval of* (α, β). *Suppose that* f *has a primitive function in* (α, β), *and let* $F: (\alpha, \beta) \to \mathbb{R}$ *be a primitive function of* f. *The improper integral* $\int_\alpha^\beta f(x)\,dx$ *is convergent if and only if the finite limits* $\lim_{x\to\alpha+0} F(x)$ *and* $\lim_{x\to\beta-0} F(x)$ *exist, and then*

$$\int_\alpha^\beta f(x)\,dx = [F]_\alpha^\beta.$$

Proof. Fix a point $x_0 \in (\alpha, \beta)$. Since f is integrable on every closed and bounded subinterval, it is easy to see that the improper integral $\int_\alpha^\beta f(x)\,dx$ is convergent if and only if the integrals $\int_\alpha^{x_0} f(x)\,dx$ and $\int_{x_0}^\beta f(x)\,dx$ are convergent by Definition 19.1 or Definition 19.3, and then

$$\int_\alpha^\beta f(x)\,dx = \int_\alpha^{x_0} f(x)\,dx + \int_{x_0}^\beta f(x)\,dx.$$

By the fundamental theorem of calculus,

$$\int_{x_0}^b f(x)\,dx = F(b) - F(x_0)$$

for all $x_0 < b < \beta$. This implies that the limits $\lim_{b\to\beta-0} [F(b) - F(x_0)]$ along with $\lim_{b\to\beta-0} \int_{x_0}^b f(x)\,dx$ either both exist or both do not exist, and if they do, then they agree. The same holds for the limits

$$\lim_{a\to\alpha+0} [F(x_0) - F(a)] \quad \text{and} \quad \lim_{a\to\alpha+0} \int_a^{x_0} f(x)\,dx$$

as well. We get that the integral $\int_\alpha^\beta f(x)\,dx$ is convergent if and only if the limits $\lim_{b\to\beta-0} [F(b) - F(x_0)]$ and $\lim_{a\to\alpha+0} [F(x_0) - F(a)]$ exist and are finite, and then the value of the integral is the sum of these two limits. This is exactly what we wanted to show. $\square$

We now turn to the rules of integration regarding improper integrals. We will see that the formulas for Riemann integration—including the rules for integration by parts and substitution—remain valid for improper integrals with almost no changes.

Theorem 19.9. *If the integrals* $\int_\alpha^\beta f\,dx$ *and* $\int_\alpha^\beta g\,dx$ *are convergent, then the integrals* $\int_\alpha^\beta c \cdot f\,dx$ $(c \in \mathbb{R})$ *and* $\int_\alpha^\beta (f+g)\,dx$ *are also convergent, and*

$$\int_\alpha^\beta c \cdot f\,dx = c \cdot \int_\alpha^\beta f\,dx \qquad (c \in \mathbb{R})$$

and

$$\int_\alpha^\beta (f+g)\,dx = \int_\alpha^\beta f\,dx + \int_\alpha^\beta g\,dx.$$

Proof. The statement is clear by the definition of the improper integral, and by Theorems 14.30, 14.31, and 10.35. $\square$

Theorem 19.10. *Let the functions f and g be differentiable, and let f' and g' be integrable on every closed and bounded subinterval of* (α, β), *and suppose that the limits* $\lim_{x \to \alpha + 0} fg$ *and* $\lim_{x \to \beta - 0} fg$ *exist and are finite. Then*

$$\int_{\alpha}^{\beta} f'g\, dx = [fg]_{\alpha}^{\beta} - \int_{\alpha}^{\beta} fg'\, dx,$$

in the sense that if one of the two improper integrals exists, then the other one does, too, and the two sides are equal.

Proof. Fix a point $x_0 \in (\alpha, \beta)$. By Theorem 15.11,

$$\int_{x_0}^{b} f'g\, dx = [fg]_{x_0}^{b} - \int_{x_0}^{b} fg'\, dx$$

for all $x_0 < b < \beta$. Thus if we let b tend to $\beta - 0$, then the left side has a limit if and only if the right side does, and if they exist, then they are equal. Similarly, if $a \to \alpha + 0$, then the left side of the equality

$$\int_{a}^{x_0} f'g\, dx = [fg]_{a}^{x_0} - \int_{a}^{x_0} fg'\, dx$$

has a limit if and only if the right side does, and if they exist, then they are equal. We can now get the statement of the theorem just as in the proof of Theorem 19.8. □

We defined Riemann integrals $\int_{a}^{b} f\, dx$ for the case $b \le a$ (see Definition 14.40). We should make this generalization for improper integrals as well.

Definition 19.11. If the improper integral $\int_{\alpha}^{\beta} f(x)\, dx$ exists, then let

$$\int_{\beta}^{\alpha} f(x)\, dx = - \int_{\alpha}^{\beta} f(x)\, dx.$$

The value of the integral $\int_{\alpha}^{\alpha} f\, dx$ is zero for all $\alpha \in \mathbb{R}$ and $\alpha = \pm\infty$, and for every function f (whether it is defined at the point α or not).

Theorem 19.12. *Let* (α, β) *be a finite or infinite open interval. Let g be strictly monotone and differentiable, and suppose that g' is integrable on every closed and bounded subinterval of* (α, β). *Let[1]* $\lim_{x \to \alpha + 0} g(x) = \gamma$ *and* $\lim_{x \to \beta - 0} g(x) = \delta$. *If f is continuous on the open interval* (γ, δ), *then*

$$\int_{\alpha}^{\beta} f(g(t)) \cdot g'(t)\, dt = \int_{\gamma}^{\delta} f(x)\, dx \tag{19.7}$$

in the sense that if one of the two improper integrals is convergent, then so is the other, and they are equal.

---

[1] These finite or infinite limits must exist by Theorem 10.68.

Proof. We suppose that g is strictly monotone increasing (we can argue similarly if g is monotone decreasing). Then $\gamma < \delta$. Fix a number $x_0 \in (\alpha, \beta)$, and let $F(x) = \int_{x_0}^{x} f(t)\, dt$ for all $x \in (\gamma, \delta)$. By the definition of the improper integral, it follows that the right-hand side of (19.7) is convergent if and only if the finite limits $\lim_{x \to \gamma+0} F(x)$ and $\lim_{x \to \delta-0} F(x)$ exist, and then the value of the integral on the right-hand side of (19.7) is $[F]_{\gamma}^{\delta}$.

By Theorem 15.5, F is a primitive function of f on the interval (γ, δ), and so by the differentiation rule for compositions of functions, $F \circ g$ is a primitive function of $(f \circ g) \cdot g'$ on (α, β). By Theorem 19.8, the left-hand side of (19.7) is convergent if and only if the finite limits $\lim_{t \to \alpha+0} F(g(t))$ and $\lim_{t \to \beta-0} F(g(t))$ exist, and then the value of the integral on the left-end side of (19.7) is $[F \circ g]_{\alpha}^{\beta}$.

Now if $\lim_{x \to \gamma+0} F(x) = A$, then by Theorem 10.41 on the limit of compositions of functions, $\lim_{t \to \alpha+0} F(g(t)) = A$. Conversely, if $\lim_{t \to \alpha+0} F(g(t)) = A$, then $\lim_{x \to \gamma+0} F(x) = A$; clearly, the function g^{-1} maps the interval (γ, δ) onto the interval (α, β) in a strictly monotone increasing way, and so $\lim_{x \to \gamma+0} g^{-1}(x) = \alpha$. Thus if $\lim_{t \to \alpha+0} F(g(t)) = A$, then $\lim_{x \to \gamma+0} F(x) = A$, since $F = (F \circ g) \circ g^{-1}$.

The same reasoning gives that the limit $\lim_{x \to \delta-0} F(x)$ exists if and only if $\lim_{t \to \beta-0} F(g(t))$ exists, and then they are equal.

Comparing all of the results above, we get the statement of the theorem we set out to prove. $\square$

The theorem above is true with much weaker conditions as well: it is not necessary to require the monotonicity of g (at least if $\gamma \neq \delta$), and instead of the continuity of f, it suffices to assume that f is integrable in the closed and bounded subintervals of the image of g. The precise theorem is as follows.

Theorem 19.13. *Let g be differentiable, suppose g' is integrable on every closed and bounded subinterval of (α, β), and suppose that the finite or infinite limits $\lim_{x \to \alpha+0} g(x) = \gamma$ and $\lim_{x \to \beta-0} g(x) = \delta$ exist. Let f be integrable on every closed and bounded subinterval of the image of g, that is, of $g((\alpha, \beta))$.*

(i) *If $\gamma \neq \delta$, then (19.7) holds in the sense that if one of the two improper integrals exists, then so does the other, and then they are equal.*

(ii) *If $\gamma = \delta$ and the left-hand side of (19.7) is convergent, then its value is zero.*

We give the proof of the theorem in the appendix of the chapter. We note that when $\gamma = \delta$, the left-hand side of (19.7) is not necessarily convergent; see Exercise 19.12.

Remark 19.14. We have seen examples in which we created improper integrals from "ordinary" Riemann integrals. The reverse case can occur as well: we can create a Riemann integral starting from an improper integral. For example, the substitution $x = \sin t$ gives us

$$\int_0^1 \frac{x^n}{\sqrt{1-x^2}}\, dx = \int_0^{\frac{\pi}{2}} \sin^n t\, dt,$$

and the substitution $x = 1/t$ gives us

$$\int_1^\infty \frac{dx}{1+x^2} = \int_1^0 \frac{1}{1+1/t^2} \cdot \frac{-1}{t^2} dt = \int_0^1 \frac{dt}{1+t^2} = \frac{\pi}{4}.$$

Exercises

19.1. Compute the values of the following integrals:

(a) $\int_1^\infty \frac{dx}{x^2+x}$;

(b) $\int_{-\infty}^\infty \frac{dx}{x^2+x+1}$;

(c) $\int_{-\infty}^\infty \frac{dx}{x^4+1}$;

(d) $\int_1^\infty \frac{x+1}{x^3+x} dx$;

(e) $\int_0^\infty \frac{x}{(1+x^2)^3} dx$;

(f) $\int_a^b \frac{dx}{\sqrt{(x-a)(b-x)}} dx$;

(g) $\int_1^\infty \frac{dx}{2^x-1} dx$;

(h) $\int_{-\infty}^\infty \frac{e^x}{e^{2x}+1} dx$;

(i) $\int_0^\infty e^{-\sqrt{x}} dx$;

(j) $\int_1^\infty \frac{\log x}{x^2} dx$;

(k) $\int_1^\infty \frac{x \log x}{(1+x^2)^2} dx$.

19.2. Prove that the integral $\int_3^\infty \frac{dx}{x \cdot \log x \cdot \log \log^c x}$ is convergent if and only if $c > 1$. How can we generalize the statement?

19.3. Compute the integral $\int_{\pi/4}^{\pi/2} dx/\sin^4 x$ with the substitution $t = \text{tg}\, x$.

19.4. Compute the integral $\int_{-\infty}^\infty dx/(1+x^2)^n$ for every positive integer n with the substitution $x = \text{tg}\, t$.

19.5. Let f be a nonnegative and continuous function in $[0, \infty)$. What simpler expression is $\lim_{h\to\infty} h \cdot \int_0^1 f(hx) dx$ equal to?

19.6. Prove that if the integral $\int_{-\infty}^\infty f(x) dx$ is convergent, then we have $\lim_{n\to\infty} \int_{-1}^1 f(nx) dx = 0$.

19.7. Prove that if f is positive and monotone decreasing in $[0, \infty)$ and the integral $\int_0^\infty f(x) dx$ is convergent, then $\lim_{x\to\infty} x \cdot f(x) = 0$. (H S)

19.8. Prove that

$$\lim_{n\to\infty} \frac{1}{\sqrt{n}} \cdot \left(\frac{1}{\sqrt{1}} + \frac{1}{\sqrt{2}} + \cdots + \frac{1}{\sqrt{n}} \right) = 2.$$

19.9. Prove that if f is monotone in $(0,1]$ and the improper integral $\int_0^1 f(x)\,dx$ is convergent, then

$$\lim_{n\to\infty} \frac{1}{n}\cdot\left(f\left(\frac{1}{n}\right)+f\left(\frac{2}{n}\right)+\cdots+f\left(\frac{n}{n}\right)\right) = \int_0^1 f(x)\,dx.$$

Is the statement true if we do not assume f to be monotone? (S)

19.10. Compute the value of $\lim_{n\to\infty}\sum_{i=1}^{n-1} 1/\sqrt{i\cdot(n-i)}$.

19.11. Let f be continuous in $[a,b]$, and let f be differentiable in $[a,b]$ except at finitely many points. Prove that if the improper integral $\int_a^b \sqrt{1+(f'(x))^2}\,dx$ is convergent and its value is I, then the graph of f is rectifiable, and its arc length is I. (S)

19.12. Let $\alpha = -\infty$, $\beta = \infty$, let $g(x) = 1/(1+x^2)$ for all $x \in \mathbb{R}$, and let $f(x) = 1/x^2$ for all $x > 0$. Check that with these assumptions, f is integrable on every closed and bounded subinterval of the image of g, but the left-hand side of (19.7) does not exist (while the right-hand side clearly exists, and its value is zero).

19.2 The Convergence of Improper Integrals

In the applications of improper integrals, the most important question is whether a given integral is convergent; computing the value of the integral if it is convergent is often only a secondary (or hopeless) task. Suppose that f is integrable on every closed and bounded subinterval of the interval $[a, \beta)$. The convergence of the improper integral $\int_a^\beta f\,dx$ depends only on the values of f close to β: for every $a < b < \beta$, the integral $\int_a^\beta f\,dx$ is convergent if and only if the integral $\int_b^\beta f\,dx$ is convergent. Clearly,

$$\int_a^\omega f\,dx = \int_a^b f\,dx + \int_b^\omega f\,dx$$

for all $b < \omega < \beta$, so if $\omega \to \beta - 0$, then the left-hand side has a finite limit if and only if the same holds for $\int_b^\omega f\,dx$.

From Examples 19.2 and 19.5, we might want to conclude that an integral $\int_b^\beta f\,dx$ can be convergent only if f tends to zero fast enough as $x \to \beta - 0$. However, this is not the case. Consider a convergent integral $\int_b^\beta f\,dx$ (for example, any of the convergent integrals appearing in the examples above), and change the values of f arbitrarily at the points of a sequence $x_n \to \beta - 0$. This does not affect the convergence of the integral $\int_a^\beta f\,dx$, since for all $a < b < \beta$, the value of the function f is changed at only finitely many points inside the interval $[a,b]$, so the Riemann integral $\int_a^b f\,dx$ does not change either. On the other hand, if at the point x_n we give f the value n for every n, then this new function does not tend to 0 at β. In fact we can find continuous functions f such that $\int_b^\beta f \, dx$ is convergent but f does not tend to 0 at β (see Exercises 19.28 and 19.29).

The converse of the (false) statement above, however, is true: if f tends to zero fast enough as $x \to \beta - 0$, then the integral $\int_b^\beta f \, dx$ is convergent. To help with the proof of this, we first give a necessary and sufficient condition for the convergence of integrals, which we call **Cauchy's criterion for improper integrals**.

Theorem 19.15 (Cauchy's Criterion). *Let f be integrable on every closed and bounded subinterval of the interval $[a, \beta)$. The improper integral $\int_a^\beta f \, dx$ is convergent if and only if for every $\varepsilon > 0$, there exists a number $a \le b < \beta$ such that $\left| \int_{b_1}^{b_2} f(x) \, dx \right| < \varepsilon$ for all $b < b_1 < b_2 < \beta$.*

Proof. The statement follows simply from Cauchy's criterion for limits of functions (Theorem 10.34) and the equality

$$\int_a^{b_2} f(x) \, dx - \int_a^{b_1} f(x) \, dx = \int_{b_1}^{b_2} f(x) \, dx.$$

□

Suppose that f is integrable on every closed and bounded subinterval of the interval $[a, \beta)$. If f is nonnegative on $[a, \beta)$, then the improper integral $\int_a^\beta f \, dx$ always exists. This is because the function $\omega \mapsto \int_a^\omega f \, dx$ ($\omega \in [a, \beta)$) is monotone increasing, and so the finite or infinite limit $\lim_{\omega \to \beta - 0} \int_a^\omega f \, dx$ must necessarily exist by Theorem 10.68.

Thus if f is integrable on every closed and bounded subinterval of the interval $[a, \beta)$, then the improper integral $\int_a^\beta |f| \, dx$ exists for sure. The question is whether its value is finite.

Definition 19.16. We call the improper integral $\int_a^\beta f \, dx$ *absolutely convergent* if f is integrable on the closed and bounded subintervals of $[a, \beta)$ and the improper integral $\int_a^\beta |f| \, dx$ is convergent.

Theorem 19.17. *If the improper integral $\int_a^\beta f \, dx$ is absolutely convergent, then it is convergent.*

Proof. The statement clearly follows from Cauchy's criterion, since

$$\left| \int_{b_1}^{b_2} f(x) \, dx \right| \le \int_{b_1}^{b_2} |f(x)| \, dx$$

for all $a \le b_1 < b_2 < \beta$.

□

The converse of the statement is generally not true; we will soon see in Examples 19.20.3 and 19.20.4 that the improper integral $\int_1^\infty (\sin x / x) \, dx$ is convergent but not absolutely convergent.

The following theorem—which is one of the most often used criteria for convergence—is called the **majorization principle**.

Theorem 19.18 (Majorization Principle). *Let f and g be integrable in every closed and bounded subinterval of $[a, \beta)$, and suppose that there exists a $b_0 \in [a, \beta)$ such that $|f(x)| \le g(x)$ for all $x \in (b_0, \beta)$. If $\int_a^\beta g(x)\, dx$ is convergent, then so is $\int_a^\beta f(x)\, dx$.*

Proof. By the "only if" part of Theorem 19.15, if $\int_a^\beta g(x)\, dx$ is convergent, then for an arbitrary $\varepsilon > 0$, there exists a $b < \beta$ such that

$$\left| \int_{b_1}^{b_2} g(x)\, dx \right| < \varepsilon$$

holds for all $b < b_1, b_2 < \beta$. Since $|f(x)| \le g(x)$ in the interval (b_0, β),

$$\left| \int_{b_1}^{b_2} f(x)\, dx \right| \le \left| \int_{b_1}^{b_2} |f(x)|\, dx \right| \le \left| \int_{b_1}^{b_2} g(x)\, dx \right| < \varepsilon$$

for all $\max(b, b_0) < b_1, b_2 < \beta$. Thus by the "if" part of Theorem 19.15, the improper integral $\int_a^\beta f(x)\, dx$ is convergent. □

Remark 19.19. The majorization principle can be used to show divergence as well: if $g(x) \ge |f(x)|$ for all $x \in (b_0, \beta)$ and $\int_a^\beta f(x)\, dx$ is divergent, then it follows that $\int_a^\beta g(x)\, dx$ is also divergent. Clearly, if this latter integral were convergent, then $\int_a^\beta f(x)\, dx$ would be convergent as well.

Such applications of the theorem are often called **minorization**.

Examples 19.20. **1.** The integrals $\int_1^\infty (\sin x / x^2)\, dx$ and $\int_1^\infty (\cos x / x^2)\, dx$ are convergent, since $\int_1^\infty x^{-2}\, dx$ is convergent, and

$$\left| \frac{\sin x}{x^2} \right| \le \frac{1}{x^2} \quad \text{and} \quad \left| \frac{\cos x}{x^2} \right| \le \frac{1}{x^2}$$

for all $x \ge 1$.

2. The integral $\int_0^1 (\sin x / x^2)\, dx$ is divergent. We know that $\lim_{x \to 0} \sin x / x = 1$, so $\sin x / x > 1/2$ if $0 < x < \delta_0$, and so $\sin x / x^2 > 1/x$ if $0 < x < \delta_0$. Thus by the divergence of $\int_0^{\delta_0} x^{-1}\, dx$, it follows that $\int_0^1 (\sin x / x^2)\, dx$ is divergent.

3. The integral $\int_1^\infty (\sin x / x)\, dx$ is convergent.

We can easily show this with the help of integration by parts. Since

$$\int_1^\omega \frac{\sin x}{x}\, dx = \left[-\frac{1}{x} \cdot \cos x \right]_1^\omega - \int_1^\omega \frac{\cos x}{x^2}\, dx$$

for all $\omega > 1$,

$$\int_1^\infty \frac{\sin x}{x}\, dx = \cos 1 - \int_1^\infty \frac{\cos x}{x^2}\, dx.$$

4. The integral $\int_1^\infty |\sin x / x|\, dx$ is divergent. Since $|\sin x| \ge \sin^2 x$ for all x, to show that the integral is divergent it suffices to show that $\int_1^\infty (\sin^2 x / x)\, dx$ is divergent. We know that

$$\int_1^\omega \frac{\sin^2 x}{x}\,dx = \frac{1}{2}\cdot\int_1^\omega \frac{1-\cos 2x}{x}\,dx = \frac{1}{2}\cdot\int_1^\omega \frac{dx}{x} - \frac{1}{2}\cdot\int_1^\omega \frac{\cos 2x}{x}\,dx.$$

Similarly to Example 3, we can show that if $\omega \to \infty$, then the second term on the right-hand side is convergent, while $\int_1^\omega dx/x \to \infty$ as $\omega \to \infty$. Thus $\int_1^\infty \sin^2 x/ x\,dx = \infty$.

5. The integral $\int_0^\infty e^{-x^2}\,dx$ is convergent, and in fact, $\int_0^\infty x^c\,e^{-x^2}\,dx$ is also convergent for all $c > 0$. Clearly, $e^{-x^2} < 1/|x|^{c+2}$ if $|x| > x_0$, so $|x^c\,e^{-x^2}| < 1/x^2$ if $|x| > x_0$.

Example 19.21. The integral $\int_0^\infty x^c \cdot e^{-x}\,dx$ is convergent if $c > -1$.

Let us inspect the integrals $\int_0^1$ and $\int_1^\infty$ separately. The integral $\int_1^\infty x^c\,e^{-x}\,dx$ is convergent for all c: we can prove this in the same way as in Example 19.20.5.

If $c \geq 0$, then $\int_0^1 x^c \cdot e^{-x}\,dx$ is an ordinary integral. If $-1 < c < 0$, then we can apply the majorization principle: since if $x \in (0,1]$, then $|x^c \cdot e^{-x}| \leq x^c$ and the integral $\int_0^1 x^c\,dx$ is convergent, we know that the integral $\int_0^1 x^c \cdot e^{-x}\,dx$ is also convergent.

The integral appearing in the example above (as a function of c) appears in applications so often that it gets its own notation. Leonhard Euler, who first worked with this integral, found it more useful to define this function on the interval $(0,\infty)$ instead of $(-1,\infty)$, so instead of notation for the integral in Example 19.21, he introduced it for the integral $\int_0^\infty x^{c-1} \cdot e^{-x}\,dx$, and this is still used today.

We use $\Gamma(c)$ to denote the value of the integral $\int_0^\infty x^{c-1} \cdot e^{-x}\,dx$ for all $c > 0$.

So for example, $\Gamma(1) = \int_0^\infty e^{-x}\,dx = 1$.

One can show that $\Gamma(c+1) = c \cdot \Gamma(c)$ for all $c > 0$. Once we know this, it easily follows that $\Gamma(n) = (n-1)!$ for every positive integer n (see Exercises 19.38 and 19.39). It is known that Γ is not an elementary function. To see more properties of the function Γ, see Exercises 19.41–19.46.

Example 19.20.3 can be greatly generalized. The following theorem gives us a convergence criterion according to which we could replace $1/x$ in the example by any monotone function that tends to zero at infinity, and we could replace $\sin x$ by any function whose integral function is bounded.

Theorem 19.22. *Suppose that*

(i) *the function f is monotone on the interval $[a,\infty)$ and $\lim_{x\to\infty} f(x) = 0$, and moreover,*

(ii) *the function g is integrable on $[a,b]$ for all $b > a$, and the integral function $G(x) = \int_a^x g(t)\,dt$ is bounded on $[a,\infty)$.*

Then the improper integral $\int_a^\infty f(x)g(x)\,dx$ is convergent.

Proof. Suppose that $|\int_a^x g(t)\,dt| \leq K$ for all $x \geq a$. Then $|\int_x^y g(t)\,dt| \leq 2K$ for all $a \leq x < y$, since $\int_x^y g(t)\,dt = \int_a^y g(t)\,dt - \int_a^x g(t)\,dt$. Let $\varepsilon > 0$ be given. Since $\lim_{x\to\infty} f(x) = 0$, there exists an x_0 such that $|f(x)| < \varepsilon$ for all $x \geq x_0$. If $x_0 \leq x < y$, then by the second mean value theorem of integration (Theorem 15.8), we get that for a suitable $\xi \in [x, y]$,

$$\left| \int_x^y f(t)g(t) \, dt \right| \leq |f(x)| \cdot \left| \int_x^\xi g(t) \, dt \right| + |f(y)| \cdot \left| \int_\xi^y g(t) \, dt \right| \leq$$

$$\leq \varepsilon \cdot 2K + \varepsilon \cdot 2K = 4K\varepsilon.$$

Since ε was arbitrary, we have shown that the improper integral $\int_a^\infty f(x)g(x) \, dx$ satisfies Cauchy's criterion, so it is convergent. □

Exercises

19.13. Are the following integrals convergent?

(a) $\displaystyle\int_0^\infty \frac{[x]!}{2^x + 3^x} \, dx;$

(b) $\displaystyle\int_{10}^\infty \frac{dx}{x \cdot \log\log x};$

(c) $\displaystyle\int_0^1 \log|\log x| \, dx;$

(d) $\displaystyle\int_0^1 |\log x|^{|\log x|} \, dx;$

(e) $\displaystyle\int_0^1 x^{\log x} \, dx;$

(f) $\displaystyle\int_0^\infty (\sqrt{x+1} - \sqrt{x}) \, dx;$

(g) $\displaystyle\int_0^1 \frac{dx}{\sin(\sin(\sin x))};$

(h) $\displaystyle\int_2^\infty \frac{x+1}{x^2\sqrt{x^2 - 1}} \, dx;$

(i) $\displaystyle\int_1^\infty \frac{x^x}{[x]!} \, dx;$

(j) $\displaystyle\int_5^\infty \frac{\sqrt[x]{2^x + 3^x}}{(x - \log x)\sqrt{x}} \, dx;$

(k) $\displaystyle\int_0^{\pi/2} \sqrt{\operatorname{tg} x} \, dx;$

(l) $\displaystyle\int_0^1 \frac{dx}{((\pi/2) - \arcsin x)^{3/2}}.$

19.14. Determine the values of A and B that make the integral

$$\int_1^\infty \left(\sqrt{1+x^2} - Ax - \frac{B}{x} \right) dx$$

convergent.

19.15. Prove that if $R = p/q$ is a rational function, $a < b$, $q(a) = 0$, and $p(a) \neq 0$, then the integral $\int_a^b R(x) \, dx$ is divergent.

19.16. Let p and q be polynomials, and suppose that q does not have a root in $[a, \infty)$. When is the integral $\int_a^\infty p(x)/q(x) \, dx$ convergent?

19.17. Let p and q be polynomials. When is the integral $\int_{-\infty}^\infty p(x)/q(x) \, dx$ convergent?

19.18. For which c are the following integrals convergent?

(a) $\int_0^{\frac{\pi}{2}} \frac{dx}{(\sin x)^c}$;

(b) $\int_0^1 \frac{1-\cos x}{x^c} dx$;

(c) $\int_0^\infty \frac{dx}{1+x^c}$;

(d) $\int_0^1 \frac{dx}{x^c - 1}$;

19.19. For which a, b is the integral $\int_0^{\pi/2} \sin^a x \cdot \cos^b x\, dx$ convergent?

19.20. (a) Prove that the improper integral $\int_0^\pi \log \sin x\, dx$ is convergent.
(b) Compute the value of the integral with the following method: use the substitution $x = 2t$, apply the identity $\sin 2t = 2\sin t \cos t$, then prove and use the fact that $\int_0^{\pi/2} \log \sin t\, dt = \int_0^{\pi/2} \log \cos t\, dt$. (S)

19.21. What does the geometric mean of the numbers

$$\sin \pi/n, \sin 2\pi/n, \ldots, \sin (n-1)\pi/n$$

tend to as $n \to \infty$? Check that the limit is smaller than the limit of the arithmetic means of the same numbers. (S)

19.22. Compute the values of the following integrals.

(a) $\int_0^{\pi/2} x \cdot \operatorname{ctg} x\, dx$;

(b) $\int_0^{\pi/2} \frac{x^2}{\sin^2 x} dx$.

19.23. Prove that $\int_0^\infty \sin x^2\, dx$ is convergent. (H)

19.24. For what values of $c \in \mathbb{R}$ are the following integrals convergent? (H)

(a) $\int_1^\infty \frac{\sin x}{x^c} dx$;

(b) $\int_1^\infty \frac{\cos x}{x^c} dx$.

19.25. Prove that

(a) $\frac{x}{\pi} \cdot \int_{-\infty}^\infty \frac{dt}{t^2 + x^2} = 1$ if $x > 0$.

(b) $\lim_{x\to 0} \frac{|x|}{\pi} \cdot \int_{-\infty}^\infty \frac{f(t)}{t^2+x^2} dt = f(0)$ if $f: \mathbb{R} \to \mathbb{R}$ is continuous and bounded.

19.26. Let f and g be positive and continuous functions on the half-line $[a, \infty)$ such that $\lim_{x\to\infty}(f/g) = 1$. Prove that the integrals $\int_a^\infty f\, dx$ and $\int_a^\infty g\, dx$ are either both convergent or both divergent.

19.27. Let f, g, and h be continuous functions in $[a, \infty)$, and suppose that $f(x) \le g(x) \le h(x)$ for all $x \ge a$. Prove that if the integrals $\int_a^\infty f\, dx$ and $\int_a^\infty h\, dx$ are convergent, then the integral $\int_0^\infty g\, dx$ is also convergent. (H)

19.28. Construct a continuous function $f: [0, \infty) \to \mathbb{R}$ such that the integral $\int_0^\infty f\, dx$ is absolutely convergent, but f does not tend to zero as $x \to \infty$. (S)

19.29. Prove that if f is uniformly continuous on the interval $[0, \infty)$ and the integral $\int_0^\infty f\, dx$ is convergent, then f tends to zero as $x \to \infty$. (S)

19.30. Construct a continuous function $f\colon [0,\infty) \to \mathbb{R}$ such that the integral $\int_0^\infty f\,dx$ is absolutely convergent, but the integral $\int_0^\infty f^2\,dx$ is divergent. (H)

19.31. Prove that if $f\colon [a,\infty) \to \mathbb{R}$ is monotone decreasing and the integral $\int_a^\infty f\,dx$ is convergent, then the integral $\int_0^\infty f^2\,dx$ is also convergent.

19.32. Let $f\colon [a,\infty) \to \mathbb{R}$ be a nonnegative and continuous function such that $\lim_{x\to\infty} \sqrt[x]{f(x)} = 1/2$. Prove that the integral $\int_a^\infty f\,dx$ is convergent.

19.33. (a) Prove that if $f\colon [a,\infty) \to \mathbb{R}$ is a nonnegative function such that the integral $\int_a^\infty f\,dx$ is convergent, then there exists a function $g\colon [a,\infty) \to \mathbb{R}$ such that $\lim_{x\to\infty} g(x) = \infty$ and the integral $\int_a^\infty g\cdot f\,dx$ is also convergent.

(b) Prove that if $f\colon [a,\infty) \to \mathbb{R}$ is a nonnegative function such that the integral $\int_a^\infty f\,dx$ exists but is divergent, then there exists a positive function $g\colon [a,\infty) \to \mathbb{R}$ such that $\lim_{x\to\infty} g(x) = 0$, and the integral $\int_a^\infty g\cdot f\,dx$ is also divergent. (H S)

19.34. Let $f\colon [a,\infty) \to \mathbb{R}$ be a nonnegative function such that the integral $\int_a^\infty f\,dx$ exists and is divergent. Is it true that in this case, the integral $\int_a^\infty f/(1+f)\,dx$ is also divergent?

19.35. Let $f\colon [3,\infty) \to \mathbb{R}$ be a nonnegative function such that $\int_3^\infty f\,dx$ is convergent. Prove that then the integral $\int_3^\infty f(x)^{1-1/\log x}\,dx$ is also convergent. ($*$ H S)

19.36. Prove that if $f\colon [a,\infty) \to \mathbb{R}$ is a decreasing nonnegative function, then the integrals $\int_a^\infty f(x)\,dx$ and $\int_a^\infty f(x)\cdot|\sin x|\,dx$ are either both convergent or both divergent.

19.37. Let f be a positive, monotone decreasing, convex, and differentiable function on $[a,\infty)$. Define the (unbounded) area of the region B_f lying under the graph of f as the integral $\int_a^\infty f(x)\,dx$, the volume of the set we obtain by rotating the region B_f about the x-axis by the integral $\pi \cdot \int_a^\infty f^2(x)\,dx$, and the surface area of the set we obtain by rotating the graph of f about the x-axis by the integral $2\pi \cdot \int_a^\infty f(x)\cdot\sqrt{1+(f'(x))^2}\,dx$. These three values can each be finite or infinite, which (theoretically) leads to eight results. Which of these outcomes can actually occur among functions with the given properties? (H S)

19.38. Prove that $\Gamma(c+1) = c\cdot\Gamma(c)$ for all $c > 0$. (H S)

19.39. Prove that $\Gamma(n) = (n-1)!$ for every positive integer n. (H)

19.40. Compute the integral $\int_0^1 \log^n x\,dx$ for every positive integer n.

19.41. Express the value of the integral $\int_0^\infty e^{-x^3}\,dx$ with the help of the function Γ. How can this result be generalized? (S)

19.42. Prove that

$$\int_0^1 (1-x)^n x^{c-1}\,dx = \frac{n!}{c(c+1)\cdots(c+n)} \tag{19.8}$$

for every $c > 0$ and every nonnegative integer n. (H)

19.43. Use the substitution $x = t/n$ in (19.8); then let n tend to infinity in the result. Prove that

$$\Gamma(c) = \lim_{n \to \infty} \frac{n^c n!}{c(c+1) \cdots (c+n)} \qquad (19.9)$$

for all $c > 0$. (∗ H S)

19.44. Apply (19.9) with $c = 1/2$. Using Wallis's formula, prove from this that $\Gamma(1/2) = \sqrt{\pi}$.

19.45. Show that $\int_0^\infty e^{-x^2} dx = \sqrt{\pi}/2$.

19.46. Prove that $\Gamma(2x) = 4^x/(2\sqrt{\pi})\Gamma(x)\Gamma(x+1/2)$ for all $x > 0$. (H)

19.3 Appendix: Proof of Theorem 19.13

Fix a point $x_0 \in (\alpha, \beta)$. We first show that

$$\int_{x_0}^\beta f(g(t)) \cdot g'(t) \, dt = \int_{g(x_0)}^\delta f(x) \, dx, \qquad (19.10)$$

in the sense that if one of the two improper integrals is convergent, then so is the other, and they are equal. We distinguish two cases.

I. First we suppose that the value δ is finite, and that g takes on the value δ in the interval $[x_0, \beta)$. Let (c_n) be a sequence such that $c_n \to \beta$ and $c_n < \beta$ for all n. Then $g(c_n) \to \delta$, and so it easily follows that the set H consisting of the numbers $g(x_0)$, δ, and $g(c_n)$ $(n = 1, 2, \ldots)$ has a greatest and a smallest element. (This is because if there is an element greater than $\max(g(x_0), \delta)$ in H, namely some $g(c_n)$, then there can be only finitely many elements greater than this, so there will be a greatest among these. This proves that H has a greatest element. We can argue similarly to show that H has a smallest element, too.) If $m = \min H$ and $M = \max H$, then by the continuity of g, we know that $[m, M]$ is a closed and bounded subinterval of the image of the function g. By the condition, f is integrable on $[m, M]$. Thus f is also integrable on the interval $[g(x_0), \delta]$ by Theorem 14.37. Let $\int_{g(x_0)}^\delta f(x) \, dx = I$. Since the integral function of f is continuous on $[m, M]$ by Theorem 15.5,

$$\lim_{n \to \infty} \int_{g(x_0)}^{g(c_n)} f(x) \, dx = I. \qquad (19.11)$$

Now applying the substitution formula (that is, Theorem 15.22) with the interval $[x_0, c_n]$, we obtain

$$\int_{x_0}^{c_n} f(g(t)) \cdot g'(t) \, dt = \int_{g(x_0)}^{g(c_n)} f(x) \, dx \qquad (19.12)$$

for all n, which we can combine with (19.11) to find that

$$\lim_{n \to \infty} \int_{x_0}^{c_n} f(g(t)) \cdot g'(t) \, dt = I. \tag{19.13}$$

Since this holds for every sequence $c_n \to \beta$, $c_n < \beta$,

$$\lim_{c \to \beta} \int_{x_0}^{c} f(g(t)) \cdot g'(t) \, dt = I, \tag{19.14}$$

which—by the definition of the improper integral—means that the integral on the left-hand side of (19.10) is convergent, and its value is I.

II. Now suppose that g does not take on the value δ in the interval $[x_0, \beta)$ (including the case $\delta = \infty$ or $\delta = -\infty$). Since g is continuous, the Bolzano–Darboux theorem implies that either $g(x) < \delta$ for all $x \in [x_0, \beta)$ (this includes the case $\delta = \infty$), or $g(x) > \delta$ for all $x \in [x_0, \beta)$ (this includes the case $\delta = -\infty$). We can assume that the first case holds (since the second one can be treated in the same way). Then $[g(x_0), \delta) \subset g([x_0, \beta))$.

Suppose that the right-hand side of (19.10) is convergent, and that its value is I. If $c_n \to \beta$ and $c_n < \beta$ for all n, then $g(c_n) \to \delta$ and $g(c_n) < \delta$ for all n. Thus (19.11) and (19.12) give us (19.13) once again. Since this holds for every sequence $c_n \to \beta$, $c_n < \beta$, we also know that (19.14) holds, that is, the integral on the left-hand side of (19.10) is convergent, and its value is I.

Now finally, let us suppose that the left-hand side of (19.10) is convergent and that its value is I. Let (d_n) be a sequence such that $d_n \to \delta$ and $g(x_0) \le d_n < \delta$ for all n. Since $[g(x_0), \delta) \subset g([x_0, \beta))$, there exist points $c_n \in [x_0, \beta)$ such that $d_n = g(c_n)$ $(n = 1, 2, \ldots)$. We now show that $c_n \to \beta$. Since $c_n < \beta$ for all n, it suffices to show that for all $b < \beta$, we have $c_n > b$ if n is sufficiently large.

We can assume that $x_0 \le b < \beta$. Since g is continuous, it must have a maximum value in $[x_0, b]$. If this is A, then $A < \delta$, since we assumed every value of g to be smaller than δ. By the condition $d_n \to \delta$, we know that $d_n > A$ if $n > n_0$. Since $d_n = g(c_n)$, if $n > n_0$, then $c_n > b$, since $c_n \le b$ would imply that $g(c_n) \le A$.

This shows that $c_n \to \beta$. This, in turn, implies that the left-hand side of (19.12) tends to I as $n \to \infty$. Since the right-hand side of (19.12) is $\int_{g(x_0)}^{d_n} f(x) \, dx$, we have shown that

$$\lim_{n \to \infty} \int_{g(x_0)}^{d_n} f(x) \, dx = I.$$

This holds for every sequence $d_n \to \delta$, $d_n < \delta$, so the integral on the right-hand side of (19.10) is convergent, and its value is I. This proves (19.10). The exact same argument shows that

$$\int_{\alpha}^{x_0} f(g(t)) \cdot g'(t) \, dt = \int_{\gamma}^{g(x_0)} f(x) \, dx \tag{19.15}$$

in the sense that if one of the two improper integrals exists, then so does the other, and they are equal.

Now we turn to the proof of the theorem. Suppose that $\gamma \neq \delta$. We can assume that $\gamma < \delta$, since the case $\gamma > \delta$ is similar. Then the interval (γ, δ) is part of the image of g, so there exists a point $x_0 \in (\alpha, \beta)$ such that $g(x_0) \in (\gamma, \delta)$. By the definition of the improper integral, it follows that the left-hand side of (19.7) is convergent if and only if the left-hand sides of (19.10) and (19.15) are convergent, and then it is equal to their sum. The same holds for the right-hand sides. Thus (19.10) and (19.15) immediately prove statement (i) of the theorem.

Now suppose that $\gamma = \delta$ and that the left-hand side of (19.7) is convergent. Then the left-hand sides of (19.10) and (19.15) are also convergent. Thus the corresponding right-hand sides are also convergent, and (19.10) and (19.15) both hold. Since the sum of the right-hand sides of (19.10) and (19.15) is $\int_\gamma^\delta f \, dx = 0$, the left-hand side of (19.7) is also zero.

Erratum:

Real Analysis

Foundations and Functions of One Variable

Fifth Edition

Miklós Laczkovich and Vera T. Sós

© Springer New York 2015
M. Laczkovich, V.T. Sós, *Real Analysis*, Undergraduate Texts in Mathematics,
DOI 10.1007/978-1-4939-2766-1

DOI 10.1007/978-1-4939-2766-1_20

The front matter was revised in the print and online versions of this book. On the title page "Fifth Edition" is incorrect—instead it should be the "First English Edition".

On the copyright page it should read as follows:

1st Hungarian edition: T. Sós, Vera, Analízis I/1 © Nemzeti Tankönyvkiadó, Budapest, 1972
2nd Hungarian edition: T. Sós, Vera, Analízis A/2 © Nemzeti Tankönyvkiadó, Budapest, 1976
3rd Hungarian edition: Laczkovich, Miklós & T. Sós, Vera: Analízis I © Nemzeti Tankönyvkiadó, Budapest, 2005
4th Hungarian edition: Laczkovich, Miklós & T. Sós, Vera: Analízis I © Typotex, Budapest, 2012
Translation from the Hungarian language 3rd edition: Valós analízis I by Miklós Laczkovich & T. Sós, Vera, © Nemzeti Tankönyvkiadó, Budapest, 2005. All rights reserved © Springer 2015.
© Springer New York 2015

The online version of the original book can be found at
http://dx.doi.org/10.1007/978-1-4939-2766-1

© Springer New York 2015
M. Laczkovich, V.T. Sós, *Real Analysis*, Undergraduate Texts
in Mathematics, DOI 10.1007/978-1-4939-2766-1_20

Hints, Solutions

Hints

Chapter 2

2.12 Draw the lines one by one. Every time we add a new line, we increase the number of regions by as many regions as the new line intersects. Show that this number is one greater than the previous number of lines.

2.16 Apply the inequality of arithmetic and geometric means with the numbers x, x, and $2 - 2x$.

2.24 Let $X = A_1 \cup \ldots \cup A_n$. Prove (using de Morgan's laws and (1.2)) that every expression $U(A_1, \ldots, A_n)$ can be reduced to the following form: $U_1 \cap U_2 \cap \ldots \cap U_N$, where for every i, $U_i = A_1^{\varepsilon_1} \cup \ldots \cup A_n^{\varepsilon_n}$. Here $\varepsilon_j = \pm 1$, $A_j^1 = A_j$, and $A_j^{-1} = X \setminus A_j$. Check that if the condition of the exercise holds for U and V of this form, then $U = V$.

Chapter 3

3.4 A finite set contains a largest element. If we add a positive number to this element, we get a contradiction.

3.10 Does the sequence of intervals $[n/x, 1/n]$ have a shared point?

3.16 Show that (a) if the number a is the smallest positive element of the set H, then $H = \{na : n \in \mathbb{Z}\}$; (b) if $H \neq \{0\}$ and H does not have a smallest positive element, then $H \cap (0, \delta) \neq \emptyset$ for all $\delta > 0$, and so H is everywhere dense.

3.25 Suppose that $H \neq \emptyset$ and H has a lower bound. Show that the least upper bound of the set of lower bounds of H is also the greatest lower bound of H.

© Springer New York 2015
M. Laczkovich, V.T. Sós, *Real Analysis*, Undergraduate Texts in Mathematics, DOI 10.1007/978-1-4939-2766-1

3.31 Since $b/a > 1$, there exists a rational number $n > 0$ such that $b/a = 1 + (1/n)$. Justify that $a = \left(1 + (1/n)\right)^n$ and $b = \left(1 + (1/n)\right)^{n+1}$. Let $n = p/q$, where p and q are relatively prime positive integers. Show that $\left((p+q)/p\right)^{p/q}$ can be rational only if $q = 1$.

3.33 Suppose first that b is rational. If $0 \le b \le 1$, then apply the inequality of arithmetic and geometric means. Reduce the case $b > 1$ to the case $0 < b < 1$. For irrational b, apply the definition of taking powers.

Chapter 4

4.2 Show that if the sequences (a_n) and (b_n) satisfy the recurrence, then for every $\lambda, \mu \in \mathbb{R}$, the sequence $(\lambda a_n + \mu b_n)$ does as well. Thus it suffices to show that if α is a root of the polynomial p, then the sequence (α^n) satisfies the recurrence.

4.3 (a) Let α and β be the roots of the polynomial $x^2 - x - 1$. According to the previous exercise, for every $\lambda, \mu \in \mathbb{R}$, the sequence $(\lambda \alpha^n + \mu \beta^n)$ also satisfies the recurrence. Choose λ and μ such that $\lambda \alpha^0 + \mu \beta^0 = 0$ and $\lambda \alpha^1 + \mu \beta^1 = 1$.

4.22 In order to construct a sequence oscillating at infinity, create a sequence that moves between 0 and 1 back and forth with each new step size getting closer to zero.

4.27 If the decimal expansion of $\sqrt{2}$ consisted of only a repeating digit from some point on, then $\sqrt{2}$ would be rational.

Chapter 5

5.4 Construct an infinite set $A \subset \mathbb{N}$ such that $A \cap \{kn : n \in \mathbb{N}\}$ is finite for all $k \in \mathbb{N}$, $k > 1$. Let $a_n = 1$ if $n \in A$ and $a_n = 0$ otherwise.

5.21 Show that the sequence $b_n = n \cdot \max\{a_i^k : 1 \le i, k \le n\}$ satisfies the conditions.

Chapter 6

6.3 Write the condition in the form $a_n - a_{n-1} \le a_{n+1} - a_n$. Show that the sequence is monotone from some point on.

6.4 (d) Separate the sequence into two monotone subsequences.

6.5 Show that $a_n \ge \sqrt{a}$ and $a_n \ge a_{n+1}$ for all $n \ge 2$.

6.7 Multiply the inequalities $1 + 1/k < e^{1/k} < 1 + 1/(k-1)$ for all $2 \leq k \leq n$.

6.9 Suppose that $(a_{n+1} - a_n)$ is monotone decreasing. Prove that $a_{2n} - a_n \geq n \cdot (a_{2n+1} - a_{2n})$ for all n.

6.17 The condition is that the finite or infinite limit $\lim_{n \to \infty} a_n = \alpha$ exists, $a_n \leq \alpha$ holds for all n, and if $a_n = \alpha$ for infinitely many n, then $a_n < \alpha$ can hold for only finitely many n.

6.22 A possible construction: let $a_n = \sqrt{k}$ if $2^{2^{k-1}} \leq n < 2^{2^k}$.

6.23 The statement is true. Use the same idea as the proof of the transference principle.

Chapter 7

7.2 (a) Give a closed form for the partial sums using the identity

$$\frac{1}{n^2 + 2n} = \frac{1}{2n} - \frac{1}{2(n+2)}.$$

A similar method can be used for the series (b), (c), and (d).

7.3 Break the rational function p/q into quotients of the form $c_i/(x - a_i)$. Show that here, $\sum_{i=1}^{k} c_i = 0$, and apply the idea used in part (c) of Exercise 7.2.

7.5 Use induction. To prove the inductive step, use the statement of Exercise 3.33.

7.8 Give the upper bound $N/10^{k-1}$ to the sum $\sum_{10^{k-1} \leq a_n < 10^k} 1/a_n$, where N denotes how many numbers there are with k digits that do not contain the digit 7.

7.10 It does not follow.

7.11 Yes, it follows. Prove that the given infinite series satisfies Cauchy's criterion.

Chapter 8

8.2 For all N, there are only finitely many sequences $(a_1, \ldots, a_k)$ such that $|a_1| + \cdots + |a_k| = N$.

8.4 Use the fact that every interval contains a rational point.

8.9 Every $x \in (0, 1]$ has a unique form $x = 2^{-a_1} + 2^{-a_2} + \cdots$, where $a_1 < a_2 < \cdots$ are natural numbers. Apply the bijections

$$x \leftrightarrow (a_1, a_2, \ldots) \leftrightarrow (a_1, a_2 - a_1, a_3 - a_2, \ldots).$$

8.11 It suffices to prove that the set of pairs (A, B) has the cardinality of the continuum, where $A, B \subset \mathbb{N}$. Find a map that maps these pairs to subsets of $\mathbb{N}$ bijectively.

Chapter 9

9.4 Such functions exist. Construct first such a function on the four-element set $a, b, -a, -b$ for all $0 < a < b$.

9.5 Let, for example, f_c be the identically 1 function for all c. A less trivial example: let $f_c(x) = x + c$ for all $c, x \in \mathbb{R}$.

The answer to the second question is no. If, for example, $g = f_{1/2}$, then $g \circ g = f_1$, and not every function f_1 has such a g. Show that if $f(1) = -1$, $f(-1) = 1$, and $f(x) = 0$ for all $x \neq \pm 1$, then there does not exist a function $g \colon \mathbb{R} \to \mathbb{R}$ such that $g \circ g = f$. (The question of which functions can be expressed in the form $g \circ g$ has been studied extensively. See, for example, the following paper: R. Isaacs, Iterates of fractional order, *Canad. J. Math.* **2** (1950), 409–416.)

9.6 (a) Let f_1 be constant and f_2 one-to-one. (c) The answer is positive. (d) The answer is positive.

Chapter 10

10.8 Show that the infimum of the set of positive periods is positive and also a period.

10.12 First show, using the statement of Exercise 3.17, that if x is irrational, then the set of numbers $\{nx\}$ is everywhere dense in $[0, 1]$.

10.15 Suppose that $\lim_{y \to x} f(y) = \infty$ for all x. Construct a sequence of nested intervals $[a_n, b_n]$ such that $f(x) > n$ for all $x \in [a_n, b_n]$.

10.16 Suppose that $\lim_{y \to x} f(y) = 0$ for all x. Construct a sequence of nested intervals $[a_n, b_n]$ such that $|f(x)| < 1/n$ for all $x \in [a_n, b_n]$.

10.20 Construct a set $A \subset \mathbb{R}$ that is not bounded from above, but for all $a > 0$, we have $n \cdot a \notin A$ if n is sufficiently large. (We can also achieve that for every $a > 0$, at most one $n \in \mathbb{N}^+$ exists such that $n \cdot a \in A$.) Let $f(x) = 1$ if $x \in A$, and let $f(x) = 0$ otherwise.

10.57 Show that if the leading coefficient of the polynomial p of degree three is positive, then $\lim_{x \to \infty} p(x) = \infty$ and $\lim_{x \to -\infty} p(x) = -\infty$. Therefore, p takes on both positive and negative values.

10.58 Apply the Bolzano–Darboux theorem to the function $f(x) - x$.

10.60 Show that if f is continuous and $f(f(x)) = -x$ for all x, then (i) f is one-to-one, (ii) f is strictly monotone, and so (iii) $f(f(x))$ is strictly monotone increasing.

10.71 (ii) First show that f and g are bounded in A; then apply the equality

$$f(y)g(y) - f(x)g(x) = f(y) \cdot (g(y) - g(x)) + g(x) \cdot (f(y) - f(x)).$$

10.76 Let $A = \{a_1, a_2, \ldots\}$. For an arbitrary real number x, let $f(x)$ be the sum of the numbers 2^{-n} for which $a_n < x$. Show that f satisfies the conditions.

10.80 Every continuous function satisfies the condition.

10.81 First show that if $p < q$ are rational numbers and n is a positive integer, then the set $\{a \in [-n, n]: \lim_{x \to a} f(x) < p < q < f(a)\}$ is countable.

10.82 Apply the ideas used in the solution of Exercise 10.17.

10.89 Not possible.

10.90 Not possible.

10.94 The function $g(x) = f(x) - f(1) \cdot x$ is additive, periodic with every rational number a period, and bounded from above on an interval. Show that $g(x) = 0$ for all x.

10.96 Apply the ideas used in the proof of Theorem 10.76.

10.102 By Exercise 10.101, it suffices to show that every point c of I has a neighborhood in which f is bounded. Let f be bounded from above on $[a, b] \subset I$. We can assume that $b < c$. Let

$$\alpha = \sup\{x \in I: x \geq a, \ f \text{ is bounded in } [a, x]\}.$$

Show (using the weak convexity of f) that $\alpha > c$.

Chapter 11

11.16 Use Exercise 6.7.

11.31 Use induction, using the identity

$$\cos(n+1)x + \cos(n-1)x = 2\cos nx \cdot \cos x.$$

11.35 Show that $f(0) = 1$. Prove, by induction, that if for some $a, c \in \mathbb{R}$ we have $f(a) = \cos(c \cdot a)$, then $f(na) = \cos(c \cdot na)$ for all n.

11.36 (a) Prove and use that $\sin^{-2}(k\pi/2n) + \sin^{-2}((n-k)\pi/2n) = 4\sin^{-2}(k\pi/n)$ for all $0 < k < n$. (b) Use induction. (c) Apply Theorem 11.26.

Chapter 12

12.9 There is no such point. Use the fact from number theory that for every irrational x there exist infinitely many rational p/q such that $|x - (p/q)| < 1/q^2$.

12.15 Show that $(f(y_n) - f(x_n))/(y_n - x_n)$ falls between the numbers

$$\min\left(\frac{f(x_n) - f(a)}{x_n - a}, \frac{f(y_n) - f(a)}{y_n - a} \right) \quad \text{and} \quad \max\left(\frac{f(x_n) - f(a)}{x_n - a}, \frac{f(y_n) - f(a)}{y_n - a} \right).$$

12.54 Suppose that $c < d$ and $f(c) > f(d)$. Let $\alpha = \sup\{x \in [c,d] : f(x) \ge f(c)\}$. Show that $\alpha < d$ and $\alpha = d$ both lead to a contradiction.

12.57 After subtracting a linear function, we can assume that $f'(c) = 0$. We can also suppose that $f''(c) > 0$. Thus f' is strictly locally increasing at c. Deduce that this means that f has a strict local minimum at c, and that for suitable $x_1 < c < x_2$, we have $f(x_1) = f(x_2)$.

12.65 Differentiate the function $(f(x) - \sin x)^2 + (g(x) - \cos x)^2$.

12.71 The statement holds for $k = n$. Prove, with the help of Rolle's theorem, that if $1 \le k \le n$ and the statement holds for k, then it holds for $k - 1$ as well.

12.76 Let $g(x) = (f(x+h) - f(x))/h$. Then

$$(f(a+2h) - 2f(a+h) + f(a))/h^2 = (g(a+h) - g(a))/h.$$

By the mean value theorem, there exists a $c \in (a, a+h)$ such that $(g(a+h) - g(a))/h = g'(c) = (f'(c+h) - f'(c))/h$. Apply Theorem 12.9 to f'.

12.81 We can suppose that $a > 1$. Prove that $a^x = x$ has a root if and only if the solution x_0 of $(a^x)' = a^x \cdot \log a = 1$ satisfies $a^{x_0} \le x_0$. Show that this last inequality is equivalent to the inequalities $1 \le x_0 \cdot \log a$ and $e \le a^{x_0} = 1/\log a$.

12.82 (a) Let $a < 1$. Show that the sequence (a_{2n+1}) is monotone increasing, the sequence (a_{2n}) is monotone decreasing, and both converge to the solution of the equation $a^x = x$. The case $a = 1$ is trivial. (b) Let $a > 1$. If the sequence is convergent, then its limit is the solution of the equation $a^x = x$. By the previous exercise, this has a solution if and only if $a \le e^{1/e}$. Show that in this case, the sequence converges (monotonically increasing) to the (smaller) solution of the equation $a^x = x$.

12.83 Apply inequality (12.32).

Chapter 13

13.2 Check that the statement follows for the polynomial x^n by the binomial theorem.

13.6 Show that the value of the equation given in Exercise 13.5 does not change if we write $-x$ in place of x. Use the fact that $\binom{n}{k} = \binom{n}{n-k}$.

13.9 Use Euler's formula (11.65) for $\cos x$.

Chapter 14

14.3 Show that $s_F(f) \leq S_F(g)$ for every partition F.

14.6 First show that the upper sums for partitions containing one base point form an interval. We can suppose that $f \geq 0$. Let $M = \sup\{f(t): t \in [a,b]\}$. Let $g(x)$ denote the upper sum corresponding to the partition $a < x < b$, and let $g(a) = g(b) = M(b-a)$. We have to see that if $a < c < b$ and $g(c) < y < M (b-a)$, then g takes on the value y. One of $M_1 = \sup\{f(t): t \in [a,c]\}$ and $M_2 = \sup\{f(t): t \in [c,b]\}$ is equal to M. By symmetry, we can assume that $M_1 = M$. If $c \leq x \leq b$, then the first term appearing in the upper sum $g(x)$, that is, $\sup\{f(t): t \in [a,x]\} \cdot (x-a) = M (x-a)$, is continuous. The second term, that is, $\sup\{f(t): t \in [x,b]\} \cdot (b-x)$, is monotone decreasing.

Thus in the interval $[c,b]$, g is the sum of a continuous and a monotone decreasing function. Moreover, $g(b) = M(b-a) = \max g$. Show that then $g([c,d])$ is an interval.

14.9 Using Exercise 14.8, construct nested intervals whose shared point is a point of continuity.

14.16 Use the equality

$$\sin \alpha + \sin 2\alpha + \cdots + \sin n\alpha = \frac{\sin(n\alpha/2)}{\sin(\alpha/2)} \cdot \sin\left((n+1)\alpha\right)/2. \qquad (1)$$

14.30 The statement is false. Find a counterexample in which f is the Dirichlet function.

14.38 Show that for every $\varepsilon > 0$, there are only finitely many points x such that $|f(x)| > \varepsilon$. Then apply the idea seen in Example 14.45.

14.42 We can assume that $c = 0$. Fix an $\varepsilon > 0$; then estimate the integrals $\int_0^\varepsilon f(tx)\,dx$ and $\int_\varepsilon^1 f(tx)\,dx$ separately.

14.44 Let $\max f = M$, and let $\varepsilon > 0$ be fixed. Show that there exists an interval $[c,d]$ on which $f(x) > M - \varepsilon$; then prove that $\sqrt[n]{\int_a^b f^n(x)\,dx} > M - 2\varepsilon$ if n is sufficiently large.

14.50 Using properties (iii), (iv), and (v), show that if $e(x) = 1$ for all $x \in (a,b)$ (the value of e can be arbitrary at the points a and b), then $\Phi(e;[a,b]) = b - a$. After this, with the help of properties (i) and (iii), show that $\Phi(f;[a,b]) = \int_a^b f(x)\,dx$ holds for every step function. Finally, use (iv) to complete the solution. (We do not need property (ii).)

14.51 Let $\alpha = \Phi(e; [0,1])$, where $e(x) = 1$ for all $x \in [0,1]$. Use (iii) and (vi) to show that if $b - a$ is rational and $e(x) = 1$ for all $x \in (a,b)$ (the value of e can be arbitrary at the points a and b), then $\Phi(e; [a,b]) = \alpha \cdot (b - a)$. Show the same for arbitrary $a < b$; then finish the solution in the same way as the previous exercise.

Chapter 15

15.3 Is it true that the function $\sqrt{x}$ is Lipschitz on $[0,1]$?

15.4 Apply the idea behind the proof of Theorem 15.1.

15.9 Not possible. See Exercise 14.9.

15.17 Compute the derivative of the right-hand side.

15.25 Use the substitution $y = \pi - x$.

15.32 Let $q = c \cdot r_1^{n_1} \cdots r_k^{n_k}$, where $r_1, \ldots, r_k$ are distinct polynomials of degree one or two. First of all, show that in the partial fraction decomposition of p/q, the numerator A of the elementary rational function $A/r_k^{n_k}$ is uniquely determined. Then show that the degree of the denominator of $(p/q) - (A/r_k^{n_k})$ is smaller than the degree of q, and apply induction.

15.37 By Exercise 15.36, it is enough to show that $\int_2^x dt/\log^{n+1} t = o(x/\log^n x)$. Use L'Hôpital's rule for this.

Chapter 16

16.13 Let the two cylinders be $\{(x,y,z) : y^2 + z^2 \leq R^2\}$ and $\{(x,y,z) : x^2 + z^2 \leq R^2\}$. Show that if their intersection is A, then the sections $A^z = \{(x,y) : (x,y,z) \in A\}$ are squares, and then apply Theorem 16.11.

16.27 Let N be the product of the first n primes. Choose a partition whose base points are the points i/N ($i = 0, \ldots, N$), as well as an irrational point between each of $(i-1)/N$ and i/N. Show that if $c \leq 2$, then the length of the corresponding inscribed polygonal path tends to infinity as $n \to \infty$. (We can use the fact that the sum of the reciprocals of the first n prime numbers tends to infinity as $n \to \infty$; see Corollary 18.16.) Thus if $c \leq 2$, then the graph is not rectifiable.

If $c > 2$, then the graph is rectifiable. To prove this, show that it suffices to consider the partitions F for which there exists an N such that the rational base points of F are exactly the points i/N ($i = 0, \ldots, N$). When finding bounds for the inscribed polygonal paths, we can use the fact that if $b > 0$, then the sums $\sum_{k=1}^{N} 1/k^{b+1}$ remain smaller than a value independent of N.

16.29 Let $F: a = t_0 < t_1 < \cdots < t_n = b$ be a fine partition, and let r_i be the smallest nonnegative number such that a disk D_i centered at $g(t_{i-1})$ with radius r_i covers the set $g([t_{i-1}, t_i])$. Show that the sum of the areas of the disks D_i is small.

16.31 (d) and (e) Show that there exist a point P and a line e such that the points of the curve are equidistant from P and e. Thus the curve is a parabola.

16.35 The graph of the function f agrees with the image of the curve if we know that $f(a \cdot \varphi \cdot \cos \varphi) = a \cdot \varphi \cdot \sin \varphi$ for all $0 \le \varphi \le \pi/4$. Check that the function $g(x) = a \cdot x \cdot \cos x$ is strictly monotone increasing on $[0, \pi/4]$. Thus $f(x) = a \cdot g^{-1}(x) \cdot \sin(g^{-1}(x))$ for all $x \in [0, a \cdot \pi \cdot \sqrt{2}/8]$. To compute the integrals $\int f \, dx$ and $\int f^2 \, dx$, use the substitution $x = g(t)$.

16.36 The graph of the function f agrees with the image of the curve if we know that $f(ax - a \sin x) = a - a \cos x$ for all $x \in [0, 2\pi]$. Check that the function $g(x) = ax - a \sin x$ is strictly monotone increasing on $[0, 2a\pi]$; then use the ideas from the previous question.

16.40 Apply the formula for the surface area of a segment of the sphere.

Chapter 17

17.2 Let $a = x_0 < x_1 < \cdots < x_n = b$ be a uniform partition of $[a, b]$ into n equal parts. Show that if $j \le k$, $c \in [x_{j-1}, x_j]$ and $d \in [x_{k-1}, x_k]$, then

$$\sum_{i=j}^{k} \omega(f; [x_{i-1}, x_i]) \cdot (x_i - x_{i-1}) \ge |f(d) - f(c)| \cdot (b-a)/n. \qquad (2)$$

17.3 Show that if $k \le (\sqrt{n})/2$, then the numbers $2/((2i+1)\pi)$ $(i = 1, \ldots, k)$ fall into different subintervals of the partition F_n. Deduce from this, using (2), that

$$\Omega_{F_n}(f) \ge \frac{1}{2} \cdot \sum_{i=1}^{[(\sqrt{n})/2]} \frac{2}{(2i+1)\pi} \cdot \frac{1}{n} \ge c \cdot \frac{\log n}{n}.$$

17.5 Suppose that the value $V_{F_0}(f)$ is the largest, where $F_0: a = c_0 < c_1 < \cdots < c_k = b$. Show that f is monotone on each interval $[c_{i-1}, c_i]$ $(i = 1, \ldots, k)$.

17.10 See Exercise 16.27.

17.14 Show that $f' \equiv 0$.

17.15 If $\alpha \ge \beta + 1$, then check that f' is bounded, and so f is Lipschitz. In the case $\alpha < \beta + 1$, show that if $0 \le x < y \le 1$, then

$$|f(x) - f(y)| \le |x^{\alpha} - y^{\alpha}| + y^{\alpha} \cdot \min\left(2, |x^{-\beta} - y^{-\beta}|\right).$$

Then prove that $y^\alpha - x^\alpha \le C \cdot (y - x)^\gamma$, and $y^\alpha \cdot \min\left(2, (x^{-\beta} - y^{-\beta})\right) \le C \cdot (y - x)^\gamma$ with a suitable constant C. In the proof of the second inequality, distinguish two cases based on whether $y - x$ is small or large compared to y.

17.16 If a function is Hölder α with some constant $\alpha \ge 1$, then f is Hölder 1, so Lipschitz, and so it has bounded variation. If $\alpha < 1$, then there exists a function that is Hölder α that does not have bounded variation. Look for the example in the form $x^n \cdot \sin x^{-n}$ where n is big.

17.19 To prove the "only if" statement, show that if f does not have bounded variation on $[a, b]$, then either there exists a point $a \le c < b$ such that f does not have bounded variation in any right-hand neighborhood of c, or there exists a point $a < c \le b$ such that f does not have bounded variation on any left-hand neighborhood of c. If, for example, f does not have bounded variation in any right-hand neighborhood of c, then look for a strictly decreasing sequence tending to c with the desired property.

Chapter 18

18.4 Let c be a shared point of discontinuity. Then there exists an $\varepsilon > 0$ such that no good δ exists for the continuity of f or g at c. Show that for every $\delta > 0$, there exists a partition with mesh smaller than δ such that the approximating sums formed with different inner points differ by at least ε^2. By symmetry, we can suppose that $c < b$ and that f is discontinuous from the right at c. Distinguish two cases based on whether g is discontinuous or continuous from the right at c.

18.5 Show that for all $\delta > 0$, there exists a partition $0 = x_0 < x_1 < \cdots < x_n = 1$ with mesh smaller than δ, and there exist inner points c_i and d_i such that

$$\sum_{i=1}^n f(c_i)\left(f(x_i) - f(x_{i-1})\right) > 1 \quad \text{and} \quad \sum_{i=1}^n f(d_i)\left(f(x_i) - f(x_{i-1})\right) < -1.$$

18.6 We give hints for two different proofs. (i) Choose a sequence of partitions F_n satisfying $\lim_{n\to\infty} \delta(F_n) = 0$, and for each n, fix the inner points. Show that the sequence of approximating sums $\sigma_{F_n}(f, g)$ is convergent. Let $\lim_{n\to\infty} \sigma_{F_n}(f, g) = I$. Show that $\int_a^b f \, dg$ exists and its value is I. (ii) Let A_n denote the set of approximating sums corresponding to partitions with mesh smaller than $1/n$ with arbitrary inner points, and let J_n be the smallest closed interval containing the set A_n. Show that $J_1 \supset J_2 \supset \ldots$, and as $n \to \infty$, the length of J_n tends to zero. Thus the intervals J_n have exactly one shared point. Let this shared point be I. Show that $\int_a^b f \, dg$ exists, and its value is I.

18.8 Use the statement of Exercise 18.4.

18.9 Use the statement of Exercise 17.19.

18.11 Suppose first that g is Lipschitz and monotone increasing, and apply the idea used in the proof of Theorem 18.10. Prove that every Lipschitz function can be expressed as the difference of two monotone increasing Lipschitz functions.

Chapter 19

19.7 By Cauchy's criterion, $\lim_{x \to \infty} \int_x^{2x} f \, dt = 0$.

19.23 First, with the help of Theorem 19.22 or by the method seen in Example 19.20.3, show that the integral $\int_1^\infty \sin x / \sqrt{x} \, dx$ is convergent; then apply the substitution $x^2 = t$.

19.24 The integrals are convergent for all $c > 0$. We can prove this with the help of Theorem 19.22 or by the method seen in Example 19.20.3. If $c \le 0$, then the integrals are divergent. Show that in this case, Cauchy's criterion is not satisfied.

19.27 Use Cauchy's criterion.

19.30 Apply the construction in the solution of Exercise 19.28, with the change of choosing the values $f_n(a_n)$ to be large.

19.33 In both exercises, choose the function g to be piecewise constant; that is, constant on the intervals (a_{n-1}, a_n), where (a_n) is a suitable sequence that tends to infinity.

19.35 Let $g(x) = f(x)^{1-1/\log x}$. Show that if at a point x, we have $f(x) < 1/x^2$, then $g(x) \le c/x^2$; if $f(x) \ge 1/x^2$, then $g(x) \le c \cdot f(x)$. Then use the majorization principle.

19.37 Only three cases are possible.

19.38 Use integration by parts.

19.39 Use induction with the help of the previous question.

19.42 Prove the statement by induction on n. (Let the nth statement be that (19.8) holds for every $c > 0$.) Prove the induction step with the help of integration by parts.

19.43 Use the fact that $(1 - t/n)^n \le e^{-t}$ for every $0 < t \le n$, and show from this that $\Gamma(c) > n^c \cdot n! / (c(c+1) \cdots (c+n))$.

Show that $e^t \cdot (1 - t/n)^n$ is monotone decreasing on $[0, n]$, and deduce that

$$e^{-t} \le (1 + \varepsilon) \cdot \left(1 - \frac{t}{n}\right)^n$$

for all $t \in [0, n]$ if n is sufficiently large. Show from this that for sufficiently large n, $\Gamma(c) < \varepsilon + (1 + \varepsilon) \cdot n^c \cdot n! / (c(c+1) \cdots (c+n))$.

19.46 Apply (19.9) and Wallis formula.

Solutions

Chapter 2

2.3 Let $\mathcal{H}_n$ denote the system of sets $H \subset \{1,\ldots,n\}$ satisfying the condition, and let a_n be the number of elements in $\mathcal{H}_n$. It is easy to check that $a_1 = 2$ and $a_2 = 3$ (the empty set works, too). If $n > 2$, then we can show that for every $H \subset \{1,\ldots,n\}$,

$$H \in \mathcal{H}_n \iff [(n \notin H) \wedge (H \cap \{1,\ldots,n-1\} \in \mathcal{H}_{n-1})] \vee$$
$$\vee [(n \in H) \wedge (H \cap \{1,\ldots,n-2\} \in \mathcal{H}_{n-2}) \wedge (n-1 \notin H)].$$

It is then clear that for $n > 2$, we have $a_n = a_{n-1} + a_{n-2}$. Thus the sequence of numbers a_n is $2, 3, 5, 8, \ldots$. This is called the Fibonacci sequence, which has an explicit formula (see Exercise 4.3).

2.11 The inductive step does not work when $n = 3$; we cannot state that $P = Q$ here.

2.14 Let $a \geq -1$. The inequality $(1+a)^n \geq 1+na$ is clearly true if $1+na < 0$, so we can assume that $1+na \geq 0$. Apply the inequality of the arithmetic and geometric means consisting of the n numbers $1+na, 1, \ldots, 1$. We get that

$$\sqrt[n]{1+na} \leq ((1+na)+n-1)/n = 1+a.$$

If we raise both sides to the nth power, then we get the inequality we want to prove.

Chapter 3

3.27 If the set $\mathbb{N}$ were bounded from above, then it would have a least upper bound. Let this be a. Then $n \leq a$ for all $n \in \mathbb{N}$. Since if $n \in \mathbb{N}$, then $n+1 \in \mathbb{N}$, we must have $n+1 \leq a$, that is, $n \leq a-1$ for all $n \in \mathbb{N}$. This, however, is impossible, since then $a-1$ would also be an upper bound. This shows that $\mathbb{N}$ is not bounded from above, so the axiom of Archimedes is satisfied.

Let $[a_n, b_n]$ be nested closed intervals. The set $A = \{a_n : n \in \mathbb{N}\}$ is bounded from above, because each b_n is an upper bound. If $\sup A = c$, then $a_n \leq c$ for all n. Since b_n is also an upper bound of A, $c \leq b_n$ for all n. Thus $c \in [a_n, b_n]$ for every n, which is Cantor's axiom.

3.32 If $b/a = c$ then $c \geq 1$. By Theorems 3.23 and 3.24, we have that $b^r/a^r = c^r \geq c^0 = 1$.

3.33 First let b be rational and $0 \le b \le 1$. Then $b = p/q$, where $q > 0$ and $0 \le p \le q$ are integers. By the inequality of arithmetic and geometric means,

$$(1+x)^b = \sqrt[q]{(1+x)^p} = \sqrt[q]{(1+x)^p \cdot 1^{q-p}} \le \frac{p(1+x) + q - p}{q} = 1 + bx. \qquad (3)$$

Now let $b > 1$ be rational. If $1 + bx \le 0$, then $(1+x)^b \ge 1 + bx$ holds. Thus we can suppose that $bx > -1$. Since $0 < 1/b < 1$, we can apply (3) to bx instead of x, and $1/b$ instead of b, to get that

$$(1+bx)^{1/b} \le 1 + (1/b) \cdot bx = 1 + x.$$

Then applying Exercise 3.32, we get that $1 + bx \le (1+x)^b$. Thus we have proved the statement for a rational exponent b.

In the proof of the general case, we can assume that $x \ne 0$ and $b \ne 0, 1$, since the statement is clear when $x = 0$ or $b = 0, 1$. Let $x > 0$ and $0 < b < 1$. If $b < r < 1$ is rational, then by Theorem 3.27, $(1+x)^b \le (1+x)^r$, and by (3), $(1+x)^r \le 1 + rx$. Thus $(1+x)^b \le 1 + rx$ for all rational $b < r < 1$. It already follows from this that $(1+x)^b \le 1 + bx$. Indeed, if $(1+x)^b > 1 + bx$, then we can chose a rational number $b < r < 1$ such that $(1+x)^b > 1 + rx$, which is impossible.

Now let $-1 < x < 0$ and $0 < b < 1$. If $0 < r < b$ is rational, then by Theorem 3.27, $(1+x)^b < (1+x)^r$, and by (3), $(1+x)^r \le 1 + rx$. Thus $(1+x)^b \le 1 + rx$ for all rational numbers $0 < r < b$. It then follows that $(1+x)^b \le 1 + bx$, since if $(1+x)^b > 1 + bx$, then we can chose a rational number $0 < r < b$ such that $(1+x)^b > 1 + rx$, which is impossible.

We can argue similarly when $b > 1$.

Chapter 4

4.1 $a_n = 1 + (-1)^n \cdot 2^{2-n}$.

4.3 (a) The roots of the polynomial $x^2 - x - 1$ are $\alpha = (1 + \sqrt{5})/2$ and $\beta = (1 - \sqrt{5})/2$. In the sense of the previous question, for every $\lambda, \mu \in \mathbb{R}$, the sequence $(\lambda \alpha^n + \mu \beta^n)$ also satisfies the recurrence. Choose λ and μ such that $\lambda + \mu = \lambda \alpha^0 + \mu \beta^0 = 0 = u_0$ and $\lambda \alpha^1 + \mu \beta^1 = 1 = u_1$ hold. It is easy to see that the choice $\lambda = 1/\sqrt{5}, \beta = -1/\sqrt{5}$ works. Thus the sequence

$$v_n = \frac{1}{\sqrt{5}} \left(\left(\frac{1+\sqrt{5}}{2} \right)^n - \left(\frac{1-\sqrt{5}}{2} \right)^n \right) \qquad (4)$$

satisfies the recurrence, and $v_0 = u_0$, $v_1 = u_1$. Then by induction, it follows that $v_n = u_n$ for all n, that is, u_n is equal to the right-hand side of (4) for all n.

4.13 For a given $\varepsilon > 0$, there exists an N such that if $n \geq N$, then $|a_n - a| < \varepsilon$. Let $|a_1 - a| + \cdots + |a_N - a| = K$. If $n \geq N$, then

$$|s_n - a| = \left| \frac{(a_1 - a) + \cdots + (a_n - a)}{n} \right| \leq \frac{|a_1 - a| + \cdots + |a_n - a|}{n} \leq \frac{K + n\varepsilon}{n} < 2\varepsilon,$$

given that $n > K/\varepsilon$. Thus $s_n \to a$. It is clear that the sequence $a_n = (-1)^n$ satisfies $s_n \to 0$.

Chapter 5

5.9 Let $\max_{1 \leq i \leq k} a_i = a$. It is clear that

$$a = \sqrt[n]{a^n} \leq \sqrt[n]{a_1^n + \cdots + a_k^n} \leq \sqrt[n]{k \cdot a^n} = \sqrt[n]{k} \cdot a \to a,$$

and so the statement follows by the squeeze theorem.

Chapter 6

6.15 If the set $\mathbb{N}$ were bounded from above, then the sequence $a_n = n$ would also be bounded, so by the Bolzano–Weierstrass theorem, it would have a convergent subsequence. This, however, is impossible, since the distance between any two terms of this sequence is at least 1. This shows that $\mathbb{N}$ is not bounded from above, so the axiom of Archimedes holds.

Let $[a_n, b_n]$ be nested closed intervals. The sequence (a_n) is bounded, since a_1 is a lower bound and every b_i is an upper bound for it. By the Bolzano–Weierstrass theorem, we can choose a convergent subsequence a_{n_k}. If $a_{n_k} \to c$, then $c \leq b_i$ for all i, since $a_{n_k} \leq b_i$ for all i and k. On the other hand, if $n_k \geq i$, then $a_{n_k} \geq a_i$, so $c \geq a_i$. Thus $c \in [a_i, b_i]$ for all i, so Cantor's axiom holds.

6.21 Consider a sequence $a_k \to \infty$ such that $a_{k+1} - a_k \to 0$. Repeating the terms of this sequence enough times gives us a suitable sequence. For example, starting from the sequence $a_k = \sqrt{k}$: (a) Let $a_n = \sqrt{k}$ if $2^{k-1} \leq n < 2^k$. (d) Let (t_k) be a strictly monotone increasing sequence of positive integers such that

$$t_{k+1} > t_k + \max_{n < t_k} s_n$$

for all k, and let $a_n = \sqrt{k}$ if $t_{k-1} \leq n < t_k$. Then (a_n) is monotone increasing and tends to infinity. If $t_{k-1} \leq n < t_k$, then $n + s_n < t_{k+1}$, so $a_{s_n} - a_n \leq \sqrt{k+1} - \sqrt{k}$, which implies $a_{s_n} - a_n \to 0$.

Chapter 7

7.2

(a) Since $1/(n^2+2n) = (1/2) \cdot (1/n - 1/(n+2))$, we have that

$$\sum_{n=1}^{N} \frac{1}{n^2+2n} = \frac{1}{2} \cdot \sum_{n=1}^{N} \left(\frac{1}{n} - \frac{1}{n+2} \right) = \frac{1}{2} \cdot \left(1 + \frac{1}{2} - \frac{1}{N+1} - \frac{1}{N+2} \right).$$

Thus the partial sums of the series tend to $3/4$, so the series is convergent with sum $3/4$.

(b) If we leave out the first term in the series in (a), then we get the series in (b). Thus the partial sums of this new series tend to $(3/4) - (1/3) = 5/12$, so it is convergent with sum $5/12$.

(c) Since $1/(n^3 - n) = (1/2) \cdot (1/(n-1) - 2/n + 1/(n+1))$, we have that

$$\sum_{n=2}^{N} \frac{1}{n^3 - n} = \frac{1}{2} \cdot \sum_{n=2}^{N} \left(\frac{1}{n-1} - \frac{2}{n} + \frac{1}{n+1} \right) = \frac{1}{2} \cdot \left(1 - \frac{1}{2} - \frac{1}{N} + \frac{1}{N+1} \right).$$

Thus the partial sums of the series tend to $1/4$, so the series is convergent with sum $1/4$.

7.5 We prove this by induction. The statement holds for $n = 1$. To prove the inductive step, we need to show that if $n \geq 1$, then

$$\frac{1}{(n+1)^{b+1}} \leq \left(1 + \frac{1}{b} - \frac{1}{b \cdot (n+1)^b} \right) - \left(1 + \frac{1}{b} - \frac{1}{b \cdot n^b} \right).$$

After multiplying this through by $b \cdot (n+1)^b$ and rearranging, this takes the form

$$1 + \frac{b}{n+1} \leq \left(1 + \frac{1}{n} \right)^b. \tag{5}$$

If $b \geq 1$, then (5) is clear from the Bernoulli inequality for real powers (the first statement of Exercise 3.33).

If $0 < b < 1$, then take the reciprocal of both sides of (5), then raise them to the power $1/b$. After some rearrangement, (5) becomes the inequality

$$\left(1 - \frac{b}{n+1+b} \right)^{1/b} \geq 1 - \frac{1}{n+1}. \tag{6}$$

Since $1/b > 1$, (6) again follows from the first statement of Exercise 3.33.

7.14 Let $\lim_{n\to\infty} s_n = A$, where s_n is the nth partial sum of the series. Since

$$a_1 + 2a_2 + \cdots + na_n = (a_1 + \cdots + a_n) + (a_2 + \cdots + a_n) + \cdots + (a_n) =$$
$$= s_n + (s_n - s_1) + \cdots + (s_n - s_{n-1}) =$$
$$= (n+1)s_n - (s_1 + \cdots + s_n),$$

we have that

$$\frac{a_1 + 2a_2 + \cdots + na_n}{n} = \frac{n+1}{n} s_n - \frac{s_1 + \cdots + s_n}{n}. \tag{7}$$

Since $s_n \to A$, we have $(s_1 + \cdots + s_n)/n \to A$ (see Exercise 4.13), and so the right-hand side of (7) tends to zero.

Chapter 9

9.6 (b), (c), (d): see the following paper: W. Sierpiński, Sur les suites infinies de fonctions définies dans les ensembles quelconques, *Fund. Math.* **24** (1935), 09–212. (See also: W. Sierpiński: *Oeuvres Choisies* (Warsaw 1976) volume III, 255–258.)

9.18 Suppose that the graph of f intersects the line $y = ax + b$ at more than two points. Then there exist numbers $x_1 < x_2 < x_3$ such that $f(x_i) = ax_i + b$ ($i = 1, 2, 3$). By the strict convexity of f, $f(x_2) < h_{x_1, x_3}(x_2)$. It is easy to see that both sides are equal to $ax_2 + b$ there, which is a contradiction.

Chapter 10

10.17 Let $\{r_n\}$ be an enumeration of the rational numbers. Let $r \in \mathbb{Q}$ and $\varepsilon = 1/2$. Since f is continuous in r, there exists a closed and bounded interval I_1 such that $\sup\{f(x) \colon x \in I_1\} < \inf\{f(x) \colon x \in I_1\} + 1$. We can assume that $r_1 \notin I_1$, since otherwise, we can take a suitable subinterval of I_1. Suppose that $n > 1$ and we have already chosen the interval I_{n-1}. Choose an arbitrary rational number r from the interior of I_{n-1}. Since f is continuous in r, there exists a closed and bounded interval $I_n \subset I_{n-1}$ such that $\sup\{f(x) \colon x \in I_n\} < \inf\{f(x) \colon x \in I_n\} + 1/n$. We can assume that $r_n \notin I_n$, since otherwise, we can choose a suitable subinterval of I_n. We can also assume that I_n is in the interior of I_{n-1} (that is, they do not share an endpoint).

Thus we have defined the nested closed intervals I_n for each n. Let $x_0 \in \bigcap_{n=1}^{\infty} I_n$. Then x_0 is irrational, since $x_0 \neq r_n$ for all n. Let $u_n = \inf\{f(x) \colon x \in I_n\}$ and $v_n = \sup\{f(x) \colon x \in I_n\}$. Clearly, $u_n \leq f(x_0) \leq v_n$ and $v_n - u_n < 1/n$ for all n. Let $\varepsilon > 0$ be fixed. If $n > 1/\varepsilon$, then $f(x_0) - \varepsilon < u_n \leq v_n < f(x_0) + \varepsilon$, from which it is clear that $|f(x) - f(x_0)| < \varepsilon$ for all $x \in I_n$. Since $x_0 \in I_{n+1}$ and I_{n+1} is in the interior of I_n, there exists a $\delta > 0$ such that $(x_0 - \delta, x + \delta) \subset I_n$. Thus $|f(x) - f(x_0)| < \varepsilon$ whenever $|x - x_0| < \delta$, which proves that f is continuous at x_0.

10.21   See the following paper: H.T. Croft, A question of limits, *Eureca* **20** (1957), 11–13. For the history and a generalization of the problem, see L. Fehér, M. Laczkovich, and G. Tardos, Croftian sequences, *Acta Math. Hung.* **56** (1990), 353–359.

10.40   Let $f(0) = 1$ and $f(x) = 0$ for all $x \neq 0$. Moreover, $g(x) = 0$ for all x. Then $\lim_{x \to 0} f(x) = \lim_{x \to 0} g(x) = 0$. On the other hand, $f(g(x)) = 1$ for all x.

10.61   If I is degenerate, then so is $f(I)$. If I is not degenerate, then let $\alpha = \inf f(I)$ and $\beta = \sup f(I)$. We show that $(\alpha, \beta) \subset f(I)$. If $\alpha < c < \beta$, then for suitable $a, b \in I$, we have $\alpha < f(a) < c < f(b) < \beta$. Since f is continuous on $[a, b]$, by Theorem 10.57 f attains the value c over $[a, b]$, that is, $c \in f(I)$. This shows that $(\alpha, \beta) \subset f(I)$.

If $\alpha = -\infty$ and $\beta = \infty$, then by the above, $f(I) = \mathbb{R}$, so the statement holds. If $\alpha \in \mathbb{R}$ and $\beta = \infty$, then $(\alpha, \infty) \subset f(I) \subset [\alpha, \infty)$, so $f(I)$ is one of the intervals (α, ∞), $[\alpha, \infty)$, so the statement holds again. We can argue similarly if $\alpha = -\infty$ and $\beta \in \mathbb{R}$. Finally, if $\alpha, \beta \in \mathbb{R}$, then by $(\alpha, \beta) \subset f(I) \subset [\alpha, \beta]$, we know that $f(I)$ can only be one of the following sets: (α, β), $[\alpha, \beta)$, $(\alpha, \beta]$, $[\alpha, \beta]$. Thus $f(I)$ is an interval.

10.88   Such is the function $-x$, for example.

10.91   It is easy to see by induction on k that

$$f\left(\frac{x_1 + \cdots + x_{2^k}}{2^k}\right) \le \frac{f(x_1) + \cdots + f(x_{2^k})}{2^k} \tag{8}$$

for all (not necessarily distinct) numbers $x_1, \ldots, x_{2^k} \in I$. If $x_1, \ldots, x_n \in I$ are fixed numbers, then let $s = (x_1 + \cdots + x_n)/n$ and $x_i = s$ for all $n < i \le 2^n$. By (8), we have

$$f(s) \le \frac{f(x_1) + \cdots + f(x_n) + (2^n - n) \cdot f(s)}{2^n},$$

and so

$$f(s) \le \frac{f(x_1) + \cdots + f(x_n)}{n}.$$

10.99   Suppose that $f(x) \le K$ for all $|x - x_0| < \delta$. If $|h| < \delta$, then

$$f(x_0) \le \frac{f(x_0 - h) + f(x_0 + h)}{2} \le \frac{K + f(x_0 + h)}{2},$$

so $f(x_0 + h) \ge 2f(x_0) - K$. This means that $2f(x_0) - K$ is a lower bound of the function f over $(x_0 - \delta, x_0 + \delta)$. Thus f is also bounded from below, and so it is bounded in $(x_0 - \delta, x_0 + \delta)$.

10.100 If we apply inequality (10.24) with the choices $a = x$ and $b = x + 2^k h$, then we get that

$$f\left(x + 2^{k-1}h\right) \le \frac{1}{2}\left[f(x) + f\left(x + 2^k h\right)\right].$$

Dividing both sides by 2^{k-1} then rearranging yields us the inequality

$$\frac{1}{2^{k-1}}f\left(x + 2^{k-1}h\right) - \frac{1}{2^k}f\left(x + 2^k h\right) \le \frac{1}{2^k}f(x).$$

If we take the sum of these inequalities for $k = 1, \ldots, n$, then the inner terms cancel out on the left-hand side, and we get that

$$f(x+h) - \frac{1}{2^n}f(x+2^n h) \le \left(\frac{1}{2} + \frac{1}{4} + \cdots + \frac{1}{2^n}\right)f(x) = \left(1 - \frac{1}{2^n}\right)f(x).$$

A further rearrangement give us the inequality

$$f(x+h) - f(x) \le \frac{1}{2^n} \cdot [f(x + 2^n h) - f(x)]. \tag{9}$$

10.101 By Exercise 10.99, f is bounded on the interval $J = (x_0 - \delta, x_0 + \delta)$. Let $|f(x)| \le M$ for all $x \in J$. Let $\varepsilon > 0$ be fixed, and choose a positive integer n such that $2^n > 2M/\varepsilon$ holds. If $|t| < \delta/2^n$, then $x_0 + 2^n t \in J$, so $|f(x_0 + 2^n t)| \le M$. Thus by (9), we have

$$f(x_0 + t) - f(x_0) \le \frac{1}{2^n} \cdot [f(x_0 + 2^n t) - f(x_0)] \le \frac{1}{2^n} \cdot 2M < \varepsilon.$$

If, however, we apply (9) with the choices $x = x_0 + t$ and $h = -t$, then we get that

$$f(x_0) - f(x_0 + t) \le \frac{1}{2^n} \cdot [f(x_0 - (2^n - 1)t) - f(x_0 + t)] \le \frac{1}{2^n} \cdot 2M < \varepsilon.$$

Finally, we conclude that $|f(x_0 + t) - f(x_0)| < \varepsilon$ for all $|t| < \delta/2^n$, which proves that f is continuous at x_0.

Chapter 11

11.36

(a) First of all, we show that

$$\sin^{-2}(k\pi/2m) + \sin^{-2}((m-k)\pi/2m) = 4\sin^{-2}(k\pi/m) \tag{10}$$

for all $0 < k < m$. This is because $\sin^{-2}((m-k)\pi/2m) = \cos^{-2}(k\pi/2m)$, and so the left-hand side of (10) is

$$\sin^{-2}(k\pi/2m) + \cos^{-2}(k\pi/2m) = \frac{\cos^2(k\pi/2m) + \sin^2(k\pi/2m)}{\sin^2(k\pi/2m) \cdot \cos^2(k\pi/2m)} =$$

$$= \frac{1}{\sin^2(k\pi/2m) \cdot \cos^2(k\pi/2m)} = \frac{4}{\sin^2(2 \cdot (k\pi/2m))} = 4 \cdot \sin^{-2}(k\pi/m).$$

Applying the equality (10) with $m = 2n$, we get that

$$\sin^{-2}(k\pi/4n) + \sin^{-2}((2n-k)\pi/4n) = 4\sin^{-2}(k\pi/2n) \qquad (11)$$

for all $0 < k < 2n$. If we now sum the equalities (11) for $k = 1, \ldots, n$, then on the left-hand side, we get every term of the sum defining A_{2n}, except that $\sin^{-2}(n\pi/4n) = 2$ appears twice, and $\sin^{-2}(2n\pi/4n) = 1$ is missing. Thus the sum of the left-hand sides is $A_{2n} + 1$. Since the sum of the right-hand sides is $4A_n$, we get that $A_{2n} = 4A_n - 1$.

(b) The statement is clear for $n = 0$, and follows for $n > 0$ by induction.

(c) The inequality is clear by Theorem 11.26. Applying this for $x = k\pi/2n$ and then summing the inequality we get for $k = 1, \ldots, n$ gives us the desired bound.

(d) By the above, the partial sum $S_{2^n} = \sum_{k=1}^{2^n}(1/k^2)$ satisfies

$$\left(\frac{\pi}{2 \cdot 2^n}\right)^2 \cdot (A_{2^n} - 2^n) \leq S_{2^n} \leq \left(\frac{\pi}{2 \cdot 2^n}\right)^2 \cdot A_{2^n},$$

and so

$$\frac{\pi^2}{6} - \frac{\pi^2 2^n}{4^{n+1}} \leq S_{2^n} \leq \frac{\pi^2}{6} + \frac{\pi^2}{3 \cdot 4^{n+1}}.$$

Then by the squeeze theorem, $S_{2^n} \to \pi^2/6$. Since the series is convergent by Example 7.11.1, we have $S_n \to \pi^2/6$, that is, $\sum_{n=1}^{\infty} 1/k^2 = \pi^2/6$.

Chapter 12

12.15 Let

$$m_n = \min\left(\frac{f(x_n) - f(a)}{x_n - a}, \frac{f(y_n) - f(a)}{y_n - a}\right)$$

and

$$M_n = \max\left(\frac{f(x_n) - f(a)}{x_n - a}, \frac{f(y_n) - f(a)}{y_n - a}\right)$$

for all n. It is clear that $m_n \to f'(a)$ and $M_n \to f'(a)$ if $n \to \infty$.
Let $p_n = (a - x_n)/(y_n - x_n)$ and $q_n = (y_n - a)/(y_n - x_n)$. Then $p_n, q_n > 0$ and $p_n + q_n = 1$. Since

$$\frac{f(y_n) - f(x_n)}{y_n - x_n} = p_n \cdot \frac{f(a) - f(x_n)}{a - x_n} + q_n \cdot \frac{f(y_n) - f(a)}{y_n - a},$$

we have $m_n \leq (f(y_n) - f(x_n))/(y_n - x_n) \leq M_n$. Then the statement follows by the squeeze theorem.

12.19 See Example 13.43.

12.23 Let (x_0, y_0) be a common point of the two graphs. Then $\sqrt{4a(a-x_0)} = \sqrt{4b(b+x_0)}$, so $a(a-x_0) = b(b+x_0)$ and $x_0 = a - b$. The slopes of the two graphs at the point (x_0, y_0) are

$$m_1 = -2a/\sqrt{4a(a-x_0)} \text{ and } m_2 = 2b/\sqrt{4b(b+x_0)}.$$

Thus

$$m_1 \cdot m_2 = \frac{-4ab}{\sqrt{4a(a-x_0)} \cdot \sqrt{4b(b+x_0)}} = \frac{-4ab}{\left(\sqrt{4a(a-x_0)}\right)^2} = \frac{-b}{a-x_0} = -1.$$

It is well known (and easy to show) that if the product of the slopes of two lines is -1, then the two lines are perpendicular.

12.25 Since $\pi - e < 1$, if $x < 0$, then $2^x < 1 < (\pi - e)^x$, and if $x > 0$, then $2^x > 1 > (\pi - e)^x$. Thus the only point of intersection of the two graphs is $(0, 1)$. At this point, the slopes of the tangent lines are $\log 2$ and $\log(\pi - e)$. This means that the angle between the tangent line of 2^x at $(0, 1)$ and the x-axis is $\operatorname{arc tg}(\log 2)$, while the angle between the tangent line of $(\pi - e)^x$ at $(0, 1)$ and the x-axis is $\operatorname{arc tg}(\log(\pi - e))$. Thus the angle between the two tangent lines is $\operatorname{arc tg}(\log 2) - \operatorname{arc tg}(\log(\pi - e))$.

12.44 If $y = \log_a x$, then $y' = 1/(x \cdot \log a)$, so xy' is constant, $(xy')' = 0$, $y' + xy'' = 0$. Thus $-1 = (-x)' = (y'/y'')' = (y''^2 - y'y''')/y''^2$, so $y'y''' - 2(y'')^2 = 0$.

12.48 Let $a \leq b$. The volume of the box is

$$K(x) = (a - 2x)(b - 2x)x = 4x^3 - 2(a+b)x^2 + abx.$$

We need to find the maximum of this function on the interval $[0, a/2]$. Since $K(0) = K(a/2) = 0$, the absolute maximum is in the interior of the interval, so it is a local extremum. The solutions of the equation $K'(x) = 12x^2 - 4(a+b)x + ab = 0$ are

$$x = \frac{a+b \pm \sqrt{a^2 + b^2 - ab}}{6}.$$

Since

$$\frac{a+b+\sqrt{a^2+b^2-ab}}{6} \geq \frac{a+b+\sqrt{a^2+b^2-b^2}}{6} = \frac{2a+b}{6} \geq \frac{a}{2}$$

and we are looking for extrema inside $(0, a/2)$, $K(x)$ has a local and therefore absolute maximum at the point

$$x = \frac{a+b-\sqrt{a^2+b^2-ab}}{6}.$$

For example, in the case $a = b$, the box has maximum volume if $x = a/6$.

12.53 $f'(x) = 1 + 4x \cdot \sin(1/x) - 2\cos(1/x)$, if $x \neq 0$ and

$$f'(0) = \lim_{x \to 0} \frac{f(x) - f(0)}{x - 0} = \lim_{x \to 0} (1 + 2x\sin(1/x)) = 1.$$

Thus $f'(0) > 0$. At the same time, f' takes on negative values in any right- or left-hand neighborhood of 0. This is because

$$f'\left(\frac{\pm 1}{2k\pi}\right) = 1 - 2 < 0$$

for every positive integer k. It follows that 0 does not have a neighborhood in which f is monotone increasing, since f is strictly locally decreasing at each of the points $1/(2k\pi)$.

12.87 Suppose that $e = p/q$, where p and q are positive integers. Then $q > 1$, since e is not an integer. Let $a_n = 1 + 1/1! + \cdots + 1/n!$. The sequence (a_n) is strictly monotone increasing, and by Exercise 12.86, $e = \lim_{n \to \infty} a_n$. Thus $e > a_n$ for all n. If $n > q$, then

$$q! \cdot (a_n - a_q) = q! \cdot \left(\frac{1}{(q+1)!} + \cdots + \frac{1}{n!}\right) =$$
$$= \frac{1}{(q+1)} + \frac{1}{(q+1)(q+2)} + \cdots + \frac{1}{(q+1) \cdot \ldots \cdot n} \leq$$
$$\leq \frac{1}{(q+1)} + \frac{1}{(q+1)^2} + \cdots + \frac{1}{(q+1)^{n-q}} =$$
$$= \frac{1}{(q+1)} \cdot \left(1 - \frac{1}{(q+1)^{n-q}}\right) / \left(1 - \frac{1}{(q+1)}\right) <$$
$$< \frac{1}{(q+1)} / \left(1 - \frac{1}{(q+1)}\right) = \frac{1}{q}.$$

This holds for all $n > q$, so $0 < q! \cdot (e - a_q) \leq 1/q < 1$. On the other hand, since $e = p/q$,

$$q! \cdot (e - a_q) = q! \cdot \left(\frac{p}{q} - 1 - \frac{1}{1!} - \frac{1}{2!} - \cdots - \frac{1}{q!}\right)$$

is an integer, which is impossible.

12.95 The function f has a strict local minimum at 0, since $f(0) = 0$ and $f(x) > 0$ if $x \neq 0$. Now

$$f'(x) = x^2 \left[4x \left(2 + \sin\frac{1}{x}\right) - \cos\frac{1}{x}\right]$$

if $x \neq 0$. We can see that f' takes on both negative and positive values in every right-hand neighborhood of 0. For example, if $k \geq 2$ is an integer, then

$$f'\left(\frac{1}{2k\pi}\right) = \frac{1}{(2k\pi)^2} \cdot \left(\frac{8}{2k\pi} - 1\right) < 0$$

and

$$f'\left(\frac{1}{(2k+1)\pi}\right) = \frac{1}{(2k+1)^2\pi^2} \cdot \left(\frac{8}{(2k+1)\pi} + 1\right) > 0.$$

Chapter 13

13.5 The interval $[0,1]$ is mapped to $[-1,1]$ by the function $2x-1$. Thus we first need to determine the nth Bernstein polynomial of the function $f(2x-1)$, which is

$$\sum_{k=0}^{n} f\left(\frac{2k}{n} - 1\right) \cdot \binom{n}{k} x^k (1-x)^{n-k}.$$

We have to transform this function back onto $[-1,1]$, that is, we need to replace x by $(1+x)/2$, which gives us the desired formula.

13.7 $B_1 = 1$, $B_2 = B_3 = (1+x^2)/2$, $B_4 = B_5 = (3+6x^2-x^4)/8$.

13.11

$$\sum_{k=0}^{n} \frac{k}{n} \binom{n}{k} x^k (1-x)^{n-k} = x \cdot \sum_{k=1}^{n} \binom{n-1}{k-1} x^{k-1} (1-x)^{n-k} =$$

$$= x \cdot \left(x + (1-x)\right)^{n-1} = x.$$

13.12

$$\sum_{k=0}^{n} \frac{k^2}{n^2} \binom{n}{k} x^k (1-x)^{n-k} = \sum_{k=1}^{n} \frac{k}{n} \binom{n-1}{k-1} x^k (1-x)^{n-k} =$$

$$= \frac{n-1}{n} \sum_{k=1}^{n} \frac{k-1}{n-1} \binom{n-1}{k-1} x^k (1-x)^{n-k} + \frac{1}{n} \sum_{k=1}^{n} \binom{n-1}{k-1} x^k (1-x)^{n-k} =$$

$$= \frac{n-1}{n} \sum_{k=2}^{n} \binom{n-2}{k-2} x^k (1-x)^{n-k} + \frac{x}{n} =$$

$$= \frac{n-1}{n} x^2 + \frac{x}{n} = x^2 + \frac{x-x^2}{n}.$$

13.32 Let $f(x) = x^2 \sin(1/x)$ if $x \neq 0$ and $f(0) = 0$. We know that f is differentiable everywhere, and $f'(x) = 2x\sin(1/x) - \cos(1/x)$ if $x \neq 0$ and $f'(0) = 0$ (see Example 13.43). The function $f' + h_2$ is continuous everywhere, so by Theorem 15.5, it has a primitive function. If $g' = f' + h_2$, then $(g-f)' = h_2$, so $g-f$ is a primitive function of h_2.

If we start with the function $f_1(x) = x^2 \cos(1/x)$, $f_1(0) = 0$, then a similar argument gives a primitive function of h_1.

13.33 The function $h_1^2 + h_2^2$ vanishes at zero, and is 1 everywhere else. Thus $h_1^2 + h_2^2$ is not Darboux, so it does not have a primitive function. On the other hand, $h_2^2 - h_1^2 = h_2(x/2)$, so $h_2^2 - h_1^2$ has a primitive function. Now the statement follows.

Chapter 14

14.6 I. The solution is based on the following statement: *Let $g = g_1 + g_2$, where $g_1: [c,b] \to \mathbb{R}$ is continuous and $g_2: [c,b] \to \mathbb{R}$ is monotone decreasing. If $g(b) = \max g$, then the image of g is an interval.*

Let $c \le d < b$ and $g(d) < y < g(b)$. We will show that if $s = \sup\{x \in [d,b]: g(t) \le y$ for all $t \in [d,x]\}$, then $g(s) = y$. To see this, suppose that $g(s) < y$. Then $s < b$ but $\lim_{x \to d+0} g_1(x) = g_1(s)$. Now $\lim_{x \to d+0} g_2(x) \le g_2(s)$ implies that $g(x) < y$ in a right-hand neighborhood of the point s, which is impossible. If $f(s) > y$, then $s > d$. Thus $\lim_{x \to d-0} g_1(x) = g_1(s)$ and $\lim_{x \to d-0} g_2(x) \ge g_2(s)$ imply that $g(x) > y$ in a left-hand neighborhood of the point s, which is also impossible. This proves the statement.

  II. Suppose first that there is only one base point. We can assume that $f \ge 0$. Let $M = \sup\{f(t): t \in [a,b]\}$. Let $g(x)$ denote the upper sum corresponding to the partition $a < x < b$, and let $g(a) = g(b) = M(b-a)$. We have to show that if $a < c < b$ and $g(c) < y < M(b-a)$, then g takes on the value y. One of the two values $M_1 = \sup\{f(t): t \in [a,c]\}$, $M_2 = \sup\{f(t): t \in [c,b]\}$ must be equal to M. By symmetry, we may assume that $M_1 = M$. If $c \le x \le b$, then the first term appearing in the upper sum of $g(x)$, which is $\sup\{f(t): t \in [a,x]\} \cdot (x-a) = M(x-a)$, is continuous. The second term, which is $\sup\{f(t): t \in [x,b]\} \cdot (b-x)$, is monotone decreasing.

  Thus over the interval $[c,b]$, the function g is the sum of a continuous and a monotone decreasing function. Moreover, $g(b) = M(b-a) = \max g$. By the above, it follows that g takes on the value of y, and so the set of upper sums corresponding to partitions with one base point forms an interval.

  III. Now let $F: a = x_0 < x_2 < \cdots < x_n = b$ be an arbitrary partition, and let $S_F < y < M(b-a)$. We have to show that there is an upper sum that is equal to y. We can assume that n is the smallest number such that there exists a partition into n parts whose upper sum is less than y. Then for the partition $F': a = x_0 < x_2 < \cdots < x_n = b$, we have $S_{F'} \ge y$. If $S_{F'} = y$, then we are done. If $S_{F'} > y$, then we can apply step II for the interval $[a, x_2]$ to find a point $a < x < x_2$ such that the partition $F'': a = x_0 < x < x_2 < \cdots < x_n = b$ satisfies $S_{F''} = y$.

14.10 The statement is not true: if f is the Dirichlet function and $F: a < b$, then the possible values of σ_F are $b - a$ and 0.

14.12 The statement is true. By Exercise 14.9, there exists a point $x_0 \in [a,b]$ where f is continuous. Since $f(x_0) > 0$, there exist points $a \le c < d \le b$ such that $f(x) > f(x_0)/2$ for all $x \in [c,d]$. It is clear that if the partition F_0 contains the points c and d, then $s_{F_0} \ge (d-c) \cdot f(x_0)/2$. By Lemma 14.4, it then follows that $S_F \ge (d-c) \cdot f(x_0)/2$ for every partition, so $\int_a^b f\,dx \ge (d-c) \cdot f(x_0)/2 > 0$.

14.33 To prove the nontrivial direction, suppose that f is nonnegative, continuous, and not identically zero. If $f(x_0) > 0$, then by a similar argument as in the solution of Exercise 14.12, we get that $\int_a^b f\,dx > 0$.

Chapter 15

15.24 We will use the notation of the proof of Stirling's formula (Theorem 15.15). Since the sequence (a_n) is strictly monotone increasing and tends to 1, we have $a_n < 1$ for all n. This proves the inequality $n! > (n/e)^n \cdot \sqrt{2\pi n}$.

By inequality (15.14),

$$\log a_{N+1} - \log a_n = \sum_{k=n}^{N} (\log a_{k+1} - \log a_k) < \sum_{k=n}^{N} \frac{1}{4k^2} < \frac{1}{4} \sum_{k=n}^{N} \frac{1}{(k-1)k} =$$

$$= \frac{1}{4(n-1)} - \frac{1}{4N}$$

for all $n < N$. Here letting N go to infinity, we get that $-\log a_n \le \frac{1}{4(n-1)}$, that is, $a_n \ge e^{-1/(4(n-1))}$, which is exactly the second inequality we wanted to prove.

15.30 $\int \sqrt{x^2 - 1}\,dx = \frac{1}{2}x\sqrt{x^2 - 1} - \frac{1}{2}\operatorname{arch} x + c$.

15.39 Start with the equality

$$\left(x^k \cdot \sqrt{f(x)}\right)' = kx^{k-1}\sqrt{f(x)} + \frac{x^k f'(x)}{2\sqrt{f(x)}} = \frac{k \cdot x^{k-1} \cdot f(x) + \frac{1}{2}x^k f'(x)}{\sqrt{f(x)}}.$$

The numerator of the fraction on the right-hand side is a polynomial of degree exactly $k+n-1$. It follows that I_{k+n-1} can be expressed as a linear combination of an elementary function and the integrals $I_0, I_1, \ldots, I_{k+n-2}$. The statement then follows.

Chapter 16

16.4 Let $\varepsilon > 0$ be given. Since A is measurable, we can find rectangles $R_1, \ldots, R_n \subset \mathbb{R}^p$ such that $A \subset \cup_{i=1}^n R_i$ and $\sum_{i=1}^n m_p(R_i) < m_p(A) + \varepsilon$. Similarly, there are

rectangles $S_1, \ldots, S_k \subset \mathbb{R}^q$ such that $B \subset \bigcup_{j=1}^k S_j$ and $\sum_{j=1}^k m_q(S_j) < m_q(B) + \varepsilon$.
Then $T_{ij} = R_i \times S_j$ is a rectangle in $\mathbb{R}^{p+q}$ and $m_{p+q}(T_{ij}) = m_p(R_i) \cdot m_q(S_j)$ for every
$i = 1, \ldots, n$, $j = 1, \ldots, k$. Clearly,

$$A \times B \subset \bigcup_{i=1}^n \bigcup_{j=1}^k T_{ij},$$

and thus

$$\overline{m}_{p+q}(A \times B) \leq \sum_{i=1}^n \sum_{j=1}^k m_{p+q}(T_{ij}) = \sum_{i=1}^n \sum_{j=1}^k m_p(R_i) \cdot m_q(S_j) =$$

$$= \left(\sum_{i=1}^n m_p(R_i) \right) \cdot \left(\sum_{j=1}^k m_q(T_{ij}) \right) <$$

$$< (m_p(A) + \varepsilon) \cdot (m_q(B) + \varepsilon).$$

This is true for every $\varepsilon > 0$, and therefore, we obtain $\overline{m}_{p+q}(A \times B) \leq m_p(A) \cdot m_q(B)$.
A similar argument gives $\underline{m}_{p+q}(A \times B) \geq m_p(A) \cdot m_q(B)$. Then we have $\underline{m}_{p+q}$ $(A \times B) = \overline{m}_{p+q}(A \times B) = m_p(A) \cdot m_q(B)$; that is, $A \times B$ is measurable, and its measure
equals $m_p(A) \cdot m_q(B)$.

16.8 The function f is nonnegative and continuous on the interval $[0, r]$, so by
Corollary 16.10, the sector $B_f = \{(x, y) : 0 \leq x \leq r, \ 0 \leq y \leq f(x)\}$ is measurable
and has area

$$\int_0^{r\cos\delta} \frac{\sin\delta}{\cos\delta} \cdot x \, dx + r^2 \int_{r\cos\delta}^r \sqrt{r^2 - x^2} \, dx =$$

$$= \frac{1}{2} r^2 \cos\delta + \sin\delta + r^2 \int_\delta^0 \sin t \cdot (-\sin t) \, dt =$$

$$= \frac{1}{2} r^2 \cos\delta \cdot \sin\delta + \frac{1}{2} r^2 \delta - \frac{r^2 \sin 2\delta}{4} = \frac{1}{2} r^2 \delta.$$

16.9 The part of the region A_u that falls in the upper half-plane is the difference
between the triangle T_u defined by the points $(0,0)$, $(u,0)$, and (u,v), and the region
$B_u = \{(x, y) : 1 \leq x \leq u, \ 0 \leq y \leq \sqrt{x^2 - 1}\}$. Thus

$$\frac{1}{2} \cdot t(A_u) = \frac{1}{2} u \sqrt{u^2 - 1} - \int_1^u \sqrt{x^2 - 1} \, dx.$$

The value of the integral, by Exercise 15.30, is $(1/2) u \sqrt{u^2 - 1} - (1/2) \operatorname{arch} u$, so
$t(A_u)/2 = (\operatorname{arch} u)/2$ and $t(A_u) = \operatorname{arch} u$.

16.27 We show that if $c > 2$, then the graph of f^c is rectifiable. Let F be an arbitrary
partition, and let N denote the least common denominator of the rational base points
of F. Since adding new base points does not decrease the length of the inscribed
polygonal path, we can assume that the points i/N $(i = 0, \ldots, N)$ are all base points.

Then all the rest of the base points of F are irrational. It is clear that in this case, the length of the inscribed polygonal path corresponding to F is at most

$$\sum_{i=1}^{N} \left(f^c((i-1)/N) + (1/N) + f^c(i/N) \right) \le 1 + 2 \cdot \sum_{i=0}^{N} f^c(i/N).$$

If $1 < k \le N$, then among the numbers of the form i/N $(i \le N)$, there are at most k with denominator k (after simplifying). Here the value of f is $1/k^c$, so

$$\sum_{i=0}^{N} f^c(i/N) \le 2 + \sum_{k=2}^{N} k \cdot (1/k^c) = 2 + \sum_{k=2}^{N} k^{1-c}.$$

We can now use the fact that if $b > 0$, then $\sum_{k=1}^{N} 1/(k^{b+1}) < (b+1)/b$ for all N (see Exercise 7.5). We get that $\sum_{k=2}^{N} k^{1-c} < 1/(c-2)$, so the length of every inscribed polygonal path is at most $5 + (2/(c-2))$.

Now we show that if $c \le 2$, then the graph of f^c is not rectifiable. Let $N > 1$ be an integer, and consider a partition

$$F_N : 0 = x_0 < y_0 < x_1 < y_1 < x_2 < \cdots < x_{N-1} < y_{N-1} < x_N = 1$$

such that $x_i = i/N$ $(i = 0, \ldots, N)$ and y_i is irrational for all $i = 0, \ldots, N-1$. If p is a prime divisor of N, then the numbers $1/p, \ldots, (p-1)/p$ appear among the base points, and the length of the segment of the inscribed polygonal path over the interval $[x_i, y_i]$ is at least $1/p^2$. It is then clear that the length of the whole inscribed polygonal path is at least $\sum_{p|N}(p-1)/p^2 \ge (1/2) \cdot \sum_{p|N} 1/p$.

Now we use the fact that the sum of the reciprocals of the first n primes tends to infinity as $n \to \infty$ (see Corollary 18.16). Thus if N is equal to the product of the first n primes, then the partition F_N above gives us an inscribed polygonal path whose length can be arbitrarily long.

16.29 Let the coordinates of g be g_1 and g_2. By Heine's theorem (Theorem 10.61), there exists a $\delta > 0$ such that if $u, v \in [a,b]$, $|u-v| < \delta$, then $|g_i(u) - g_i(v)| < \varepsilon/2$ $(i = 1, 2)$, and so $|g(u) - g(v)| < \varepsilon$.

Let the arc length of the curve be L, and let $F : a = t_0 < t_1 < \cdots < t_n = b$ be a partition with mesh smaller than δ. If $r_i = \sup\{|g(t) - g(t_{i-1})| : t \in [t_{i-1}, t_i]\}$, then by choosing δ, we have $r_i \le \varepsilon$ for all i. Then the disks $D_i = \{x \in \mathbb{R}^2 : |x - g(t_{i-1})| \le r_i\}$ $(i = 1, \ldots, n)$ cover the set $g([a,b])$.

Choose points $u_i \in [t_{i-1}, t_i]$ such that $|g(u_i) - g(t_{i-1})| \ge r_i/2$ $(i = 1, \ldots, n)$. Since the partition with base points t_i and u_i has an inscribed polygonal path of length at most L, we have $\sum_{i=1}^{n} r_i \le 2 \cdot \sum_{i=1}^{n} |g(u_i) - g(t_{i-1})| \le 2L$. Thus the area of the union of the discs D_i is at most

$$\sum_{i=1}^{n} \pi r_i^2 \le \pi \cdot \sum_{i=1}^{n} \varepsilon \cdot r_i \le 2L\pi \cdot \varepsilon.$$

16.40 Cover the disk D by a sphere G of the same radius. For every planar strip S, consider the (nonplanar) strip S' in space whose projection onto the plane is S. If the strips S_i cover the disk, then the corresponding strips S'_i cover the sphere in space. Each strip S'_i cuts out a strip of the sphere with surface area $2\pi r \cdot d_i$ from G, where d_i is the width of the strips S_i and S'_i. Since these strips of the sphere cover G, we have $\sum 2\pi r \cdot d_i \geq 4r^2\pi$, so $\sum d_i \geq 2r$.

Chapter 17

17.1 Let $F: a = x_0 < x_1 < \cdots < x_n = b$. By the Lipschitz condition, we know that $\omega_i = \omega(f; [x_{i-1}, x_i]) \leq K \cdot (x_i - x_{i-1}) \leq K \cdot \delta(F)$ for all i. Thus

$$\Omega_F(f) = \sum_{i=1}^{n} \omega_i \cdot (x_i - x_{i-1}) \leq K \cdot \delta(F) \cdot \sum_{i=1}^{n} (x_i - x_{i-1}) = K \cdot (b-a) \cdot \delta(F).$$

17.12 Since every integrable function is already bounded, there exists a K such that $|f'(x)| \leq K$ for all $x \in [a,b]$. By the mean value theorem, it follows that f is Lipschitz, so by statement (ii) of Theorem 17.3, f is of bounded variation.

Let $F: a = x_0 < \cdots < x_n = b$ be an arbitrary partition of the interval $[a,b]$. By the mean value theorem, there exist points $c_i \in (x_{i-1}, x_i)$ such that $|f(x_i) - f(x_{i-1})| = |f'(c_i)|(x_i - x_{i-1})$ for all $i = 1, \ldots, n$, and so $V_F(f)$ is equal to the Riemann sum $\sigma_F(|f'| : (c_i))$.

Let $\int_a^b |f'| dx = I$. For each $\varepsilon > 0$, there exists a partition F such that every Riemann sum of the function $|f'|$ corresponding to the partition F differs from I by at most ε. Thus $V_F(f)$ also differs from I by less than ε, so it follows that $V(f; [a,b]) \geq I$.

On the other hand, an arbitrary partition F has a refinement F' such that every Riemann sum of the function $|f'|$ corresponding to F' differs from I by less than ε. Then $V_{F'}(f)$ also differs from I by less than ε, so $V_{F'}(f) < I + \varepsilon$. Since $V_F(f) \leq V_{F'}(f)$, we have $V_F(f) < I + \varepsilon$. This holds for every partition F and every $\varepsilon > 0$, so $V(f; [a,b]) \leq I$.

17.13 Let $0 \leq x < y \leq 1$. If $\alpha \geq 1$, then by the mean value theorem, for a suitable $z \in (x,y)$, we have

$$y^\alpha - x^\alpha = \alpha \cdot z^{\alpha-1} \cdot (y-x) \leq \alpha \cdot (y-x) = \alpha \cdot (y-x)^\beta.$$

If $\alpha < 1$, then $y^\alpha - x^\alpha \leq (y-x)^\alpha = (y-x)^\beta$.

17.15 If $\alpha \geq \beta + 1$, then it is easy to check that f' is bounded in $(0, 1]$. Then f is Lipschitz, that is, Hölder 1 on the interval $[0, 1]$. We can then suppose that $\alpha < \beta + 1$ and $\gamma = \alpha/(\beta + 1) < 1$. Let $0 \leq x < y \leq 1$ be fixed. Then

$$|f(x) - f(y)| = \left| x^\alpha \cdot \sin x^{-\beta} - y^\alpha \cdot \sin y^{-\beta} \right| \leq$$

$$\leq \left| x^\alpha \cdot \sin x^{-\beta} - y^\alpha \cdot \sin x^{-\beta} \right| + y^\alpha \cdot \left| \sin x^{-\beta} - \sin y^{-\beta} \right| \leq$$

$$\leq |x^\alpha - y^\alpha| + y^\alpha \cdot \min \left(2, \left| x^{-\beta} - y^{-\beta} \right| \right),$$

so it suffices to show that $y^\alpha - x^\alpha \leq C \cdot (y - x)^\gamma$ and

$$y^\alpha \cdot \min \left(2, (x^{-\beta} - y^{-\beta}) \right) \leq C \cdot (y - x)^\gamma \tag{12}$$

with a suitable constant C. By the previous exercise, there is a constant C depending only on α such that $y^\alpha - x^\alpha \leq C \cdot (y - x)^{\min(1, \alpha)} \leq C \cdot (y - x)^\gamma$, since $\gamma < \min(1, \alpha)$.

We distinguish two cases in the proof of the inequality (12). If $y - x \geq y^{\beta+1}/2$, then

$$y^\alpha \cdot 2 = 2 \cdot \left(y^{\beta+1} \right)^{\alpha/(\beta+1)} \leq 2 \cdot 2^{\alpha/(\beta+1)} \cdot (y - x)^{\alpha/(\beta+1)} < 4 \cdot (y - x)^\gamma.$$

Now suppose that $y - x < y^{\beta+1}/2$. Then on the one hand, $x > y/2$, and on the other hand, $y > (2(y-x))^{1/(\beta+1)}$. By the mean value theorem, for a suitable $u \in (x, y)$, we have

$$y^\alpha \cdot (x^{-\beta} - y^{-\beta}) = y^\alpha \cdot (-\beta) \cdot u^{-\beta-1} \cdot (x - y) =$$

$$= y^\alpha \cdot \beta \cdot u^{-\beta-1} \cdot (y - x) <$$

$$< y^\alpha \cdot \beta \cdot (y/2)^{-\beta-1} \cdot (y - x) \leq C \cdot y^{\alpha-\beta-1} \cdot (y - x) <$$

$$< C \cdot (2(y - x))^{(\alpha-\beta-1)/(\beta+1)} \cdot (y - x) < C \cdot (y - x)^\gamma,$$

where $C = \beta \cdot 2^{\beta+1}$.

Chapter 18

18.4 Let c be a shared point of discontinuity. We can assume that $c < b$, and f is discontinuous from the right at c (since the proof is similar when $c > a$ and f is discontinuous from the left at c). We distinguish two cases. First, we suppose that g is discontinuous from the right at c. Then there exists an $\varepsilon > 0$ such that every right-hand neighborhood of c contains points x and y such that $|f(x) - f(c)| \geq \varepsilon$ and $|g(y) - g(c)| \geq \varepsilon$. It follows that for every $\delta > 0$, we can find a partition $a = x_0 < x_1 < \cdots < x_n = b$ with mesh smaller than δ such that c is one of the base points, say $c = x_{k-1}$, and $|g(x_k) - g(c)| \geq \varepsilon$.

Let $c_i = x_{i-1}$ for all $i = 1,\ldots,n$, let $d_i = x_{i-1}$ for all $i \neq k$, and let $d_k \in (x_{k-1}, x_k]$ be a point such that $|f(d_k) - f(c_k)| = |f(d_k) - f(c)| \geq \varepsilon$. Then the sums $S_1 = \sum_{i=1}^n f(c_i) \cdot (g(x_i) - g(x_{i-1}))$ and $S_2 = \sum_{i=1}^n f(d_i) \cdot (g(x_i) - g(x_{i-1}))$ differ only in the kth term, and

$$|S_1 - S_2| = |(f(c_k) - f(d_k)) \cdot (g(x_k) - g(x_{k-1}))| \geq \varepsilon^2.$$

This means that we cannot find a δ for ε^2 such that the condition in Definition 18.3 is satisfied, that is, the Stieltjes integral $\int_a^b f \, dg$ does not exist.

Now suppose that g is continuous from the right at c. Then $c \in (a,b)$, and g is discontinuous from the left at c. It follows that for every $\delta > 0$, we can find a partition $a = x_0 < x_1 < \cdots < x_n = b$ with mesh smaller than δ such that c is not a base point, say $x_{k-1} < c < x_k$, and $|g(x_k) - g(x_{k-1})| \geq \varepsilon$. Let $c_i = d_i = x_{i-1}$ for all $i \neq k$. Also, let $c_k = c$ and $d_k \in (c, x_k]$ be a point such that $|f(d_k) - f(c_k)| = |f(d_k) - f(c)| \geq \varepsilon$. Then (just as in the previous case) the sums S_1 and S_2 differ by at least ε^2 from each other, and so the Stieltjes integral $\int_a^b f \, dg$ does not exist.

18.5 Let $x_i = 2/((2i+1)\pi)$ for all $i = 0, 1 \ldots$. Then $f(x_i) = (-1)^i \cdot \sqrt{2/((2i+1)\pi)}$ ($i \in \mathbb{N}$). Let $\delta > 0$ be given. Fix an integer $N > 1/\delta$, and let $x_N = y_0 < y_1 < \cdots < y_n = 1$ be a partition of $[x_N, 1]$ with mesh smaller than δ. Then

$$F_M : 0 < x_M < x_{M-1} < \cdots < x_N < y_1 < \cdots < y_n = 1$$

is a partition of $[0, 1]$ with mesh smaller than δ for all $M > N$. We show that if M is sufficiently large, then there exists a approximating sum that is greater than 1, and there also exists one that is less than -1.

In each of the intervals $[0, x_M]$ and $[y_{j-1}, y_j]$ ($j = 1, \ldots, n$), let the inner point be the left endpoint of the interval. Let $c_i = x_i$ ($N \leq i \leq M - 1$). Then $f(c_i) = f(x_i) = (-1)^i \cdot \sqrt{2/((2i+1)\pi)}$ for all $N \leq i \leq M - 1$, so

$$\sum_{i=N}^{M-1} f(c_i)(f(x_i) - f(x_{i+1})) = \sum_{i=N}^{M-1} \sqrt{\frac{2}{(2i+1)\pi}} \cdot \left(\sqrt{\frac{2}{(2i+1)\pi}} + \sqrt{\frac{2}{(2i+3)\pi}} \right) >$$

$$> \sum_{i=N}^{M-1} \frac{2}{(2i+2)\pi} = \frac{1}{\pi} \cdot \sum_{i=N+1}^{M} \frac{1}{i}.$$

We get the approximating sum corresponding to the partition F_M by taking the above sum and adding the terms corresponding to $[y_{j-1}, y_j]$ ($j = 1, \ldots, n$). Note that the sum of these new terms does not depend on M. Since $\lim_{M \to \infty} \sum_{i=N+1}^{M} (1/i) = \infty$, choosing M sufficiently large gives us a approximating sum that is arbitrarily large. Similarly, we can show that with the choice $c_i = x_{i+1}$ ($N \leq i \leq M - 1$), we can get arbitrarily small approximating sums for sufficiently large M.

Chapter 19

19.7 By Cauchy's criterion, $\lim_{x\to\infty} \int_x^{2x} f \, dt = 0$. Since if $t \in [x, 2x]$ then $f(t) \geq f(2x)$, we have that

$$0 \leq x \cdot f(2x) \leq \lim_{x\to\infty} \int_x^{2x} f \, dt,$$

and so by the squeeze theorem, $x \cdot f(2x) \to 0$ if $x \to \infty$.

19.9

(i) We can suppose that f is monotone decreasing and nonnegative.

Let $\varepsilon > 0$ be fixed, and choose a $0 < \delta < 1$ such that $\left| \int_x^1 f \, dt - I \right| < \varepsilon$ holds for all $0 < x \leq \delta$, where $I = \int_0^1 f \, dx$. Then $\int_0^x f \, dt < \varepsilon$ also holds for all $0 < x \leq \delta$. Fix an integer $n > 1/\delta$. If $k/n \leq \delta < (k+1)/n$, then the partition $F: k/n < \cdots < n/n = 1$ gives us intervals of length $1/n$, so by (14.19), the lower sum

$$s_F = \frac{1}{n} \cdot \sum_{i=k+1}^{n} f\left(\frac{i}{n}\right)$$

corresponding to F differs from the integral $\int_{k/n}^1 f \, dx$ by less than $(f(k/n) - f(1))/n$, so it differs from the integral I by less than $\varepsilon + (f(k/n) - f(1))/n$. Now by $k/n \leq \delta$, it follows that

$$\frac{1}{n} \cdot \sum_{i=1}^{k} f\left(\frac{i}{n}\right) \leq \int_0^{k/n} f \, dx < \varepsilon,$$

and thus $f(1/n)/n < \varepsilon$. Therefore,

$$\left| \frac{1}{n} \cdot \sum_{i=1}^{n} f\left(\frac{i}{n}\right) - I \right| \leq \varepsilon + |s_F - I| \leq$$

$$\leq \varepsilon + \varepsilon + \frac{1}{n} \cdot (f(k/n) - f(1)) \leq 2\varepsilon + \frac{1}{n} \cdot f(1/n) < 3\varepsilon.$$

This holds for all $n > 1/\delta$, which concludes the solution of the first part of the exercise.

(ii) If the function is not monotone, the statement is not true. The function $f(1/n) = n^2$ $(n = 1, 2, \ldots)$, $f(x) = 0$ $(x \neq 1/n)$ is a simple counterexample.

19.11 We can assume that f is differentiable on $[a, b)$ and that f' is integrable on $[a, x]$ for all $a < x < b$.

We show that the length of every inscribed polygonal path is $\leq I$. Let $F: a = x_0 < x_1 < \ldots < x_n = b$ be a partition, and let $\varepsilon > 0$ be given. By the continuity of f and the convergence of the improper integral, there exists $0 < \delta < \varepsilon$ such that $|f(x) - f(b)| < \varepsilon$ and $|\int_a^x f \, dt - I| < \varepsilon$ for all $b - \delta < x < b$. Since adding new base points does not decrease the length of the inscribed polygonal path, we can assume that $x_{n-1} > b - \delta$. By Remark 16.21, the graph of f over the interval $[a, x_{n-1}]$ is rectifiable, and its arc length is $\int_a^{x_{n-1}} f\,dx < I + \varepsilon$. Thus the inscribed polygonal path corresponding to the partition $F_1: a = x_0 < x_1 < \cdots < x_{n-1}$ has length $< I + \varepsilon$, and so the inscribed polygonal path corresponding to the partition F has length less than

$$I + \varepsilon + \sqrt{(b - x_{n-1})^2 + (f(b) - f(x_{n-1}))^2} \le$$
$$\le I + \varepsilon + |b - x_{n-1}| + |f(b) - f(x_{n-1})| \le I + 3\varepsilon.$$

This is true for every ε, which shows that the graph of f is rectifiable, and its arc length is at most I. On the other hand, for every $a < x < b$, there exists a partition of $[a, x]$ such that the corresponding inscribed polygonal path gets arbitrarily close to the value of the integral $\int_a^x f\,dt$. Thus it is clear that the arc length of the graph of f is not less than I.

19.20

(a) Since $\sin x$ is concave on $[0, \pi/2]$, we have that $\sin x \ge 2x/\pi$ for all $x \in [0, \pi/2]$. Thus $|\log \sin x| = -\log \sin x \le -\log(2x/\pi) = |\log x| + \log(\pi/2)$. By Example 19.5, $\int_0^1 |\log x|\,dx$ is convergent (and its value is 1), so applying the majorization principle, we get that $\int_0^{\pi/2} \log \sin x\,dx$ is convergent. The substitution $x - \pi - t$ gives that $\int_{\pi/2}^\pi \log \sin x\,dx$ is also convergent.

(b) Let $\int_0^{\pi/2} \log \sin x\,dx = \int_{\pi/2}^\pi \log \sin x\,dx = I$. Applying the substitution $x = (\pi/2) - t$ gives us that $\int_0^{\pi/2} \log \cos x\,dx = I$. Now apply the substitution $x = 2t$:

$$2I = \int_0^\pi \log \sin x\,dx = \int_0^{\pi/2} \log \sin(2t) \cdot 2\,dt =$$
$$= 2 \cdot \int_0^{\pi/2} (\log 2 + \log \sin t + \log \cos t)\,dt =$$
$$= \log 2 \cdot \pi + 4I,$$

so $I = -\log 2 \cdot \pi/2$ and $\int_0^\pi \log \sin x\,dx = -\log 2 \cdot \pi$.

19.21 The function $\log \sin x$ is monotone on the intervals $[0, \pi/2]$ and $[\pi/2, \pi]$. Thus by the statement of Exercise 19.9,

$$\lim_{n \to \infty} \frac{\pi}{n} \cdot \sum_{i=1}^{n-1} \log \sin \frac{i\pi}{n} = \int_0^\pi \log \sin x\,dx = -\log 2 \cdot \pi,$$

so

$$\lim_{n \to \infty} \frac{1}{n-1} \cdot \sum_{i=1}^{n-1} \log \sin \frac{i\pi}{n} = -\log 2.$$

If we raise e to the power of the expressions present on each side, we get that if $n \to \infty$, then the geometric mean of the numbers $\sin \pi/n, \sin 2\pi/n, \ldots, \sin(n-1)\pi/n$ tends to $e^{-\log 2} = 1/2$.

The arithmetic mean clearly tends to $\frac{1}{\pi} \cdot \int_0^\pi \sin x \, dx = 2/\pi$. The inequality $(1/2) <$ $2/\pi$ is obviously true.

19.28 For every $n \in \mathbb{N}^+$, let $f_n \colon [n-1,n] \to \mathbb{R}$ be a nonnegative continuous function such that $f_n(n-1) = f_n(n) = 0$, $\max f_n \geq 1$, and $\int_{n-1}^n f_n \, dx \leq 1/2^n$. (We may take the function for which $f_n(x) = 0$ if $|x - a_n| \geq \varepsilon_n$, $f_n(a_n) = 1$, and f_n is linear in the intervals $[a_n - \varepsilon_n, a_n]$ and $[a_n, a_n + \varepsilon_n]$, where $a_n = (2n-1)/2$ and $\varepsilon_n = 2^{-n}$.)

Let $f(x) = f_n(x)$ if $x \in [n-1, n]$ and $n \in \mathbb{N}^+$. Clearly, f is continuous. Since f is also nonnegative, the function $x \mapsto \int_0^x f \, dt$ is monotone increasing, and so the limit $\lim_{x \to \infty} \int_0^x f \, dt$ exists. On the other hand, $\int_0^n f \, dt \leq 2^{-1} + \cdots + 2^{-n} < 1$ for all n, so the limit is finite, and the improper integral $\int_0^\infty f \, dt$ is convergent.

19.29 Let $\varepsilon > 0$ be given. By the uniform continuity of f, there exists a $\delta > 0$ such that $|x - y| < \delta$ implies $|f(x) - f(y)| < \varepsilon$. By Cauchy's criterion, for the convergence of the integral, we can pick a number $K > 0$ such that if $K < x < y$, then $|\int_x^y f \, dt| < \varepsilon \cdot \delta$. We show that $|f(x)| < 2\varepsilon$ for all $x > K$. Suppose that $x > K$ and $f(x) \geq 2\varepsilon$. Then by the choice of δ, we have $f(t) \geq \varepsilon$ for all $t \in (x, x + \delta)$, and so $\int_x^{x+\delta} f \, dt \geq \varepsilon \cdot \delta$, which contradicts the choice of K. Thus $f(x) < 2\varepsilon$ for all $x > K$, and we can similarly show that $f(x) > -2\varepsilon$ for all $x > K$.

19.33

(a) Let $a_0 = a$. If $n > 0$ and we have already chosen the number $a_{n-1} > a$, then let $a_n > \max(n, a_{n-1})$ be such that $\int_{a_n}^\infty f \, dx < 1/(n \cdot 2^n)$. Thus we have chosen numbers a_n for all $n = 0, 1, \ldots$. Let $g(x) = n - 1$ for all $x \in (a_{n-1}, a_n)$ and $n = 1, 2, \ldots$. Then $\lim_{x \to \infty} g(x) = \infty$. Now

$$\int_a^{a_n} g \cdot f \, dx = \sum_{i=0}^{n-1} \int_{a_i}^{a_{i+1}} g \cdot f \, dx = \sum_{i=1}^{n-1} i \cdot \int_{a_i}^{a_{i+1}} f \, dx < \sum_{i=1}^{n-1} i \cdot \frac{1}{i \cdot 2^i} < 1,$$

and so the function $x \mapsto \int_a^x g \cdot f \, dt$ is bounded. Since it is also monotone, its limit is finite, and so the improper integral is convergent.

(b) Let $a_0 = a$. If $n > 0$ and we have already chosen the number $a_{n-1} > a$, then let $a_n > \max(n, a_{n-1})$ be such that $\int_{a_{n-1}}^{a_n} f \, dx > n$. Thus we have chosen numbers a_n for all $n = 0, 1, \ldots$. Let $g(x) = 1/n$ for all $x \in (a_{n-1}, a_n)$ and $n = 1, 2, \ldots$. Then $\lim_{x \to \infty} g(x) = 0$. Now

$$\int_a^{a_n} g \cdot f \, dx = \sum_{i=1}^n \int_{a_{i-1}}^{a_i} g \cdot f \, dx = \sum_{i=1}^n \frac{1}{i} \cdot \int_{a_{i-1}}^{a_i} f \, dx > \sum_{i=1}^n \frac{1}{i} \cdot i = n,$$

so $\int_a^\infty g \cdot f \, dx$ is divergent.

19.35 We will apply the majorization principle. Let $x \geq 3$ be fixed. If we have $0 \leq f(x) < 1/x^2$, then

$$\log\left(f(x)^{1-1/\log x}\right) = (1-1/\log x)\cdot \log f(x) \le$$

$$\le \left(1-\frac{1}{\log x}\right)\cdot(-2\log x) = 2-2\log x,$$

so

$$f(x)^{1-1/\log x} \le e^{2-2\log x} = e^2/x^2.$$

If, however, $f(x) \ge 1/x^2$, then

$$f(x)^{1-1/\log x} = f(x)\cdot f(x)^{-1/\log x} \le f(x)\cdot x^{2/\log x} = e^2\cdot f(x).$$

We get that $f(x)^{1-1/\log x} \le \max\left(e^2/x^2, e^2\cdot f(x)\right) \le e^2\cdot\left((1/x^2)+f(x)\right)$ for all $x \ge 3$. Since the integral $\int_3^\infty \left((1/x^2)+f(x)\right)dx$ is convergent, $\int_3^\infty f(x)^{1-1/\log x}\,dx$ must also be convergent.

19.37 Since f is decreasing and convex, f' is increasing and nonpositive, so $|f'|$ is bounded. It is then clear that the integrals $\int_a^\infty f(x)\cdot\sqrt{1+(f'(x))^2}\,dx$ and $\int_a^\infty f(x)\,dx$ are either both convergent or both divergent. By Exercise 19.31, if $\int_a^\infty f\,dx$ is convergent, then $\int_a^\infty f^2\,dx$ is also convergent. Thus only three configurations are possible: All three integrals are convergent, $\int_a^\infty f(x)\cdot\sqrt{1+(f'(x))^2}\,dx$ and $\int_a^\infty f(x)\,dx$ are divergent while $\int_a^\infty f^2\,dx$ is convergent, or all three integrals are divergent. Examples for each three cases are given by the functions $1/x^2$, $1/x$, and $1/\sqrt{x}$ over the interval $[1,\infty)$.

19.38 Applying integration by parts gives

$$\Gamma(c+1) = \int_0^\infty x^c\cdot e^{-x}\,dx = \left[x^c\cdot(-e^{-x})\right]_0^\infty + \int_0^\infty c\cdot x^{c-1}\cdot e^{-x}\,dx = 0+c\cdot\Gamma(c).$$

19.41 Using the substitution $x^3 = t$ gives us that

$$\int_0^\infty e^{-x^3}\,dx = \int_0^\infty e^{-t}\cdot\frac{1}{3}\cdot t^{-2/3}\,dt = \frac{1}{3}\cdot\Gamma(1/3).$$

We similarly get that $\int_0^\infty e^{-x^s}\,dx = \Gamma(1/s)/s$ for all $s > 0$.

19.43 Use the substitution $x = t/n$ in (19.8). We get that

$$\int_0^n \left(1-\frac{t}{n}\right)^n\cdot t^{c-1}\,dt = \frac{n^c\cdot n!}{c(c+1)\cdots(c+n)} \tag{13}$$

for all $n = 1, 2, \ldots$. Since $(1-t/n)^n \le e^{-t}$ for all $0 < t \le n$, by (13) we get that $\Gamma(c) > \left(n^c\cdot n!\right)/\left(c(c+1)\cdots(c+n)\right)$.

On the other hand, for a given n, the function $e^t \cdot (1 - t/n)^n$ is monotone decreasing on the interval $[0, n]$, since its derivative there is

$$e^t \cdot \left(1 - \frac{t}{n}\right)^n - e^t \cdot \left(1 - \frac{t}{n}\right)^{n-1} \leq 0.$$

Let $\varepsilon > 0$ be fixed. Choose a number $K > 0$ such that $\int_K^\infty e^{-t} \cdot t^{c-1} \, dt < \varepsilon$ holds, and n_0 such that

$$e^K \cdot \left(1 - \frac{K}{n}\right)^n > \frac{1}{1+\varepsilon}$$

holds for all $n > n_0$. If $n \geq \max(n_0, K)$ and $0 < t < K$, then

$$e^t \cdot \left(1 - \frac{t}{n}\right)^n \geq e^K \cdot \left(1 - \frac{K}{n}\right)^n \geq \frac{1}{1+\varepsilon},$$

so

$$e^{-t} \leq (1 + \varepsilon) \cdot \left(1 - \frac{t}{n}\right)^n,$$

and thus

$$\Gamma(c) < \int_K^\infty e^{-t} \cdot t^{c-1} + (1+\varepsilon) \int_0^K \left(1 - \frac{t}{n}\right)^n \cdot t^{c-1} \, dt <$$

$$< \varepsilon + (1+\varepsilon) \frac{n^c \cdot n!}{c(c+1) \cdots (c+n)}. \tag{14}$$

Since $\varepsilon > 0$ was arbitrary and (14) holds for every sufficiently large n, (19.9) also holds.

Notation

© Springer New York 2015
M. Laczkovich, V.T. Sós, *Real Analysis*, Undergraduate Texts
in Mathematics, DOI 10.1007/978-1-4939-2766-1

References

1. Davidson, K.R., Dosig, A.P.: Real Analysis and Applications. Theory in Practice. Springer, New York (2010)
2. Erdős, P., Surányi, J.: Topics in the Theory of Numbers. Springer, New York (2003)
3. Euclid: The Thirteen Books of the Elements [Translated with introduction and commentary by Sir Thomas Heath]. Second Edition Unabridged. Dover, New York (1956)
4. Hewitt, E., Stromberg, K.: Real and Abstract Analysis. Springer, New York (1975)
5. Niven, I., Zuckerman, H.S., Montgomery, H.L.: An Introduction to the Theory of Numbers, 5th edn. Wiley, New York (1991)
6. Rademacher, H., Toeplitz, O.: Von Zahlen und Figuren. Springer, Berlin (1933) [English translation: The Enjoyment of Mathematics]. Dover, New York (1990)
7. Rudin, W.: Principles of Mathematical Analysis, 3rd edn. McGraw-Hill, New York (1976)
8. Zaidman, S.: Advanced Calculus. An Introduction to Mathematical Analysis. World Scientific, Singapore (1997)

© Springer New York 2015
M. Laczkovich, V.T. Sós, *Real Analysis*, Undergraduate Texts
in Mathematics, DOI 10.1007/978-1-4939-2766-1

Index

© Springer New York 2015
M. Laczkovich, V.T. Sós, *Real Analysis*, Undergraduate Texts
in Mathematics, DOI 10.1007/978-1-4939-2766-1

Printed in the United States
By Bookmasters